PEARSON EDEXCEL INTERNATIONAL GCSE (9–1)

MATHEMATICS A

Student Book 2

David Turner
Ian Potts

Published by Pearson Education Limited, 80 Strand, London, WC2R 0RL.

www.pearsonglobalschools.com

Copies of official specifications for all Pearson Edexcel qualifications may be found on the website: https://qualifications.pearson.com

Text © Pearson Education Limited 2017
Edited by Lyn Imeson
Answers checked by Laurice Suess
Designed by Cobalt id
Typeset by Cobalt id
Original illustrations © Pearson Education Limited 2017
Illustrated by © Cobalt id
Cover design by Pearson Education Limited
Picture research by Andreas Schindler
Cover photo/illustration © Shutterstock.com: Grey Carnation

The rights of David Turner and Ian Potts to be identified as authors of this work have been asserted by them in accordance with the Copyright, Designs and Patents Act 1988.

First published 2017

22 21 20 19
10 9 8 7 6 5

British Library Cataloguing in Publication Data
A catalogue record for this book is available from the British Library

ISBN 978 0 435 18305 9

Printed in Slovakia by Neografia

Dedicated to Viv Hony who started the whole project.

Grateful for contributions from Jack Barraclough, Chris Baston, Ian Bettison, Sharon Bolger, Phil Boor, Ian Boote, Linnet Bruce, Andrew Edmondson, Keith Gallick, Rachel Hamar, Kath Hipkiss, Sean McCann, Diane Oliver, Harry Smith, Robert Ward-Penny and our Development Editor: Gwen Burns.

Endorsement Statement

In order to ensure that this resource offers high-quality support for the associated Pearson qualification, it has been through a review process by the awarding body. This process confirms that this resource fully covers the teaching and learning content of the specification or part of a specification at which it is aimed. It also confirms that it demonstrates an appropriate balance between the development of subject skills, knowledge and understanding, in addition to preparation for assessment.

Endorsement does not cover any guidance on assessment activities or processes (e.g. practice questions or advice on how to answer assessment questions), included in the resource nor does it prescribe any particular approach to the teaching or delivery of a related course.

While the publishers have made every attempt to ensure that advice on the qualification and its assessment is accurate, the official specification and associated assessment guidance materials are the only authoritative source of information and should always be referred to for definitive guidance.

Pearson examiners have not contributed to any sections in this resource relevant to examination papers for which they have responsibility.

Examiners will not use endorsed resources as a source of material for any assessment set by Pearson. Endorsement of a resource does not mean that the resource is required to achieve this Pearson qualification, nor does it mean that it is the only suitable material available to support the qualification, and any resource lists produced by the awarding body shall include this and other appropriate resources.

CONTENTS

UNIT 9

UNIT 10

ABOUT THIS BOOK

This two-book series is written for students following the Pearson Edexcel International GCSE (9–1) Maths A Higher Tier specification. There is a Student Book for each year of the course.

The course has been structured so that these two books can be used in order, both in the classroom and for independent learning.

Each book contains five units of work. Each unit contains five sections in the topic areas: *Number, Algebra, Sequences, Graphs, Shape and Space, Sets* and *Handling Data*.

Each unit contains concise explanations and worked examples, plus numerous exercises that will help you build up confidence.

Non-starred and starred parallel exercises are provided, to bring together basic principles before being challenged with more difficult questions. These are supported by parallel revision exercises at the end of each chapter.

Challenges, which provide questions applying the basic principles in unusual situations, feature at the back of the book along with *Fact Finders* which allow you to practise comprehension of real data.

Points of Interest put the maths you are about to learn in a real-world context.

Learning Objectives show what you will learn in each lesson.

Basic Principles outline assumed knowledge and key concepts from the beginning.

Transferable Skills are highlighted to show what skill you are using and where.

Activities are a gentle way of introducing a topic.

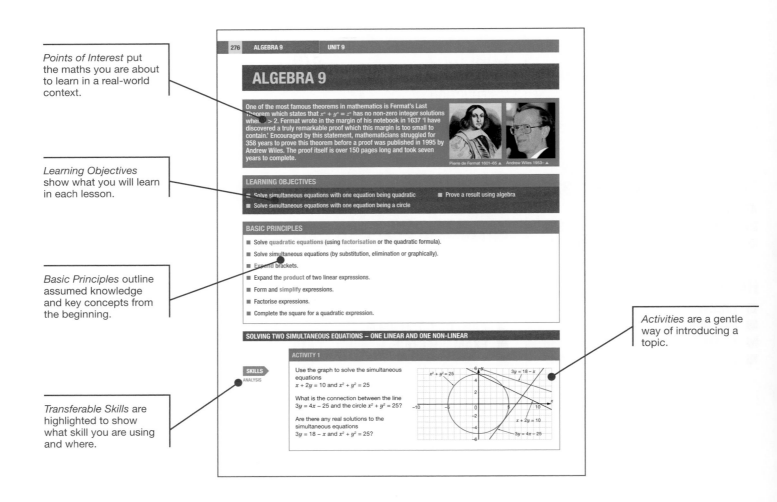

276   ALGEBRA 9   UNIT 9

ALGEBRA 9

One of the most famous theorems in mathematics is Fermat's Last Theorem which states that $x^n + y^n = z^n$ has no non-zero integer solutions when $n > 2$. Fermat wrote in the margin of his notebook in 1637 'I have discovered a truly remarkable proof which this margin is too small to contain.' Encouraged by this statement, mathematicians struggled for 358 years to prove this theorem before a proof was published in 1995 by Andrew Wiles. The proof itself is over 150 pages long and took seven years to complete.

Pierre de Fermat 1601–65 ▲   Andrew Wiles 1953– ▲

LEARNING OBJECTIVES

■ Solve simultaneous equations with one equation being quadratic   ■ Prove a result using algebra
■ Solve simultaneous equations with one equation being a circle

BASIC PRINCIPLES

■ Solve quadratic equations (using factorisation or the quadratic formula).
■ Solve simultaneous equations (by substitution, elimination or graphically).
■ Expand brackets.
■ Expand the product of two linear expressions.
■ Form and simplify expressions.
■ Factorise expressions.
■ Complete the square for a quadratic expression.

SOLVING TWO SIMULTANEOUS EQUATIONS – ONE LINEAR AND ONE NON-LINEAR

ACTIVITY 1

SKILLS
ANALYSIS

Use the graph to solve the simultaneous equations
$x + 2y = 10$ and $x^2 + y^2 = 25$

What is the connection between the line $3y = 4x - 25$ and the circle $x^2 + y^2 = 25$?

Are there any real solutions to the simultaneous equations
$3y = 18 - x$ and $x^2 + y^2 = 25$?

Key Points boxes summarise the essentials.

Questions have been given a *Pearson step* from 1 to 12. This tells you how difficult the question is. The higher the step, the more challenging the question.

Examples provide a clear, instructional framework.

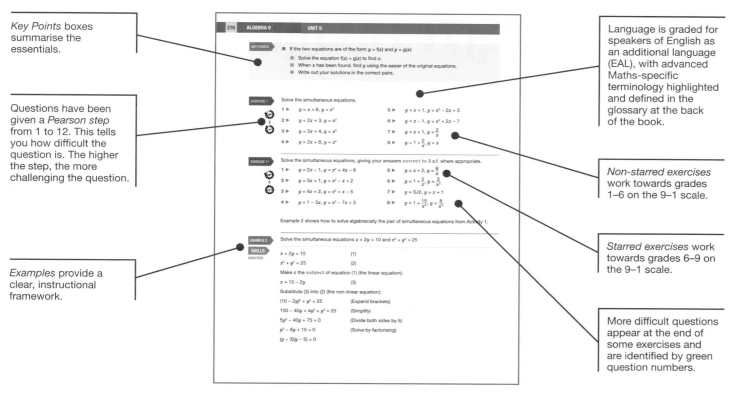

278   ALGEBRA 9   UNIT 9

KEY POINTS

■ If the two equations are of the form $y = f(x)$ and $y = g(x)$:
  ■ Solve the equation $f(x) = g(x)$ to find x.
  ■ When x has been found, find y using the easier of the original equations.
  ■ Write out your solutions in the correct pairs.

EXERCISE 1

Solve the simultaneous equations.

1 ▶ $y = x + 6, y = x^2$ 5 ▶ $y = x + 1, y = x^2 - 2x + 3$
2 ▶ $y = 2x + 3, y = x^2$ 6 ▶ $y = x - 1, y = x^2 + 2x - 7$
3 ▶ $y = 3x + 4, y = x^2$ 7 ▶ $y = x + 1, y = \frac{2}{x}$
4 ▶ $y = 2x + 8, y = x^2$ 8 ▶ $y = 1 + \frac{2}{x}, y = x$

EXERCISE 1*

Solve the simultaneous equations, giving your answers correct to 3 s.f. where appropriate.

1 ▶ $y = 2x - 1, y = x^2 + 4x - 6$ 5 ▶ $y = x + 2, y = \frac{8}{x}$
2 ▶ $y = 3x + 1, y = x^2 - x + 2$ 6 ▶ $y = 1 + \frac{2}{x}, y = \frac{3}{x^2}$
3 ▶ $y = 4x + 2, y = x^2 + x - 5$ 7 ▶ $y = 3\sqrt{x}, y = x + 1$
4 ▶ $y = 1 - 3x, y = x^2 - 7x + 3$ 8 ▶ $y = 1 + \frac{15}{x^2}, y = \frac{8}{x}$

Example 2 shows how to solve algebraically the pair of simultaneous equations from Activity 1.

EXAMPLE 2

SKILLS
ANALYSIS

Solve the simultaneous equations $x + 2y = 10$ and $x^2 + y^2 = 25$

$x + 2y = 10$ (1)
$x^2 + y^2 = 25$ (2)

Make x the subject of equation (1) (the linear equation):

$x = 10 - 2y$ (3)

Substitute (3) into (2) (the non-linear equation):

$(10 - 2y)^2 + y^2 = 25$ (Expand brackets)
$100 - 40y + 4y^2 + y^2 = 25$ (Simplify)
$5y^2 - 40y + 75 = 0$ (Divide both sides by 5)
$y^2 - 8y + 15 = 0$ (Solve by factorising)
$(y - 3)(y - 5) = 0$

Language is graded for speakers of English as an additional language (EAL), with advanced Maths-specific terminology highlighted and defined in the glossary at the back of the book.

Non-starred exercises work towards grades 1–6 on the 9–1 scale.

Starred exercises work towards grades 6–9 on the 9–1 scale.

More difficult questions appear at the end of some exercises and are identified by green question numbers.

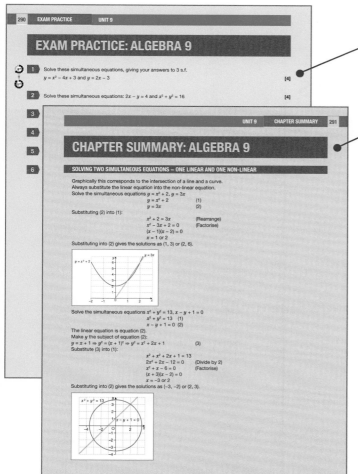

290   EXAM PRACTICE   UNIT 9

EXAM PRACTICE: ALGEBRA 9

1   Solve these simultaneous equations, giving your answers to 3 s.f.
    $y = x^2 - 4x + 3$ and $y = 2x - 3$ [4]

2   Solve these simultaneous equations: $2x - y = 4$ and $x^2 + y^2 = 16$ [4]

3

4

5

6

UNIT 9   CHAPTER SUMMARY   291

CHAPTER SUMMARY: ALGEBRA 9

SOLVING TWO SIMULTANEOUS EQUATIONS – ONE LINEAR AND ONE NON-LINEAR

Graphically this corresponds to the intersection of a line and a curve.
Always substitute the linear equation into the non-linear equation.
Solve the simultaneous equations $y = x^2 + 2, y = 3x$
$y = x^2 + 2$ (1)
$y = 3x$ (2)

Substituting (2) into (1):
$x^2 + 2 = 3x$ (Rearrange)
$x^2 - 3x + 2 = 0$ (Factorise)
$(x - 1)(x - 2) = 0$
$x = 1$ or 2
Substituting into (2) gives the solutions as (1, 3) or (2, 6).

Solve the simultaneous equations $x^2 + y^2 = 13, x - y + 1 = 0$
$x^2 + y^2 = 13$ (1)
$x - y + 1 = 0$ (2)
The linear equation is equation (2).
Make y the subject of equation (2):
$y = x + 1 \Rightarrow y^2 = (x + 1)^2 \Rightarrow y^2 = x^2 + 2x + 1$ (3)
Substitute (3) into (1):
$x^2 + x^2 + 2x + 1 = 13$
$2x^2 + 2x - 12 = 0$ (Divide by 2)
$x^2 + x - 6 = 0$ (Factorise)
$(x + 3)(x - 2) = 0$
$x = -3$ or 2
Substituting into (2) gives the solutions as $(-3, -2)$ or $(2, 3)$.

Exam Practice tests cover the whole chapter and provide quick, effective feedback on your progress.

Chapter Summaries state the most important points of each chapter.

EXTRA RESOURCES

Interactive practice activities and teacher support are provided online as part of Pearson's ActiveLearn Digital Service. This includes downloadable materials in the Teacher's Resource Pack for Student Books 1 and 2:

• 150 lesson plans
• 100 prior knowledge presentations and worksheets
• 90 starter activities, presentations and worksheets
• 200 videos and animations
• Pearson progression self-assessment charts.

ASSESSMENT OBJECTIVES

The following tables give an overview of the assessment for this course.

We recommend that you study this information closely to help ensure that you are fully prepared for this course and know exactly what to expect in the assessment.

PAPER 1	PERCENTAGE	MARK	TIME	AVAILABILITY
HIGHER TIER MATHS A Written examination paper Paper code 4MA1/3H Externally set and assessed by Edexcel	50%	100	2 hours	January and June examination series First assessment June 2018
PAPER 2	**PERCENTAGE**	**MARK**	**TIME**	**AVAILABILITY**
HIGHER TIER MATHS A Written examination paper Paper code 4MA1/4H Externally set and assessed by Edexcel	50%	100	2 hours	January and June examination series First assessment June 2018

ASSESSMENT OBJECTIVES AND WEIGHTINGS

ASSESSMENT OBJECTIVE	DESCRIPTION	% IN INTERNATIONAL GCSE
AO1	Demonstrate knowledge, understanding and skills in number and algebra: • numbers and the numbering system • calculations • solving numerical problems • equations, formulae and identities • sequences, functions and graphs	57–63%
AO2	Demonstrate knowledge, understanding and skills in shape, space and measures: • geometry and trigonometry • vectors and transformation geometry	22–28%
AO3	Demonstrate knowledge, understanding and skills in handling data: • statistics • probability	12–18%

ASSESSMENT SUMMARY

The Edexcel International GCSE (9–1) in Mathematics (Specification A) **Higher Tier** requires students to demonstrate application and understanding of the following topics.

NUMBER
- Use numerical skills in a purely mathematical way and in real-life situations.

ALGEBRA
- Use letters as equivalent to numbers and as variables.
- Understand the distinction between expressions, equations and formulae.
- Use algebra to set up and solve problems.
- Demonstrate manipulative skills.
- Construct and use graphs.

GEOMETRY
- Use the properties of angles.
- Understand a range of transformations.
- Work within the metric system.
- Understand ideas of space and shape.
- Use ruler, compasses and protractor appropriately.

STATISTICS
- Understand basic ideas of statistical averages.
- Use a range of statistical techniques.
- Use basic ideas of probability.

Students should also be able to demonstrate **problem-solving skills** by translating problems in mathematical or non-mathematical contexts into a process or a series of mathematical processes.

Students should be able to demonstrate **reasoning skills** by
- making deductions and drawing conclusions from mathematical information
- constructing chains of reasoning
- presenting arguments and proofs
- interpreting and communicating information accurately.

CALCULATORS

Students will be expected to have access to a suitable electronic calculator for both examination papers. The electronic calculator to be used by students attempting **Higher Tier** examination papers (3H and 4H) should have these functions as a minimum:

$+, -, \times, \div, x^2, \sqrt{x}$, memory, brackets, $x^y, x^{\frac{1}{y}}, \bar{x}, \Sigma x, \Sigma fx$, standard form, sine, cosine, tangent and their inverses.

PROHIBITIONS

Calculators with any of the following facilities are prohibited in all examinations:
- databanks
- retrieval of text or formulae
- QWERTY keyboards
- built-in symbolic algebra manipulations
- symbolic differentiation or integration.

UNIT 6

The probability of throwing a six when you roll a dice is $\frac{1}{6}$

The cells of a beehive are hexagons (a polygon with 6 sides).

6 is the smallest perfect number: a number whose divisors add up to that number ($1 \times 2 \times 3 = 6$ and $1 + 2 + 3 = 6$).

The product of any three consecutive integers is always divisible by 6.

NUMBER 6

Until the early part of the 17th century, all calculations were done by long multiplication and division, which was very laborious. In 1614, Scottish mathematician John Napier showed how multiplication could be simplified using special tables called 'logarithms'. These changed the tasks of multiplying and dividing into addition and subtraction. They also simplified the calculation of powers and roots. The ideas behind these tables come from the index laws $a^m \times a^n = a^{m+n}$ and $a^m \div a^n = a^{m-n}$. Nowadays calculators can do all these tasks much faster and more accurately!

John Napier 1550–1617 ▶

LEARNING OBJECTIVES

- Recognise and use direct proportion
- Recognise and use inverse proportion

- Use index laws to simplify numerical expressions involving negative and fractional indices

BASIC PRINCIPLES

- Plot curved graphs.

- Make a table of values, plotting the points and joining them using a smooth curve.

- Rules of **indices**:
    - When multiplying, add the indices: $x^m \times x^n = x^{m+n}$
    - When dividing, subtract the indices: $x^m \div x^n = x^{m-n}$
    - When raising to a **power**, multiply the indices: $(x^m)^n = x^{mn}$

- Express numbers in **prime factor** form: $72 = 8 \times 9 = 2^3 \times 3^2$

- Convert metric units of length.

- Understand **ratio**.

DIRECT PROPORTION

If two quantities are in **direct proportion**, then when one is multiplied or divided by a number, so is the other.

For example, if 4 kg of apples cost $6, then 8 kg cost $12, 2 kg cost $3 and so on.

This relationship produces a straight-line graph through the origin.

When two quantities are in direct proportion, the graph of the relationship will always be a straight line through the origin.

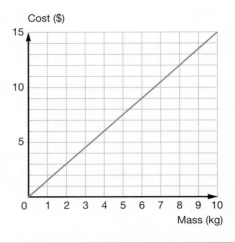

EXAMPLE 1

SKILLS

PROBLEM SOLVING

The cost of a phone call is directly proportional to its length. A three-minute call costs $4.20.

a What is the cost of an eight-minute call?   **b** A call costs $23.10. How long is it?

a 3 minutes costs $4.20
  $\Rightarrow$ 1 minute costs $4.20 \div 3 = \$1.40$
  $\Rightarrow$ 8 minutes costs $8 \times \$1.40 = \$11.20$

b 1 minute costs $1.40
  $\Rightarrow$ \$1 gives $\frac{1}{1.40}$ minutes
  $\Rightarrow$ \$23.10 gives $23.10 \times \frac{1}{1.40}$ minutes = 16.5 minutes

In Example 1, the graph of cost plotted against the number of minutes is a straight line through the origin.

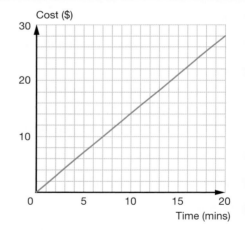

KEY POINTS

If two quantities are in direct proportion:
- when one is multiplied or divided by a number, so is the other
- their ratio stays the same as they increase or decrease
- the graph of the relationship will always be a straight line through the origin.

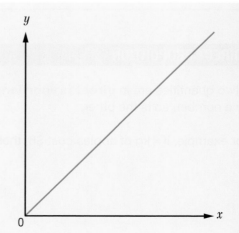

EXERCISE 1

1 ▶ Are these pairs of quantities in direct proportion?
Give reasons for your answers.

  a 12 hotdogs cost £21.60, 15 hotdogs cost £27.00
  b 5 apples cost £1.60, 9 apples cost £2.97
  c Tom took 45 minutes to run 10 km, Ric took 1 hour 2 minutes to run 14 km

2 ▶ The table shows the distance travelled by a car over a period of time.

DISTANCE, s (km)	3	6	15	24
TIME, t (minutes)	4	8	20	32

  a Is s in direct proportion to t? Give a reason for your answer.
  b What is the formula connecting s and t?
  c Work out the distance travelled after 24 minutes.

3 ▶ A band of four people can play a song in 4 minutes 30 seconds. How long
does it take a band of eight people to play the same song?

4 ▶ In a science experiment, the speed of a ball bearing is measured at different times as it rolls
down a slope. The table shows the results.

TIME, t (s)	1	2	3	4	5
SPEED, v (m/s)	0.98	1.96	2.94	3.92	4.9

Are time and speed in direct proportion? Give reasons for your answer.

5 ▶ The time it takes a kettle to boil some water, t, is directly proportional to the volume, V, of water in the kettle.

a Copy and complete the table.

t (seconds)	60		140		260
V (cm³)		500	700	1000	

b What is the formula connecting V and t?

c How many minutes will it take to boil 1.5 litres of water?

6 ▶ A queen termite lays 1 million eggs in 18 days.

a How many eggs does the queen lay in one day?

b How many eggs does the queen lay in one minute?

State any assumptions you make.

7 ▶ 4 tickets to a theme park cost $104.

a How much will 18 tickets cost?

b Jo pays $182 for some tickets. How many did she buy?

8 ▶ The cost of wood is directly proportional to its length. A 3000 mm length costs $58.50.

a Find the cost of a 5 m length of this wood.

b Find the length of a piece of this wood that costs $44.85.

EXERCISE 1*

1 ▶ Seven girls take 3 min 30 s to do a dance routine for an exam.
How long will it take a group of nine girls to do the same routine?

2 ▶ The table shows some lengths in both miles and kilometres.

MILES	2.5	8	12.5	14	17.5
KILOMETRES	4	12.8	20	22.4	28

a Are miles and kilometres in direct proportion? Give a reason for your answer.

b What is the ratio of miles : kilometres in the form $1 : n$?

c How many miles is 100 kilometres?

3 ▶ Kishan measures the extension of a spring when different weights are added to it. The table shows the results.

WEIGHT, w (N)	2.5	3.5	4.5	5.5	6.5
EXTENSION, e (mm)	11.25	15.75	22.05	24.75	29.25

The weight and extension are in direct proportion, but Kishan made one mistake in recording his results. Find and correct the mistake.

4 ▶ The local supermarket is offering an exchange rate of 1.2585 euros for £1. Chas receives 377.55 euros in a transaction. How many £ did Chas exchange?

5 ▶ The cost of a newspaper advertisement is directly proportional to the area of the advertisement.

a Copy and complete the table.

AREA, A (cm²)		30		70	100
COST, C ($)	1125		1800	3150	

b Find the formula connecting C and A.
c An advertisement costs $3825. What is the area of the advertisement?

6 ▶ The length of the shadow of an object is directly proportional to its height. A 4.8 m tall lamp post has a shadow 2.1 m long.

a Find the height of a nearby bus stop with a shadow 1.05 m long.
b A nearby church spire is 30.4 m tall. Find the length of its shadow.

7 ▶ A recipe for chocolate cheesecake that serves 6 people uses 270 g of chocolate.

a Naomi wants to make a cheesecake for 8 people. How much chocolate should she buy?
b The chocolate is only available in 450 g bars. If she uses all the chocolate in the cheesecake, and scales the other ingredients accordingly, how many people will the cheesecake serve?

8 ▶ The resistance, R ohms, of some wire is directly proportional to its length, l mm. A piece of wire 600 mm long has a resistance of 2.1 ohms.

a Find the resistance of a 1.5 m length of this wire.
b Find the formula connecting R and l.
c Tamas needs to make a 7.7 ohm resistor. How many metres of wire does he need?

INVERSE PROPORTION

ACTIVITY 1

SKILLS

MODELLING

Ethan and Mia are planning a journey of 120 km. They first work out the time it will take them travelling at various speeds.

Copy and complete the table showing their results.

TIME, t (hours)	2	3	4	5		8
SPEED, v (km/hr)		40			20	
$t \times v$		120				

Plot their results on a graph of v km/hr against t hours for $0 \le t \le 8$

Use your graph to find the speed required to do the journey in $2\frac{1}{2}$ hours.

Is there an easier way to find this speed?

The quantities (time and speed) in Activity 1 are in **inverse proportion**.

The graph you drew in Activity 1 is called a **reciprocal graph**.

In Activity 1 when the two quantities are multiplied together the answer is always 120. The **product** of two quantities is always constant if the quantities are in inverse proportion. This makes it easy to calculate values.

KEY POINTS

When two quantities are in inverse proportion:
- the graph of the relationship is a reciprocal graph
- one quantity increases at the same **rate** as the other quantity decreases, for example, as one doubles (× 2) the other halves (÷ 2)
- their product is constant.

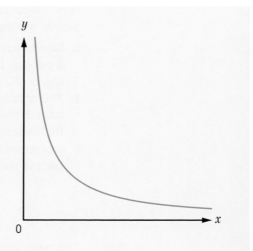

EXAMPLE 2

SKILLS

MODELLING

The emergency services are planning to pump out a flooded area. The number of pumps needed, n, is in inverse proportion to the time taken in days, t. It will take 6 days using two pumps.

a How long will it take using 4 pumps?

b How many pumps are needed to do the job in 2 days?

The quantities are in inverse proportion, so $n \times t$ is constant.
When $n = 2$, $t = 6 \Rightarrow nt = 12$

a If $n = 4$ then $t = 3$ (as $4 \times 3 = 12$) so it will take 3 days.

b If $t = 2$ then $n = 6$ (as $6 \times 2 = 12$) so six pumps are needed.

If the number of pumps is plotted against time a reciprocal graph will be produced.

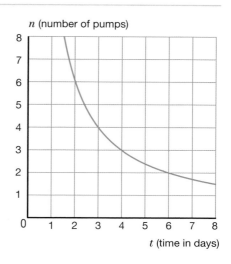

n (number of pumps)

t (time in days)

EXERCISE 2

1 ▶ The time it takes an object to travel a fixed distance is shown in the table.

TIME, t (seconds)	4	8	10	15
SPEED, s (m/s)	150	75	60	40

a Are time and speed inversely proportional? Give a reason for your answer.

b If the time taken is 12 seconds, what is the speed?

2 ▶ The number of burgers sold each day by Ben's Burger Bar is inversely proportional to the temperature.

a Copy and complete this table.

TEMPERATURE, T (°C)	10			25	30
NUMBER SOLD, N		375	300	240	

b Find the formula connecting T and N.

c Plot a graph to show this data.

d Comment on predicted sales as the temperature approaches 0 °C.

3 ▶ It has been estimated that it took 4000 men 30 years to build the largest pyramid at Giza, in Egypt, over 4500 years ago.

Assuming the number of men and the time taken are in inverse proportion,

a How long would it have taken with 6000 men?

b If a **similar** pyramid needed to be built in 10 years, how many men would be required?

4 ▶ Ava finds the time, t seconds, taken to download a music video is inversely proportional to her internet connection speed, s Mb/s. It takes 2 seconds when the speed is 10 Mb/s.

a How long will it take if the internet connection speed is 16 Mb/s?

b What must the internet connection speed be to download it in 0.5 s?

c Find the formula connecting s and t.

5 ▶ The Forth Bridge is a famous bridge over the Firth of Forth in Scotland. It was recently repainted, and it took a team of 400 painters 10 years to complete.

   a If 250 painters had been used, how long would it have taken?

   b How many painters would have been needed to complete the work in 8 years?

6 ▶ The food in Luke's bird feeder lasts for 3 days when 200 birds visit it per day. Luke thinks these two quantities are inversely proportional.

   a Assuming Luke is correct, how many days will the food last if only 150 birds visit it per day?

   b When 300 birds visit per day, the food lasts for 2 days. Is Luke's assumption correct? Give a reason for your answer.

EXERCISE 2*

1 ▶ Two quantities, A and B, are in inverse proportion.

   a Copy and complete the table.

A	3		5		32
B		10	9.6	8	

   b Find the formula connecting A and B.

2 ▶ There are 3 trillion trees on Earth and the concentration of carbon dioxide in the atmosphere is 400 parts per million. Assuming the number of trees is inversely proportional to the concentration of carbon dioxide, find

   a the concentration of carbon dioxide if the number of trees falls to 2 trillion

   b the number of extra trees needed to reduce the concentration of carbon dioxide to 300 parts per million.

3 ▶ Bella's food mixer has 10 speed settings. When the setting is at 6, it takes her 4 minutes to whip some cream. Bella thinks the speed setting is inversely proportional to the time taken.

   a If Bella is correct, how long will it take her to whip the cream at a speed setting of 8?

   b When Bella whips cream at a speed setting of 4, it takes 5 minutes. Is the speed setting inversely proportional to the time taken? Give a reason for your answer.

4 ▶ In an electrical circuit, the current (A amps) is inversely proportional to the resistance (R ohms). The current is 9 amps when the resistance is 16 ohms.

   a Find the current when the resistance is 12 ohms.

   b Find the formula connecting A and R.

   c The current must be limited to 6 amps. What resistance will achieve this?

5 ▶ The Shard in London is 306 m high, making it the tallest building in Europe. It has 11 000 panes of glass which take a team of 6 window cleaners 20 days to clean. The window cleaners have to abseil down the outside of the building to do this.

    a If one of the cleaners falls ill, how long will it take to clean the windows?

    b How many cleaners would be needed to clean the windows in 15 days?

6 ▶ Sienna is training for a charity cycle ride. When she averages 24 km/hr she takes 2 hrs 15 mins to do the course.

    a When she started training she took 2 hrs 42 mins. What was her average speed?

    b When Sienna finally did the ride she averaged 27.5 km/hr. How long, to the nearest minute, did it take her?

ACTIVITY 2

SKILLS

MODELLING

The speed of a professional tennis player's serve is around 250 km/hr. Could you return it?

The speed of the serve is inversely related to the reaction time needed. When the speed of serve is 250 km/hr, the reaction time needed is 0.25 s.

To work out your reaction time, do the following experiment with a 30 cm ruler.

- Ask a friend to hold a ruler near the 30 cm mark, with the ruler hanging vertically.
- Without touching the ruler, place your index finger and thumb either side of the 0 cm mark.
- Your friend must now let go of the ruler without warning, and you must catch it as soon as possible.
- Record the reading (x cm) where you catch the ruler.

1 ▶ Use the formula $t = \sqrt{\dfrac{x}{490}}$ seconds, to convert x into your reaction time.

2 ▶ Calculate a better estimate of your reaction time by repeating the experiment five times to find the **mean** time.

3 ▶ What speed of serve could you return?

FRACTIONAL INDICES

The laws of indices can be extended to fractional indices.

ACTIVITY 3

SKILLS

REASONING

a Use your calculator to work out:

   i $4^{\frac{1}{2}}$ **ii** $9^{\frac{1}{2}}$ **iii** $16^{\frac{1}{2}}$ **iv** $25^{\frac{1}{2}}$ **v** $36^{\frac{1}{2}}$

What does this suggest $a^{\frac{1}{2}}$ means?

Check your theory with some other numbers.

b Use your calculator to work out:

   i $8^{\frac{1}{3}}$ **ii** $27^{\frac{1}{3}}$ **iii** $64^{\frac{1}{3}}$

What does this suggest $a^{\frac{1}{3}}$ means?

Check your theory with some other numbers.

Using the laws of indices, $4^{\frac{1}{2}} \times 4^{\frac{1}{2}} = 4^{\left(\frac{1}{2}+\frac{1}{2}\right)} = 4^1 = 4$

But $\sqrt{4} \times \sqrt{4} = 4$

This means that $4^{\frac{1}{2}} = \sqrt{4} = 2$ so $4^{\frac{1}{2}}$ means the square root of 4

Similarly, $8^{\frac{1}{3}} \times 8^{\frac{1}{3}} \times 8^{\frac{1}{3}} = 8^{\left(\frac{1}{3}+\frac{1}{3}+\frac{1}{3}\right)} = 8^1 = 8$

But $\sqrt[3]{8} \times \sqrt[3]{8} \times \sqrt[3]{8} = 8$

This means that $8^{\frac{1}{3}} = \sqrt[3]{8} = 2$ so $8^{\frac{1}{3}}$ means the cube root of 8

In a similar way it can be shown that $5^{\frac{1}{4}} = \sqrt[4]{5}$, $7^{\frac{1}{5}} = \sqrt[5]{7}$ and so on.

Writing numbers as powers can simplify the **working**.

EXAMPLE 3

Work out $8^{\frac{1}{3}}$

Method 1: Using $x^{\frac{1}{m}} = \sqrt[m]{x}$

$8^{\frac{1}{3}} = \sqrt[3]{8} = 2$

Method 2: Write 8 as a power of 2, then use the index laws to simplify.

$8^{\frac{1}{3}} = (2^3)^{\frac{1}{3}} = 2^{3 \times \frac{1}{3}} = 2^1 = 2$

EXAMPLE 4

Work out $8^{\frac{2}{3}}$

Method 1: Using $x^{\frac{1}{m}} = \sqrt[m]{x}$

$8^{\frac{2}{3}} = \left(8^{\frac{1}{3}}\right)^2 = \left(\sqrt[3]{8}\right)^2 = 2^2 = 4$

Method 2: Write 8 as a power of 2, then use the index laws to simplify.

$8^{\frac{2}{3}} = (2^3)^{\frac{2}{3}} = 2^{3 \times \frac{2}{3}} = 2^2 = 4$

As $8^{\frac{2}{3}} = \left(8^{\frac{1}{3}}\right)^2$, $8^{\frac{2}{3}}$ can be thought of as the square of the cube root of 8.

Alternatively, as $8^{\frac{2}{3}} = (8^2)^{\frac{1}{3}}$ it can be thought of as the cube root of 8 squared.

EXAMPLE 5   Work out $\sqrt[3]{27^2}$

Method 1: Using $x^{\frac{n}{m}} = \left(\sqrt[m]{x}\right)^n$

$\sqrt[3]{27^2} = (27^2)^{\frac{1}{3}} = 27^{\frac{2}{3}} = \left(\sqrt[3]{27}\right)^2 = 3^2 = 9$

Method 2: Write 27 as a power of 3 and write the cube root as a fractional **index**.

$\sqrt[3]{27^2} = [(3^3)^2]^{\frac{1}{3}} = 3^{3 \times 2 \times \frac{1}{3}} = 3^2 = 9$

EXAMPLE 6   Work out $\left(\dfrac{4}{25}\right)^{\frac{3}{2}}$

Method 1: Using $x^{\frac{n}{m}} = \left(\sqrt[m]{x}\right)^n$

$\left(\dfrac{4}{25}\right)^{\frac{3}{2}} = \left(\sqrt{\dfrac{4}{25}}\right)^3 = \left(\dfrac{2}{5}\right)^3 = \dfrac{8}{125}$

Method 2: Write the numbers as powers, then use the index laws to **simplify**.

$\left(\dfrac{4}{25}\right)^{\frac{3}{2}} = \left(\dfrac{2^2}{5^2}\right)^{\frac{3}{2}} = \dfrac{2^{2 \times \frac{3}{2}}}{5^{2 \times \frac{3}{2}}} = \dfrac{2^3}{5^3} = \dfrac{8}{125}$

ACTIVITY 4

SKILLS

REASONING

Use your calculator to check Examples 3, 4, 5 and 6.

KEY POINTS

- Use the rules:
  - $x^{\frac{1}{m}} = \sqrt[m]{x}$
  - $x^{\frac{n}{m}} = \left(\sqrt[m]{x}\right)^n = \sqrt[m]{x^n}$

- Or, write the numbers in prime factor form first.

EXERCISE 3

Work out

1 ▶ $25^{\frac{1}{2}}$

2 ▶ $27^{\frac{1}{3}}$

3 ▶ $16^{\frac{1}{4}}$

4 ▶ $\left(\dfrac{1}{4}\right)^{\frac{1}{2}}$

5 ▶ $9^{\frac{3}{2}}$

6 ▶ $8^{\frac{5}{3}}$

7 ▶ $\left(\dfrac{16}{25}\right)^{\frac{1}{2}}$

8 ▶ $\left(\dfrac{4}{9}\right)^{\frac{3}{2}}$

9 ▶ $\left(\dfrac{27}{8}\right)^{\frac{2}{3}}$

10 ▶ $4^{\frac{5}{2}} \times 64^{\frac{1}{3}}$

11 ▶ $3^{\frac{1}{3}} \times 9^{\frac{1}{3}}$

12 ▶ $4^{\frac{2}{3}} \div 2^{\frac{1}{3}}$

Solve for x

13 ▶ $\sqrt[3]{20} = 20^x$

14 ▶ $3^x = \left(\sqrt{3}\right)^5$

EXERCISE 3*

Work out

1 ▶ $144^{\frac{1}{2}}$

4 ▶ $\left(\dfrac{1}{32}\right)^{\frac{1}{5}}$

7 ▶ $\left(\dfrac{64}{49}\right)^{\frac{1}{2}}$

10 ▶ $16^{\frac{2}{5}} \times 4^{\frac{1}{5}}$

2 ▶ $900^{\frac{1}{2}}$

5 ▶ $81^{\frac{3}{4}}$

8 ▶ $\left(\dfrac{81}{256}\right)^{\frac{3}{4}}$

11 ▶ $25^{\frac{3}{5}} \div 5^{\frac{1}{5}}$

3 ▶ $\left(\dfrac{1}{8}\right)^{\frac{1}{3}}$

6 ▶ $(-125)^{\frac{2}{3}}$

9 ▶ $\left(-3\dfrac{3}{8}\right)^{\frac{4}{3}}$

12 ▶ $8^{\frac{5}{3}} \div \left(\dfrac{16}{25}\right)^{\frac{3}{2}}$

Solve for x

13 ▶ $7^x = \left(\sqrt[4]{7}\right)^3$

14 ▶ $\sqrt[5]{5} = 25^x$

NEGATIVE INDICES

The rules of indices can also be extended to work with negative indices.

In an earlier number chapter you used $10^{-2} = \dfrac{1}{10^2}$, $10^{-3} = \dfrac{1}{10^3}$ and so on.

This also works with numbers other than 10.

EXAMPLE 7

Show that $2^{-2} = \dfrac{1}{2^2}$

$2^2 \div 2^4 = 2^{-2}$ (Using $x^m \div x^n = x^{m-n}$)

But $2^2 \div 2^4 = \dfrac{2^2}{2^4} = \dfrac{1}{2^2} \Rightarrow 2^{-2} = \dfrac{1}{2^2}$

In a similar way it can be shown that $2^{-3} = \dfrac{1}{2^3}$, $5^{-4} = \dfrac{1}{5^4}$ and so on.

Note that $2^{-1} = \dfrac{1}{2^1} = \dfrac{1}{2}$

EXAMPLE 8

Work out $8^{-\frac{2}{3}}$

$8^{-\frac{2}{3}} = \dfrac{1}{8^{\frac{2}{3}}} = \dfrac{1}{4}$ (Using the result of Example 4)

EXAMPLE 9

Work out $\left(\dfrac{2}{3}\right)^{-2}$

$\left(\dfrac{2}{3}\right)^{-2} = \dfrac{1}{\left(\dfrac{2}{3}\right)^2} = 1 \times \left(\dfrac{3}{2}\right)^2 = \dfrac{9}{4}$

Example 9 shows that $\left(\dfrac{x}{y}\right)^{-n} = \dfrac{1}{\left(\dfrac{x}{y}\right)^n} = \left(\dfrac{y}{x}\right)^n$

EXAMPLE 10

Work out $\left(\dfrac{16}{81}\right)^{-\frac{3}{4}}$

$\left(\dfrac{16}{81}\right)^{-\frac{3}{4}} = \left(\dfrac{81}{16}\right)^{\frac{3}{4}} = \dfrac{\left(\sqrt[4]{81}\right)^3}{\left(\sqrt[4]{16}\right)^3} = \dfrac{3^3}{2^3} = \dfrac{27}{8}$

EXAMPLE 11

Show that $2^0 = 1$

$2^3 \div 2^3 = 2^0$ (Using $x^m \div x^n = x^{m-n}$)

But $2^3 \div 2^3 = \dfrac{2^3}{2^3} = 1 \Rightarrow 2^0 = 1$

In a similar way it can be shown that $3^0 = 1$, $4^0 = 1$, and so on.

KEY POINTS

- $x^{-n} = \dfrac{1}{x^n}$ for any number n, $x \neq 0$

- $\left(\dfrac{x}{y}\right)^{-n} = \dfrac{1}{\left(\dfrac{x}{y}\right)^n} = \left(\dfrac{y}{x}\right)^n$, $x, y \neq 0$

- $x^{-1} = \dfrac{1}{x}$, $x \neq 0$

- $x^0 = 1$, $x \neq 0$

EXERCISE 4

Work out

1 ▶ 3^{-2}

2 ▶ 2^{-3}

3 ▶ 4^{-1}

4 ▶ $(3^{-2})^{-1}$

5 ▶ $3^{-2} \div 3^{-1}$

6 ▶ 5^0

7 ▶ $\left(\dfrac{1}{2}\right)^{-1}$

8 ▶ $\left(\dfrac{3}{4}\right)^{-1}$

9 ▶ $\left(\dfrac{1}{3}\right)^{-4}$

10 ▶ $\left(\dfrac{2}{3}\right)^{-3}$

11 ▶ $\left(\dfrac{5}{3}\right)^{-2}$

12 ▶ $8^{-\frac{1}{3}}$

13 ▶ $25^{-\frac{1}{2}}$

14 ▶ $16^{-\frac{3}{4}}$

15 ▶ $1^{-\frac{3}{2}}$

16 ▶ $\left(\dfrac{1}{4}\right)^{-\frac{1}{2}}$

17 ▶ $\left(\dfrac{1}{8}\right)^{-\frac{2}{3}}$

18 ▶ $\left(\dfrac{4}{9}\right)^{-\frac{1}{2}}$

19 ▶ $\dfrac{4^{-3} \times 4^{-1}}{4^{-4}}$

20 ▶ $27^{-\frac{1}{3}} \times 9^{\frac{3}{2}}$

Solve for x

21 ▶ $\dfrac{1}{7} = 7^x$

22 ▶ $\left(\dfrac{1}{4}\right)^x = 4$

23 ▶ $\left(\dfrac{3}{2}\right)^x = \dfrac{2}{3}$

24 ▶ $\left(\dfrac{2}{3}\right)^x = \dfrac{9}{4}$

25 ▶ $\left(\dfrac{16}{25}\right)^x = \dfrac{5}{4}$

EXERCISE 4*

Work out

1 ▶ 5^{-2}

2 ▶ 4^{-3}

3 ▶ 12^{-1}

4 ▶ $\left(\dfrac{1}{4}\right)^{-1}$

5 ▶ $\left(\dfrac{2}{7}\right)^{-2}$

6 ▶ $\left(\dfrac{4}{5}\right)^{-3}$

7 ▶ $\left(1\dfrac{2}{3}\right)^{-2}$

8 ▶ $64^{-\frac{1}{2}}$

9 ▶ $16^{-\frac{1}{4}}$

10 ▶ $8^{-\frac{5}{3}}$

11 ▶ $1^{-\frac{4}{3}}$

12 ▶ $\left(\dfrac{1}{81}\right)^{-\frac{1}{2}}$

13 ▶ $\left(\dfrac{1}{125}\right)^{-\frac{2}{3}}$

14 ▶ $\left(\dfrac{27}{8}\right)^{-\frac{4}{3}}$

15 ▶ $\left(\dfrac{162}{32}\right)^{-\frac{3}{4}}$

16 ▶ $2 \div 32^{-\frac{1}{5}}$

17 ▶ $\left[\left(\frac{5}{7}\right)^{-\frac{2}{3}}\right]^{-3}$ **18 ▶** $\dfrac{7^{-2} \times 7^{-3}}{7^{-5}}$ **19 ▶** $\left(\frac{4}{25}\right)^{-\frac{1}{2}} \times \left(\frac{8}{27}\right)^{\frac{1}{3}}$ **20 ▶** $\left(\frac{27}{125}\right)^{-\frac{2}{3}} \div \left(\frac{81}{64}\right)^{-\frac{1}{2}}$

Solve for x

21 ▶ $\left(\frac{4}{5}\right)^{x} = \frac{25}{16}$ **23 ▶** $\left(\sqrt[3]{5}\right)^{4} = 25^{x}$ **25 ▶** $8^{-\frac{8}{3}} = \left(\frac{1}{4}\right)^{x}$

22 ▶ $\left(\frac{49}{64}\right)^{x} = \frac{8}{7}$ **24 ▶** $\left(\frac{1}{8}\right)^{12} = 16^{x}$

EXERCISE 5   REVISION

1 ▶ The table gives the relationship between two **variables**.

VARIABLE A	3	5	8	9	13
VARIABLE B	12	20	32	36	51

Are A and B in direct proportion? Give a reason for your answer.

2 ▶ The pressure of water on an object is directly proportional to its depth.

a Copy and complete the table.

DEPTH, d (metres)	5		12		40
PRESSURE, P (bars)		0.8	1.2	2.5	

b Find the formula connecting pressure and depth.
c A diver's watch has been guaranteed to work at a pressure up to 8.5 bars. A diver takes the watch down to 75 m. Will the watch still work? Give a reason for your answer.

3 ▶ The weight of a steel pipe is directly proportional to its length. A 3 m length of pipe weighs 45.75 kg.

a Find the weight of a 10 m length of this pipe.
b A piece of this pipe weighs 122 kg. How long is it?

4 ▶ The number of workers needed to install the seats in a stadium is inversely proportional to the number of days it takes.

a Copy and complete the table.

NUMBER OF WORKERS, w	4	8		2
NUMBER OF DAYS, d		6	8	

b Find the formula connecting d and w.
c How many workers are needed to install the seats in one day?

5 ▶ When making jewellery, Erin has to glue the stones into the setting. She finds that the time taken for the glue to set is inversely proportional to the temperature. When the temperature is 20 °C, the glue takes 24 hours to set fully.

a One day the heating in Erin's workshop fails and the temperature drops to 15 °C. How long will it take the glue to set?
b Erin needs to finish some jewellery in a hurry and needs the glue to set in 8 hours. What temperature will she need to achieve this?

6 ▶ Chuck has a large farm growing corn. When he uses all three of his combine harvesters it takes 4 days to harvest the corn. Assuming the number of combine harvesters and the days taken to harvest are in inverse proportion, find

    a how long it will take to harvest if one of his combine harvesters breaks down

    b how many harvesters are needed to harvest the corn in $2\frac{1}{2}$ days.

Work out

7 ▶ $36^{\frac{1}{2}}$ **12 ▶** $16^{\frac{3}{2}}$ **17 ▶** $\left(\frac{2}{3}\right)^{-2}$ **22 ▶** $(2^2)^{-1}$

8 ▶ $8^{\frac{1}{3}}$ **13 ▶** $1^{\frac{2}{3}}$ **18 ▶** $125^{-\frac{1}{3}}$ **23 ▶** $2^2 \times 2^{-1}$

9 ▶ $81^{\frac{1}{4}}$ **14 ▶** $\left(\frac{64}{125}\right)^{\frac{1}{3}}$ **19 ▶** $27^{-\frac{2}{3}}$ **24 ▶** $\dfrac{3^{-1} \times 3^{-2}}{3^{-4}}$

10 ▶ $\left(\frac{1}{9}\right)^{\frac{1}{2}}$ **15 ▶** 4^{-3} **20 ▶** $\left(\frac{1}{64}\right)^{-\frac{1}{2}}$ **25 ▶** $\frac{1}{5} \div 125^{-\frac{1}{3}}$

11 ▶ $1000^{\frac{2}{3}}$ **16 ▶** $\left(\frac{1}{2}\right)^{-3}$ **21 ▶** $\left(\frac{1}{32}\right)^{-\frac{3}{5}}$

Solve for x

26 ▶ $\left(\sqrt[3]{2}\right)^3 = 2^x$ **27 ▶** $\left(\frac{1}{3}\right)^x = 3$ **28 ▶** $\left(\frac{25}{4}\right)^x = \frac{2}{5}$

EXERCISE 5*   **REVISION**

1 ▶ The force on a mass is directly proportional to the **acceleration** of the mass.

    a Copy and complete the table.

FORCE, F (N)	1.8			10.8	18
ACCELERATION, a (m/s²)		0.7	1.2	3	

    b Find the formula connecting F and a.

    c A force of 90 N is applied. What is the acceleration?

2 ▶ Mira is downloading some computer games. The download time is directly proportional to the file size. A 45 MB file takes $3\frac{1}{3}$ seconds to download.

   a How long will it take her to download an 81 MB file?
   b One game takes 12 seconds to download. How large is the file?

3 ▶ The table gives the relationship between two variables.

VARIABLE X	40	7.2	5	4.8	2.4
VARIABLE Y	3.6	20	28.8	30	60

Are the variables in inverse proportion? Give a reason for your answer.

4 ▶ The time taken to fill a tank with water is inversely proportional to the number of pipes filling it.

   a Copy and complete the table.

NUMBER OF PIPES, n		4		8	10
TIME, t (hrs)	18		3	2.25	

   b Find the formula connecting n and t.
   c The tank must be filled in 1.5 hrs. How many pipes are needed?

5 ▶ A world record pizza was just over 40 m in **diameter**. If it had been cut into 80 cm² pieces it would have fed 157 700 people.

   a If it had been cut into 100 cm² pieces, how many people could have been fed?
   b What area of piece would be needed to feed 252 320 people?

6 ▶ The size of the **exterior angle** of a regular **polygon** is inversely proportional to the number of sides. The exterior angle of a square is 90°.

   a The exterior angle of a regular polygon is 9°. How many sides does it have?
   b If the number of sides of a regular polygon is doubled, by what factor does the exterior angle change?

Work out

7 ▶ $121^{\frac{1}{2}}$ **12 ▶** $\left(\frac{27}{64}\right)^{\frac{1}{3}}$ **17 ▶** 5^{-3} **22 ▶** $\left(-\frac{1}{27}\right)^{-\frac{1}{3}}$

8 ▶ $(-125)^{\frac{1}{3}}$ **13 ▶** $1^{\frac{2}{5}}$ **18 ▶** $\left(\frac{1}{4}\right)^{-4}$ **23 ▶** $\left(\frac{1}{81}\right)^{-\frac{3}{4}}$

9 ▶ $256^{\frac{1}{4}}$ **14 ▶** π^{0} **19 ▶** $81^{-\frac{1}{4}}$ **24 ▶** $\left(\frac{81}{16}\right)^{\frac{3}{4}} \times \left(\frac{9}{25}\right)^{-\frac{3}{2}}$

10 ▶ $10000^{\frac{3}{4}}$ **15 ▶** $\left(1\frac{61}{64}\right)^{\frac{2}{3}}$ **20 ▶** $(-8)^{-\frac{1}{3}}$ **25 ▶** $\left(\frac{27}{8}\right)^{-\frac{2}{3}} \div \left(\frac{81}{256}\right)^{-\frac{3}{4}}$

11 ▶ $8^{\frac{4}{3}}$ **16 ▶** 2^{-5} **21 ▶** $27^{-\frac{4}{3}}$

Solve for x

26 ▶ $\frac{4}{25} = \left(\frac{5}{2}\right)^{x}$ **27 ▶** $\left(\frac{125}{64}\right)^{x} = \frac{4}{5}$ **28 ▶** $\left(\frac{3}{2}\right)^{18} = \left(\frac{8}{27}\right)^{x}$

EXAM PRACTICE: NUMBER 6

1   The cost of fuel is directly proportional to the volume bought.

   a   Copy and complete the table.

VOLUME, v (litres)		15		45
COST, C (£)	8.5	17	34	

   b   Find the formula connecting C and v.   **[4]**

2   Oliver finds that the number of pages of homework he writes is inversely proportional to the loudness of his music. When the volume control is set at 12 he writes 4 pages.

   a   If the volume control is set at 16, how many pages will he write?

   b   Oliver needs to write 6 pages to keep his teacher happy. What volume control setting should he use?

[4]

3   Work out

   a $64^{\frac{1}{2}}$ **e** $9^{\frac{3}{2}}$ **i** $\left(\frac{1}{3}\right)^{-1}$ **m** $\left(\frac{1}{36}\right)^{-\frac{1}{2}}$

   b $64^{\frac{1}{3}}$ **f** $16^{\frac{3}{4}}$ **j** $\left(\frac{3}{2}\right)^{-3}$ **n** $\left(\frac{81}{16}\right)^{-\frac{3}{4}}$

   c $32^{\frac{3}{5}}$ **g** $\left(2\frac{1}{4}\right)^{\frac{1}{2}}$ **k** $27^{-\frac{1}{3}}$ **o** $1 \div 16^{-\frac{1}{4}}$

   d $\left(\frac{8}{27}\right)^{\frac{1}{3}}$ **h** 3^{-4} **l** $8^{-\frac{2}{3}}$ **p** $16^{-\frac{1}{2}} \div 64^{-\frac{2}{3}}$

   Solve for x

   q $\left(\frac{8}{27}\right)^{x} = \frac{3}{2}$ **[17]**

                                                       [Total 25 marks]

CHAPTER SUMMARY: NUMBER 6

DIRECT PROPORTION

If two quantities are in direct proportion
- the graph of the relationship will always be a straight line through the origin
- when one is multiplied or divided by a number, so is the other
- their ratio stays the same as they increase or decrease.

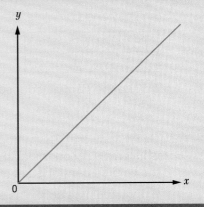

The cost of ribbon is directly proportional to its length.

A 3.5 m piece of ribbon costs £2.38. What is the cost of 8 m of this ribbon?

1 m of ribbon costs £2.38 ÷ 3.5 = £0.68 ⇒ 8 m costs 8 × £0.68 = £5.44

INVERSE PROPORTION

If two quantities are in inverse proportion
- the graph of the relationship is a reciprocal graph
- one quantity increases at the same rate as the other quantity decreases, for example, as one doubles (× 2) the other halves (÷ 2)
- their product is constant.

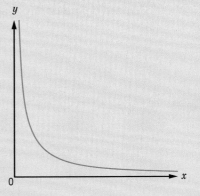

A farmer has enough food for 250 chickens (c) for 24 days (d).

He buys 50 more chickens. For how many days will the food last?

The quantities are in inverse proportion
so $c \times d$ will be constant and equal to $250 \times 24 = 6000$

The number of chickens is now 300 ⇒ $300 \times d = 6000 \Rightarrow d = 20$ so it will last for 20 days.

FRACTIONAL INDICES

Write the numbers in prime factor form first:

- $x^{\frac{1}{m}} = \sqrt[m]{x}$ $\quad 8^{\frac{1}{3}} = \sqrt[3]{8} = 2$ $\quad$ or $\quad 8^{\frac{1}{3}} = (2^3)^{\frac{1}{3}} = 2^{3 \times \frac{1}{3}} = 2^1 = 2$

- $x^{\frac{n}{m}} = \left(\sqrt[m]{x}\right)^n = \sqrt[m]{x^n}$ $\quad 8^{\frac{2}{3}} = \left(\sqrt[3]{8}\right)^2 = 2^2 = 4$ $\quad$ or $\quad 8^{\frac{2}{3}} = (2^3)^{\frac{2}{3}} = 2^{3 \times \frac{2}{3}} = 2^2 = 4$

NEGATIVE INDICES

- $x^{-n} = \frac{1}{x^n}$ for any number n, $x \neq 0$ $\qquad\qquad 2^{-3} = \frac{1}{2^3} = \frac{1}{8}$

- $\left(\frac{x}{y}\right)^{-n} = \frac{1}{\left(\frac{x}{y}\right)^n} = \left(\frac{y}{x}\right)^n$, $x, y \neq 0$ $\qquad \left(\frac{3}{5}\right)^{-2} = \frac{1}{\left(\frac{3}{5}\right)^2} = \left(\frac{5}{3}\right)^2 = \frac{25}{9}$

- $x^{-1} = \frac{1}{x}$, $x \neq 0$ $\qquad\qquad 5^{-1} = \frac{1}{5}$

- $x^0 = 1$, $x \neq 0$ $\qquad\qquad 2^0 = 1$

ALGEBRA 6

The Vitruvian Man is a drawing by Leonardo da Vinci from around 1490. For example, the head measured from the forehead to the chin was exactly one tenth of the total height. It was found with notes based on the studies of the architect Vitruvius who tried to find connections between proportions found in science, nature, art and the human body.

Leonardo da Vinci 1452–1519 ▶

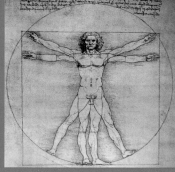

The Vitruvian Man ▶

LEARNING OBJECTIVES

■ Write and use formulae to solve problems involving direct proportion

■ Write and use formulae to solve problems involving inverse proportion

■ Use index notation involving fractional, negative and zero powers

BASIC PRINCIPLES

■ Use formulae relating one **variable** to another, for example $A = \pi r^2$

■ Substitute numerical values into formulae and relate the answers to applied real situations.

■ Derive simple formulae (linear, quadratic and cubic).

■ Know and use the laws of **indices**:
$a^m \times a^n = a^{m+n}$
$a^m \div a^n = a^{m-n}$
$(a^m)^n = a^{mn}$

■ Know and use the laws of indices involving fractional, negative and zero **powers** in a number context.

PROPORTION

If two quantities are related to each other, given enough information, it is possible to write a formula describing this relationship.

ACTIVITY 1

Copy and complete this table to show which paired items are related.

VARIABLES	RELATED? (YES OR NO)
Area of a circle A and its **radius** r	Y
Circumference of a circle C and its radius r	
Distance travelled, D, at a constant speed, x	
Mathematical ability, M, and a person's height, h	
Cost of a tin of paint, P, and its density, d	
Weight of water, W, and its volume, v	
Value of a painting, V, and its area, a	
Swimming speed, S, and collar size, x	

DIRECT PROPORTION – LINEAR

When water is poured into an empty cubical fish tank, each litre that is poured in increases the depth by a fixed amount.

 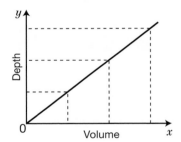

A graph of depth, y, against volume, x, is a straight line through the origin, showing a linear relationship.

In this case, y is **directly proportional** to x. If y is doubled, so is x. If y is halved, so is x, and so on.

This relationship can be expressed in *any* of these ways:

- y is directly proportional to x.
- y varies directly with x.
- y varies as x.

All these statements have the same meaning.

In symbols, direct proportion relationships can be written as $y \propto x$. The $\propto$ **sign** can then be replaced by '$= k$' to give the formula $y = kx$, where k is the constant of proportionality.

The graph of $y = kx$ is the equation of a straight line through the origin, with **gradient** k.

EXAMPLE 1

SKILLS

PROBLEM
SOLVING

The extension, y cm, of a spring is directly proportional to the **mass**, x kg, hanging from it.

If $y = 12$ cm when $x = 3$ kg, find

a the formula for y in terms of x

b the extension y cm when a 7 kg mass is attached

c the mass x kg that produces a 20 cm extension.

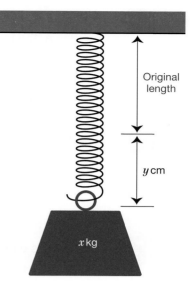

a y is directly proportional to x, so $y \propto x \Rightarrow y = kx$
$y = 12$ when $x = 3$, so $12 = k \times 3$
therefore, $k = 4$
Hence the formula is $y = 4x$

b When $x = 7$, $y = 4 \times 7 = 28$
The extension produced from a 7 kg mass is 28 cm.

c When $y = 20$ cm, $20 = 4x$
therefore, $x = 5$

A 5 kg mass produces a 20 cm extension.
The graph of extension plotted against mass would look like this.

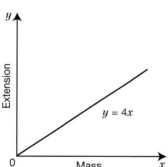

KEY POINTS

■ y is directly proportional to x is written as $y \propto x$ and this means $y = kx$, for some constant k.

■ The graph of y against x is a straight line through the origin.

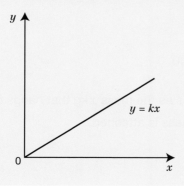

EXERCISE 1

1 ▶ y is directly proportional to x. If $y = 10$ when $x = 2$, find

    a the formula for y in terms of x

    b y when $x = 6$

    c x when $y = 25$

    d Sketch the graph of y against x.

2 ▶ d is directly proportional to t. If $d = 100$ when $t = 25$, find

    a the formula for d in terms of t

    b d when $t = 15$

    c t when $d = 180$

3 ▶ y is directly proportional to x.

    $y = 15$ when $x = 3$

    a Express y in terms of x.

    b Find y when $x = 10$

    c Find x when $y = 65$

4 ▶ y is directly proportional to x.

    $y = 52$ when $x = 8$

    a Write a formula for y in terms of x.

    b Find y when $x = 14$

    c Find x when $y = 143$

    d Sketch the graph of y against x.

5 ▶ y is directly proportional to x.

    a $y = 6$ when $x = 4$. Find x when $y = 7.5$

    b $y = 31.5$ when $x = 7$. Find x when $y = 45.5$

    c $y = 8$ when $x = 5$. Find x when $y = 13$

6 ▶ y is directly proportional to x.

    When $x = 600$, $y = 10$

    a Find a formula for y in terms of x.

    b Calculate the value of y when $x = 540$

7 ▶   An elastic string's extension y cm varies as the mass x kg that hangs from it.

    The string extends 4 cm when a 2 kg mass is attached.

    a Find the formula for y in terms of x.

    b Find y when $x = 5$

    c Find x when $y = 15$

8 ▶ A bungee jumping rope's extension e m varies as the mass M kg of the person attached to it.

If $e = 4$ m when $M = 80$ kg, find

a the formula for e in terms of M

b the extension for a person with a mass of 100 kg

c the mass of a person when the extension is 6 m.

9 ▶ An ice-cream seller discovers that, on any particular day, the number of sales (I) is directly proportional to the temperature (t °C). 1500 sales are made when the temperature is 20 °C.

a Find the formula for I in terms of t.

b How many sales might be expected on a day with a temperature of 26 °C?

10 ▶ The number of people in a swimming pool (N) varies with the daily temperature (t °C). 175 people swim when the temperature is 25 °C.

a Find the formula for N in terms of t.

b The pool's capacity is 200 people. Will people have to queue and wait if the temperature reaches 30 °C?

EXERCISE 1*

1 ▶ The speed of a stone, v m/s, falling off a cliff is directly proportional to the time, t seconds, after release. Its speed is 4.9 m/s after 0.5 s.

a Find the formula for v in terms of t.

b What is the speed after 5 s?

c At what time is the speed 24.5 m/s?

2 ▶ The cost, c cents, of a tin of salmon varies directly with its mass, m g.

The cost of a 450 g tin is 150 cents.

a Find the formula for c in terms of m.

b How much does a 750 g tin cost?

c What is the mass of a tin costing $2?

3 ▶ The distance a honey bee travels, d km, is directly proportional to the mass of the honey, m g, that it produces. A bee travels 150 000 km to produce 1 kg of honey.

a Find the formula for d in terms of m.

b What distance is travelled by a bee to produce 10 g of honey?

c What mass of honey is produced by a bee travelling once around the world – a distance of 40 000 km?

4 ▶ The mass of sugar, m g, used in making chocolate cookies varies directly with the number of cookies, n. 3.25 kg of sugar is used to make 500 cookies.

 a Find the formula for m in terms of n.

 b What mass of sugar is needed for 150 cookies?

 c How many cookies can be made using 10 kg of sugar?

5 ▶ The height of a tree, h m, varies directly with its age, y years.

 A 9 m tree is 6 years old.

 a Find the formula for h in terms of y.

 b What height is a tree that is 6 months old?

 c What is the age of a tree that is 50 cm tall?

6 ▶ The distance, d (in km), covered by an aeroplane is directly proportional to the time taken, t (in hours).

 The aeroplane covers a distance of 1600 km in 3.2 hours.

 a Find the formula for d in terms of t.

 b Find the value of d when $t = 5$

 c Find the value of t when $d = 2250$

 d What happens to the distance travelled, d, when the time, t, is

  i doubled   **ii** halved?

7 ▶ The cost, C (in £), of a newspaper advert is directly proportional to the area, A (in cm²), of the advert.

 An advert with an area of 40 cm² costs £2000.

 a Sketch a graph of C against A.

 b Write a formula for C in terms of A.

 c Use your formula to work out the cost of an 85 cm² advert.

8 ▶ y is directly proportional to x.

 $y = 46$ when $x = 6$

 a Write a formula for y in terms of x.

 b Find y when $x = 24$

 c Find x when $y = 161$

9 ▶ The distance, d (in km), covered by a long-distance runner is directly proportional to the time taken, t (in hours).

 The runner covers a distance of 42 km in 4 hours.

 a Find a formula for d in terms of t.

 b Find the value of d when $t = 8$

 c Find the value of t when $d = 7.7$

 d What happens to the distance travelled, d, when the time, t, is

  i trebled   **ii** divided by 3?

10 ▶ The amount, C (in £), that a plumber charges is directly proportional to the time, t (in hours), that the plumber works. A plumber earns £247.50 when she works 5.5 hours.

    a Sketch a graph of C against t.

    b Write a formula for C in terms of t.

    c Use your formula to work out how many hours the plumber has worked when she earns £1035.

DIRECT PROPORTION – NONLINEAR

Water is poured into an empty inverted cone. Each litre poured in will result in a different depth increase.

A graph of volume, y, against depth, x, will illustrate a direct **nonlinear relationship**.

 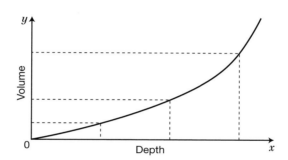

This is the graph of a **cubic** relationship.
This relationship can be expressed in *either* of these ways:

- y is directly proportional to x cubed.

- y varies as x cubed.

Both statements have the same meaning.
In symbols, this relationship is written as $y \propto x^3$
The $\propto$ sign can then be replaced by '$= k$' to give the formula $y = kx^3$, where k is the constant of proportionality.
A quantity can also be directly proportional to the square or the square root of another quantity.

EXAMPLE 2

SKILLS

ANALYSIS

Express these relationships as equations with constants of proportionality.

    a y is directly proportional to x squared.

    b m varies directly with the cube of n.

    c s is directly proportional to the square root of t.

    d v squared varies as the cube of w.

    a $y \propto x^2 \Rightarrow y = kx^2$

    b $m \propto n^3 \Rightarrow m = kn^3$

    c $s \propto \sqrt{t} \Rightarrow s = k\sqrt{t}$

    d $v^2 \propto w^3 \Rightarrow v^2 = kw^3$

KEY POINTS

- If y is proportional to x squared then $y \propto x^2$ and $y = kx^2$, for some constant k.
- If y is proportional to x cubed then $y \propto x^3$ and $y = kx^3$, for some constant k.
- If y is proportional to the square root of x then $y \propto \sqrt{x}$ and $y = k\sqrt{x}$, for some constant k.

EXAMPLE 3

SKILLS

PROBLEM
SOLVING

The cost of Luciano's take-away pizzas (C cents) is directly
proportional to the square of the **diameter** (d cm) of the pizza.
A 30 cm pizza costs 675 cents.

a Find a formula for C in terms of d and use it to find
the price of a 20 cm pizza.

b What size of pizza can you expect for $4.50?

675¢

30 cm

---

a C is proportional to d^2, so $C \propto d^2$ $C = kd^2$
$C = 675$ when $d = 30$ $675 = k(30)^2$
$k = 0.75$

The formula is therefore $C = 0.75d^2$
When $d = 20$ $C = 0.75(20)^2$
$C = 300$

The cost of a 20 cm pizza is 300 cents ($3).

b When $C = 450$ $450 = 0.75d^2$
$d^2 = 600$
$d = \sqrt{600} = 24.5$ (3 s.f.)

A $4.50 pizza should be 24.5 cm in diameter.

EXERCISE 2

1 ▶ y is directly proportional to the square of x. If $y = 100$ when $x = 5$, find
a the formula for y in terms of x
b y when $x = 6$
c x when $y = 64$

2 ▶ p varies directly as the square of q. If $p = 72$ when $q = 6$, find
a the formula for p in terms of q
b p when $q = 3$
c q when $p = 98$

3 ▶ v is directly proportional to the cube of w. If $v = 16$ when $w = 2$, find
a the formula for v in terms of w
b v when $w = 3$
c w when $v = 128$

4 ▶ m varies directly as the square root of n. If $m = 10$ when $n = 1$, find
a the formula for m in terms of n
b m when $n = 4$
c n when $m = 50$

5 ▶   The distance fallen by a parachutist, y m, is directly proportional to the square of the time
taken, t secs. If 20 m are fallen in 2 s, find
a the formula expressing y in terms of t
b the distance fallen through in 3 s
c the time taken to fall 100 m.

6 ▶ 'Espirit' perfume is available in bottles of different volumes of **similar** shapes. The price, P, is directly proportional to the cube of the bottle height, h cm. A 10 cm high bottle is $50. Find
   a the formula for P in terms of h
   b the price of a 12 cm high bottle
   c the height of a bottle of 'Espirit' costing $25.60

7 ▶ The kinetic energy, E (in joules, J), of an object varies in direct proportion to the square of its speed, s (in m/s). An object moving at 5 m/s has 125 J of kinetic energy.
   a Write a formula for E in terms of s.
   b How much kinetic energy does the object have if it is moving at 2 m/s?
   c What speed is the object moving at if it has 192.2 J of kinetic energy?
   d What happens to the kinetic energy, E, if the speed of the object is doubled?

8 ▶ The cost of fuel per hour, C (in £), to move a boat through the water is directly proportional to the cube of its speed, s (in mph).
A boat travelling at 10 mph uses £50 of fuel per hour.
   a Write a formula for C in terms of s.
   b Calculate C when the boat is travelling at 5 mph.

EXERCISE 2*

1 ▶ f is directly proportional to g^2. Copy and complete this table.

g	2	4	
f	12		108

2 ▶ m is directly proportional to n^3. Copy and complete this table.

n	1		5
m	4	32	

3 ▶ In a factory, chemical reactions are carried out in spherical containers. The time, T (in minutes), the chemical reaction takes is directly proportional to the square of the radius, R (in cm), of the spherical container.

When $R = 120$, $T = 32$
   a Write a formula for T in terms of R.
   b Find the value of T when $R = 150$

4 ▶ When an object accelerates steadily from rest, the distance, d (in metres), that it travels varies in direct proportion to the square of the time, t (in seconds), that it has been travelling.

An object moves 176.4 m in 6 seconds.
   a Write a formula for d in terms of t.
   b How far does an object move if it accelerates like this for 10 seconds from rest?
   c How many seconds has an object been accelerating for, if it has moved 1102.5 m?
   d What happens to the distance moved, d, if the time for which an object has been accelerating is doubled?

5 ▶ The volume, V (in cm³), of a sphere is directly proportional to the cube of its radius, r (in cm). A sphere with a radius of 5 cm has a volume of 523.5 cm³.

    a Write a formula for V in terms of r.
    b Calculate V when the radius is 20 cm.
    c Sketch the graph of V against r.

6 ▶ The resistance to motion, R newtons, of the 'Storm' racing car is directly proportional to the square of its speed, s km/hour. When the car travels at 160 km/hour it experiences a 500 newton resistance.

    a Find the formula for R in terms of s.
    b What is the car's speed when it experiences a resistance of 250 newtons?

7 ▶ The height of giants, H metres, is directly proportional to the cube root of their age, y years. An 8-year-old giant is 3 m tall.

    a Find the formula for H in terms of y.
    b What age is a 12 m tall giant?

8 ▶ The surface area of a sphere is directly proportional to the square of its radius.
A sphere of radius 10 cm must be increased to a radius of x cm if its surface area is to be doubled. Find x.

ACTIVITY 2

SKILLS

ANALYSIS

The German astronomer Kepler (1571–1630) created three astronomical laws. Kepler's third law gives the relationship between the orbital period, t days, of a planet around the Sun, and its **mean** distance, d km, from the Sun.

In simple terms, this law states that t^2 is directly proportional to d^3.

Find a formula relating t and d, given that the Earth is 150 million km from the Sun.

Copy and complete the table.

PLANET	d (MILLION KM)	ORBITAL PERIOD AROUND SUN t (Earth days)
Mercury	57.9	
Jupiter		4315

Find the values of d and t for other planets in the Solar System, and see if they fit the same relationship.

INVERSE PROPORTION

The temperature of a cup of coffee decreases as time increases.

A graph of temperature (T) against time (t) shows an **inverse** relationship.

This can be expressed as: 'T is inversely proportional to t'.

In symbols, this is written as $T \propto \dfrac{1}{t}$

The $\propto$ sign can then be replaced by '$= k$' to give the formula $T = \dfrac{k}{t}$ where k is the constant of proportionality.

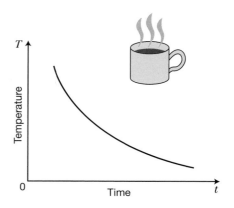

EXAMPLE 4

SKILLS

ANALYSIS

Express these equations as relationships with constants of proportionality.

a y is inversely proportional to x squared.

b m varies inversely as the cube of n.

c s is inversely proportional to the square root of t.

d v squared varies inversely as the cube of w.

a $y \propto \dfrac{1}{x^2} \Rightarrow y = \dfrac{k}{x^2}$

b $m \propto \dfrac{1}{n^3} \Rightarrow m = \dfrac{k}{n^3}$

c $s \propto \dfrac{1}{\sqrt{t}} \Rightarrow s = \dfrac{k}{\sqrt{t}}$

d $v^2 \propto \dfrac{1}{w^3} \Rightarrow v^2 = \dfrac{k}{w^3}$

KEY POINT

■ If y is inversely proportional to x then $y \propto \dfrac{1}{x}$ and $y = \dfrac{k}{x}$, for some constant k. The graph of y plotted against x looks like this.

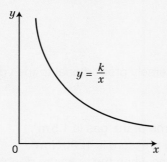

$y = \dfrac{k}{x}$

EXAMPLE 5

SKILLS

PROBLEM SOLVING

Sound intensity, I dB (decibels), is inversely proportional to the square of the distance, d m, from the source. At a music festival, it is 110 dB, 3 m away from a speaker.

a Find the formula relating I and d.
b Calculate the sound intensity 2 m away from the speaker.
c At what distance away from the speakers is the sound intensity 50 dB?

a I is inversely proportional to d^2 $I = \dfrac{k}{d^2}$

$I = 110$ when $d = 3$ $110 = \dfrac{k}{3^2}$

$k = 990$

The formula is therefore $I = \dfrac{990}{d^2}$

b When $d = 2$ $I = \dfrac{990}{2^2} = 247.5$

The sound intensity is 247.5 dB, 2 m away. (This is enough to cause deafness.)

c When $I = 50$ $50 = \dfrac{990}{d^2}$

$d^2 = 19.8$

$d = 4.45$ (3 s.f.)

The sound intensity is 50 dB, 4.45 m away from the speakers.

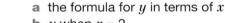

EXERCISE 3

1 ▶ y is inversely proportional to x. If $y = 4$ when $x = 3$, find

  a the formula for y in terms of x
  b y when $x = 2$
  c x when $y = 3$
  d Sketch the graph of y against x.

2 ▶ d varies inversely with t. If $d = 10$ when $t = 25$, find

  a the formula for d in terms of t
  b d when $t = 2$
  c t when $d = 50$

3 ▶ The pressure, P (in N/m²), of a gas is inversely proportional to the volume, V (in m³).
$P = 1500\,\text{N/m}^2$ when $V = 2\,\text{m}^3$

  a Write a formula for P in terms of V.
  b Work out the pressure when the volume of the gas is 1.5 m³.
  c Work out the volume of gas when the pressure is 1200 N/m².
  d What happens to the volume of the gas when the pressure doubles?

4 ▶ The time taken, t (in seconds), to boil water in a kettle is inversely proportional to the power, p (in watts) of the kettle.
A full kettle of power 1500 W boils the water in 400 seconds.

  a Write a formula for t in terms of p.
  b A similar kettle has a power of 2500 W. Can this kettle boil the same amount of water in less than 3 minutes?

5 ▶ m varies inversely with the square of n. If $m = 4$ when $n = 3$, find

  a the formula for m in terms of n
  b m when $n = 2$
  c n when $m = 1$

6 ▶ V varies inversely with the cube of w. If $V = 12.5$ when $w = 2$, find

     a the formula for V in terms of w
     b V when $w = 1$
     c w when $V = 0.8$

7 ▶ Light intensity, I candle-power, is inversely proportional to the square of the distance, $d\,$m, of an object from this light source. If $I = 10^5$ when $d = 2\,$m, find

     a the formula for I in terms of d
     b the light intensity at $2\,$km.

8 ▶ The life-expectancy, L days, of a cockroach varies inversely with the square of the density, d people/m², of the human population near its habitat. If $L = 100$ when $d = 0.05$, find

     a the formula for L in terms of d
     b the life-expectancy of a cockroach in an area where the human population density is 0.1 people/m².

EXERCISE 3*

1 ▶ The cost of Mrs Janus's electricity bill, $\$C$, varies inversely with the average temperature, $t\,$°C, over the period of the bill. If the bill is \$200 when the temperature is 25 °C, find

     a the formula expressing C in terms of t
     b the bill when the temperature is 18 °C
     c the temperature generating a bill of \$400.

2 ▶ As a balloon is blown up, the thickness of its walls, t (mm), decreases and its volume, V (cm³), increases. V is inversely proportional to t. When V is 15 000 cm³, t is 0.05 mm.

     a Write a formula for V in terms of t.
     b When the thickness of the wall of the balloon is 0.03 mm, the balloon will pop. Is it possible to blow up this balloon to a volume of 30 000 cm³?

3 ▶ y is inversely proportional to x.

x	0.25	0.5	1	2	4	5	10	20
y	48	24	12	6	3	2.4	1.2	0.6

     a Draw a graph of y against x.
       What type of graph is this?

     b $y = \dfrac{k}{x}$ where k is the constant of proportionality. Find k.

     c Work out $x \times y$ for each pair of values in the table. What do you notice?

4 ▶ a is inversely proportional to b^2. Copy and complete this table.

b	2	5	
a	50		2

5 ▶ A scientist gathers this data.

t	1	4		10
r	20		4	2

a Which of these relationships describes the collected data?

$$r \propto \frac{1}{\sqrt{t}} \qquad r \propto \frac{1}{t} \qquad r \propto \frac{1}{t^2}$$

b Copy and complete the table.

6 ▶ The electrical resistance, R ohm, of a fixed length of wire is inversely proportional to the square of its radius, r mm. If $R = 0.5$ when $r = 2$, find

a the formula for R in terms of r
b the resistance of a wire of 3 mm radius.

7 ▶ The number of people shopping at Tang's corner shop per day, N, varies inversely with the square root of the average outside temperature, t °C.

a Copy and complete this table.

DAY	N	t
Monday	400	25
Tuesday		20
Wednesday	500	

b For the rest of the week (Thursday to Saturday) the weather is hot, with a constant daily average temperature of 30 °C. What is the average number of people per day who shop at Tang's for that week? (The shop is closed on Sundays.)

8 ▶ The time for a pendulum to swing, T s, is inversely proportional to the square root of the acceleration due to gravity, g m/s². On Earth, $g = 9.8$ m/s², but on the Moon, $g = 1.9$ m/s². Find the time of swing on the Moon of a pendulum when the time that it takes to swing on Earth is 2 s.

ACTIVITY 3

SKILLS

ANALYSIS

This graph shows an inverse relationship between the body mass, M kg, of mammals and their average heart rate B beats/min.

Use the graph to complete this table.

	B (BEATS/MIN)	M (KG)
RABBIT		5
DOG	135	
MAN		70
HORSE	65	

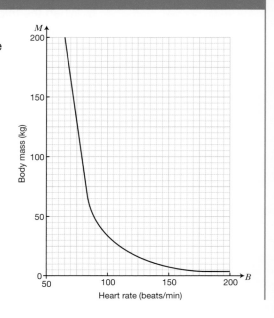

An unproven theory in biology states that the hearts of all mammals beat the same number of beats in an average life-span.

A man lives on average for 75 years. Calculate the total number of heart beats in a man's average life-span.

Test out this theory by calculating the expected life-span of the creatures in the table above.

INDICES

NEGATIVE INDICES AND FRACTIONAL INDICES INCLUDING ZERO

The laws of indices can also be applied to algebraic expressions with fractional or negative powers.

EXAMPLE 6

Work out the value of

a $27^{\frac{2}{3}}$

b $16^{-\frac{3}{4}}$

a $27^{\frac{2}{3}} = \left(27^{\frac{1}{3}}\right)^2 = 3^2 = 9$

b $16^{-\frac{3}{4}} = \dfrac{1}{\left(16^{\frac{1}{4}}\right)^3} = \dfrac{1}{2^3} = \dfrac{1}{8}$

HINT
Use the rule $(x^m)^n = x^{mn}$. Work out the cube root of 27 first. Then square the answer.

HINT
Use $x^{-n} = \dfrac{1}{x^n}$

EXAMPLE 7

Simplify

a $a^{\frac{1}{2}} \times a^{\frac{1}{2}} \times a$

a $a^{\frac{1}{2}} \times a^{\frac{1}{2}} \times a = a^{\left(\frac{1}{2}+\frac{1}{2}+1\right)} = a^2$

b $3a^2 \times 2a^{-3}$

b $3a^2 \times 2a^{-3} = 6 \times a^{2+(-3)} = 6a^{-1} = \dfrac{6}{a}$

HINT
An expression containing a negative power is not fully simplified:
x^{-1} should be written as $\dfrac{1}{x}$

c $(6a^{-2}) \div (2a^2)$

c $(6a^{-2}) \div (2a^2) = 3a^{(-2-2)} = 3a^{-4} = \dfrac{3}{a^4}$

d $\left(\dfrac{a^2}{b^4}\right)^{-\frac{1}{2}}$

d $\left(\dfrac{a^2}{b^4}\right)^{-\frac{1}{2}} = \dfrac{1}{\left(\dfrac{a^2}{b^4}\right)^{\frac{1}{2}}} = \left(\dfrac{b^4}{a^2}\right)^{\frac{1}{2}} = \dfrac{b^2}{a}$

KEY POINTS

- $a^{\frac{1}{m}} = \sqrt[m]{a}$

- $a^{\frac{n}{m}} = \left(\sqrt[m]{a}\right)^n = \sqrt[m]{a^n}$

- $a^0 = 1, \; a \neq 0$

- $a^{-1} = \dfrac{1}{a}, \; a \neq 0$

- $a^{-n} = \dfrac{1}{a^n}$ for any number n, $a \neq 0$

- $\left(\dfrac{a}{b}\right)^{-n} = \dfrac{1}{\left(\dfrac{a}{b}\right)^n} = \left(\dfrac{b}{a}\right)^n$ for any number n, $a, b \neq 0$

EXERCISE 4

Simplify

1 ▶ $a^2 \times a^{-1}$

2 ▶ $b^4 \times b^{-2}$

3 ▶ $c^{-1} \div c^2$

4 ▶ $d^3 \div d^{-2}$

5 ▶ $e^2 \times e^3 \times e^{-4}$

6 ▶ $f^3 \times f^2 \times f^{-4}$

7 ▶ $(a^2)^{-1}$

8 ▶ $(g^{-1})^2$

9 ▶ $a^{2\frac{1}{2}} \times a^{-\frac{1}{2}}$

10 ▶ $b^{-\frac{1}{3}} \times b^{-\frac{1}{3}} \times b^{-\frac{1}{3}}$

11 ▶ $c^{2\frac{1}{2}} \div c^{-\frac{1}{2}}$

12 ▶ $d^{\frac{2}{3}} \div d^{-\frac{1}{3}}$

13 ▶ $\left(e^{\frac{2}{3}}\right)^3$

14 ▶ $\left(f^2\right)^{-\frac{1}{2}}$

15 ▶   **a**   Copy and complete. $2^3 \div 2^3 = 2^{\square}$

      b   Write down 2^3 as a whole number.

      c   Copy and complete $2^3 \div 2^3 = 8 \div \square = \square$

      d   Copy and complete using parts **a** and **c**. $2^3 \div 2^3 = 2^{\square} = \square$

      e   Repeat parts **a** and **b** for $7^5 \div 7^5$

      f   Write down a rule for a^0, where a is any number.

16 ▶   Work out

      a 3^{-1}

      b 2^{-4}

      c 10^{-5}

      d $\left(\dfrac{3}{4}\right)^{-1}$

      e $\left(\dfrac{4}{5}\right)^{-3}$

      f $\left(1\dfrac{1}{4}\right)^{-1}$

      g $\left(2\dfrac{3}{4}\right)^{-2}$

      h $(0.7)^{-1}$

      i $(0.1)^{-5}$

      j $(0.4)^{-3}$

      k $(5^{-1})^0$

      l $(7^{-1})^{-1}$

EXERCISE 4*

Simplify

1 ▶ $a^{-2} \times a^2 \div a^{-2}$

2 ▶ $b^4 \times b^{-3} \div b^{-2}$

3 ▶ $2(c^2)^{-2}$

4 ▶ $(2c^{-1})^2$

5 ▶ $3a^{-2} \times 4a$

6 ▶ $4b^2 \times 2b^{-3}$

7 ▶ $a^{\frac{1}{2}} \times a^{\frac{1}{2}}$

8 ▶ $b^{\frac{1}{4}} \div b^{\frac{1}{4}}$

9 ▶ $(c^{-2})^{\frac{1}{2}}$

10 ▶ $\left(d^{\frac{1}{3}}\right)^{-3}$

11 ▶ $(-3a)^3 \div (3a^{-3})$

12 ▶ $a^{\frac{1}{2}} \times a^{-2\frac{1}{2}}$

13 ▶ $c^{-2\frac{1}{3}} \div c^{-\frac{1}{3}}$

14 ▶ $\left(e^{-\frac{1}{2}}\right)^{-2}$

15 ▶   Find k if $x^k = \sqrt[3]{x} \div \dfrac{1}{x^2}$

16 ▶   Find k if $x^{k-1} = \dfrac{(x^2)^{-2}}{x^3}$

Find the values of x and y in these equations.

17 ▶ $(a^x b^y)^{\frac{1}{6}} = a^{\frac{1}{2}} \times b^{-\frac{1}{3}}$

18 ▶ $\sqrt[5]{a^4 b^{-3}} = a^x b^y$

19 ▶ Work out

  a $27^{-\frac{1}{3}} \times 9^{\frac{3}{2}}$

  b $\left(\dfrac{4}{25}\right)^{-\frac{3}{2}} \times \left(\dfrac{8}{27}\right)^{\frac{1}{3}}$

  c $\left(\dfrac{81}{16}\right)^{\frac{3}{4}} \times \left(\dfrac{9}{25}\right)^{-\frac{3}{2}}$

20 ▶ Find the value of n.

  a $16 = 2^n$

  b $\sqrt[3]{27} = 27^n$

  c $\dfrac{1}{100} = 10^n$

  d $\sqrt{\dfrac{4}{9}} = \left(\dfrac{9}{4}\right)^n$

  e $(\sqrt{3})^7 = 3^n$

  f $(\sqrt[4]{8})^7 = 8^n$

EXERCISE 5

REVISION

1 ▶ y is directly proportional to x. If $y = 12$ when $x = 2$, find
  a the formula for y in terms of x
  b y when $x = 7$
  c x when $y = 66$

2 ▶ p varies as the square of q. If $p = 20$ when $q = 2$, find
  a the formula for p in terms of q
  b p when $q = 10$
  c q when $p = 605$

3 ▶ The cost, $\$c$, of laying floor tiles is directly proportional to the square of the area, $a\,\mathrm{m}^2$, to be covered. If a $40\,\mathrm{m}^2$ kitchen floor costs $\$1200$ to cover with tiles, find
  a the formula for c in terms of a
  b the cost of tiling a floor of area $30\,\mathrm{m}^2$
  c the area of floor covered by these tiles costing $\$600$.

4 ▶ The time taken, t hours, to make a set of 20 curtains is inversely proportional to the number of people, n, who work on them. One person would take 80 hours to finish the task.
Copy and complete this table.

n	1	2	4	
t	80			10

5 ▶ In an experiment, measurements of g and h were taken.

h	2	5	7
g	24	375	1029

Which of these relationships fits the result?
$g \propto h \quad g \propto h^2 \quad g \propto h^3 \quad g \propto \sqrt{h}$

6 ▶ The graph shows two variables that are inversely proportional to each other.
Find the values for a and b.

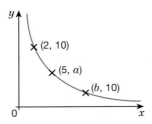

7 ▶ A farmer employs fruit pickers to harvest his apple crop.
The fruit pickers work in different-sized teams.
The farmer records the times it takes different teams to harvest the apples from 10 trees.

NUMBER OF PEOPLE IN TEAM, n	3	2	5	8	9	10	6	4	7
TIME TAKEN, t (minutes)	95	155	60	40	30	25	40	85	40

Q7a HINT

A curve of best fit should be drawn.

a Draw the graph of t against n.
b Given that t is inversely proportional to n, find the formula for t in terms of n.
c Use your formula to estimate the time it would take a team of 15 people to harvest the apples from 10 trees.

8 ▶ Here are four graphs of y against x.

A 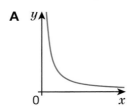   **B** 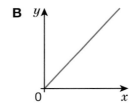   **C** 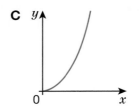   **D**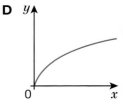

Match each of these proportionality relationships to the correct graph.

a $y \propto x$ **b** $y \propto \sqrt{x}$ **c** $y \propto \dfrac{1}{x}$ **d** $y \propto x^2$

Simplify

9 ▶ $a^3 \times a^{-1}$

10 ▶ $a^3 \div a^{-1}$

11 ▶ $(d^{-1})^2$

12 ▶ $b^{\frac{1}{2}} \times b^{-2}$

13 ▶ $\left(c^{-\frac{1}{2}}\right)^{-2}$

14 ▶ Evaluate

a $100^{\frac{1}{2}}$

b $8^{\frac{1}{3}}$

c $\left(\dfrac{4}{9}\right)^{\frac{1}{2}}$

d $\left(\dfrac{25}{49}\right)^{\frac{1}{2}}$

e $-27^{\frac{1}{3}}$

f $-\left(\dfrac{1}{27}\right)^{\frac{1}{3}}$

g $-\left(\dfrac{4}{9}\right)^{\frac{1}{2}}$

h $\left(\dfrac{1}{1000}\right)^{\frac{1}{3}}$

15 ▶ Evaluate $(8ab^2)^0$

EXERCISE 5*

REVISION

1 ▶ y squared varies as x cubed. If $y = 20$ when $z = 2$, find

    a the formula relating y to z
    b y when $z = 4$
    c z when $y = 100$

2 ▶ The frequency of radio waves, f MHz, varies inversely as their wavelength, μ metres.
If Radio 1 has $f = 99$ MHz and $\mu = 3$ m, what is the wavelength of the BBC World Service on 198 kHz?

3 ▶ m is inversely proportional to the square root of n.
If $m = 2.5 \times 10^7$ when $n = 1.25 \times 10^{-7}$, find

    a the formula for m in terms of n
    b m when $n = 7.5 \times 10^{-4}$
    c n when m is one million.

4 ▶ If y is inversely proportional to the nth power of x, copy and complete this table.

x	0.25	1	4	25
y		10	5	

Find the formula for y in terms of x.

5 ▶ In an experiment, measurements of t and w are taken. The table shows some results.

t	0.9	5	12
w	400	12.96	2.25

The variables, t and w, are connected by one of these rules.

A $w \propto \dfrac{1}{t}$ **B** $w \propto \dfrac{1}{t^2}$ **C** $w \propto \dfrac{1}{\sqrt{t}}$

Which is the correct rule?
Show **working** to justify your answer.

6 ▶ When 20 litres of water are poured into any **cylinder**, the depth, D (in cm), of the water is inversely proportional to the square of the radius, r (in cm), of the cylinder. When $r = 15$ cm, $D = 28.4$ cm.

    a Write a formula for D in terms of r.
    b Find the depth of water when the radius of the cylinder is 25 cm.
    c Find the radius of the cylinder when the depth is 64 cm.
    d Cylinder A has radius x cm and is filled with water to a depth of d cm.
       This water is poured into cylinder B with radius $2x$ cm.
       What is the depth of water in cylinder B?

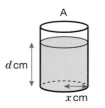
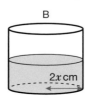

7 ▶ The gravitational force between two objects, F (in newtons, N), is inversely proportional to the square of the distance, d (in metres), between them.

A satellite orbiting the Earth is 4.2×10^7 m from the centre of the Earth.
The force between the satellite and the Earth is 60 N.

a Write a formula for F in terms of d.
b The force between two objects is 16 N. What is the value of the force when the distance between the objects doubles?

8 ▶ w is inversely proportional to the square of p.
Copy and complete this table for values of p and w.

p	2		6
w	7	$1\frac{3}{4}$	

Simplify

9 ▶ $a^3 \times a^{-2} \div a^{-1}$

10 ▶ $3 \times (c^{-1})^2$

11 ▶ $2\left(a^{\frac{1}{3}}\right)^{-3}$

12 ▶ $d^{\frac{1}{3}} \times d^{\frac{2}{3}}$

13 ▶ $(-3a)^3 \div (3a^{-3})$

14 ▶ $(2b^2)^{-1} \div (-2b)^{-2}$

15 ▶ $(3x^2 y^7)^0$

16 ▶ $\sqrt{25 x^4 y^6}$

EXAM PRACTICE: ALGEBRA 6

1 y is directly proportional to x.
If $x = 10$ when $y = 5$, find

   a   a formula for y in terms of x **[1]**

   b y when $x = 5$ **[1]**

   c x when $y = \dfrac{1}{2}$ **[1]**

2 p is directly proportional to q squared.
If $q = 10$ when $p = 20$, find

   a   a formula for p in terms of q **[1]**

   b p when $q = 20$ **[1]**

   c q when $p = 180$ **[2]**

3   A machine produces coins of a fixed thickness from a given volume of metal.

   The number of coins, N, produced is inversely proportional to the square of the diameter, d.

   a   4000 coins are made of diameter 1.5 cm. Find the value of the constant of proportionality, k.   **[2]**

   b   Find the formula for N in terms of d.   **[2]**

   c   Find the number of coins that can be produced of diameter 2 cm.   **[2]**

   d   If 1000 coins are produced, find the diameter.   **[2]**

4   When 30 litres of water are poured into any cylinder, the depth, D (in cm), of the water is inversely proportional to the square of the radius, r (in cm), of the cylinder.

   When $r = 30$ cm, $D = 10.6$ cm.

   a   Write a formula for D in terms of r.   **[1]**

   b   Find the depth of the water when the radius of the cylinder is 15 cm.   **[1]**

   c   Find the radius of the cylinder (to 1 **decimal place**) when the depth is 60 cm.   **[2]**

   d   Cylinder P has radius x cm and is filled with water to a depth of d cm.

     This water is poured into cylinder Q and fills it to a depth of $3d$ cm. What is the radius of cylinder Q?   **[2]**

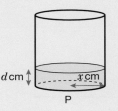
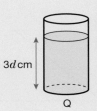

5   Simplify

   a $3a^{-2} \times 4a^3 \div 6a^{-1}$ **[1]**

   b $(b^{-4})^{\frac{1}{2}}$ **[1]**

   c $c^{\frac{5}{2}} \times \left(c^{\frac{1}{4}}\right)^{-2} \div c^{-2}$ **[1]**

   d   Find x if $\left(\dfrac{16}{9}\right)^x = \dfrac{3}{4}$ **[1]**

[Total 25 marks]

CHAPTER SUMMARY: ALGEBRA 6

PROPORTION

DIRECT PROPORTION

All these statements have the same meaning:

- y is directly proportional to x.
- y varies directly with x.
- y varies as x.

y is directly proportional to x means $y = kx$, for some constant k (constant of proportionality).

The graph of y against x is a straight line through the origin.

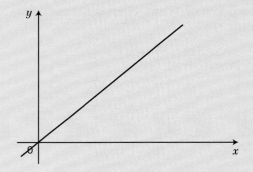

If $y = 12$ when $x = 3$, then $12 = k \times 3 \Rightarrow k = 4$
So the equation is $y = 4x$

For some constant k:

- When y is directly proportional to x^2 then $y \propto x^2$ and $y = kx^2$
- When y is directly proportional to x^3 then $y \propto x^3$ and $y = kx^3$
- When y is directly proportional to $\sqrt{x}$ then $y \propto \sqrt{x}$ and $y = k\sqrt{x}$

INVERSE PROPORTION

y is inversely proportional to x means $y = \dfrac{k}{x}$ for some constant k.

The graph of y plotted against x looks like this.

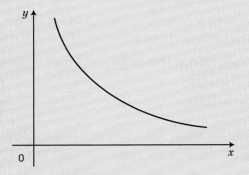

As x increases, y decreases.

If $y = 3$ when $x = 4$, then $3 = \dfrac{k}{4} \Rightarrow k = 12$

So the equation is $y = \dfrac{12}{x}$

For some constant k:

- When y is inversely proportional to x^2 then $y \propto \dfrac{1}{x^2}$ and $y = \dfrac{k}{x^2}$
- When y is inversely proportional to x^3 then $y \propto \dfrac{1}{x^3}$ and $y = \dfrac{k}{x^3}$
- When y is inversely proportional to $\sqrt{x}$ then $y \propto \dfrac{1}{\sqrt{x}}$ and $y = \dfrac{k}{\sqrt{x}}$

INDICES

For a negative or fractional index, the laws of indices still apply:

- $a^{\frac{1}{m}} = \sqrt[m]{a}$
- $a^{\frac{n}{m}} = \left(\sqrt[m]{a}\right)^n = \sqrt[m]{a^n}$
- $a^0 = 1,\ a \neq 0$
- $a^{-1} = \dfrac{1}{a},\ a \neq 0$
- $a^{-n} = \dfrac{1}{a^n}$ for any number n, $a \neq 0$
- $\left(\dfrac{a}{b}\right)^{-n} = \dfrac{1}{\left(\dfrac{a}{b}\right)^n} = \left(\dfrac{b}{a}\right)^n$ for any number n, $a, b \neq 0$

SEQUENCES

Carl Friedrich Gauss (1777–1855) is one of the greatest mathematicians of all time, showing amazing mathematical talent from an early age. When Carl was eight years old, his teacher asked his class to add together all the numbers from 1 to 100, hoping this would keep them quiet for some time. Gauss gave the correct answer after a few seconds. Gauss had summed the terms of the arithmetic sequence to find the correct answer.

LEARNING OBJECTIVES

- Find a general formula for the nth term of an arithmetic sequence
- Determine whether a particular number is a term of a given arithmetic sequence
- Find the sum of an arithmetic series

BASIC PRINCIPLES

- These are examples of important sequences:

  Natural numbers: 1, 2, 3, 4, …

  Even numbers: 0, 2, 4, 6, …

  Odd numbers: 1, 3, 5, 7, …

  Triangle numbers: 1, 3, 6, 10, …

  Square numbers: 1, 4, 9, 16, …   (The squares of the natural numbers)

  Powers of 2: 1, 2, 4, 8, …   (Numbers of the form 2^n)

  Powers of 10: 1, 10, 100, …   (Numbers of the form 10^n)

  Prime numbers: 2, 3, 5, 7, …   (Note that 1 is not a prime number)

- Solve two-step linear equations.
- Solve linear **inequalities**.

CONTINUING SEQUENCES

A set of numbers that follows a definite pattern is called a sequence. You can continue a sequence if you know how the terms are related.

ACTIVITY 1

Seema is decorating the walls of a hall with balloons in preparation for a disco. She wants to place the balloons in a triangular pattern.

To make a 'triangle' with one row she needs one balloon.

To make a triangle with two rows she needs three balloons.

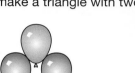

To make a triangle with three rows she needs six balloons.

The number of balloons needed form the sequence 1, 3, 6, …
Describe in words how to continue the sequence and find the next three terms.

Seema thinks she can work out how many balloons are needed for n rows by using the formula $\frac{1}{2}n(n+1)$. Her friend Julia thinks the formula is $\frac{1}{2}n(n-1)$.

Who is correct?

If Seema has 100 balloons find, using 'trial and improvement', how many rows she can make, and how many balloons will be left over (remaining).

Write down the first five terms of these sequences.

a Starting with −8, keep adding 4

b Starting with 1, keep multiplying by 3

a −8, −4, 0, 4, 8

b 1, 3, 9, 27, 81

EXAMPLE 2

SKILLS

PROBLEM
SOLVING

For these sequences, describe a rule for going from one term to the next.

a 5, 2, –1, –4, …

b 288, 144, 72, 36, …

---

a Subtract 3

b Divide by 2

EXERCISE 1

For Questions 1–5, write down the first four terms of the sequence.

1 ▶   Starting with 2, keep adding 2

2 ▶   Starting with –9, keep adding 3

3 ▶   Starting with 15, keep subtracting 5

4 ▶   Starting with 2, keep multiplying by 2

5 ▶   Starting with 12, keep dividing by 2

For Questions 6–10, describe the rule for going from one term to the next, and write down the next three numbers in the sequence.

6 ▶   3, 7, 11, 15, …, …, …

7 ▶   13, 8, 3, –2, …, …, …

8 ▶   3, 6, 12, 24, …, …, …

9 ▶   64, 32, 16, 8, …, …, …

10 ▶   0.2, 0.5, 0.8. 1.1, …, …, …

EXERCISE 1*

For Questions 1–4, write down the first four terms of the sequence.

1 ▶   Starting with –1, keep adding 1.5

2 ▶   Starting with 3, keep subtracting 1.25

3 ▶   Starting with 1, keep multiplying by 2.5

4 ▶   Starting with 3, keep dividing by –3

5 ▶   The first two terms of a sequence are 1, 1. Each successive term is found by adding together the previous two terms. Find the next four terms.

For Questions 6–10, describe the rule for going from one term to the next, and write down the next three numbers in the sequence.

6 ▶ $3, 5\frac{1}{2}, 8, 10\frac{1}{2}, …, …, …$

7 ▶   243, 81, 27, 9, …, …, …

8 ▶   2, 4, 16, 256, …, …, …

9 ▶ $1, -\frac{1}{2}, \frac{1}{4}, -\frac{1}{8}, …, …, …$

10 ▶   1, 3, 7, 15, 31, …, …, …

FORMULAE FOR SEQUENCES

Sometimes the sequence is given by a formula. This means that any term can be found without working out all the previous terms.

EXAMPLE 3

SKILLS

ANALYSIS

Find the first four terms of the sequence given by the nth term = $2n - 1$

Find the 100th term.

Substituting $n = 1$ into the formula gives the first term as	$2 \times 1 - 1 = 1$
Substituting $n = 2$ into the formula gives the second term as	$2 \times 2 - 1 = 3$
Substituting $n = 3$ into the formula gives the third term as	$2 \times 3 - 1 = 5$
Substituting $n = 4$ into the formula gives the fourth term as	$2 \times 4 - 1 = 7$
Substituting $n = 100$ into the formula gives the 100th term as	$2 \times 100 - 1 = 199$

EXAMPLE 4

SKILLS

ANALYSIS

A sequence is given by the nth term = $4n + 2$. Find the value of n for which the nth term equals 50.

$4n + 2 = 50 \Rightarrow 4n = 48 \Rightarrow n = 12$

So the 12th term equals 50.

KEY POINT

■ When a sequence is given by a formula, any term can be worked out.

EXERCISE 2

In Questions 1–6, find the first four terms of the sequence.

1 ▶ nth term = $2n + 1$

2 ▶ nth term = $5n - 1$

3 ▶ nth term = $33 - 3n$

4 ▶ nth term = $n^2 + 1$

5 ▶ nth term = $3n$

6 ▶ nth term = $\dfrac{n + 1}{n}$

In Questions 7–9, find the value of n for which the nth term has the value given in brackets.

7 ▶ nth term = $4n + 4$ (36)

8 ▶ nth term = $6n - 12$ (30)

9 ▶ nth term = $22 - 2n$ (8)

10 ▶ If the nth term = $\dfrac{1}{n - 1}$, which is the first term less than $\dfrac{1}{20}$?

EXERCISE 2*

In Questions 1–6, find the first four terms of the sequence.

1 ▶ nth term $= 5n - 6$

2 ▶ nth term $= 100 - 3n$

3 ▶ nth term $= \frac{1}{2}(n + 1)$

4 ▶ nth term $= n^2 + n + 1$

5 ▶ nth term $= n^2 + 2$

6 ▶ nth term $= \frac{2n + 1}{2n - 1}$

In Questions 7–9, find the value of n for which the nth term has the value given in brackets.

7 ▶ nth term $= 7n + 9$ (65)

8 ▶ nth term $= 3n - 119$ (−83)

9 ▶ nth term $= 12 - 5n$ (−38)

10 ▶ If the nth term $= \frac{1}{2n - 1}$, which is the first term less than 0.01?

ACTIVITY 2

SKILLS

ANALYSIS

A sequence is given by nth term $= an$. What is the connection between a and the numbers in the sequence?

A sequence is given by nth term $= 3n + 2$. Copy and fill in the table.

HINT

Try nth term $= 2n$, nth term $= 3n$, …

n	1	2	3	4	10
$3n + 2$	5				

What is the connection between the numbers 3 and 2 and the numbers in the sequence?

A sequence is given by nth term $= an + b$

What is the connection between a and b and the numbers in the sequence?

THE DIFFERENCE METHOD

When it is difficult to spot a pattern in a sequence, the difference method can often help. Under the sequence write down the differences between each pair of terms. If the differences show a pattern then the sequence can be extended.

EXAMPLE 5

Find the next three terms in the sequence 2, 5, 10, 17, 26, …

SKILLS

ANALYSIS

Sequence: 2   5   10   17   26

Differences: 3   5   7   9

$= 5 - 2$

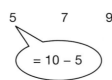
$= 10 - 5$

The differences increase by 2 each time so the table can now be extended.

Sequence:	2		5		10		17		26		**37**		**50**		**65**
Differences:		3		5		7		9		**11**		**13**		**15**		

If the pattern in the differences is not clear, add a third row giving the differences between the terms in the second row. More rows can be inserted until a pattern is found but remember not all sequences will result in a pattern.

EXAMPLE 6

SKILLS

ANALYSIS

Find the next three terms in the sequence 0, 3, 16, 45, 96, …

Sequence:	0		3		16		45		96
Differences:		3		13		29		51	
			10		16		22		
				6		6			

Now the table can be extended to give:

Sequence:	0		3		16		45		96		**175**		**288**		**441**
Differences:		3		13		29		51		**79**		**113**		**153**	
			10		16		22		**28**		**34**		**40**		
				6		6		**6**		**6**		**6**			

KEY POINT

■ The difference method finds patterns in sequences when the patterns are not obvious.

EXERCISE 3

Find the next three terms of the following sequences using the difference method.

1 ▶ 2, 5, 8, 11, 14, … 　　　　4 ▶ 1, 6, 14, 25, 39, …

2 ▶ 8, 5, 2, −1, −4, … 　　　　5 ▶ 5, −2, −6, −7, −5, …

3 ▶ 4, 5.5, 7, 8.5, 10, … 　　　6 ▶ 1, 4, 5, 4, 1, …

EXERCISE 3*

Find the next three terms of the following sequences using the difference method.

1 ▶   1, 3, 8, 16, 27, 41, …   **4 ▶**   1, 3, 6.5, 11.5, 18, 26, …

2 ▶   1, 6, 9, 10, 9, 6, …   **5 ▶**   1, 3, 6, 11, 19, 31, 48, …

3 ▶   1, −4, −7, −8, −7, −4, …   **6 ▶**   1, 2, 4, 6, 7, 6, 2, …

ACTIVITY 3

SKILLS

PROBLEM
SOLVING

Use a spreadsheet to find the next three terms of these sequences.

2, 4, 10, 20, 34, 52, …

3, 5, 10, 20, 37, 63, 100, …

2, 1, 1, 0, −4, −13, −29, …

FINDING A FORMULA FOR A SEQUENCE

ACTIVITY 4

SKILLS

REASONING

Seema decides to decorate the walls of the hall
with different patterns of cylindrical balloons.
She starts with a triangular pattern.

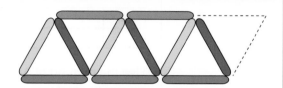

Seema wants to work out how many balloons she will need to make 100 triangles.

If t is the number of triangles and b is the number of balloons, copy and fill in the
following table.

t	1	2	3	4	5	6
b						

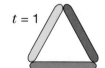 $t = 1$  $t = 2$

As the sequence in the row headed b increases by 2 each time add another row to
the table headed $2t$.

t	1	2	3	4	5	6
b						
$2t$	2	4	6	8	10	12

Write down the formula that connects b and $2t$. How many balloons does Seema need to
make 100 triangles?

Use paper clips (or other small objects) to make up some other patterns that Seema might use and find a formula for the number of balloons needed. Some possible patterns are given below.

ACTIVITY 5

SKILLS

REASONING

12 is a square **perimeter** number because 12 stones can be arranged as the perimeter of a square.

The first square perimeter number is 4.

Copy and complete the following table where n is the square perimeter number and s is the number of stones needed.

n	1	2	3	4	5	6
s	4					

Use the method of Activity 4 to find a formula for the number of pebbles in the nth square perimeter number.

KEY POINT

■ If the first row of differences is constant and equal to a then the formula for the nth term will be nth term $= an + b$ where b is another constant.

EXAMPLE 7

SKILLS

ANALYSIS

Find a formula for the nth term of the sequence 40, 38, 36, 34 ...

Sequence: 40   38   36   34

Differences: −2   −2   −2

The first row of differences is constant and equal to −2 so formula is $-2n + b$

When $n = 1$ the formula must give the first term as 40.

So $-2 \times 1 + b = 40 \Rightarrow b = 42$

The formula for the nth term is $42 - 2n$

Sometimes it is easy to find the formula.

EXAMPLE 8

Find a formula for the nth term of the sequence $\dfrac{1}{3}, \dfrac{2}{4}, \dfrac{3}{5}, \dfrac{4}{6}, \dfrac{5}{7}, \ldots$

SKILLS

ANALYSIS

The **numerator** is given by n

The **denominator** is always 2 more than the numerator, so is given by $n + 2$

So the nth term $= \dfrac{n}{n + 2}$

EXERCISE 4

In Questions 1–4, find a formula for the nth term of the sequence.

1 ▶ $1, \dfrac{1}{2}, \dfrac{1}{3}, \dfrac{1}{4}, \dfrac{1}{5}, \ldots$ **3 ▶**   4, 7, 10, 13, …

2 ▶ $1, \dfrac{1}{2}, \dfrac{1}{4}, \dfrac{1}{6}, \dfrac{1}{8}, \ldots$ **4 ▶**   30, 26, 22, 18, …

5 ▶   Anna has designed a range of candle decorations using a triangle of wood and some candles. She makes them in various sizes. The one shown here is the 3-layer size because it has three layers of candles.

     a   Copy and complete this table where l is the number of layers and c is the number of candles.

l	1	2	3	4	5	6
c						

     b   Find a formula connecting l and c.
     c   Mr Rich wants a decoration with exactly 100 candles. Explain why this is impossible. What is the largest number of layers than can be made if 100 candles are available?

6 ▶   Julia is investigating rectangle perimeter numbers with a constant width of three stones.

     a   Copy and complete the following table where n is the number in the sequence and s is the number of stones.

n	1	2	3	4	5	6
s	8					

     b   Find a formula connecting n and s.
     c   Given only 100 stones, what is the largest rectangle in her sequence that Julia can construct?

In Questions 1–4, find a formula for the nth term of the sequence.

1 ▶ $\dfrac{2}{1}, \dfrac{3}{2}, \dfrac{4}{3}, \dfrac{5}{4}, \dfrac{6}{5}, \ldots$

3 ▶ $3, 7, 11, 15, \ldots$

2 ▶ $\dfrac{1}{3}, \dfrac{3}{5}, \dfrac{5}{7}, \dfrac{7}{9}, \dfrac{9}{11}, \ldots$

4 ▶ $6, 3, 0, -3, \ldots$

5 ▶   Pippa has designed a tessellation (pattern with shapes) based on this shape.

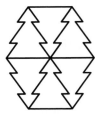

Here are the first three terms of the tessellation sequence.

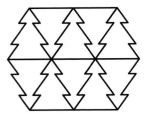
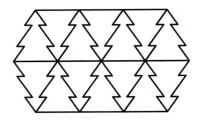

a Copy and complete this table where n is the number in the sequence and s is the number of shapes used.

n	1	2	3	4	5	6
s	6					

b Find a formula giving s in terms of n.
c How many shapes will be needed to make the 50th term of the sequence?

6 ▶   Marc is using stones to investigate rectangle perimeter numbers where the inner rectangle is twice as long as it is wide.

a Copy and complete this table where n is the number in the sequence and s is the number of stones.

n	1	2	3	4	5	6
s	10					

b Find a formula connecting n and s.
c What is the largest term of the sequence that Marc can build with only 200 stones?

ARITHMETIC SEQUENCES

When the difference between any two **consecutive** terms in a sequence is the same, the sequence is called an **arithmetic sequence**. The difference between the terms is called the **common difference**. The common difference can be positive or negative.

2   5   8   11   14   ...   is an arithmetic sequence with a common difference of 3
1   2   3   4   5   ...   is an arithmetic sequence with a common difference of 1
5   3   1   −1   −3   ...   is an arithmetic sequence with a common difference of −2
2   3   5   7   11   ...   is NOT an arithmetic sequence.

EXAMPLE 9

SKILLS

ANALYSIS

Show that **a** and **b** are arithmetic sequences.

a 5, 8, 11, 14, ...   **b** 21, 19, 17, 15, ...

---

a Sequence: 5   8   11   14
Differences: 3   3   3

As the differences are constant, the sequence is an arithmetic sequence with common difference 3.

b Sequence: 21   19   17   15
Differences: −2   −2   −2

As the differences are constant, the sequence is an arithmetic sequence with common difference −2.

NOTATION

The letter a is used for the first term, and the letter d is used for the common difference.

1st term	2nd term	3rd term	4th term	...	nth term
a	$a + d$	$a + 2d$	$a + 3d$	...	$a + (n - 1)d$

An arithmetic sequence is given by $a, a + d, a + 2d, a + 3d, ..., a + (n - 1)d$

EXAMPLE 10

SKILLS

PROBLEM SOLVING

a Find the nth term of the sequence 5, 8, 11, 14, 17, ...
b Use your answer to part **a** to find the 100th term.
c Is 243 a term of the sequence?

Sequence: 5   8   11   14   17
Differences: 3   3   3   3

So $a = 5$ and $d = 3$

a nth term is $a + (n - 1)d = 5 + 3(n - 1) = 5 + 3n - 3 = 3n + 2$

b   100th term is $3 \times 100 + 2 = 302$

c $3n + 2 = 243 \Rightarrow 3n = 241 \Rightarrow n = 80\frac{1}{3}$

243 cannot be in the sequence as $80\frac{1}{3}$ is not an **integer**.

EXAMPLE 11   The first term of an arithmetic sequence is 8 and the 20th term is 65. Find the 50th term.

SKILLS

PROBLEM SOLVING

The first term is 8 so $a = 8$.

The 20th term is $8 + (20 - 1)d \Rightarrow$ $8 + (20 - 1)d = 65$

$8 + 19d = 65$

$19d = 57$

$d = 3$

The 50th term is $8 + (50 - 1) \times 3 = 155$

KEY POINTS   In an arithmetic sequence:

■ The first term is a

■ The common difference is d

■ The nth term is $a + (n - 1)d$

EXERCISE 5

1 ▶   Write down the first four terms and the 100th term of sequences with these nth terms.

    a $3n$ **c** $7n - 4$ **e** $-4n + 8$

    b $6n + 2$ **d** $21 - 3n$

2 ▶   Find the nth term for each sequence.

    a 7, 12, 17, 22, 27, ...   **c** 19, 17, 15, 13, 11, ...

    b 2, 6, 10, 14, 18, ...

3 ▶   For each sequence, explain whether the number in the bracket is a term of the sequence or not.

    a 1, 5, 9, 13, ... (101)   **c** 40, 35, 30, 25, ... (−20)

    b 4, 9, 14, 19, ... (168)

4 ▶   Find the first term over 100 for each sequence.

    a 9, 18, 27, 36, ...   **c** 3, 8, 13, 18, ...

    b 7, 10, 13, 16, ...   **d** 10, 16, 22, 28, ...

5 ▶   The first term of an arithmetic sequence is 3 and the 10th term is 39.
Find the 30th term.

6 ▶   The first term of an arithmetic sequence is 101 and the 30th term is −44.
Find the 20th term.

7 ▶ The 11th term of an arithmetic sequence is 42 and the common difference is 3.
Find the first term.

8 ▶ The 10th term of an arithmetic sequence is 3 and the common difference is –3.
Find the 3rd term.

9 ▶ The 3rd term of an arithmetic sequence is 13 and the 6th term is 28.
Find the first term and the common difference.

10 ▶ The 2nd term of an arithmetic sequence is 18 and the 10th term is 2.
Find a and d.

11 ▶ After an injury, you only jog for 10 minutes each day for the first
week. Each week after, you increase that time by 5 minutes.
After how many weeks will you be jogging for an hour a day?

12 ▶ Tonie started receiving weekly pocket money of $5 from her mathematical father on her
9th birthday. The amount increases by 10 cents every week. How much will she receive
on the week of her 15th birthday?

EXERCISE 5*

1 ▶ Write down the first three terms and the 100th term of sequences with these nth terms.

   a $9n - 7$ **c** $3n - 30$ **e** $0.5 - 0.25n$
   b $7 - 9n$ **d** $2 + 0.5n$

2 ▶ For each sequence, explain whether each number in the brackets is a term of the sequence
or not.

   a 10, 7, 4, 1, … (–230)   **c** –31, –26, –21, –16, … (285)
   b 4, 5.5, 7, 8.5, … (99.5)

3 ▶ Find the first term over 1000 for each sequence.

   a 3, 10, 17, 24, …   **c** $0, \frac{1}{3}, \frac{2}{3}, 1, …$
   b –62, –53, –44, –35, …

4 ▶ The 9th term of an arithmetic sequence is 29 and the 27th term is 83.
Find a and d.

5 ▶ The first term of an arithmetic sequence is 10 and the 15th term is –32.
Find the 35th term.

6 ▶ The 18th term of an arithmetic sequence is 134 and the common difference is 7.
Find and **simplify** an expression for the nth term.

7 ▶ The 10th term of an arithmetic sequence is 33 and the 20th term is 63.
Find the 40th term.

8 ▶ The 8th term of an arithmetic sequence is –42 and the 16th term is –74.
Find and simplify an expression for the nth term.

9 ▶ The 7th term of one sequence is 7 and the 17th term is 37. The 9th term of another sequence is −8, and the 29th term is −88. The two sequences have an equal term. Find this term.

10 ▶ The decimal expansion of π has been memorised to 67 000 digits. One day Casper decides to do the same memory challenge. He memorises 10 digits on the first day and a further 10 digits every day after that. How long will it take him? (Give your answer in days and years to 2 s.f.)

11 ▶ One day Amanda received 3 junk emails and every day after that the number increased by 5. She deletes her junk emails every day. The limit for her inbox is 5000 emails. How long will it take to go over her limit?

12 ▶ Jay's dog weighs 35 kilograms and is overweight. The dog is put on a diet, losing 0.4 kg every week. How long does it take to reach the target weight of 28 kilograms?

SUM OF AN ARITHMETIC SEQUENCE

ACTIVITY 6

SKILLS

PROBLEM
SOLVING

In this activity you will discover how Gauss added up the numbers from 1 to 100.

A good strategy with mathematical problems is to start with a simpler version of the problem.

To add up the numbers from 1 to 5 write the series forwards and backwards and add.

$1 + 2 + 3 + 4 + 5$
                              +
$5 + 4 + 3 + 2 + 1$
_____

$6 + 6 + 6 + 6 + 6$

Adding five lots of six is done by calculating $5 \times 6 = 30$

This is twice the series so $1 + 2 + 3 + 4 + 5 = \dfrac{30}{2} = 15$

Use this technique to show that $1 + 2 + 3 + ... + 9 + 10 = 55$

Show that the answer to Gauss' problem $1 + 2 + 3 + ... + 99 + 100$ is 5050

Find a formula to add $1 + 2 + 3 + ... + (n - 1) + n$. Check that your formula works with Gauss' problem.

Gauss was adding up an arithmetic sequence. This is called an **arithmetic series**.

$4 + 7 + 10 + 13 + ...$ is an arithmetic series because the terms are an arithmetic sequence.

The general terms in an arithmetic sequence are $a, a + d, a + 2d, a + 3d, ..., a + (n - 1)d$

The notation S_n is used to mean 'the **sum** to n terms'.

S_6 means the sum of the first six terms, S_9 means the sum of the first nine terms and so on.

The sum of an arithmetic series is

$S_n = a + (a + d) + (a + 2d) + (a + 3d) + ... + (a + (n - 1)d)$

$S_n = \dfrac{n}{2}[2a + (n - 1)d]$

The proof of this follows Example 14.

EXAMPLE 12

SKILLS

PROBLEM
SOLVING

Note: This is Gauss' problem.

Show that $1 + 2 + 3 + ... + 99 + 100 = 5050$ using the formula

$$S_n = \frac{n}{2}[2a + (n-1)d]$$

$a = 1$, $d = 1$ and $n = 100 \Rightarrow S_{100} = \frac{100}{2}[2 + (100 - 1)1] = 50 \times 101 = 5050$

EXAMPLE 13

SKILLS

PROBLEM
SOLVING

Find $4 + 7 + 10 + 13 + ... + 151$

$a = 4$ and $d = 3$.

The nth term is $a + (n-1)d \Rightarrow 4 + (n-1) \times 3 = 151$

$$4 + 3n - 3 = 151$$
$$3n = 150$$
$$n = 50$$

$S_{50} = \frac{50}{2}[2 \times 4 + (50 - 1)3] = 3875$

EXAMPLE 14

SKILLS

PROBLEM
SOLVING

The first term of an arithmetic series is 5 and the 20th term is 81.

Find the sum of the first 30 terms.

$a = 5$

The 20th term is $5 + (20 - 1)d \Rightarrow 5 + (20 - 1)d = 81$

$$5 + 19d = 81$$
$$19d = 76$$
$$d = 4$$

$S_{30} = \frac{30}{2}[2 \times 5 + (30 - 1)4] = 1890$

PROOF OF $S_n = \frac{n}{2}[2a + (n-1)d]$

The proof uses the same method as Activity 6. The series is written forwards and backwards and then added.

$$S_n = a \qquad\qquad + \qquad (a + d) \qquad + ... + (a + (n-2)d) + (a + (n-1)d)$$
$$S_n = (a + (n-1)d) \quad + \quad (a + (n-2)d) \qquad + ... + (a + d) \qquad + \quad a$$

$$2S_n = [2a + (n-1)d] + [2a + (n-1)d] \qquad + ... + [2a + (n-1)d] + [2a + (n-1)d]$$

There are n terms that are all the same in the square brackets.

$$\Rightarrow 2S_n = n[2a + (n-1)d]$$
$$\Rightarrow S_n = \frac{n}{2}[2a + (n-1)d]$$

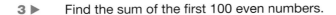

KEY POINTS

- $a + (a + d) + (a + 2d) + (a + 3d) + \ldots + (a + (n - 1)d)$ is an arithmetic series.

- The sum to n terms of an arithmetic series is $S_n = \frac{n}{2}[2a + (n - 1)d]$

EXERCISE 6

1 ▶ Find $1 + 2 + 3 + 4 + 5 + \ldots + 1000$.

2 ▶ Given $5 + 9 + 13 + 17 + \ldots$, find the sum of the first 40 terms.

3 ▶ Find the sum of the first 100 even numbers.

4 ▶ Find the sum of the first 200 odd numbers.

5 ▶ The first term of an arithmetic series is 8 and the 10th term is 44. Find the sum of the first 20 terms.

6 ▶ The third term of an arithmetic sequence is 12 and the common difference is 3. Find the sum of the first 80 terms.

7 ▶ The second term of an arithmetic series is 12 and the third term is 17. Find the sum of the first 30 terms.

8 ▶ The first term of an arithmetic series is 9 and the sum to 10 terms is 225. Find the sum of the first 20 terms.

9 ▶ Logs are stored in a pile of 20 rows, with each row having one fewer log than the one below it. If there are 48 logs on the bottom row, how many logs are in the pile?

10 ▶ Each hour a clock strikes the number of times that corresponds to the time of day. For example, at 5 o'clock, it will strike 5 times. How many times does the clock strike in a 12-hour day?

11 ▶ Zoe is saving for a holiday. In the first week she saves \$100. Each week after that she increases the amount she saves by \$10. How much has she saved after 15 weeks?

12 ▶ Bees make their honeycomb by starting with a single hexagonal cell, then forming ring after ring of hexagonal cells around the initial cell, as shown. The numbers of cells in successive rings (not including the initial cell) form an arithmetic sequence.

  a Find the formula for the number of cells in the nth ring.
  b What is the total number of cells (including the initial cell) in the honeycomb after the 20th ring is formed?

Initial cell

First ring

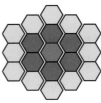

Second ring

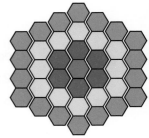

Third ring

EXERCISE 6*

1 ▶ Given 15 + 18 + 21 + 24 + …, find the sum of the first 50 terms.

2 ▶ Given 32 + 28 + 24 + 20 + …, find the sum of the first 100 terms.

3 ▶ Find 2 + 5 + 8 + 11 + … + 119

4 ▶ Find the sum of the first 50 **multiples** of 5.

5 ▶ The 19th term of an arithmetic sequence is 132 and the common difference is 7. Find the sum of the first 200 terms.

6 ▶ The 12th term of an arithmetic series is 2 and the 30th term is 38. Find the sum of the first 21 terms.

7 ▶ The 5th term of an arithmetic series is 13 and the 15th term is 33. Find and simplify an expression for the sum of n terms.

8 ▶ The common difference of an arithmetic series is 8 and the sum of the first 20 terms is 1720. Find the sum of the first 40 terms.

9 ▶ The 10th term of an arithmetic series is 48 and the 20th term is 88. Find the sum from the 15th to the 30th terms.

Q10 HINT
Use trial and improvement.

10 ▶ The first term of an arithmetic series is 3 and the common difference is 2. The sum to n terms is 675. Find n.

11 ▶ The Ancient Egyptian Rhind papyrus from around 1650 BCE contains the following problem: 'Divide 10 hekats of barley among 10 men so that the common difference between the amount each man receives is $\frac{1}{8}$ of a hekat of barley.' (A hekat was an ancient Egyptian measurement of volume used for grain, bread and beer.) Find the smallest and the largest amounts received.

12 ▶ A concert hall has 25 rows of seats. There are 22 seats on the first row, 24 seats on the second row, 26 seats on the third row, and so on. Each seat in rows 1 to 10 costs $40, each seat in rows 11 to 18 costs $30, and each seat in rows 19 to 25 costs $20.

　　a How many seats are in the hall?
　　b How much money does the concert hall make if an event is sold out?

ACTIVITY 7

SKILLS

PROBLEM
SOLVING

Ancient legend has it that when the Emperor asked the man who invented chess to name his reward he replied 'Give me one grain of rice for the first square on the chess board, two grains for the second square, four grains for the third square, eight grains for the fourth square and so on, doubling each time until the 64th square.' The Emperor thought this was a pretty trivial reward until he asked the court mathematician to work out how much rice was involved!

Copy and complete the table to show how much rice is needed.

NUMBER OF SQUARES	1	2	3	4	5	6	7	64
NUMBER OF GRAINS	$1 = 2^0$	$2 = 2^1$	$4 = 2^2$	$8 = 2^3$				
TOTAL NUMBER	1	3	7	15				

A typical grain of rice measures 6 mm by 2 mm by 2 mm. Assuming that the grain is a **cuboid**, calculate the volume of rice needed for the inventor's reward. Give your answer in mm^3 and km^3 **correct to** 3 s.f.

The squares on a typical chess board measure 4 cm by 4 cm. Work out the height of the pile of rice on the last square. Give your answer in mm and km correct to 3 s.f.

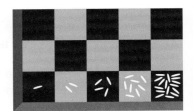

The distance from the Earth to the Sun is around 1.5×10^8 km. Work out the **ratio** of the height of the pile to the distance from the Earth to the Sun. Give your answer to 2 s.f.

EXERCISE 7

REVISION

1 ▶   Find the next three terms of these sequences.

     a 2, 4, 8, 14, 22, …   **b** 10, 20, 28, 34, 38, …

2 ▶   Phil is training for a long distance run.
On the first day of his training he runs 1000 m.
Each day after that he runs an extra 200 m.

     a How far does he run on the fifth day?

     b How far does he run on the nth day?

     c One day he runs 8 km. How many days has he been training?

3 ▶   The first four terms of two sequences are given in the following tables.

n	1	2	3	4
s	6	12	24	48

n	1	2	3	4
s	3	7	11	15

     a A formula for one of these sequences is 3×2^n
       Which sequence is this? What is the tenth term of the sequence?

     b Find a formula for the nth term of the other sequence.

4 ▶　**a** What is the name given to this sequence: 1, 3, 5, 7, 9, …?

　　　b Copy and complete:
　　　　$1 + 3 = \ldots$
　　　　$1 + 3 + 5 = \ldots$
　　　　$1 + 3 + 5 + 7 = \ldots$
　　　　$1 + 3 + 5 + 7 + 9 = \ldots$

　　　c Find the sum $1 + 3 + 5 + \ldots + 19 + 21$

　　　d Find a formula for the sum of the first n terms of the original sequence.

　　　e The sum of m terms of this sequence is 841. What is m?

5 ▶　The 4th term of an arithmetic sequence is 7 and the 8th term is 19. Find a and d.

6 ▶　The 3rd term of an arithmetic sequence is 19 and the 9th term is −5. Find a and d.

7 ▶　Find the first term over 200 for the arithmetic sequence 11, 14, 17, 20, …

8 ▶　The first term of an arithmetic sequence is 5 and the 11th term is 45. Find the 25th term.

9 ▶　Wilf is learning to lay bricks. On his first day he manages to lay 90 bricks and a further 15 bricks every day after that. How long will it take him to reach his target of 750 bricks in a day?

10 ▶　The 16th term of an arithmetic sequence is 127 and the common difference is 8. Find the sum of the first 16 terms.

11 ▶　The 4th term of an arithmetic sequence is 20 and the 8th term is 44. Find the sum of the first 24 terms.

12 ▶　A mosaic is made by starting with a central tile measuring 2 cm by 1 cm, then surrounding it with successive rings of tiles as shown.

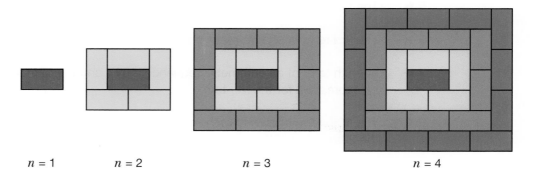

　　　$n = 1$　　　$n = 2$　　　$n = 3$　　　$n = 4$

　　　a Find an expression for the number of tiles in the nth ring.

　　　b After 70 rings have been added, how many tiles have been used?

REVISION

1 ▶   Find the next three terms of the following sequences.

     a   10, 8, 8, 10, 14, …   **b**   0, 8, 13, 15, 14, …

2 ▶   On Jamie's fifth birthday, she was given pocket money of 50p per month, which increased by 20p each month.

     a   How much pocket money does Jamie get on her sixth birthday?

     b   Find a formula to find Jamie's pocket money on the nth month after her fifth birthday.

     c   How old was Jamie when her pocket money became £17.30 per month?

3 ▶   **a**   Find the next four terms in the sequence 1, 5, 9, 13, …

     b   The first and third terms of this sequence are square numbers.
Find the positions of the next two terms of the sequence that are square numbers.

     c   Form a new sequence from the numbers giving the positions of the square numbers (i.e. starting 1, 3, …). Use this sequence to find the position of the fifth square number in the original sequence.

4 ▶   **a**   For this sequence of stones, copy and complete the table, where n is the number in the sequence and s is the number of pebbles.

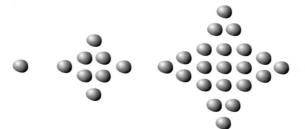

n	1	2	3	4	5
s	1	8			

     b   By adding a row for $3n^2$, find a formula connecting n and s.

     c   A pattern uses 645 pebbles. Which term of the sequence is this?

5 ▶   The 5th term of an arithmetic sequence is 8 and the 10th term is 28. Find a and d.

6 ▶   The 3rd term of an arithmetic sequence is 5 and the 7th term is −5. Find a and d.

7 ▶   The 8th term of an arithmetic sequence is 171 and the 12th term is 239. Find and simplify an expression for the nth term.

8 ▶   The 10th term of an arithmetic sequence is 37 and the 15th term is 62. Find the sum from the 20th to the 30th terms.

9 ▶ $1000 is to be divided between 8 children so that the common difference between the amount each child receives is $15. Write down the sequence of amounts in full.

10 ▶   The sum of the first 11 terms of an arithmetic sequence is 319 and the sum of the first 21 terms is 1029. Find and simplify an expression for the sum of n terms.

11 ▶ On a sponsored walk of 200 km, Karen walks 24 km the first day, but due to tiredness she walks 0.5 km less on each day after. Use trial and improvement to find how long it takes her to complete the walk.

12 ▶ A paper manufacturer sells paper rolled onto cardboard tubes. The thickness of the paper is 0.1 mm, the **diameter** of the tube is 75 mm and the overall diameter of the roll is 200 mm. Assuming the layers of paper are concentric rings, find the total length of paper in a roll. Give your answer in metres correct to 3 s.f.

EXAM PRACTICE: SEQUENCES

1 Find the first three terms of the sequences with the given nth term.

   a $84 - 4n$ **b** $n(2n + 1)$ **c** $\dfrac{n - 1}{n + 1}$ **[3]**

2 Find the next three terms of the following sequences using the difference method.

   a 11, 7, 3, –1, –5, …   **b** 2, 8, 16, 26, 38, …   **c** 0, 5, 12, 21, 32, …   **[3]**

3 The 4th term of an arithmetic sequence is 20 and the 9th term is 50. Find a and d.   **[4]**

4 The 5th term of an arithmetic sequence is 16 and the 20th term is 61. Find the sum of the first 20 terms.   **[5]**

5 The sum of the first 20 terms of an arithmetic sequence is 990. The common difference is 5.
Find the first term.   **[5]**

6 Kris is building the end wall of a house. The first row has 38 bricks, and each successive row has one fewer brick than the row below. The top row has only one brick. How many bricks are needed to build the wall?   **[5]**

[Total 25 marks]

CHAPTER SUMMARY: SEQUENCES

■ A set of numbers that follows a definite pattern is called a **sequence**.

■ When a sequence is given by a formula, any term can be worked out.

The nth term of a sequence is $3n - 2$. Find the 2nd and 100th terms.

Substituting $n = 2$ gives the 2nd term as 4

Substituting $n = 100$ gives the 100th term as 298

■ The difference method finds patterns in sequences when the patterns are not obvious.

Sequence: 5   7   11   17   25

Differences: 2   4   6   8

              2   2   2

■ If the first row of differences is constant and equal to a then the formula for the nth term will be nth term $= an + b$ where b is another constant.

Find a formula for the nth term of the sequence 5, 8, 11, 14, ...

Sequence: 5   8   11   14

Differences: 3   3   3

The first row of differences is constant and equal to 3 so formula is $3n + b$.

When $n = 1$ the formula must equal 5

The formula for the nth term is $3n + 2$

■ The formula can be obvious.

Find a formula for the nth term of the sequence 1, 4, 9, 16, ...

The sequence is square numbers so the nth term is n^2.

■ When the difference between any two consecutive terms in a sequence is the same, the sequence is called an **arithmetic sequence**. The difference between the terms is called the **common difference**. The common difference can be positive or negative.

■ In an arithmetic sequence:
  ■ The first term is a
  ■ The common difference is d
  ■ The nth term is $a + (n - 1)d$

■ $a + (a + d) + (a + 2d) + (a + 3d) + ... + (a + (n - 1)d)$ is an **arithmetic series**.

The sum to n terms of an arithmetic series is
$$S_n = \frac{n}{2}[2a + (n - 1)d]$$

Find the 20th term and the sum to 20 terms of the arithmetic sequence 7, 10, 13, 16, ...

$a = 7$, $d = 3$ and $n = 20$

The 20th term is $7 + (20 - 1) \times 3 = 64$

The sum to 20 terms is $\frac{20}{2}[2 \times 7 + (20 - 1) \times 3] = 710$

SHAPE AND SPACE 6

Crop circles have appeared in farmers' fields all over the world. These crop circles show circle geometry in use. Some people believe that they are created by aliens. However, the majority of scientists agree that they are man-made.

LEARNING OBJECTIVES

- Understand and use the alternate segment theorem
- Understand and use the internal and intersecting chord properties
- Solve angle problems using circle theorems

BASIC PRINCIPLES

- Triangle OAB is **isosceles**.

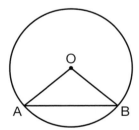

- The angle at the centre of a circle is twice the angle at the **circumference** when both are **subtended** by the same **arc**.

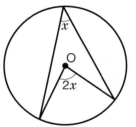

- A **tangent** is a straight line that touches a circle at one point only.

  The angle between a tangent and the **radius** is 90°

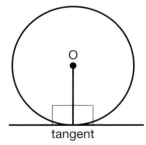

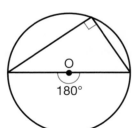

- An angle in a **semicircle** is always a **right angle**.

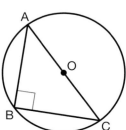

- A figure is **cyclic** if a circle can be drawn through its vertices. The vertices are **concyclic** points.

■ Opposite angles of a **cyclic quadrilateral** sum to 180°.

$a° + b° = 180°$

$x° + y° = 180°$

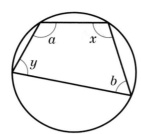

■ Angles in the same **segment** are equal.

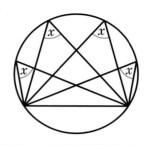

CIRCLE THEOREMS 2

The basic circle theorems are frequently combined when solving circle geometry questions. These questions will allow you to revise your understanding of these basic circle theorems.

EXAMPLE 1

SKILLS

PROBLEM SOLVING

Find ∠PNM.

∠NPM = 35° (Angles in the same segment are equal)
∠LNM = 90° (Angle in a semicircle is 90°)
∠MNQ = 90° (Angles on a straight line total 180°)
So ∠PNM = 30° (Angle sum of ΔPNQ is 180°)

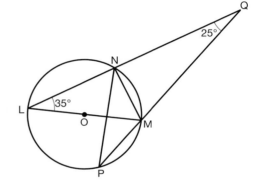

EXAMPLE 2

SKILLS

PROBLEM SOLVING

Prove that XY meets OZ at right angles.

∠OXY = 30° (**Alternate angle** to ∠ZYX)
∠OYX = 30° (Base angles in an isosceles triangle are equal)
∠YZO = 60° (Base angles in an isosceles triangle are equal)
∠ZNY = 90° (Angle sum of ΔZNY is 180°)

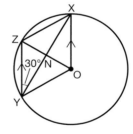

KEY POINTS

When trying to find angles or lengths in circles:

■ Always draw a neat diagram, and include all the facts. Use a pair of **compasses** to draw all circles.

■ Give a reason, in brackets, after each statement.

EXERCISE 1

For Questions 1–10, find the coloured angles. Explain your reasoning.

1 ▶

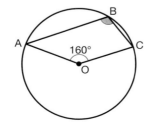

2 ▶

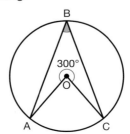

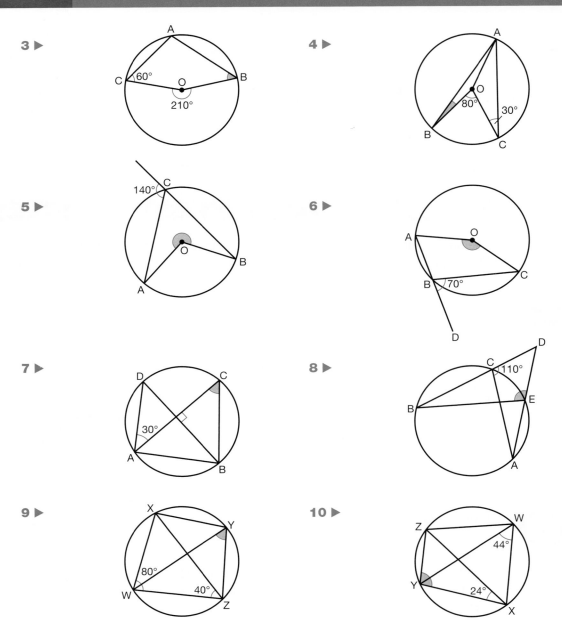

3 ▶

4 ▶

5 ▶

6 ▶

7 ▶

8 ▶

9 ▶

10 ▶

11 ▶ A and B are points on the
circumference of a circle, centre O.
BC is a tangent to the circle.
AOC is a straight line.
Angle ABO = 35°

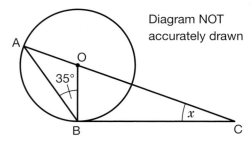

Diagram NOT
accurately drawn

Work out the size of the angle marked x.
Give reasons for your answer.

12 ▶ The diagram shows a circle, centre O.
CB and CD are tangents to the circle.

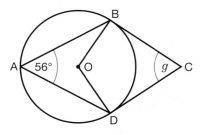

Find angle g. Explain your reasoning.

13 ▶ Pria says that just from knowing the angle GHO, she can work out all the angles inside triangles GFO and HGO.

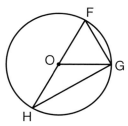

a Prove that Pria is correct.
b If angle GHO = 41°, work out the angles in triangles GFO and HGO.

14 ▶ Work out the sizes of angles a, b and c. Give reasons for your answers.

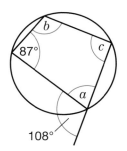

For Questions 15 and 16, prove that the points ABCD are concyclic.

HINT

A figure is cyclic if a circle can be drawn through its vertices. The vertices are concyclic points.

15 ▶

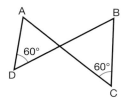

16 ▶

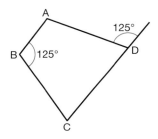

EXERCISE 1* For Questions 1–8, find the coloured angles, fully explaining your reasoning.

1 ▶

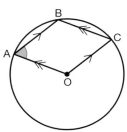

2 ▶

3 ▶

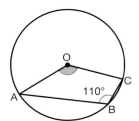

4 ▶

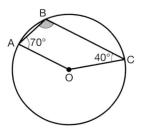

5 ▶

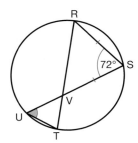

6 ▶

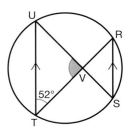

7 ▶

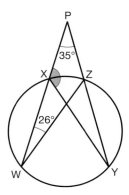

8 ▶

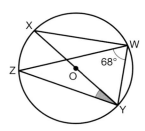

9 ▶ Find, in terms of x, ∠AOX.

10 ▶ Find, in terms of x, ∠BCD.

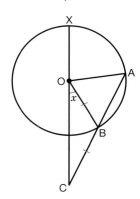

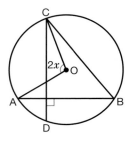

Q11 HINT

A figure is cyclic if a circle can be drawn through its vertices. The vertices are concyclic points.

11 ▶ ABCD is a **quadrilateral** in which AB = AD and BD = CD.
Let ∠DBA = x° and ∠DBC = $2x$°. Prove that A, B, C and D are concyclic.

12 ▶ ABCDEF is a hexagon inscribed in a circle.
By joining AD, prove that ∠ABC + ∠CDE + ∠EFA = 360°

13 ▶ Prove that ∠CEA = ∠BDA.

14 ▶ In the figure, OX is the **diameter** of the smaller circle, which cuts XY at A.
Prove that AX = AY.

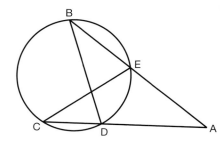

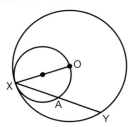

15 ▶ WXYZ is a cyclic quadrilateral. The sides XY and WZ produced meet at Q. The sides XW and YZ produced meet at P. $\angle WPZ = 30°$ and $\angle YQZ = 20°$
Find the angles of the quadrilateral.

16 ▶ PQ and PR are any two **chords** of a circle, centre O. The diameter, **perpendicular** to PQ, cuts PR at X. Prove that the points Q, O, X and R are concyclic.

ALTERNATE SEGMENT THEOREM

ACTIVITY 1

SKILLS

ANALYSIS

Find the angles in circle C_1 and circle C_2.

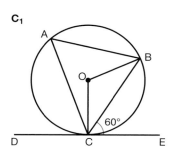 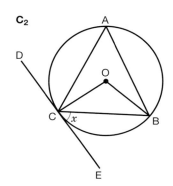

Copy and complete the table for C_1 and C_2.

CIRCLE	$\angle$ECB	$\angle$OCB	$\angle$OBC	$\angle$BOC	$\angle$BAC
C_1	60°				
C_2	$x°$				

What do you notice about angles ECB and angles BAC?

The row for angles in C_2 gives the structure for a formal proof.

A full proof requires reasons for every stage of the calculation.

Calculations	**Reasons**
$\angle ECB = x°$	General angle chosen.
$\angle OCB = (90 - x)°$	Radius is perpendicular to tangent.
$\angle OBC = (90 - x)°$	Base angles in an isosceles triangle are equal.
$\angle BOC = 2x°$	Angle sum of a triangle = 180°
$\angle BAC = x°$	Angle at centre = 2 × angle at circumference.
$\angle BCE = \angle BAC$	

Note: A formal proof of the theorem is not required by the specification.

■ The angle between a chord and a tangent is equal to the angle in the alternate segment.

This is called the 'Alternate Segment Theorem'.

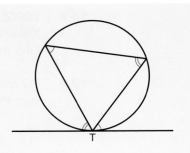

EXAMPLE 3

In the diagram, PAQ is the tangent at A to a circle with centre O.

Angle BOC = 112°

Angle CAQ = 73°

Work out the size of angle OBA.

Show your **working**, giving reasons for any statements you make.

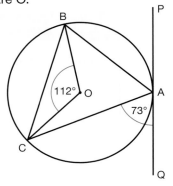

Not drawn accurately

∠OBC = 34° (ΔOBC is isosceles)

∠ABC = 73° (Alternate segment theorem)

∠OBA = 73° − 34° = 39°

EXERCISE 2

In this exercise, O represents the centre of a circle and T represents a tangent to the circle.

For Questions 1–8, find the coloured angles, fully explaining your reasons.

1 ▶

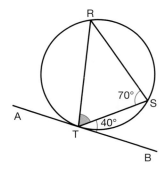

2 ▶

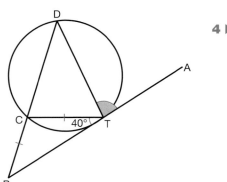

3 ▶

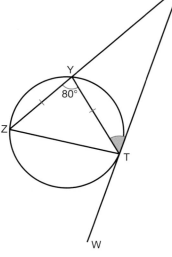

4 ▶

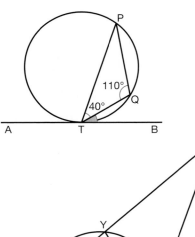

5 ▶

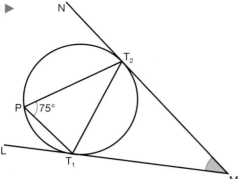

6 ▶

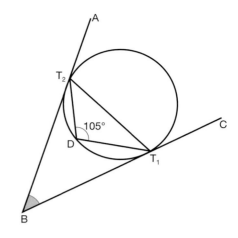

7 ▶

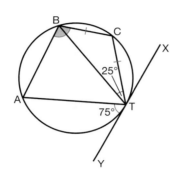

8 ▶

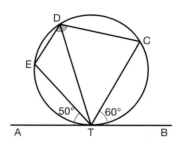

9 ▶ Find
 a ∠OTX
 b ∠TOB
 c ∠OBT
 d ∠ATY

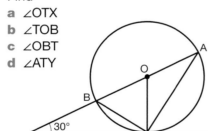

10 ▶ Find
 a ∠OTB
 b ∠OTC
 c ∠OCT
 d ∠DTA

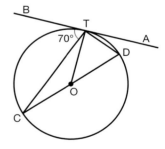

11 ▶ Copy and complete these two statements to prove ∠NPT = ∠PLT.

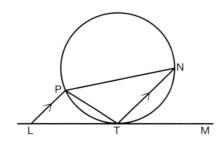

∠NTM = ... (Alternate segment)
∠PLT = ... (**Corresponding angles**)

12 ▶ Copy and complete these two statements to prove ∠ATF = ∠BAF.

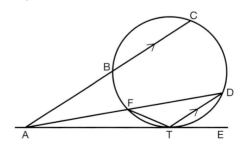

∠ATF = ∠FDT (.......)
∠FDT = ∠BAF (.......)

13 ▶ Copy and complete these two statements to prove ∠ATC = ∠BTD.

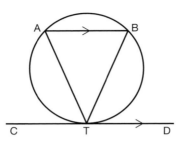

∠ATC = ... (Alternate segment theorem)

∠ABT = ∠BTD (.........)

14 ▶ Copy and complete these two statements to prove that triangles BCT and TBD are similar.

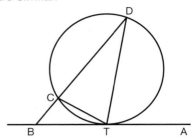

∠CTB = ... (Alternate segment theorem)

.... is common.

EXERCISE 2*
In this exercise, O represents the centre of a circle and T represents a tangent to the circle.

For Questions 1–4, find the coloured angles, fully explaining your reasons.

1 ▶

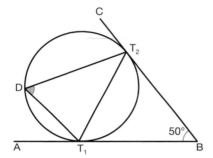

2 ▶

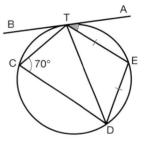

3 ▶

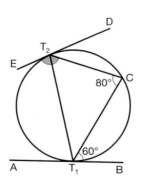

4 ▶

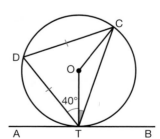

5 ▶ Prove that AB is the diameter.

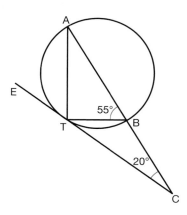

6 ▶ Given that ∠BCT = ∠TCD, prove that ∠TBC = 90°

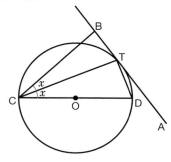

7 ▶ Work out the value of x in this diagram.

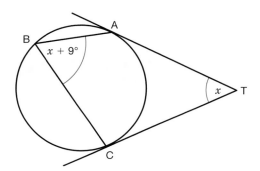

8 ▶ **a** Find ∠DAE.
    b Find ∠BED.
    c Prove that triangle ACD is isosceles.

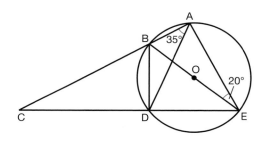

9 ▶ **a** Find ∠DTA.
    b Find ∠BCT.
    c Prove that triangles BCT and BTD are **similar**.

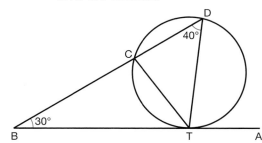

10 ▶ CD and AB are tangents at T. Find ∠ETF.

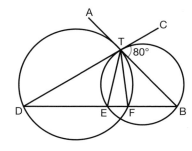

11 ▶ AB and CD are tangents at T and ∠DTB is less than 90°. Find ∠DEB.

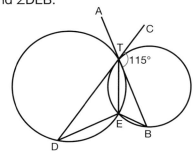

12 ▶ **a** Explain why ∠ACG = ∠ABF = 15°
    b Prove that the points CFGB are concyclic.

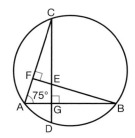

13 ▶ AB and DE are tangents to the circle, centre O.

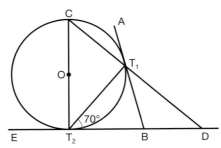

a Write down all the angles equal to
  i 70°   **ii** 20°
b Prove that B is the **mid-point** of DT$_2$.

14 ▶ **a** Giving reasons, find, in terms of x, the angles EOC and CAE.
b Use your answers to show that triangle ABE is isosceles.
c If BE = CE prove that BE will be the tangent to the larger circle at E.

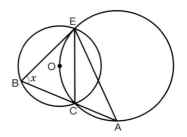

INTERSECTING CHORDS THEOREMS

ACTIVITY 2

SKILLS

ANALYSIS

O is the centre of a circle.

OA and OB are radii. OM is perpendicular to AB.

Prove that triangles OAM and OBM are **congruent** (the same).

Show that M is the mid-point of AB.

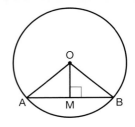

KEY POINTS

■ A **chord** is a straight line connecting two points on a circle.

■ The perpendicular from the centre of a circle to a chord bisects the chord and the line drawn from the centre of a circle to the mid-point of a chord is at right angles to the chord.

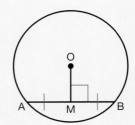

EXAMPLE 4

SKILLS

PROBLEM SOLVING

O is the centre of a circle.
The length of chord AB is 18 cm.
OM is perpendicular to AB.
Work out the length of AM.
State any circle theorems that you use.

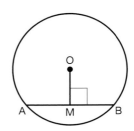

AB = 18 cm

So, AM = $\frac{18}{2}$ = 9 cm (The perpendicular from the centre of a circle to a chord bisects the chord.

So, the length of AM will be exactly half the length of AB.)

■ Two chords **intersect** inside a circle.

$AP \times BP = CP \times DP$

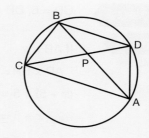

EXAMPLE 5

SKILLS

ANALYSIS

$AP = 10$, $PD = 4$ and $PB = 6$. Find CP.

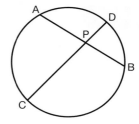

If $CP = x$

$AP \times PB = CP \times PD$

$10 \times 6 = x \times 4$

$60 = 4x$

$x = 15$

EXAMPLE 6

$AP = 4$, $BP = 3$ and $CD = 8$. Find DP.

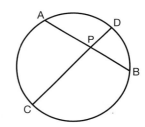

If $DP = x$

$CP = 8 - x$

$AP \times BP = CP \times DP$

$4 \times 3 = (8 - x)x$

$12 = 8x - x^2$

$x^2 - 8x + 12 = 0$

$(x - 2)(x - 6) = 0$

$x = 2$ or 6

KEY POINT

■ Two chords intersecting outside a circle.

$AP \times BP = CP \times DP$

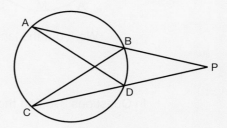

EXAMPLE 7

SKILLS

ANALYSIS

CD = 6, DP = 4 and BP = 5. Find AB.

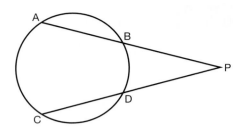

If AB = x

$AP \times BP = CP \times DP$

$(x + 5) \times 5 = 10 \times 4$

$x + 5 = 8$

$x = 3$

EXAMPLE 8

AB = 8, BP = 6 and CD = 5. Find DP.

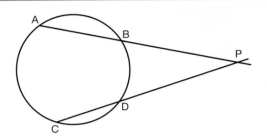

If DP = x

$CP = 5 + x$

$AP \times BP = CP \times DP$

$14 \times 6 = (x + 5)x$

$84 = x^2 + 5x$

$0 = x^2 + 5x - 84$

$0 = (x - 7)(x + 12)$

$x = 7$ or -12

DP = 7

EXERCISE 3

1 ▶ O is the centre of a circle.
OA = 17 cm and AB = 16 cm.
M is the mid-point of AB.

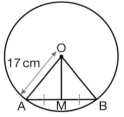

Work out the length of OM.

2 ▶ O is the centre of a circle.
M is the mid-point of chord AB.
Angle OAB = 25°

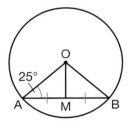

a What is angle AMO?
b Work out angle AOM.
c Work out angle AOB.

In Questions 3–8 find the length marked x.

3 ▶

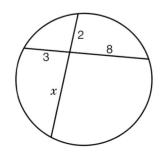

4 ▶

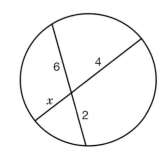

5 ▶

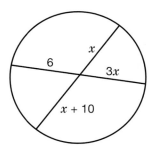

6 ▶

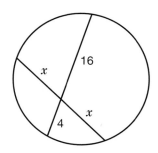

7 ▶

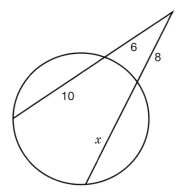

8 ▶

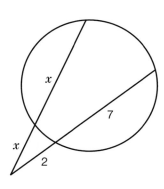

ACTIVITY 3

SKILLS

ANALYSIS

Two chords intersecting inside a circle

Show that triangles APD and CPB are similar and use this fact to prove that AP × BP = CP × DP

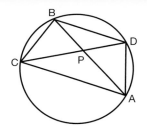

Note: A formal proof of the theorem is not required by the specification.

ACTIVITY 4

SKILLS

ANALYSIS

Two chords intersecting outside a circle

Show that triangles APD and CPB are similar and use this fact to prove that AP × BP = CP × DP

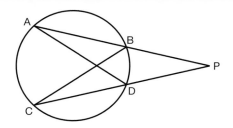

Note: A formal proof of the theorem is not required by the specification.

EXERCISE 3*

1 ▶ O is the centre of a circle.
M is a point on chord AB.
The length of chord AB is 12 cm.
OM is perpendicular to AB. OM is 8 cm.

    a Work out the length of AM. State any circle theorems that you use.
    b What is the length of the radius of the circle?

2 ▶ O is the centre of a circle. The radius of the circle is 26 cm. The distance from O to the mid-point of chord AB is 24 cm. Work out the length of chord AB.

In Questions 3–8 find the length marked x.

3 ▶

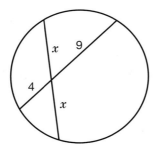

4 ▶

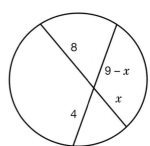

5 ▶

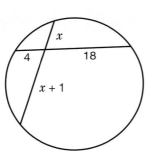

6 ▶

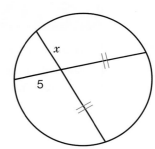

7 ▶

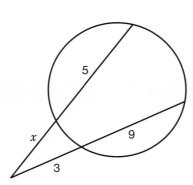

8 ▶

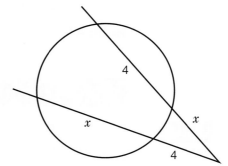

ACTIVITY 5

ANALYSIS AND CRITICAL THINKING

Constructing the circumcircle of a triangle

- Draw any triangle ABC.
- Construct the perpendicular **bisector** of AB, BC and CA.
- The point of intersection of these three lines is at point P. This is the circumcentre of a circle passing through points A, B and C.
- The radius (PA, PB and PC) of the circumcircle is called the circumradius enabling the circumcircle of triangle ABC to be drawn.

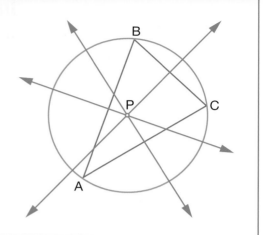

EXERCISE 4   REVISION

1 ▶ O is the centre of a circle.
OA = 20 cm and AB = 24 cm.
M is the mid-point of AB.
Work out the length of OM.

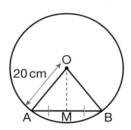

2 ▶ AP = x, AB = 16, CP = 4 and DP = 7.
Find x.

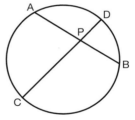

3 ▶ In each diagram, AT is a tangent to the circle.

Work out the size of each angle marked with a letter. Give reasons for each step in your working.

a

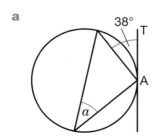

b

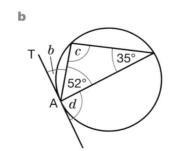

c

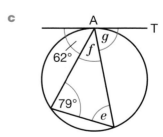

4 ▶ Work out the size of each angle marked with a letter. Give reasons for each step in your working.

a

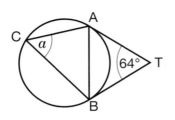

b

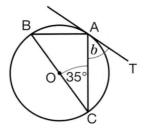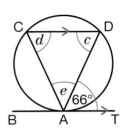

c

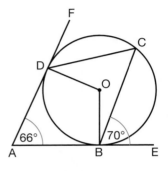

5 ▶ O is the centre of the circle. DAT and BT are tangents to the circle.
Angle CAD = 62° and angle ATB = 20°.

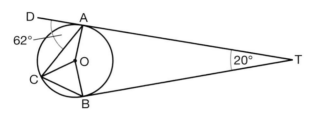

Work out the size of

a angle CAO   d angle COB
b angle AOB   e angle CBO.
c angle AOC

Give reasons for each step in your working.

6 ▶ B, C and D are points on the circumference of a circle, centre O. ABE and ADF are tangents to the circle.

Angle DAB = 66°
Angle CBE = 70°
Work out the size of angle ODC.

REVISION

1 ▶ O is the centre of a circle. M is the mid-point of chord AB.
Angle OAB = 45°

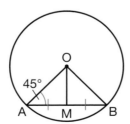

a Work out angle AMO.
b Work out angle AOM.
c Work out angle AOB.
d Which of the triangles AMO, BOM and ABO are similar?

2 ▶ AB = 9, BP = x, CD = 9 and DP = 3. Find x.

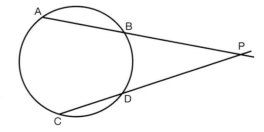

3 ▶ In the diagram, ABCD is a cyclic quadrilateral.
Prove that $x + y = 180°$

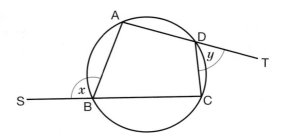

4 ▶ Work out the size of each angle marked with a letter.
Give a reason for each step of your working.

a

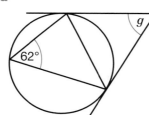

b

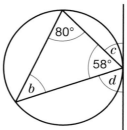

c

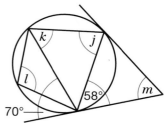

d

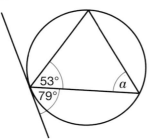

e

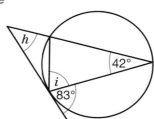

f

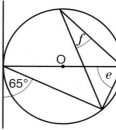

5 ▶ Points A, B, C and D lie on a circle
with centre O.
CT is a tangent to the circle at C.
CA bisects angle BAD.

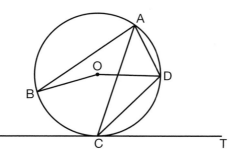

Prove that angle DCT is $\frac{1}{4}$ of angle BOD.

6 ▶ O is the centre of the circle. A, C and E
are all points on the circumference.
BCD is a tangent touching the circle at
point C.
DEF is a tangent touching the circle at
point E.

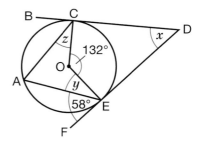

Work out the sizes of angles x, y and z.
Give reasons for any statements you
make.

EXAM PRACTICE: SHAPE AND SPACE 6

1 O is the centre of a circle with radius 6.5 cm.
AB is a chord with length 12 cm. Angle OMB = 90°

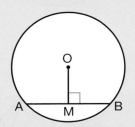

    a Write down the length of OA.
    b What is the length of AM?
    c Work out the length of OM.　　　　**[3]**

2 In each part, find x.

    a

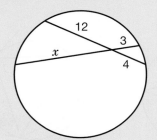

    b

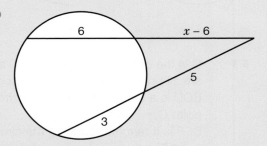

　　　　　　　　　　　　　　　　　　[6]

3 In each part, find the angles x, y and z. Give reasons for each step of your working.

    a

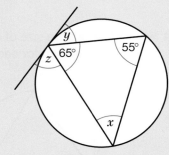

    b

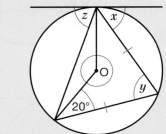

　　　　　　　　　　　　　　　　　　[12]

4 TBP and TCQ are tangents to the circle with centre O.
Point A lies on the circumference of the circle.

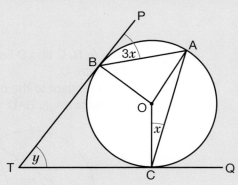

Prove that $y = 4x$
Give reasons for any statements you make.　　**[4]**

[Total 25 marks]

CHAPTER SUMMARY: SHAPE AND SPACE 6

ALTERNATE SEGMENT THEOREM

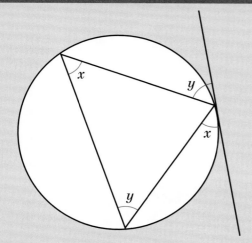

The angle between a chord and a tangent is equal to the angle in the alternate segment.

INTERSECTING CHORDS THEOREMS

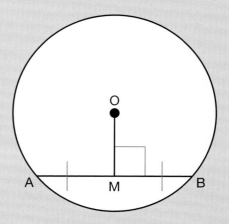

A **chord** is a straight line connecting two points on a circle.

The perpendicular from the centre of a circle to a chord bisects the chord and the line drawn from the centre of a circle to the mid-point of a chord is at right angles to the chord.

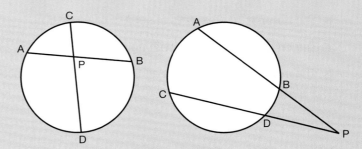

For two chords intersecting inside or outside a circle:

$AP \times BP = CP \times DP$

SETS 2

Cantor (1845–1918) was a German mathematician whose work on set theory was controversial. He managed to prove that there are an infinite number of fractions, but if numbers like $\sqrt{2}$ and π are included then this set (called the set of real numbers) cannot be counted. In other words, there are more than an infinite number of real numbers! This proof is quite simple and is called 'Cantor's Diagonal Proof'.

```
6  .  2  8  3  1  8  5  3  0  7  .  .  .
2  .  7  1  8  2  8  1  8  2  8  .  .  .
1  .  4  1  4  2  1  3  5  6  2  .  .  .
4  .  6  9  2  0  1  6  0  9  .  .  .
0  .  5  7  7  2  1  5  6  6  4  .  .  .
0  .  6  9  3  1  4  7  1  8  0  .  .  .
?  .  ?  ?  ?  ?  ?  ?  ?  ?  ?  .  .  .
?  .  ?  ?  ?  ?  ?  ?  ?  ?  ?  .  .  .
```

LEARNING OBJECTIVES

- Use Venn diagrams to represent three sets
- Solve problems involving sets
- Use set-builder notation

BASIC PRINCIPLES

- A **set** is a collection of objects, described by a list or a rule. $A = \{1, 3, 5\}$

- Each object is an **element** or **member** of the set. $1 \in A, 2 \notin A$

- Sets are **equal** if they have exactly the same elements. $B = \{5, 3, 1\}, B = A$

- The **number of elements** of set A is given by $n(A)$. $n(A) = 3$

- The **empty set** is the set with no members. $\{ \}$ or $\varnothing$

- The **universal set** contains all the elements being discussed in a particular problem. $\mathscr{E}$

- B is a **subset** of A if every member of B is a member of A. $B \subset A$

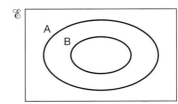

- The **complement** of set A is the set of all elements not in A. A'

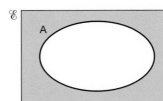

■ The **intersection** of A and B is the set of elements which are in both A and B.

$A \cap B$

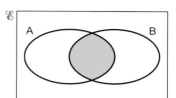

■ The **union** of A and B is the set of elements which are in A or B or both.

$A \cup B$

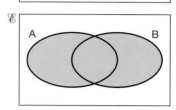

THREE-SET PROBLEMS

Questions involving three sets are more involved than two-set questions. The **Venn diagram** must show all the possible intersections of the sets.

EXAMPLE 1

SKILLS

CRITICAL THINKING

There are 80 students studying either French (F), Italian (I) or Spanish (S) in a sixth form college. 7 of the students study all three languages. 15 study French and Italian, 26 study French and Spanish and 17 study Italian and Spanish. 43 study French and 52 study Spanish.

a Draw a Venn diagram to show this information.
b How many study only Italian?

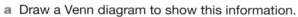

a Start by drawing three intersecting ovals labelled F, I and S.

As 7 students study all three subjects, put the number 7 in the intersection of all three circles.

15 students study French and Italian, so $n(F \cap I) = 15$
This number includes the 7 studying all three, so
$15 - 7 = 8$ must go in the region marked R.

The other numbers are worked out in a similar way.
This shows, for example, that 19 students study French and Spanish only.

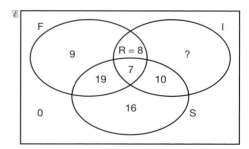

b As there are 80 students, the numbers must all add up to 80. The numbers shown add up to 69, so the number doing Italian only (the region marked '?') is $80 - 69 = 11$

KEY POINTS

■ A ∩ B ∩ C is the intersection of A, B and C i.e. where all three sets intersect.

■ A ∪ B ∪ C is the union of A, B and C, that is, all three sets combined.

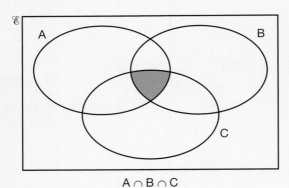

A ∩ B ∩ C

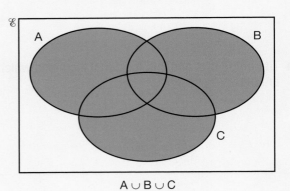

A ∪ B ∪ C

EXERCISE 1

6th

7th

1 ▶ In the Venn diagram
ℰ = {people in a night club}
P = {people who like pop music}
C = {people who like classical music}
J = {people who like jazz}

a How many people liked pop music only?
b How many liked pop music and classical music?
c How many liked jazz and classical music, but not pop music?
d How many liked all three types of music?
e How many people were in the night club?

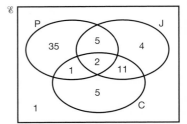

2 ▶ ℰ = {positive **integers** ≤ 24}
D = {factors of 20}
E = {factors of 24}
F = {first seven even numbers}

a Copy and complete the Venn diagram for these sets.
b List the elements of F ∪ D.
c How many elements are there in E′?

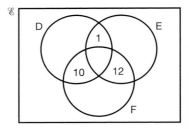

3 ▶ ℰ = {pack of 52 playing cards}
B = {Black cards}
C = {Clubs}
K = {Kings}

a Draw a Venn diagram to show the sets B, C and K.
b Describe the set B ∪ K.
c Describe the set B ∪ K ∪ C.
d Describe the set B′ ∪ K.

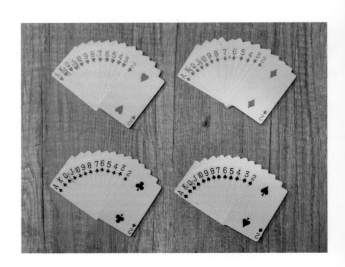

4 ▶ In the village of Cottersock, not all the houses are
connected to electricity, water or gas.

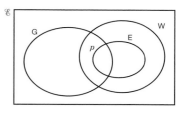

ℰ = {houses in Cottersock}
E = {houses with electricity}
W = {houses with water}
G = {houses with gas}
The relationship of these sets is shown in the Venn diagram.

a Express both in words and in set notation the relationship between E and W.
b Give all the information you can about house p shown on the Venn diagram.
c Copy the diagram and shade the set G ∩ E.
d Mark on the diagram house q that is connected to water but not to gas or electricity.

5 ▶ ℰ = {letters of the alphabet}
V = {vowels}
A = {a, b, c, d, e}
B = {d, e, u}

a Draw a Venn diagram to show the sets V, A and B.
b List the set V ∪ A.
c Describe the set V′.
d Is B ⊂ (V ∪ A)?

6 ▶ ℰ = {all triangles}
E = {**equilateral triangles**}
I = {**isosceles triangles**}
R = {**right-angled triangles**}

a Draw a Venn diagram to show the sets E, I and R.
b **Sketch** a member of I ∩ R.
c Describe the sets I ∪ E and I ∪ R.
d Describe the sets I ∩ E and E ∩ R.

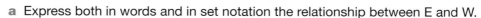

EXERCISE 1*

1 ▶ In the Venn diagram
ℰ = {ice-creams in a shop}
C = {ice-creams containing chocolate}
N = {ice-creams containing nuts}
R = {ice-creams containing raisins}

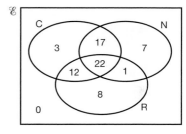

a How many ice-creams contain both chocolate and nuts?
b How many ice-creams contain all three ingredients?
c How many ice-creams contain just raisins?
d How many ice-creams contain chocolate and raisins but not nuts?
e How many different types of ice-creams are there in the shop?

2 ▶ ℰ = {positive integers less than 10}
P = {**prime numbers**}
E = {even numbers}
F = {factors of 6}

a Illustrate this information on a Venn diagram.
b List P′ ∩ E, E ∩ F, P ∩ F′.
c Describe the set P ∩ E ∩ F.

3 ▶   The universal set, $\mathscr{E}$ = {2, 3, 4, …, 12}.
      A = {factors of 24}
      B = {**multiples** of 3}
      C = {even numbers}

      On a copy of the diagram, draw a ring to represent the set C,
      and write the members of $\mathscr{E}$ in the appropriate regions.

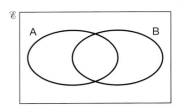

4 ▶   All boys in a class of 30 study at least one of the three sciences:
      Physics, Chemistry and Biology.

      14 study Biology.
      15 study Chemistry.
      6 study Physics and Chemistry.
      7 study Biology and Chemistry.
      8 study Biology and Physics.
      5 study all three.

      Use the Venn diagram to work out how many boys study Physics.

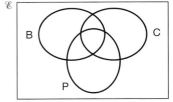

5 ▶   Draw a Venn diagram to show the intersections of three sets A, B and C.
      Given that $n(A) = 18$, $n(B) = 15$, $n(C) = 16$, $n(A \cup B) = 26$, $n(B \cup C) = 23$, $n(A \cap C) = 7$
      and $n(A \cap B \cap C) = 1$, find $n(A \cup B \cup C)$.

6 ▶   Show that a set of 3 elements has 8 subsets including $\varnothing$. Find a rule giving the number of
      subsets (including $\varnothing$) for a set of n elements.

PRACTICAL PROBLEMS

Entering the information from a problem into a Venn diagram usually means the numbers in the sets
can be worked out. Sometimes it is easier to use some algebra as well.

EXAMPLE 2

In a class of 23 students, 15 like coffee and 13 like tea. 4 students don't like either drink. How many
like **a** both drinks, **b** tea only, **c** coffee only?

SKILLS

CRITICAL
THINKING

Enter the information into a Venn diagram in stages. Let C be the set
of coffee drinkers and T the set of tea drinkers. Let x be the number
of students who like both.

The 4 students who don't like either drink can be put in, along with x
for the students who like both.

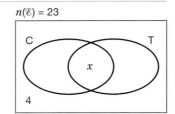

As 15 students like coffee, $15 - x$ students like coffee only and so
$15 - x$ goes in the region shown. Similarly, 13 like tea so $13 - x$ goes
in the region shown.

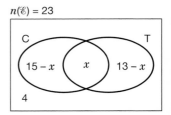

a The number who like coffee, tea or both is 23 − 4 = 19

This means $n(C \cup T) = 19$

So $(15 − x) + x + (13 − x) = 19$

$\Rightarrow 28 − x = 19$

$\Rightarrow x = 9$

So 9 students like both.

b The number who like tea only is $13 − x = 4$
c The number who like coffee only is $15 − x = 6$

EXERCISE 2

1 ▶ In a class of 40 pupils, 18 watched 'Next Door' last night and 23 watched 'Westenders'. 7 watched both programmes. How many students did not watch either programme?

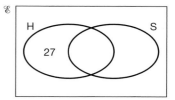

2 ▶ There are 182 spectators at a football match. 79 are wearing a hat, 62 are wearing scarves and 27 are wearing a hat but not a scarf. How many are wearing neither a hat nor a scarf?

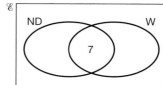

3 ▶ In one class in a school, 13 students are studying Media and 12 students are studying Sociology. 8 students are studying both, while 5 students are doing neither subject. How many students are there in the class?

4 ▶ In a town, 9 shops sell magazines and 12 shops sell sweets. 7 shops sell both, while 27 shops sell neither magazines nor sweets. How many shops are there in the town?

5 ▶ A youth group has 31 members. 15 like skateboarding, 13 like roller skating and 8 don't like either. How many like **a** skateboarding only, **b** roller skating only, **c** both?

6 ▶ 52 students are going on a skiing trip. 28 have skied before, 30 have snowboarded before while 12 have done neither. How many have done both sports before?

EXERCISE 2*

1 ▶ In a class of 30 girls, 18 play basketball, 12 do gymnastics and 4 don't do sports at all.
a How many girls do both gymnastics and basketball?
b How many girls play basketball but do not do gymnastics?

2 ▶ A social club has 40 members. 18 like singing. The same number like both singing and dancing, dancing only and neither singing nor dancing. How many like dancing?

3 ▶ At Tom's party there were both pizzas and burgers to eat. Some people ate one of each. The number of people who ate a burger only was seven more than the number who ate both. The number of people who ate pizza only was two times the number of people who ate both pizzas and burgers. The number of people who ate nothing was the same as the number of people who ate both pizza and burgers. If there were 57 people at the party, how many people ate both?

4 ▶ During the final year exams, 68 students took Mathematics, 72 took Physics and 77 took Chemistry. 44 took Mathematics and Physics, 55 took Physics and Chemistry, 50 took Mathematics and Chemistry while 32 took all three subjects. Draw a Venn diagram to represent this information and hence calculate how many students took these three exams.

Q5 HINT

In the Venn diagram let x be the number who chose all three fillings.

5 ▶ At Billy's Baguette Bar there is a choice of up to three fillings: salad, chicken or cheese. One afternoon there were 80 customers. 44 chose salad, 46 chose chicken and 35 chose cheese. 22 chose salad and chicken, 14 chose chicken and cheese while 17 chose salad and cheese. How many chose all three ingredients?

6 ▶ Sonia asked 19 friends if they liked the singer Abbey or the singer Boston. The number who liked neither was twice the number who liked both. The number who liked only Boston was the same as the number who liked both. 7 liked Abbey.
  a How many liked both?
  b How many liked Abbey only?

7 ▶ In a class of 25 pupils, 19 have scientific calculators and 14 have graphic calculators. If x pupils have both and y pupils have neither, what are the largest and smallest possible values of x and y?

8 ▶ It is claimed that 75% of teenagers can ride a bike and 65% can swim. What can be said about the percentage who do both?

SHADING SETS

Sometimes it can be difficult to find the intersection or union of sets in a Venn diagram. If one set is shaded in one direction and the other set in another direction, then the intersection is given wherever there is cross shading; the union is given by all the areas that are shaded.

The diagrams show first the sets A and B, then the set A′ shaded one way, then the set B shaded another way.

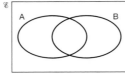
Sets A and B

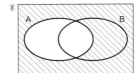
Set A′ shaded one way

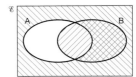
Set B shaded the other way

EXAMPLE 3

SKILLS

ANALYSIS

Show on a Venn diagram

a A′ ∩ B   **b** A′ ∪ B

a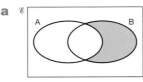
Shading shows A′ ∩ B

b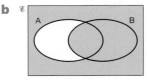
Shading shows A′ ∪ B

KEY POINT

■ Shading sets differently makes it easier to find the intersection or union.

EXERCISE 3

1 ▶ On copies of diagram 1 shade these sets.

    a $A \cap B'$　　　　**d** $A' \cup B'$
    b $A \cup B'$　　　　**e** $(A \cup B)'$
    c $A' \cap B'$

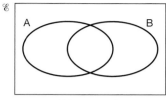

Diagram 1

2 ▶ On copies of diagram 2 shade these sets.

    a $A \cap B'$　　　　**c** $A' \cap B'$
    b $A \cup B'$　　　　**d** $A' \cup B'$

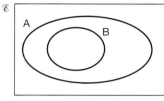

Diagram 2

3 ▶ On copies of diagram 3 shade these sets.

    a $A \cap B \cap C$
    b $A' \cup (B \cap C)$

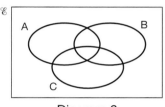

Diagram 3

4 ▶ On copies of diagram 3 shade these sets.

    a $(A \cup B') \cap C$　　　　**b** $A \cup B \cup C'$

5 ▶ Describe the shaded sets using set notation.

    a　　　　　　　　　　**b**　　　　　　　　　　**c**

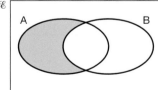 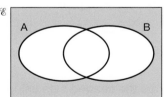

6 ▶ Describe the shaded sets using set notation.

    a　　　　　　　　　　**b**　　　　　　　　　　**c**

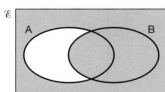 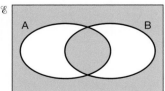 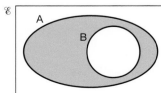

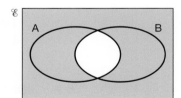

EXERCISE 3*

1 ▶   On copies of diagram 1 shade these sets.
   a $(A' \cap B)'$ **c** $(A \cap B')'$
   b $(A \cup B')'$ **d** $(A \cup B)'$

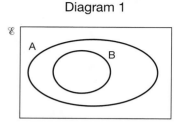

Diagram 1

2 ▶   On copies of diagram 2 shade these sets.
   a $(A \cap B')'$ **c** $(A \cap B)'$
   b $(A \cup B')'$ **d** $(A' \cup B)'$

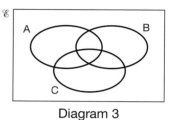

Diagram 2

3 ▶   On copies of diagram 3 shade these sets.
   a $A \cap B \cap C$ **c** $(A \cap B') \cup C$
   b $(A \cap B \cap C)'$ **d** $(A \cup B)' \cap C$

Diagram 3

4 ▶   On copies of diagram 3 shade these sets.
   a $A \cup (B' \cap C)$ **c** $A \cup (B \cap C)$
   b $(A \cap B) \cup C'$ **d** $(A \cup B) \cap C$

5 ▶   Describe the shaded sets using set notation.

   a   **b**   **c**

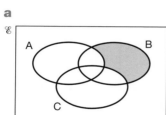   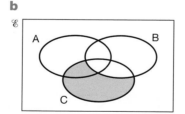   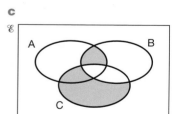

6 ▶   Describe the shaded sets using set notation.

   a   **b**   **c**

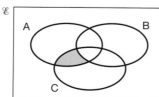   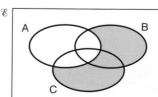   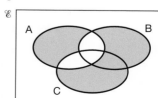

ACTIVITY 1

This activity is about De Morgan's Laws. One law states that $(A \cup B)' = A' \cap B'$.

Shade copies of Diagram 4 to show the following sets: $A \cup B$, $(A \cup B)'$, A', B' and $A' \cap B'$ and therefore prove the law.

Another law states $(A \cap B)' = A' \cup B'$. Use copies of Diagram 4 to prove this law.

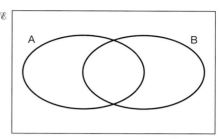

Diagram 4

SET-BUILDER NOTATION

Sets can be described using **set-builder notation**:
A = {x such that $x > 2$} means 'A is the set of all x such that x is greater than 2'.
Rather than write '**such that**' the notation A = {x: $x > 2$} is used.

B = {x: $x > 2$, x is positive integer} means the set of positive integers x such that x is greater than 2.
This means B = {3, 4, 5, 6, ...}.

C = {x: $x < 2$ or $x > 2$} can be written as {x: $x < 2$} $\cup$ {x: $x > 2$}.
The symbol $\cup$ (union) means that the set includes all values satisfied by either **inequality**.

Certain sets of numbers are used so frequently that they are given special symbols.
- $\mathbb{N}$ is the set of **natural numbers** or positive integers { 1, 2, 3, 4, ...}.
- $\mathbb{Z}$ is the set of integers { ..., −2, −1, 0, 1, 2, ...}.
- $\mathbb{Q}$ is the set of **rational numbers**. These are numbers that can be written as **recurring** or **terminating** decimals. $\mathbb{Q}$ does not contain numbers like $\sqrt{2}$ or π.
- $\mathbb{R}$ is the set of **real numbers**. This contains $\mathbb{Q}$ and numbers like $\sqrt{2}$ or π.

Express in set-builder notation

a the set of natural numbers which are greater than 2
b the set of all real numbers greater than 2.

a {x: $x > 2$, $x \in \mathbb{N}$ }
b {x: $x > 2$, $x \in \mathbb{R}$ }

Note: a ≠ b since, for example, 3.2 is a real number but it is not a natural number.

EXAMPLE 5

List these sets.

a $\{x: x \text{ is even}, x \in \mathbb{N}\}$
b $\{x: x = 3y, y \in \mathbb{N}\}$

a $\{2, 4, 6, \ldots\}$
b $\{3, 6, 9, 12, \ldots\}$

EXERCISE 4

7th

1 ▶   List these sets.

a $\{x: x \text{ is a weekday beginning with T}\}$

c $\{x: x < 7, x \in \mathbb{N}\}$

b $\{z: z \text{ is a colour in traffic lights}\}$

d $\{x: -2 < x < 7, x \in \mathbb{Z}\}$

2 ▶   List these sets.

a $\{x: x \text{ is a continent}\}$

c $\{x: x \leq 5, x \in \mathbb{N}\}$

b $\{y: y \text{ is a Mathematics teacher in your school}\}$

d $\{x: -4 < x \leq 2, x \in \mathbb{Z}\}$

3 ▶   Express in set-builder notation the set of natural numbers which are

a less than 7
b greater than 4
c between 2 and 11 inclusive

d between −3 and 3
e odd
f prime.

4 ▶   Express in set-builder notation the set of natural numbers which are

a greater than −3
b less than or equal to 9
c between 5 and 19

d between −4 and 31 inclusive
e multiples of 5
f factors of 48

EXERCISE 4*

7th

10th

1 ▶ $A = \{x: x \leq 6, x \in \mathbb{N}\}$, $B = \{x: x = 2y, y \in A\}$, $C = \{1, 3, 5, 7, 9, 11\}$
List these sets.

a B
b $\{x: x = 2y + 1, y \in C\}$

c $A \cap B$
d Give a rule to describe $B \cup C$.

2 ▶ $A = \{x: -2 \leq x \leq 2, x \in \mathbb{Z}\}$, $B = \{x: x = y^2, y \in A\}$, $C = \{x : x = 2^y, y \in A\}$
List these sets.

a B
b C

c $A \cap B \cap C$
d $\{(x, y): x = y, x \in A, y \in C\}$

3 ▶   List these sets.

    a $\{x : 2^x = -1, x \in \mathbb{R}\}$ **c** $\{x : x^2 + x - 6 = 0, x \in \mathbb{N}\}$

    b $\{2^{-x} : 0 \leq x \leq 5, x \in \mathbb{Z}\}$ **d** $\{x : x^2 + x - 6 = 0, x \in \mathbb{Z}\}$

4 ▶   List these sets.

    a $\{x : x^2 + 1 = 0, x \in \mathbb{R}\}$ **c** $\{x : x^2 + 2x - 6 = 0, x \in \mathbb{Q}\}$

    b $\{2^x : 0 \leq x < 5, x \in \mathbb{Z}\}$ **d** $\{x : x^2 + 2x - 6 = 0, x \in \mathbb{R}\}$

5 ▶   Show the sets $\mathbb{N}$, $\mathbb{Z}$, $\mathbb{Q}$ and $\mathbb{R}$ in a Venn diagram.

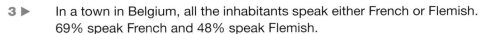

EXERCISE 5   REVISION

1 ▶ $n(A) = 20$, $n(A \cap B) = 7$ and $n(A' \cap B) = 10$.

    a Draw a Venn diagram to show   **b** Find $n(B)$

       this information.   **c** Find $n(A \cup B)$

2 ▶   In a class of 20 students, 16 drink tea, 12 drink coffee and 2 students drink neither.

    a How many drink tea only?   **c** How many drink both?

    b How many drink coffee only?

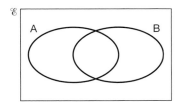

3 ▶   In a town in Belgium, all the inhabitants speak either French or Flemish.
    69% speak French and 48% speak Flemish.

    a What percentage speak both languages?

    b What percentage speak French only?

    c What percentage speak Flemish only?

4 ▶   On copies of the diagram, shade these sets.

    a $A' \cap B$ **b** $(A \cup B)'$

5 ▶ Describe the shaded set using set notation.

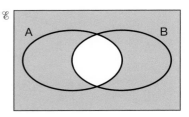

6 ▶ List these sets.

    a $\{x: -3 < x < 4, x \in \mathbb{Z}\}$ **b** $\{x: x < 5, x \in \mathbb{N}\}$ **c** $\{x: -6 < x < -1, x \in \mathbb{N}\}$

7 ▶ Express in set-builder notation the set of natural numbers which are

    a even   **b** factors of 24   **c** between −1 and 4 inclusive.

8 ▶ Solve the inequality $3(x - 1) \leq 4x + 1$ where x is an integer. Write your answer in set-builder notation.

EXERCISE 5*   **REVISION**

1 ▶ $n(A \cap B \cap C) = 2$, $n(A \cup B \cup C)' = 5$, $n(A) = n(B) = n(C) = 15$ and $n(A \cap B) = n(A \cap C) = n(B \cap C) = 6$. How many are in the universal set?

2 ▶ A youth club has 140 members. 80 listen to pop music, 40 listen to rock music, 75 listen to heavy metal and 2 members don't listen to any music. 15 members listen to pop and rock only. 12 listen to rock and heavy metal only, while 10 listen to pop and heavy metal only. How many listen to all three types of music?

3 ▶ In a group of 50 students at a summer school, 15 play tennis, 20 play cricket, 20 swim and 7 students do nothing. 3 students play tennis and cricket, 6 students play cricket and swim, while 5 students play tennis and swim. How many students do all three sports?

4 ▶   On copies of the diagram, shade these sets.

    a $(A' \cap B) \cup C$ **b** $A \cap (B \cup C)$

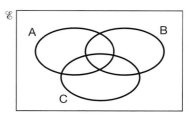

5 ▶   Describe the shaded sets using set notation.

    a 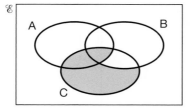   **b**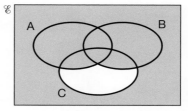

6 ▶   List these sets.

    a $\{x: x^2 - 1 = 0, x \in \mathbb{R}\}$ **b** $\{x: x^2 + 4x = 0, x \in \mathbb{Q}\}$ **c** $\{x: x^2 + 2x + 2 = 0, x \in \mathbb{R}\}$

7 ▶   Express in set-builder notation the set of natural numbers which are

    a greater than 5   **b** between 4 and 12   **c** multiples of 3

8 ▶   Solve the inequality $x < 5x + 1 \le 4x + 5$ where x is a real number. Write your answer in set-builder notation.

EXAM PRACTICE: SETS 2

1 In a group of 20 children, 11 have a scooter and 13 have a pair of roller skates. 4 children have neither. Use a Venn diagram to find how many children have both. **[4]**

2 On copies of the following diagram, shade these sets.

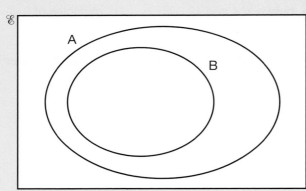

a (A ∩ B)′
b (A′ ∪ B)′   **[6]**

3 Describe the shaded set using set notation.   **[3]**

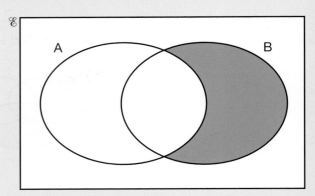

4 List the set {x: −2 < x ≤ 2, x ∈ ℤ}.   **[3]**

5 Express in set-builder notation the set of natural numbers which are multiples of four.   **[3]**

6 In a group of 100 students, 40 play tennis, 55 play football, 30 play squash and 8 do none of these. 12 students play tennis and football, 18 play football and squash and 10 play squash and tennis. Use a Venn diagram to find out how many students

a play all three sports
b play only squash.   **[6]**

[Total 25 marks]

CHAPTER SUMMARY: SETS 2

THREE-SET PROBLEMS

A ∩ B ∩ C is the intersection of A, B and C i.e. where all three sets intersect.

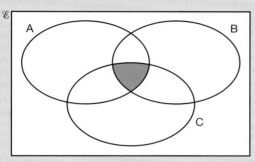

A ∩ B ∩ C

A ∪ B ∪ C is the union of A, B and C i.e. all three sets combined.

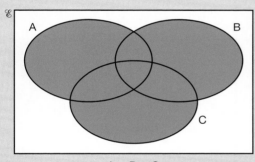

A ∪ B ∪ C

PRACTICAL PROBLEMS

Enter as much information as possible in the Venn diagram. Sometimes letting the number in a set or subset be x can help solve the problem.

$n(E) = 30$, $n(A) = 20$, $n(B) = 16$ and $n(A \cup B)' = 0$. Find $n(A \cap B)$.

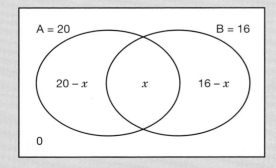

$(20 - x) + x + (16 - x) = 30 \Rightarrow x = 6$ so $n(A \cap B) = 6$

SHADING SETS

If one set is shaded in one direction and the other set in another direction, then the intersection is given wherever there is cross shading; the union is given by any shading at all.

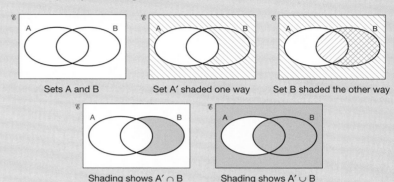

Sets A and B   Set A′ shaded one way   Set B shaded the other way

Shading shows A′ ∩ B   Shading shows A′ ∪ B

SET-BUILDER NOTATION

Certain sets of numbers are used so frequently that they are given special symbols.

- $\mathbb{N}$ is the set of natural numbers or positive integers { 1, 2, 3, 4, ...}.
- $\mathbb{Z}$ is the set of integers { ..., −2, −1, 0, 1, 2, ...}.
- $\mathbb{Q}$ is the set of rational numbers. These are numbers that can be written as recurring or terminating decimals. $\mathbb{Q}$ does not contain numbers like $\sqrt{2}$ or π.
- $\mathbb{R}$ is the set of real numbers. This contains $\mathbb{Q}$ and numbers like $\sqrt{2}$ or π.

The set A = $\{x: x \geq 5, x \in \mathbb{R}\}$ is the set of all real numbers greater than or equal to five.

B = $\{x: x \geq 5, x \in \mathbb{N}\}$ is the set { 5, 6, 7, 8, ...}.

Note that, for example, $6.2 \in A$ but $6.2 \notin B$ so A ≠ B.

The set $\{x: x$ is odd, $x \in \mathbb{N}\}$ is the set {1, 3, 5, 7, ...}.

The set $\{x: x = 2y, y \in \mathbb{N}\}$ is the set {2, 4, 6, 8, ...}.

The set C = $\{x: x < 2$ or $x > 2\}$ can be written as $\{x: x < 2\} \cup \{x: x > 2\}$.

The symbol ∪ (union) means that the set includes all values satisfied by either inequality.

UNIT 7

A rainbow has seven colours (blue, green, indigo, orange, red, violet and yellow).

The opposite sides of a cubical die add up to 7.

No perfect square ends in a 7. It is the smallest prime happy number.

NUMBER 7

The Egyptians used fractions as early as 1800 BC. They had a number system similar to the modern-day decimal system in base 10 and adopted symbols (hieroglyphs) to represent the numbers 1, 10, 100, 1000, 10 000, 100 000 and 1 000 000.

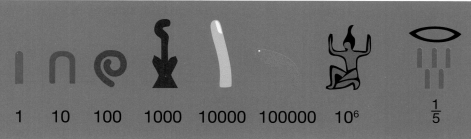

1   10   100   1000   10000   100000 10^6 $\frac{1}{5}$

LEARNING OBJECTIVES

- Convert recurring decimals to fractions
- Use a calculator for more complex calculations

BASIC PRINCIPLES

- **Simplify** fractions.

- Use a scientific calculator to work out arithmetic calculations (including use of the memory, sign change and power keys).

- All fractions can be written as either terminating decimals or decimals with a set of recurring digits.

- Fractions that produce terminating decimals have, in their simplest form, denominators with only 2 or 5 as factors. This is because 2 and 5 are the only factors of 10 (decimal system).

- The dot notation is used to indicate which digits recur. For example,

  - $0.\dot{3}$ = 0.333 …

  - $0.\dot{2}\dot{3}$ = 0.232 323 …

  - $0.\dot{0}5\dot{6}$ = 0.056 056 056 …

  - $1.2\dot{3}\dot{4}$ = 1.234 343 4 …

RECURRING DECIMALS

Fractions that have an exact decimal equivalent are called **terminating decimals**.

Fractions that have a decimal equivalent that repeats itself are called recurring decimals.

<table>
<tr><td>**EXAMPLE 1**</td><td>Change $0.\dot{5}$ to a fraction.</td></tr>
</table>

EXAMPLE 1

Change $0.\dot{5}$ to a fraction.

SKILLS

ANALYSIS

Let $x = 0.555\,555\ldots$ (Multiply both sides by 10 as one digit recurs)

$10x = 5.555\,555\ldots$ (Subtract the value of x from the value of $10x$)

$9x = 5$ (Divide both sides by 9)

$x = \dfrac{5}{9}$

EXAMPLE 2

Change $0.\dot{7}\dot{9}$ to a fraction.

SKILLS

ANALYSIS

Let $x = 0.797\,979\ldots$ (Multiply both sides by 100 as two digits recur)

$100x = 79.797\,979\ldots$ (Subtract the value of x from the value of $100x$)

$99x = 79$ (Divide both sides by 99)

$x = \dfrac{79}{99}$

KEY POINT

■ To change a recurring decimal to a fraction, first form an equation by putting x equal to the recurring decimal. Then multiply both sides of the equation by 10 if one digit recurs, by 100 if two digits recur, and by 1000 if 3 digits recur etc.

NO. OF REPEATING DIGITS	MULTIPLY BY
1	10
2	100
3	1000

EXERCISE 1

For Questions 1–4, *without* doing any calculation, write down the fractions that produce terminating decimals.

1 ▶ $\dfrac{5}{11}, \dfrac{9}{16}, \dfrac{2}{3}, \dfrac{5}{6}, \dfrac{2}{15}$

3 ▶ $\dfrac{2}{19}, \dfrac{3}{20}, \dfrac{5}{48}, \dfrac{5}{64}, \dfrac{13}{22}$

2 ▶ $\dfrac{5}{7}, \dfrac{4}{33}, \dfrac{5}{32}, \dfrac{7}{30}, \dfrac{3}{8}$

4 ▶ $\dfrac{3}{40}, \dfrac{5}{17}, \dfrac{7}{80}, \dfrac{9}{25}, \dfrac{9}{24}$

For Questions 5–18, change each recurring decimal to a fraction in its simplest form.

5 ▶ $0.\dot{3}$

12 ▶ $0.0\dot{1}$

6 ▶ $0.\dot{4}$

13 ▶ $0.0\dot{3}$

7 ▶ $0.\dot{5}$

14 ▶ $0.0\dot{2}$

8 ▶ $0.\dot{6}$

15 ▶ $0.0\dot{5}$

9 ▶ $0.\dot{7}$

16 ▶ $0.0\dot{6}$

10 ▶ $0.\dot{9}$

17 ▶ $0.\dot{7}\dot{3}$

11 ▶ $0.0\dot{7}$

18 ▶ $0.\dot{1}\dot{5}$

EXAMPLE 3

SKILLS

ANALYSIS

Change $0.\dot{1}2\dot{3}$ to a fraction.

Let $x = 0.123\,123...$ (Multiply both sides by 1000 as three digits recur)

$1000x = 123.123\,123...$ (Subtract the value of x from the value of $1000x$)

$999x = 123$ (Divide both sides by 999)

$x = \dfrac{123}{999}$

$x = \dfrac{41}{333}$ (In its simplest form)

EXAMPLE 4

SKILLS

ANALYSIS

Change $0.3\dot{4}\dot{5}$ to a fraction.

Let $x = 0.3\dot{4}\dot{5}$ (Multiply by 10 to create recurring digits after the decimal point)

$10x = 3.454\,545...$ (Multiply by 100 as there are two recurring digits after the decimal point)

$1000x = 345.454\,545...$ (Subtract $10x$ from $1000x$)

$990x = 342$

$x = \dfrac{342}{990} = \dfrac{19}{55}$ (Write answer in its simplest form)

EXERCISE 1*

For Questions 1–12, change each recurring decimal to a fraction in its simplest form.

1 ▶ $0.\dot{2}\dot{4}$ **4 ▶** $0.\dot{9}\dot{3}$ **7 ▶** $0.0\dot{2}\dot{7}$ **10 ▶** $0.\dot{1}0\dot{1}$

2 ▶ $0.\dot{3}\dot{8}$ **5 ▶** $9.0\dot{1}\dot{9}$ **8 ▶** $0.0\dot{3}\dot{6}$ **11 ▶** $0.\dot{3}8\dot{4}$

3 ▶ $0.\dot{3}\dot{0}$ **6 ▶** $8.0\dot{2}\dot{9}$ **9 ▶** $0.\dot{4}1\dot{2}$ **12 ▶** $0.\dot{4}7\dot{4}$

For Questions 13–16, change each recurring decimal to a fraction.

13 ▶ $0.1\dot{2}$ **14 ▶** $0.8\dot{6}$ **15 ▶** $0.0\dot{5}\dot{6}$ **16 ▶** $0.1\dot{5}\dot{6}$

For Questions 17–18, write each answer as a recurring decimal.

17 ▶ $0.\dot{7}\dot{3} \times 0.\dot{0}\dot{5}$ **18 ▶** $0.0\dot{7} \times 0.\dot{2}14285\dot{7}$

ADVANCED CALCULATOR PROBLEMS

The efficient use of a calculator is key to obtaining accurate solutions to more complex calculations. Scientific calculators automatically apply the operations in the correct order. However, the use of extra brackets may be needed in some calculations. The calculator's memory may also be a help with more complicated numbers and calculations.

The instruction manual should be kept and studied with care.

EXAMPLE 5

SKILLS

ANALYSIS

Calculate the following, giving answers **correct to 3 significant figures**.

a $\left(\dfrac{1.76 \times 10^3}{\sqrt{5.3 \times 10^{-2}}}\right)^4$ b $\sqrt[5]{\dfrac{5.4 \times 10^5}{\pi^3}}$

a   (1.76 ×10ˣ 3 ▼ √ 5.3 ×10ˣ –2) xᵃ 4 =

$3.41585... \times 10^{15} = 3.42 \times 10^{15}$ (3 s.f.)

b   ⁿ√ 5   5.4 ×10ˣ 5 ▼ π xᵃ 3 =

$7.04999... = 7.05$ (3 s.f.)

Become familiar with the function keys on your own calculator as they may be different from those shown here.

EXERCISE 2

Calculate the following, giving answers correct to 3 significant figures.

1 ▶ $\dfrac{2.157 \times 6.871}{1.985}$

2 ▶ $\dfrac{5.679 + 7.835}{3.873 - 0.7683}$

3 ▶ $\dfrac{3.457 \times 10^5}{7.321 \times 10^3 - 3.578 \times 10^2}$

4 ▶ $\dfrac{5.325 \times 10^4 - 3.567 \times 10^3}{7.215 \times 10^7}$

5 ▶ $\dfrac{1}{9} + \dfrac{3}{11} - \dfrac{2}{17}$

6 ▶ $(3.257 + 1.479 \times 10^3)^2$

7 ▶ $\dfrac{1.25}{1.17^2} + \dfrac{5.63}{3.86^2}$

8 ▶ $\left(\dfrac{5.85}{1.31} - \dfrac{4.82}{2.35}\right)^2$

9 ▶ $\left(\dfrac{1.75 \times 10^3}{3.52 \times 10^2}\right)^2$

10 ▶ $\left(\dfrac{10\pi^2}{5.75 \times 10^2 - \pi^2}\right)^2$

11 ▶ $\left(\dfrac{1}{5.68 \times 10^{-3}}\right)^2$

12 ▶ $\dfrac{\sqrt{\pi} + \sqrt{2}}{\pi^2 - \sqrt{2}}$

13 ▶ If $s = ut + \dfrac{1}{2}at^2$, find s if $u = 4$, $t = 5$ and $a = 10$

14 ▶ If $E = (a^2 + b^2 + c^2)^d$, find E if $a = 2.3$, $b = 4.1$, $c = 1.5$ and $d = 0.5$

15 ▶ If $E = \sqrt{a + \sqrt{b + \sqrt{c + \sqrt{d}}}}$, find E if $a = 2$, $b = 10$, $c = -1$ and $d = 8$

16 ▶ The formula $F = \dfrac{9C}{5} + 32$ is used to convert temperatures from degrees Celsius to degrees Fahrenheit.
   a Convert 28 °C into degrees Fahrenheit.
   b Make C the **subject** of the formula.
   c Convert 104 °F into degrees Celsius.

EXERCISE 2*

Calculate the following, giving answers correct to 3 significant figures.

1 ▶ $\left(\dfrac{1.5 \times 5^3}{1.1 \times 3^5}\right)^5$

2 ▶ $\left(\dfrac{7}{2.3^7} + \dfrac{2.3^7}{7}\right)^{-7}$

3 ▶ $\sqrt{\dfrac{7.5 \times 10^2}{5.7 \times 10^{-2}}}$

4 ▶ $\sqrt[3]{\dfrac{\pi^3}{3\pi}} + \sqrt[3]{\dfrac{3\pi}{\pi^3}}$

5 ▶ $\sqrt{\dfrac{\sin45°}{\sqrt{3}} + \sqrt{\dfrac{\sqrt{7}}{\sin60°}}}$

7 ▶ $\dfrac{1}{\sqrt[4]{7.5 \times 10^{-4}}}$

6 ▶ $\sqrt[5]{\dfrac{7.5 \times 10^{-3}}{1.5 \times 10^{-5}}}$

8 ▶ $\left(\pi + (5.75 \times \sqrt{23})^{-2}\right)^{-3}$

9 ▶ $\sqrt{5 + \sqrt{4 + \sqrt{3 + \sqrt{2 + \sqrt{1}}}}}$

10 ▶ $\sqrt[5]{\dfrac{\dfrac{1}{2^3} + \dfrac{3}{4^3} + \tan30°}{7.5 \times \pi^{-2}}}$

11 ▶ $2\pi\sqrt{\dfrac{2.5^3 + 1.5^3}{3.5 \times \sqrt[3]{\pi}}}$

12 ▶ $\sqrt[5]{5 + \sqrt[4]{4 + \sqrt[3]{3 + \sqrt[2]{2 + \sqrt{1}}}}}$

13 ▶ If $x = \dfrac{-b \pm \sqrt{b^2 - 4ac}}{2a}$, find x if $a = 3.2$, $b = -7.3$ and $c = 1.4$

14 ▶ The escape **velocity**, v (m/s) of an object from a planet of **mass** M kg, **radius** R m is given by the formula

$v = \sqrt{\dfrac{2GM}{R}}$, where $G = 6.673 \times 10^{-11}$ Nm²/kg²

Find the escape velocity in m/s to 4 s.f. from

a Earth ($M = 5.97 \times 10^{24}$ kg , $R = 6.37 \times 10^6$ m)

b Mars ($M = 6.39 \times 10^{23}$ kg , $R = 3.39 \times 10^6$ m)

15 ▶ Newton's Law of Universal Gravitation can be used to calculate the force (F) between two objects.

$F = \dfrac{Gm_1m_2}{r^2}$, where G is the gravitational constant (6.67×10^{-11} Nm² kg⁻²),

m_1 and m_2 are the masses of the two objects (kg) and r is the distance between them (km).

a Rearrange the formula to make r the subject.

The gravitational force between the Earth and the Sun is 3.52×10^{22} N.

The mass of the Sun is 1.99×10^{30} kg and the mass of the Earth is 5.97×10^{24} kg.

b Work out the distance between the Sun and the Earth.

16 ▶ A circle with centre (h, k) and radius r has equation

$(x - h)^2 + (y - k)^2 = r^2$

Find the radius of a circle with centre (2.5, 3) which has the point (7, 15.5) on its **circumference**. Give your answer to 3 s.f.

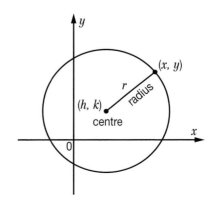

SKILLS

ADAPTIVE
LEARNING

ACTIVITY 1

Calculations such as $3 \times 2 \times 1$ can be represented by the symbol $3!$

Similarly $4! = 4 \times 3 \times 2 \times 1$

So $1! = 1$, $2! = 2$, $3! = 6$, $4! = 24$ and so on...

The $x!$ symbol is called the **factorial** of x.

Thus $x! = x \times (x - 1) \times (x - 2) \times \ldots \times 3 \times 2 \times 1$

The factorial function is very useful in mathematics when considering patterns.

Scientific calculators have this function represented by the button $\boxed{x!}$

Note that $\dfrac{4!}{3!} = \dfrac{4 \times 3 \times 2 \times 1}{3 \times 2 \times 1} = 4$

1 ▶ Find the values of the following expressions without the use of a calculator.

    a $5!$ **b** $6!$ **c** $7!$ **d** $\dfrac{5!}{4!}$ **e** $\dfrac{10!}{8!}$ **f** $\dfrac{100!}{98!}$

2 ▶ Find the values of the following with the use of a calculator.

    a $\dfrac{10!}{7!5!}$ **d** $\sqrt[5!]{\pi^{5!}}$

    b $\left(\dfrac{12!}{9!} - \dfrac{6!}{3!} \right)^{\frac{3!}{2!}}$ **e** $1\dfrac{1}{1!} + 1\dfrac{1}{2!} + 1\dfrac{1}{3!} + 1\dfrac{1}{4!} + 1\dfrac{1}{5!}$

    c $\left(\dfrac{\sin(10 \times 3!)}{\cos(10 \times 3!)} \right)^{3!}$ **f** $\left(1\dfrac{1}{1!} - 1\dfrac{1}{2!} + 1\dfrac{1}{3!} - 1\dfrac{1}{4!} + 1\dfrac{1}{5!} \right)^{\frac{10!}{5 \times 9!}}$

3 ▶ Solve these equations.

    a $x! = 3\,628\,800$ **c** $2^{x!} = 16\,777\,216$

    b $x^2 - x + 4! = 12 + 3!x$ **d** $(x!)^2 = 1\,625\,702\,400$

4 ▶ Show that $\dfrac{n!}{(n - 2)!} = n^2 - n$

EXERCISE 3

REVISION

1 ▶ Write these recurring decimals as fractions in their simplest form.

    a $0.\dot{2}$ **b** $0.0\dot{7}$ **c** $0.\dot{2}\dot{3}$

2 ▶ Show that $0.\dot{9} = 1$

3 ▶ Show that $0.\dot{x} = \dfrac{x}{9}$

4 ▶ Show that $0.0\dot{x} = \dfrac{x}{90}$

5 ▶ Calculate these, giving answers correct to 3 significant figures.

    a $\dfrac{3.61 + 7.41}{1.1^2}$ **b** $\dfrac{7.1^3}{4.5 \times 10^{-3}}$ **c** $\sqrt{\dfrac{7.7 \times 10^7}{5 \times 10^5 - 3 \times 10^3}}$

6 ▶ If $v^2 = u^2 + 2as$, find the value of v if $u = 10$, $a = 4$ and $s = 5.5$

7 ▶   **a** Make T the subject of the formula $S = \dfrac{D}{T}$

   b Sometimes the distance between the Earth and Mars is about 57.6 million kilometres. The speed of light is approximately 3×10^8 m/s.

   Calculate the time taken for light to travel from Mars to the Earth.

8 ▶   The formula, $d = \sqrt{2Rh}$, where $R \simeq 6.37 \times 10^6$ metres is the radius of the Earth, gives the approximate distance to the horizon of someone whose eyes are h metres above sea level.

   Use this formula to calculate the distance (to the nearest metre) to the horizon of someone who stands

   a at sea level and is 1.7 m tall

   b on the summit of Mount Taranaki, New Zealand, which is 2518 m above sea level.

EXERCISE 3*

REVISION

1 ▶   Write these recurring decimals as fractions in their simplest form.

   a $0.\dot{7}$ **b** $0.0\dot{1}$ **c** $0.\dot{6}\dot{7}$ **d** $3.04\dot{5}$

2 ▶   Show that $0.\dot{x}\dot{y} = \dfrac{xy}{99}$

3 ▶   Show that $0.\dot{x}y\dot{z} = \dfrac{xyz}{999}$

4 ▶   Show that $1.0\dot{x}\dot{y} = \dfrac{10xy - 10}{990}$

5 ▶   Calculate these, giving answers correct to 3 significant figures.

   a $\sqrt{\dfrac{\pi^3}{3^\pi}}$ **b** $\sqrt[3]{\dfrac{5.7 \times 10^7}{\sin 60°}}$ **c** $\sqrt[5]{\dfrac{\sqrt{7.5 \times 10^5}}{\dfrac{2}{\pi} + \dfrac{\pi}{2}}}$

6 ▶   If $T = 2\pi\sqrt{\dfrac{a+b}{g}}$ and $a = 5$, $b = 12$ and $g = 9.8$, find the value of T to 3 s.f.

7 ▶   The deposit, D, needed when booking a skiing holiday is in two parts:
   ■ a non-returnable booking fee, B
   ■ one-tenth of the total cost of the holiday, which is worked out by multiplying the price per person, P, by the number of people, N, in the group.

   $D = B + \dfrac{NP}{10}$

   a Find the deposit needed to book a holiday for four people when the cost per person is £2000 and the booking fee is £150.

   b Make P the subject of the formula.

   c What is the price per person when $D = £500$, $B = £150$ and $N = 5$?

8 ▶   The formula gives the monthly repayments, £M, needed to pay off a mortgage over n years when the amount borrowed is £P and the interest rate is $r\%$.

   $$M = \dfrac{Pr\left(1 + \dfrac{1}{100}r\right)^n}{1200\left[\left(1 + \dfrac{1}{100}r\right)^n - 1\right]}$$

   Calculate the monthly repayments when the amount borrowed is £250 000 over 25 years and the interest rate is 5%.

EXAM PRACTICE: NUMBER 7

1 Convert the following recurring decimals to fractions in their simplest form.

a $0.\dot{8}$ **b** $0.8\dot{5}$ **c** $0.\dot{7}5\dot{4}$ **d** $0.0\dot{2}3\dot{7}$ **e** $0.7\dot{3}$ **f** $3.0\dot{2}\dot{1}$ **[12]**

2 Calculate these, giving answers correct to 3 significant figures.

a $\dfrac{1}{2^2} + \dfrac{1}{3^2} + \dfrac{1}{4^2} + \dfrac{1}{5^2}$ **b** $\dfrac{7.35 \times 10^5}{9.78 \times 10^6}$ **c** $\sqrt[7]{\dfrac{5^3 + 3.5 \times 10^5}{4.9 \times 10^{-3}}}$ **[6]**

3 Find the value of z to 3 sig. figs given that $z = \sqrt[3]{\dfrac{a^2 + b^2 + c^2}{\tan(b - c)^\circ}}$, where $a = 60$, $b = 30$, $c = 15$ **[4]**

4 Show that $3.\dot{x}\dot{y} = \dfrac{xy + 297}{99}$ **[3]**

[Total 25 marks]

CHAPTER SUMMARY: NUMBER 7

RECURRING DECIMALS

Fractions that have a decimal equivalent that repeats itself are called recurring decimals.

The dot notation of recurring decimals should be clearly understood.

$0.\dot{3}$ $= 0.333\ 333...$

$0.\dot{3}\dot{2}$ $= 0.323\ 232...$

$0.\dot{3}2\dot{1}$ $= 0.321\ 321...$

All recurring decimals can be written as exact fractions. To change a recurring decimal to a fraction, first form an equation by putting x equal to the recurring decimal. Then multiply both sides of the equation by 10 if one digit recurs, by 100 if two digits recur, and by 1000 if 3 digits recur etc.

NO. OF REPEATING DIGITS	MULTIPLY BY
1	10
2	100
3	1000

Change $0.\dot{1}$ to a fraction.

Let $x = 0.111\ ...$

$10x = 1.111...$

$9x = 1$

$x = \dfrac{1}{9}$

Change $0.\dot{6}\dot{3}$ to a fraction.

Let $x = 0.636363...$

$100x = 63.636363...$

$99x = 63$

$x = \dfrac{63}{99} = \dfrac{7}{11}$

Change $0.7\dot{4}\dot{5}$ to a fraction.

Let $x = 0.7\dot{4}\dot{5}$

$10x = 7.454545...$

$1000x = 745.454545...$

$990x = 738$

$x = \dfrac{738}{990} = \dfrac{41}{55}$

ADVANCED CALCULATOR FUNCTIONS

It is important to learn how to operate your calculator efficiently.

The instruction manual will give clear steps for how to achieve best practice.

ALGEBRA 7

It was not possible to solve all types of quadratic equation until around 800 AD, cubic equations were first solved by Tartaglia during a mathematical contest in 1535, and the solution to quartic equations followed soon after in 1540. In 1823 the Norwegian mathematician, Abel, gave a proof showing that solving quintic equations (highest power is x^5) was impossible, but the proof was dismissed by the great mathematician Gauss. Today we can solve any equation numerically, but we can't necessarily find the exact solutions.

Niels Henrik Abel 1802–1829 ▶

LEARNING OBJECTIVES

■ Solve quadratic equations of the form $ax^2 + bx + c = 0$ by factorisation

■ Complete the square for a quadratic expression

■ Solve quadratic equations by completing the square

■ Solve quadratic equations by using the quadratic formula

■ Solve problems involving quadratic equations

■ Solve quadratic inequalities

BASIC PRINCIPLES

■ **Expand** brackets.

■ Expand the **product** of two linear expressions.

■ Substitute into formulae.

■ **Factorise** quadratic expressions of the form $x^2 + bx + c$

■ Solve linear equations.

■ Solve **quadratic equations** of the form $x^2 + bx + c = 0$ by factorising

■ Solve problems by setting up and solving quadratic equations of the form $x^2 + bx + c = 0$

SOLVING QUADRATIC EQUATIONS BY FACTORISING

KEY POINTS

■ There are three types of quadratic equations with $a = 1$

■ If $b = 0$ $x^2 - c = 0$ (Rearrange)
    $\Rightarrow$ $x^2 = c$ (Square root both sides)
    $\Rightarrow$ $x = \pm\sqrt{c}$

■ If $c = 0$ $x^2 + bx = 0$ (Factorise)
    $\Rightarrow$ $x(x + b) = 0$ (Solve)
    $\Rightarrow$ $x = 0$ or $x = -b$

■ If $b \neq 0$ and $c \neq 0$ $x^2 + bx + c = 0$ (Factorise)
    $\Rightarrow$ $(x + p)(x + q) = 0$ (Solve)
    $\Rightarrow$ $x = -p$ or $x = -q$
    where $p \times q = c$ and $p + q = b$

■ If c is negative then p and q have opposite signs to each other.

■ If c is positive then p and q have the same sign as b.

EXAMPLE 1

Solve these quadratic equations.

a $x^2 - 64 = 0$ **b** $x^2 - 81 = 0$ **c** $x^2 - 7x = 0$ **d** $x^2 - 10x + 21 = 0$

a $x^2 = 64$
   $x = 8$ or $x = -8$

b $(x - 9)(x + 9) = 0$
   $x = 9$ or $x = -9$

c $x(x - 7) = 0$
   $x = 0$ or $x = 7$

d $(x - 7)(x - 3) = 0$
   $x = 7$ or $x = 3$

Part **b** shows that if c is a square number then using the difference of two squares gives the same answers.

EXERCISE 1

Solve these equations by factorising.

1 ▶ $x^2 + 3x + 2 = 0$ **5 ▶** $x^2 + 9 = 6x$ **9 ▶** $x^2 - 4 = 0$

2 ▶ $x^2 + x = 6$ **6 ▶** $z^2 + 4z - 12 = 0$ **10 ▶** $p^2 + 5p - 84 = 0$

3 ▶ $x^2 + 7x + 10 = 0$ **7 ▶** $x^2 + x = 0$

4 ▶ $x^2 - 2x = 15$ **8 ▶** $t^2 = 4t$

EXERCISE 1*

Solve these equations by factorising.

1 ▶ $x^2 + 6x + 5 = 0$ **5 ▶** $49 + p^2 = 14p$ **9 ▶** $x^2 - 169 = 0$

2 ▶ $x^2 + 4 = 5x$ **6 ▶** $t^2 - 3t = 40$ **10 ▶** $x^2 + 2x - 143 = 0$

3 ▶ $z^2 + 15z + 56 = 0$ **7 ▶** $x^2 = 13x$

4 ▶ $x^2 = 4x + 45$ **8 ▶** $x^2 + 17x = 0$

FURTHER FACTORISATION

When $a \neq 1$ the factorisation may take longer.

KEY POINT

■ Always take out any **common factor** first.

EXAMPLE 2

Solve these quadratic equations.

a $9x^2 - 25 = 0$ **b** $3x^2 - 12x = 0$ **c** $12x^2 - 24x - 96 = 0$

a $9x^2 - 25 = 0$

$\Rightarrow 9x^2 = 25$

$\Rightarrow x^2 = \dfrac{25}{9}$

$\Rightarrow x = \pm\dfrac{5}{3}$

b $3x^2 - 12x = 0$

$\Rightarrow 3x(x - 4) = 0$

$\Rightarrow x = 0$ or $x = 4$

c $12x^2 - 24x - 96 = 0$

$\Rightarrow 12(x^2 - 2x - 8) = 0$

$\Rightarrow 12(x + 2)(x - 4) = 0$

$\Rightarrow x = -2$ or $x = 4$

If there is no simple number **factor**, then the factorisation takes more time.

EXAMPLE 3

Solve $x(3x - 13) = 10$

Expand the brackets and rearrange into the form $ax^2 + bx + c = 0$

$3x^2 - 13x - 10 = 0$

Factorise $3x^2 - 13x - 10$

The first terms in each bracket multiply together to give $3x^2$ so the factorisation starts as

$(3x \ldots\ldots)(x \ldots\ldots)$

The last terms in each bracket multiply together to give –10. The possible pairs are –1 and 10, 1 and –10, 2 and –5, –2 and 5.

The outside and inside terms **multiply out** and **sum** to $-13x$

So the factorisation is $(3x + 2)(x - 5)$

$\Rightarrow \quad (3x + 2)(x - 5) = 0$

$\Rightarrow \quad 3x + 2 = 0 \text{ or } x - 5 = 0$

$\Rightarrow \quad x = -\frac{2}{3} \text{ or } x = 5$

EXERCISE 2

Solve these equations by factorising.

1 ▶ $16x^2 = 81$
2 ▶ $3x^2 + 6x = 0$
3 ▶ $5x^2 = 5x$
4 ▶ $2x^2 - 10x + 12 = 0$

5 ▶ $z(2z - 5) + 2 = 0$
6 ▶ $3t^2 + 9t + 6 = 0$
7 ▶ $x(3x - 5) = 0$
8 ▶ $4(x^2 - x) = 24$

9 ▶ $3x^2 + 8x + 4 = 0$
10 ▶ $3z^2 + 10z = 8$

EXERCISE 2*

Solve these equations by factorising.

1 ▶ $128 - 18x^2 = 0$
2 ▶ $6x^2 = 9x$
3 ▶ $2z^2 + 6 = 7z$
4 ▶ $x(6x - 7) = 3$

5 ▶ $2p(4p + 3) + 1 = 0$
6 ▶ $5x^2 + 10 = 27x$
7 ▶ $3x^2 = 17x + 28$
8 ▶ $4x^2 + 40x + 100 = 0$

9 ▶ $4t^2 = 29t - 7$
10 ▶ $9x^2 + 25 = 30x$

SOLVING QUADRATIC EQUATIONS BY COMPLETING THE SQUARE

If a quadratic equation cannot be factorised, then the method of completing the square can be used to solve it. This involves writing one side of the equation as a **perfect square** and a constant. A perfect square is an expression like $(x + 2)^2$, which is equal to $x^2 + 4x + 4$

	x	2
x	x^2	$2x$
2	$2x$	4

EXAMPLE 4

Solve $x^2 + 4x - 3 = 0$

$$x^2 + 4x - 3 = 0 \Rightarrow \quad x^2 + 4x + 4 = 7 \qquad \text{(Add 7 to both sides to make LHS a perfect square)}$$
$$\Rightarrow \quad (x + 2)^2 = 7 \qquad \text{(Square root both sides)}$$
$$\Rightarrow \quad x + 2 = \pm\sqrt{7}$$
$$\Rightarrow \quad x = -2 + \sqrt{7} \text{ or } -2 + \sqrt{7}$$

We want to write $x^2 + bx + c$ in the form $(x + p)^2 + q$ to have a perfect square.

$$x^2 + bx + c = (x + p)^2 + q \qquad (1)$$
$$x^2 + bx + c = x^2 + 2px + p^2 + q \quad \text{(Multiplying out)}$$

So if we take $b = 2p$ and $c = p^2 + q$ then the expressions will be equal.

This means $p = \dfrac{b}{2}$

Also as $c = p^2 + q$ then $q = -p^2 + c$ or $q = -\left(\dfrac{b}{2}\right)^2 + c$

So substituting for p and q in (1) gives

$$x^2 + bx + c = \left(x + \frac{b}{2}\right)^2 - \left(\frac{b}{2}\right)^2 + c$$

EXAMPLE 5

Write these quadratic expressions in the form $(x + p)^2 + q$

a $x^2 + 4x + 5$ b $x^2 + 5x - 1$

a $x^2 + 4x + 5 = (x + 2)^2 - (2)^2 + 5 = (x + 2)^2 + 1$

$\quad\quad\;\; \uparrow \;\; \uparrow \quad\quad\; \uparrow \quad\; \uparrow \quad\;\; \uparrow$

$\quad\quad\;\; b \;\;\; c \quad\quad \dfrac{b}{2} \quad \dfrac{b}{2} \quad\; c$

b $x^2 + 5x - 1 = \left(x + \dfrac{5}{2}\right)^2 - \left(\dfrac{5}{2}\right)^2 - 1 = \left(x + \dfrac{5}{2}\right)^2 - \dfrac{29}{4}$

$\quad\quad\;\; \uparrow \;\; \uparrow \quad\quad\; \uparrow \quad\; \uparrow \quad\;\; \uparrow$

$\quad\quad\;\; b \;\;\; c \quad\quad \dfrac{b}{2} \quad \dfrac{b}{2} \quad\; c$

If b is negative then care is needed with the signs.

EXAMPLE 6

Write $x^2 - 8x + 7$ in the form $(x + p)^2 + q$

Treat $x^2 - 8x + 7$ as $x^2 + (-8)x + 7$ so $\dfrac{b}{2} = -4$

Then $x^2 + (-8)x + 7 \quad = (x + (-4))^2 - (-4)^2 + 7$
$$= (x - 4)^2 - 16 + 7$$
$$= (x - 4)^2 - 9$$

EXERCISE 3

Write these in the form $(x + p)^2 + q$

1 ▶ $x^2 + 2x + 3$	**5 ▶** $x^2 + 3x + 1$
2 ▶ $x^2 + 6x - 4$	**6 ▶** $x^2 + 5x - 3$
3 ▶ $x^2 - 4x + 2$	**7 ▶** $x^2 - 7x - 1$
4 ▶ $x^2 - 10x - 3$	**8 ▶** $x^2 - 9x + 2$

EXERCISE 3*

Write these in the form $(x + p)^2 + q$

1 ▶ $x^2 - 6x + 1$	**5 ▶** $x(x - 9) - 1$
2 ▶ $x^2 + 3 - 12x$	**6 ▶** $x^2 - 11(x - 2)$
3 ▶ $5x + x^2 - 12$	**7 ▶** $15x - (8 - x^2)$
4 ▶ $x^2 - 13 - 7x$	**8 ▶** $x^2 + 2px + 3$

If the **coefficient** of $x^2 \neq 1$ then continue as shown in Example 7.

EXAMPLE 7

Write $4x^2 + 12x - 5$ in the form $a(x + p)^2 + q$

$4x^2 + 12x - 5 = 4(x^2 + 3x) - 5$

Now $x^2 + 3x = \left(x + \dfrac{3}{2}\right)^2 - \left(\dfrac{3}{2}\right)^2 = \left(x + \dfrac{3}{2}\right)^2 - \dfrac{9}{4}$

So $4(x^2 + 3x) - 5 = 4\left[\left(x + \dfrac{3}{2}\right)^2 - \dfrac{9}{4}\right] - 5$

$\qquad\qquad\qquad = 4\left(x + \dfrac{3}{2}\right)^2 - 9 - 5$

$\qquad\qquad\qquad = 4\left(x + \dfrac{3}{2}\right)^2 - 14$

EXAMPLE 8

Write $-x^2 + 4x + 1$ in the form $a(x + p)^2 + q$

$-x^2 + 4x + 1 = -1[x^2 - 4x] + 1$

$\qquad\qquad\qquad = -1[(x - 2)^2 - 2^2] + 1$

$\qquad\qquad\qquad = -(x - 2)^2 + 5$

EXERCISE 4

Write these in the form $a(x + p)^2 + q$

1 ▶ $3x^2 + 6x - 4$	**3 ▶** $6x^2 - 12x - 8$	**5 ▶** $2x - x^2 + 4$
2 ▶ $2x^2 + 6x + 1$	**4 ▶** $2x^2 - 10x + 5$	**6 ▶** $3 - 3x - x^2$

EXERCISE 4*

Write these in the form $a(x + p)^2 + q$

1 ▶ $2x^2 + 16x + 4$	**3 ▶** $x(5x - 12) + 5$	**5 ▶** $3 - 16x - 4x^2$
2 ▶ $2x^2 - 5x - 7$	**4 ▶** $5 - x^2 + 6x$	**6 ▶** $9(1 + 2x) - 6x^2$

Any quadratic equation can be written in the form $p(x + q)^2 + r = 0$ by completing the square. It can then be solved.

EXAMPLE 9

By completing the square solve

a $x^2 + 2x - 2 = 0$　　　**b** $2x^2 - 6x = 4$

a $x^2 + 2x - 2 = 0 \Rightarrow (x + 1)^2 - 1 - 2 = 0$　(Rearrange)

$\Rightarrow (x + 1)^2 = 3$　　　(Square root both sides)

$\Rightarrow x + 1 = \pm\sqrt{3}$

$\Rightarrow x = -1 + \sqrt{3}$ or $x = -1 - \sqrt{3}$

b $2x^2 - 6x = 4 \Rightarrow x^2 - 3x = 2$　　　(Divide both sides by 2)

$\Rightarrow \left(x - \dfrac{3}{2}\right)^2 - \left(\dfrac{-3}{2}\right)^2 = 2$

$\Rightarrow \left(x - \dfrac{3}{2}\right)^2 = \dfrac{17}{4}$

$\Rightarrow x - \dfrac{3}{2} = \dfrac{\pm\sqrt{17}}{2}$

$\Rightarrow x = \dfrac{3}{2} \pm \dfrac{\sqrt{17}}{2}$

KEY POINTS

■ $x^2 + bx + c = \left(x + \dfrac{b}{2}\right)^2 - \left(\dfrac{b}{2}\right)^2 + c$

■ Take care with the signs if b is negative.

■ $ax^2 + bx + c$ can be written as $a\left(x^2 + \dfrac{b}{a}x\right) + c$ before completing the square for $x^2 + \dfrac{b}{a}x$

■ Give your answers in **surd** form (unless told otherwise) as these answers are exact.

EXERCISE 5

Solve these equations by completing the square.
For odd number questions, give answers to 3 s.f.
For even number questions, give answers in surd form.

1 ▶ $x^2 + 2x - 5 = 0$ 　　　　**5 ▶** $x^2 + 3x - 3 = 0$ 　　　　**9 ▶** $3x^2 + 6x + 1 = 0$

2 ▶ $x^2 - 2x - 6 = 0$ 　　　　**6 ▶** $x^2 - 7x + 5 = 0$ 　　　　**10 ▶** $10x + 5 - 2x^2 = 0$

3 ▶ $x^2 + 4x = 8$ 　　　　　　**7 ▶** $3x^2 - 12x = 60$

4 ▶ $x^2 + 10x + 15 = 0$ 　　　**8 ▶** $6x = 4 - 2x^2$

EXERCISE 5*

Solve these equations by completing the square.
For odd number questions, give answers to 3 s.f.
For even number questions, give answers in surd form.

1 ▶ $x^2 - 6x + 1 = 0$ 　　　　**5 ▶** $x(x - 9) = 1$ 　　　　**9 ▶** $7x^2 = 4(1 + x)$

2 ▶ $x^2 + 3 = 12x$ 　　　　　**6 ▶** $16x - 2x^2 - 4 = 0$ 　　　**10 ▶** $2x^2 + 4 = 5x$

3 ▶ $12 - x^2 = 5x$ 　　　　　**7 ▶** $2x^2 - 5x = 6$

4 ▶ $x^2 - 13 = 7x$ 　　　　　**8 ▶** $x(5x + 12) = -5$

SOLVING QUADRATIC EQUATIONS USING THE QUADRATIC FORMULA

The quadratic formula is used to solve quadratic equations that may be difficult to solve using other methods. Proof of the formula follows later in the chapter.

KEY POINT

- The quadratic formula

  If $ax^2 + bx + c = 0$ then $x = \dfrac{-b \pm \sqrt{b^2 - 4ac}}{2a}$

The quadratic formula is very important and is given on the formula sheet. However, it is used so often that you should memorise it.

EXAMPLE 10

Solve $3x^2 - 8x + 2 = 0$ giving answers **correct to** 3 s.f.

Here $a = 3$, $b = -8$, $c = 2$

Substituting into the formula gives $x = \dfrac{-(-8) \pm \sqrt{(-8)^2 - 4 \times 3 \times 2}}{2 \times 3} = \dfrac{8 \pm \sqrt{64 - 24}}{6}$

so $x = \dfrac{8 + \sqrt{40}}{6} = 2.39$ or $x = \dfrac{8 - \sqrt{40}}{6} = 0.279$ to 3 s.f.

Note how the negative number, −8, is handled in Example 10. Be very careful when substituting negative numbers into the formula; this is usually the reason for any mistakes.

EXAMPLE 11

Solve $3.5x + 2.3x^2 = 4.8$ giving answers correct to 3 s.f.

$2.3x^2 + 3.5x - 4.8 = 0$ (Rearrange the equation into the form $ax^2 + bx + c = 0$)

$a = 2.3$, $b = 3.5$, $c = -4.8$ (Write down the numbers to substitute in the formula)

$x = \dfrac{-3.5 \pm \sqrt{3.5^2 - 4 \times 2.3 \times (-4.8)}}{2 \times 2.3} = \dfrac{-3.5 \pm \sqrt{56.41}}{4.6}$ (Substitute the numbers)

so $x = \dfrac{-3.5 + \sqrt{56.41}}{4.6} = 0.872$ or $x = \dfrac{-3.5 - \sqrt{56.41}}{4.6} = -2.39$ to 3 s.f.

KEY POINTS

- a is the coefficient of x^2, it is not necessarily the first number in the equation. Always rearrange the equation into the form $ax^2 + bx + c = 0$ before using the formula.
- If the question asks for answers to a number of s.f. or d.p. then it is fairly certain that the quadratic formula is required.

EXERCISE 6

Solve these equations using the quadratic formula, giving answers correct to 3 s.f.

1 ▶ $x^2 + 3x + 2 = 0$ 5 ▶ $4x + x^2 = 8$ 9 ▶ $2.4x^2 - 4.5x + 1.7 = 0$

2 ▶ $x^2 + 2x - 5 = 0$ 6 ▶ $15 = 10x - x^2$ 10 ▶ $0.7 + 1.6x = 8.3x^2$

3 ▶ $x^2 - 6x + 3 = 0$ 7 ▶ $3x^2 = 4x + 20$

4 ▶ $x^2 - 4x - 2 = 0$ 8 ▶ $3x + 5x^2 = 4$

EXERCISE 6*

Solve these equations using the quadratic formula, giving answers correct to 3 s.f.

1 ▶ $x^2 - 6x + 2 = 0$

2 ▶ $x^2 + 5x - 12 = 0$

3 ▶ $x(4 + x) = 8$

4 ▶ $2x^2 + 5 = 10x$

5 ▶ $2x^2 = 13x + 45$

6 ▶ $3t^2 - 5t = 1$

7 ▶ $(q + 3)(q - 2) = 5$

8 ▶ $2.3x^2 - 12.6x + 1.3 = 0$

9 ▶ $4.7z^2 + 1.4z = 7.5$

10 ▶ $x(x + 1) + (x - 1)(x + 2) = 3$

PROOF OF THE QUADRATIC FORMULA

The following proof is for the case when $a = 1$, i.e. for the equation $x^2 + bx + c = 0$

$x^2 + bx + c = 0 \Rightarrow \quad \left(x + \dfrac{b}{2}\right)^2 - \left(\dfrac{b}{2}\right)^2 + c = 0$ (completing the square)

$\Rightarrow \quad \left(x + \dfrac{b}{2}\right)^2 = \left(\dfrac{b}{2}\right)^2 - c$ (rearranging)

$\Rightarrow \quad \left(x + \dfrac{b}{2}\right)^2 = \dfrac{b^2}{4} - c$ (squaring the fraction)

$\Rightarrow \quad \left(x + \dfrac{b}{2}\right)^2 = \dfrac{b^2 - 4c}{4}$ (**simplifying**)

$\Rightarrow \quad x + \dfrac{b}{2} = \pm\sqrt{\dfrac{b^2 - 4c}{4}}$ (square rooting both sides)

$\Rightarrow \quad x + \dfrac{b}{2} = \dfrac{\pm\sqrt{b^2 - 4c}}{2}$ (square rooting fraction)

$\Rightarrow \quad x = \dfrac{-b \pm \sqrt{b^2 - 4c}}{2}$ (rearranging and simplifying)

ACTIVITY 1

Change the proof above to prove the quadratic formula for $ax^2 + bx + c = 0$

Start by dividing both sides of the equation by a.

PROBLEMS LEADING TO QUADRATIC EQUATIONS

Example 12 revises the process of setting up and solving a quadratic equation to solve a problem. The steps to be followed are:

■ Where relevant, draw a clear diagram and put all the information on it.

■ Let x stand for what you are trying to find.

■ Form a quadratic equation in x and simplify it.

■ Solve the equation by factorising, completing the square or by using the formula.

■ Check that the answers make sense.

EXAMPLE 12

The width of a rectangular photograph is 4 cm more than the height.

The area is 77 cm².
Find the height of the photograph.

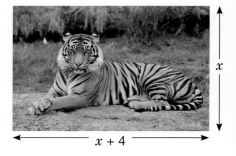

Let x be the height in cm.
Then the width is $x + 4$ cm.
As the area is 77 cm²,

$x(x + 4) = 77$ (Multiply out the brackets)
$x^2 + 4x = 77$ (Rearrange into the form $ax^2 + bx + c = 0$)
$x^2 + 4x - 77 = 0$ (Factorise)
$(x - 7)(x + 11) = 0$

So, $x = 7$ or -11 cm
The height cannot be negative, so the height is 7 cm.

EXAMPLE 13

The **chords** of a circle **intersect** as shown.

Find the value of x.

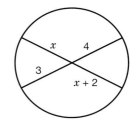

Using the intersecting chords theorem:

$x(x + 2) = 3 \times 4$ (Multiply out the brackets)
$x^2 + 2x = 12$ (Rearrange into the form $ax^2 + bx + c = 0$)
$x^2 + 2x - 12 = 0$

This expression does not factorise.

Using the quadratic formula with $a = 1$, $b = 2$ and $c = -12$ gives

$$x = \frac{-2 \pm \sqrt{2^2 - 4 \times (-12)}}{2}$$

$x = 2.61$ or -4.61 (3 s.f.)

As x cannot be negative, $x = 2.61$ to 3 s.f.

EXAMPLE 14

A rectangular fish pond is 6 m by 9 m. The pond is surrounded by a concrete path of constant width. The area of the pond is the same as the area of the path. Find the width of the path.

Let x be the width of the path.

The area of the path is $(2x + 9)(2x + 6) - 9 \times 6$
$\qquad\qquad\qquad\qquad = 4x^2 + 30x + 54 - 54$
$\qquad\qquad\qquad\qquad = 4x^2 + 30x$

The area of the pond is $9 \times 6 = 54$ m².

As the area of the path equals the area of the pond,

$4x^2 + 30x = 54$ (Rearrange into the form $ax^2 + bx + c = 0$)
$4x^2 + 30x - 54 = 0$ (Divide both sides of the equation by 2)
$2x^2 + 15x - 27 = 0$ (Factorise)
$(2x - 3)(x + 9) = 0$

So, $x = 1.5$ or -9 m
As x cannot be negative, the width of the path is 1.5 m.

EXERCISE 7

Where appropriate give answers to 3 s.f.

1 ▶ The height, h m, of a firework rocket above the ground after t seconds is given by $h = 35t - 5t^2$.

    a Show that when the rocket is 50 m above the ground $t^2 - 7t + 10 = 0$
    b Solve for t to find when the rocket is 50 m above the ground.
    c When does the rocket land?

2 ▶ Two **integers** differ by six and their product is 216. Find the two integers.

3 ▶ The height of a rectangle is 3 cm more than the width. The area is 30 cm². Find the dimensions of the rectangle.

4 ▶ The chords of a circle intersect as shown. Find x.

5 ▶ The sum of an integer and its square is 210. Find the two possible values of the integer.

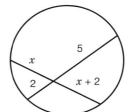

6 ▶ The **hypotenuse** of a **right-angled triangle** is 10 cm long. The other two sides differ by 3 cm. What are the lengths of the other two sides?

7 ▶ A rectangular picture, 12 cm by 18 cm, is surrounded by a frame of constant width. The area of the frame is the same as the area of the picture. What is the width of the frame?

8 ▶ The sum of the squares of two **consecutive** integers is 113.

    a If x is one of the integers show that $x^2 + x - 56 = 0$
    b Solve $x^2 + x - 56 = 0$ to find the two consecutive integers.

9 ▶ A **rhombus** has a side length equal to one of the diagonals. The other diagonal is 4 cm longer. Find the area of the rhombus.

10 ▶ The sum of the first n integers $1 + 2 + 3 + \ldots + n = \dfrac{n(n + 1)}{2}$

    a How many numbers must be taken to have a sum greater than one million?
    b Why can the sum never equal 1 000 000?

EXERCISE 7*

Where appropriate give answers to 3 s.f.

1 ▶ Dylan kicks a rugby ball. Its height, h m, is given by $h = 20t - 5t^2 + 1$ where t is the time in seconds after the kick.

    a When is the ball 10 m above the ground?
    b When does the ball land?

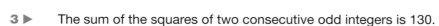

2 ▶ The height of a triangle is 3 cm more than the base. The area is 14 cm². Find the base of the triangle.

3 ▶ The sum of the squares of two consecutive odd integers is 130.

    a If x is one of the integers show that $x^2 + 2x - 63 = 0$
    b Solve $x^2 + 2x - 63 = 0$ to find the two consecutive integers.

4 ▶ Jos is 33 years younger than his father. The product of their ages is 658.
How old is Jos?

5 ▶ A farmer uses 21 m of fencing to make all four sides of a rectangular area of $27\,m^2$.
What are the dimensions of the area?

6 ▶ A swimming pool is in the shape of a rectangle 10 m by 8 m with two semi-circular ends.
The paved area around the pool is of constant width. The paved area is double the area of
the pool. What is the width of the paved area around the pool?

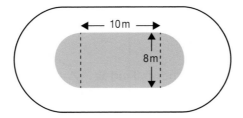

Q7 HINT
The surface area
of a sphere of
radius r is $4\pi r^2$

7 ▶ As a weather balloon rises, its **radius** increases
because the air pressure is lower. A spherical weather
balloon's surface area doubles when its radius
increases by 2 m. What was the original radius?

8 ▶ Two cars start from the same point on flat ground. The first car drives off directly North at
$25\,ms^{-1}$. Two seconds later the second car drives off directly East at $20\,ms^{-1}$. After how
many seconds are they 1 km apart?

9 ▶ An n-sided polygon has $\dfrac{n(n-3)}{2}$ diagonals.

a How many sides has a polygon with 665 diagonals?
b Why is it impossible for a polygon to have 406 diagonals?

10 ▶ Lee spent $1200 on holiday. If he had spent $50 less per day, he could have stayed an
extra two days. How long was his holiday?

SOLVING QUADRATIC INEQUALITIES

Squares and square roots involving **inequalities** need care.

EXAMPLE 15 ▶ Solve $x^2 - 4 < 0$

First **sketch** $y = x^2 - 4$

To do this, find where the graph intersects the x-axis by solving
the equation $x^2 - 4 = 0$.
$x^2 - 4 = 0 \Rightarrow x^2 = 4 \Rightarrow x = -2$ or $x = 2$.

So the graph intersects the x-axis at $x = -2$ and $x = 2$.

Also, when $x = 0$, $y = -4$, so the graph cuts the y-axis at -4.

The graph is a **parabola**, which is U-shaped.

The required region is below the x-axis. As this is one region,
the answer is one inequality.

The solution is $-2 < x < 2$.

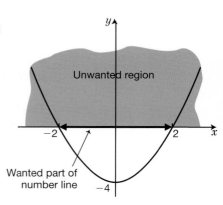

Unwanted region

Wanted part of
number line

EXAMPLE 16   Solve $x^2 - x - 2 \leq 0$, showing the solution set on a number line.

First sketch $y = x^2 - x - 2$

To do this, find where the graph intersects the x-axis by solving the equation $x^2 - x - 2 = 0$

$x^2 - x - 2 = 0 \Rightarrow (x - 2)(x + 1) = 0 \Rightarrow x = 2$ or $x = -1$
so the graph intersects the x-axis at $x = 2$ and $x = -1$

When $x = 0$, $y = -2$

The required region is below the x-axis. As this is one region, the answer is one inequality.

The solution is $-1 \leq x \leq 2$.

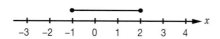

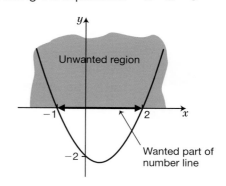

EXAMPLE 17   Solve $x^2 - 5x + 6 \geq 0$

First sketch $y = x^2 - 5x + 6$

To do this, find where the graph intersects the x-axis by solving the equation $x^2 - 5x + 6 = 0$

$x^2 - 5x + 6 = 0 \Rightarrow (x - 2)(x - 3) = 0 \Rightarrow x = 2$ or $x = 3$
so the graph intersects the x-axis at $x = 2$ and $x = 3$

When $x = 0$, $y = 6$

The required region is above the x-axis. As this has two components, the answer is two inequalities.

The solution is $x \leq 2$ or $x \geq 3$

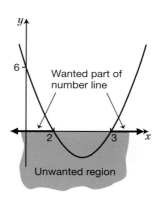

KEY POINT

■ To solve a quadratic inequality, sketch the graph of the quadratic function.

EXERCISE 8   Solve these inequalities, and show the solution set on a number line.

1 ▶ $x^2 < 16$ 5 ▶ $4x^2 + 3 > 67$ 9 ▶ $x^2 + 7x + 10 \geq 0$

2 ▶ $2x^2 \geq 50$ 6 ▶ $(x - 1)(x + 3) \leq 0$ 10 ▶ $x^2 + 2x - 15 \leq 0$

3 ▶ $x^2 + 3 \leq 84$ 7 ▶ $(x + 3)(x + 4) > 0$

4 ▶ $3x^2 < 75$ 8 ▶ $(2x - 1)(x + 1) < 0$

EXERCISE 8*   In Questions 1–6 solve the inequalities, and show the solution set on a number line.

1 ▶ $5x^2 + 3 \geq 23$ 3 ▶ $3x^2 - 7x - 20 \geq 0$ 5 ▶ $(3x - 1)(x + 1) \geq x(2x - 3) - 5$

2 ▶ $(x - 5)^2 > 4$ 4 ▶ $(x + 1)^2 \leq 5x^2 + x + 1$ 6 ▶ $(x - 5)^2 + 5(2x - 3) > 2x(x + 3) - 6$

7 ▶   Two numbers differ by seven. The product of the two numbers is less than 78.

Find the possible **range** of values for the smaller number.

8 ▶ The area of the right-angled triangle A is greater than the area of rectangle B. Find the range of possible values of x.

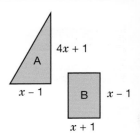

9 ▶ The **perimeter** of a rectangle is 28 cm. Find the range of possible values of the width of the rectangle if the diagonal is less than 10 cm.

10 ▶ The area of a rectangle is 12 cm². Find the range of possible values of the width of the rectangle if the diagonal is more than 5 cm.

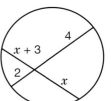

EXERCISE 9

REVISION

1 ▶ Solve these equations.
  a $x^2 - 25 = 0$ **b** $2x^2 + 8x = 0$

2 ▶ Solve these equations by factorisation.
  a $x^2 + x - 12 = 0$
  b $5x^2 - 5x - 30 = 0$
  c $3x^2 + x - 2 = 0$

3 ▶ By writing the equations in completed square form, solve the equations. Give your answers in surd form.
  a $x^2 + 4x - 3 = 0$ **b** $2x^2 - 8x - 2 = 0$
  c $3x^2 + 18x = 12$

4 ▶ Solve these equations giving your answers correct to 3 s.f.
  a $x^2 - 2x - 4 = 0$ **b** $3x^2 - 5x + 1 = 0$

5 ▶ The chords of a circle intersect as shown in the diagram. Find x.

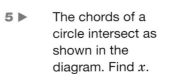

6 ▶ The sum of the squares of two consecutive integers is 145.
  a If x is one of the integers show that $x^2 + x - 72 = 0$
  b Solve $x^2 + x - 72 = 0$ to find the two consecutive integers.

7 ▶ A piece of wire 60 cm long is bent to form the perimeter of a rectangle of area 210 cm². Find the dimensions of the rectangle.

8 ▶ Solve these inequalities and represent the solution set on a number line.
  a $3x^2 \geq 12$ **b** $x^2 + 2x - 15 < 0$

EXERCISE 9*

REVISION

1 ▶ Solve these equations.
  a $x^2 - 20 = 0$
  b $x^2 - 9x = 0$

2 ▶ Solve these equations by factorisation.
  a $x^2 + x - 72 = 0$
  b $7x^2 = 14x + 168$
  c $2x(4x + 7) = 15$

3 ▶ By writing the equations in completed square form, solve the equations. Give your answers in surd form.
  a $x^2 - 6x - 2 = 0$
  b $4x^2 + 10x + 1 = 0$
  c $3x^2 = 21x + 3$

4 ▶ Solve these equations giving your answers correct to 3 s.f.
  a $3x^2 = 7x + 5$
  b $2.1x^2 + 8.4x - 4.3 = 0$

5 ▶ The area of a rectangular grass surface is 30 m². During planning, the length was decreased by 1 m and the width increased by 1 m, but the area did not change. Find the original dimensions of the grass surface.

6 ▶ Two chords of a circle intersect as shown in the diagram. Find x.

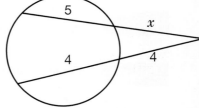

7 ▶ When a sheet of A4 paper is cut in half as shown, the result is a sheet of A5 paper. The two rectangles formed by A4 and A5 paper are **similar**. Find x.

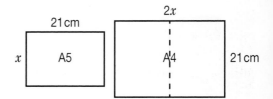

8 ▶ Solve the quadratic inequalities and represent the solution set on a number line.
  a $x^2 - 12x + 32 < 0$
  b $x^2 - 2x \geq 10$

EXAM PRACTICE: ALGEBRA 7

1 Solve the following equations by factorising.

a $5x^2 - 125 = 0$

b $3x^2 + x - 2 = 0$

c $x(2x - 5) = 3$ **[5]**

2 a Complete the square for $x^2 - 8x + 11$

b Therefore solve $x^2 - 8x + 11 = 0$ giving your answers in surd form.   **[4]**

3 Use the quadratic formula to solve these equations.

a $x^2 + x - 1 = 0$

b $2x^2 - 4x + 1 = 0$ **[4]**

4 Solve the quadratic inequalities and represent the solution set on a number line.

a $2x^2 > 18$

b $x^2 + x - 2 \leq 0$ **[4]**

5 a Use the intersecting chords theorem to form an equation in x from the diagram.

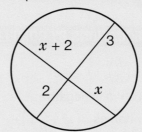

b Solve your equation to find x, giving your answer correct to 3 s.f.   **[4]**

6 The width of a rectangular photograph is 5 cm more than the height. The area is 80 cm². Find the height of the photograph.   **[4]**

[Total 25 marks]

CHAPTER SUMMARY: ALGEBRA 7

SOLVING QUADRATIC EQUATIONS BY FACTORISING

The different types of quadratic equation are shown, with examples, below.

No x term: $\qquad\qquad$ $x^2 - 4 = 0 \Rightarrow x^2 = 4 \Rightarrow x = \pm 2$

No number term: $\qquad$ $x^2 + 4x = 0 \Rightarrow x(x + 4) = 0 \Rightarrow x = 0$ or -4

Simple factorising: $\qquad$ $x^2 - x - 2 = 0 \Rightarrow (x + 1)(x - 2) = 0 \Rightarrow x = -1$ or 2

Number factor: $\qquad$ $3x^2 - 3x - 6 = 0 \Rightarrow 3(x^2 - x - 2) = 0 \Rightarrow 3(x + 1)(x - 2) = 0 \Rightarrow x = -1$ or 2

More complex factorising: $2x^2 + 5x + 2 = 0 \Rightarrow (2x + 1)(x + 2) = 0 \Rightarrow x = -\dfrac{1}{2}$ or -2

SOLVING QUADRATIC EQUATIONS BY COMPLETING THE SQUARE

$$x^2 + bx + c = \left(x + \frac{b}{2}\right)^2 - \left(\frac{b}{2}\right)^2 + c$$

Take care with the signs if b is negative.

$ax^2 + bx + c$ can be written as $a\left(x^2 + \dfrac{b}{a}x\right) + c$ before completing the square for $x^2 + \dfrac{b}{a}x$

Give your answers in surd form (unless told otherwise) as these are exact.

Complete the square for $x^2 - 4x + 1$

$$x^2 - 4x + 1 = (x - 2)^2 - (-2)^2 + 1 = (x - 2)^2 - 3$$

Therefore solve $x^2 - 4x + 1 = 0$ giving exact answers.

$$x^2 - 4x + 1 = 0 \Rightarrow (x - 2)^2 - 3 = 0$$
$$\Rightarrow (x - 2)^2 = 3$$
$$\Rightarrow x - 2 = \pm\sqrt{3}$$
$$\Rightarrow x = 2 + \sqrt{3} \text{ or } 2 - \sqrt{3}$$

SOLVING QUADRATIC EQUATIONS USING THE QUADRATIC FORMULA

If $ax^2 + bx + c = 0$ then $x = \dfrac{-b \pm \sqrt{b^2 - 4ac}}{2a}$

Write down the values of a, b and c. Take care with the signs. If $b = -3$ then $-b = +3$, and b^2 must be positive. If one of a or c is negative then $-4ac$ will be positive.

Solve $2x^2 - 5x - 1 = 0$

$a = 2$, $b = -5$ and $c = -1$

Substituting into the formula: $\quad x = \dfrac{5 \pm \sqrt{5^2 - 4 \times 2 \times -1}}{2 \times 2}$

$$x = \frac{5 \pm \sqrt{33}}{4}$$

SOLVING QUADRATIC INEQUALITIES

To solve a quadratic inequality, sketch the graph of the quadratic function.

Solve $\quad$ **a** $x^2 - x - 2 < 0$ $\quad$ **b** $x^2 - x - 2 \geq 0$

First sketch $y = x^2 - x - 2$ by finding where the graph intersects the x-axis.

$x^2 - x - 2 = 0 \Rightarrow (x + 1)(x - 2) = 0 \Rightarrow x = -1$ and $x = 2$

So the graph intersects the x-axis at $x = -1$ and $x = 2$

The graph is a positive parabola, which is U-shaped.

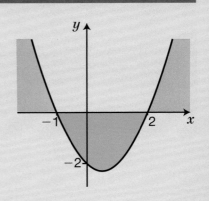

a The required region is below the x-axis. As this is one region, the answer is one inequality, so $-1 < x < 2$

b The required region is above the x-axis. As this has two components, the answer is two inequalities, so $x \leq -1$ or $x \geq 2$

GRAPHS 6

The architect Zaha Hadid has designed many amazing buildings around the world. The Heydar Aliyev Cultural Centre in Azerbaijan's capital city, Baku, is a stunning structure consisting of cubic and reciprocal curves and flowing lines.

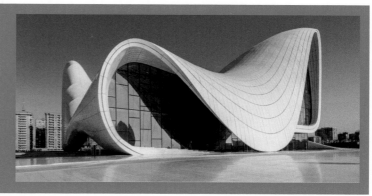

LEARNING OBJECTIVES

■ Recognise and draw graphs of cubic functions   ■ Recognise and draw graphs of reciprocal functions

BASIC PRINCIPLES

■ Substitute positive and negative numbers (including fractions) into algebraic expressions (including those involving squares and cubes) and those of the form $\frac{1}{x}$

■ Recognise linear and **quadratic graphs**.

■ Draw graphs (linear and quadratic) using tables of values.

■ A graph of p against q implies that p is on the vertical axis and q is on the horizontal axis. These values are **variables** and can represent physical quantities such as distance, speed, time and many others. Often a **range** of values is stated where the graph is a 'good' mathematical model and the values can be trusted to produce accurate results.

CUBIC GRAPHS $y = ax^3 + bx^2 + cx + d$

A **cubic function** is one in which the highest **power** of x is x^3.

All cubic functions can be written in the form $y = ax^3 + bx^2 + cx + d$ where a, b, c and d are constants.

The graphs of cubic functions have distinctive shapes and can be used to model real-life situations.

> **EXAMPLE 1**
>
> Plot the graph of $y = x^3 + 1$ for $-2 \le x \le 2$

Method 1 The complete table is shown with all cells filled in to find y.

x	−2	−1	0	1	2
x^3	−8	−1	0	1	8
+1	+1	+1	+1	+1	+1
y	−7	0	1	2	9

Method 2 The table is produced showing only x and y values.
This can be generated from functions in many calculators.

x	−2	−1	0	1	2
y	−7	0	1	2	9

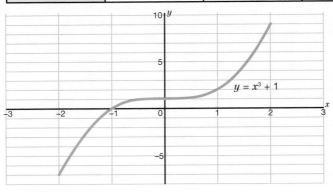

$y = x^3 + 1$

EXAMPLE 2

SKILLS

REASONING

Draw the graph of $y = 2x^3 + 2x^2 - 4x$ for $-2 \le x \le 2$ by creating a suitable table of values.

x	−2	−1	0	$\dfrac{1}{2}$	1	2
$2x^3$	−16	−2	0	$\dfrac{1}{4}$	2	16
$2x^2$	8	2	0	$\dfrac{1}{2}$	2	8
$-4x$	8	4	0	−2	−4	−8
y	0	4	0	$-1\dfrac{1}{4}$	0	16

As $y = 0$ for both $x = 0$ and $x = 1$, $x = \dfrac{1}{2}$ is used to find the value of y between $x = 0$ and $x = 1$

Plot the points and draw a smooth curve through all the points.

Note: Modern calculators often have a function that enables tables of graphs to be produced accurately and quickly so that only the x and y values are shown. However, you will still be expected to be able to fill in tables like the one above to plot some graphs.

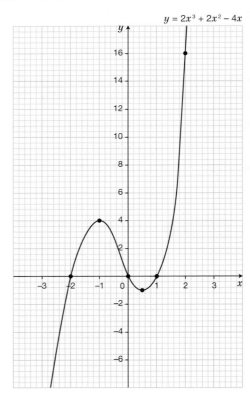

$y = 2x^3 + 2x^2 - 4x$

■ A cubic function is one whose highest power of x is x^3
It is written in the form $y = ax^3 + bx^2 + cx + d$

When $a > 0$ the function looks like this or

When $a < 0$ the function looks like this 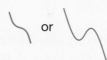 or

The graph **intersects** the y-axis at the point $y = d$

EXERCISE 1

Create a suitable table of values for these equations, between the stated x values, and then use these to draw the graphs of each.

1 ▶ $y = x^3 + 2$ $-3 \leq x \leq 3$ **4 ▶** $y = x^3 - 3x$ $-3 \leq x \leq 3$

2 ▶ $y = x^3 - 2$ $-3 \leq x \leq 3$ **5 ▶** $y = x^3 + x^2 - 2x$ $-3 \leq x \leq 3$

3 ▶ $y = x^3 + 3x$ $-3 \leq x \leq 3$

6 ▶   Copy and complete the table and use it to draw the graph of $y = x^3 - 3x^2 + x$ for $-2 \leq x \leq 3$

x	-2	-1	0	1	2	3
y	-22		0		-2	

7 ▶   This water tank has dimensions as shown, in metres.

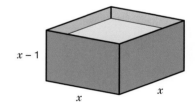

a Show that the volume of the tank, $V\text{m}^3$, is given by the formula $V = x^3 - x^2$

b Draw the graph of V against x for $2 \leq x \leq 5$ by first copying and completing the table.

x	2	2.5	3	3.5	4	4.5	5
V	4		18		48		100

c Use your graph to estimate the volume of a tank for which the base area is $16\,\text{m}^2$.

d What are the dimensions of a tank of volume $75\,\text{m}^3$?

8 ▶   The **cross-section** of a hilly region can be drawn as the graph of $y = x^3 - 8x^2 + 16x + 8$ for $0 \leq x \leq 5$, where x is measured in kilometres and y is the height above sea level in metres.

a Draw the graph of $y = x^3 - 8x^2 + 16x + 8$ for $0 \leq x \leq 5$.

b The peak is called Triblik and at the base of the valley is Vim Tarn. Mark these two features on your graph (cross-section), and estimate the height of Triblik above Vim Tarn.

EXERCISE 1*

Draw the graphs of these equations between the stated x values after creating a suitable table of values.

1 ▶ $y = 2x^3 - x^2 + x - 3$ $-3 \leq x \leq 3$

2 ▶ $y = 2x^3 - 2x^2 - 24x$ $-4 \leq x \leq 4$

3 ▶ $y = -2x^3 + 3x^2 + 4x$ $-3 \leq x \leq 3$

4 ▶ Copy and complete the table of values and use it to draw the graph of $y = -x^3 - 2x^2 + 11x + 12$ for $-4 \leq x \leq 4$

x	−4	−3	−2	−1	0	1	2	3	4
y	0		−10		12		18		−40

5 ▶ The equation $v = 27t - t^3$ (valid for $0 \leq t \leq 5$) gives the **velocity**, in metres per second, of a firework moving through the air t seconds after it was set off .

  a Draw the graph of v against t after first creating a table of values for the given values of t.

  b Use your graph to estimate the greatest velocity of the firework, and the time at which this occurs.

  c For how long does the firework travel faster than 30 m/s?

6 ▶ A toy is made that consists of a cylinder of **diameter** $2x$ cm and height x cm upon which is fixed a right circular cone of base **radius** x centimetres and height 6 cm.

  a Volume of a right circular cone = $\frac{1}{3}$ × base area × height.

  Show that the total volume V, in cubic centimetres, of the toy is given by $V = \pi x^2(x + 2)$

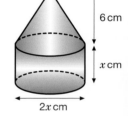

  b Draw the graph of V against x for $0 \leq x \leq 5$ after creating a suitable table of values.

  c Use your graph to find the volume of a toy of diameter 7 cm.

  d What is the curved surface area of the cylinder if the total volume of the toy is 300 cm³? ($A = 2\pi rh$)

7 ▶ A closed cylindrical can of height h cm and radius r cm is made from a thin sheet of metal. The *total* surface area is 100π cm².

  a Show that $h = \dfrac{50}{r} - r$

  Hence, show that the volume of the can, V cm³, is given by $V = 50\pi r - \pi r^3$

  b Draw the graph of V against r for $0 \leq r \leq 7$ by first creating a suitable table of values.

  c Use the graph to estimate the greatest possible volume of the can.

  d What are the diameter and height of the can of maximum volume?

8 ▶ An open box is made from a thin square metal sheet measuring 10 cm by 10 cm. Four squares of side x centimetres are cut away, and the remaining sides are folded upwards to make the box of depth x centimetres.

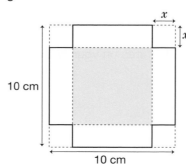
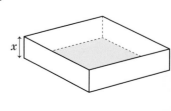

   a Show that the side length of the box is $(10 - 2x)$ cm.

   b Show that the volume V in cm³ of the box is given by the formula

   $V = 100x - 40x^2 + 4x^3$ for $0 \leq x \leq 5$

   c Draw the graph of V against x by first constructing a suitable table of values.

   d Use your graph to estimate the maximum volume of the box, and state its dimensions.

ACTIVITY 1

SKILLS

ANALYSIS

A forest contains F foxes and R rabbits. Their numbers change throughout a given year as shown in the graph of F against R. t is the number of months after 1 January.

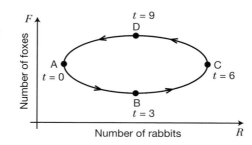

Copy and complete this table.

YEAR INTERVAL	FOX NUMBERS, F	RABBIT NUMBERS, R	REASON
JAN–MAR (A–B)	Decreasing	Increasing	Fewer foxes to eat rabbits
APR–JUN (B–C)			
JUL–SEP (C–D)			
OCT–DEC (D–A)			

Sketch two graphs of F against t and R against t for the interval $0 \leq t \leq 12$, placing the horizontal axes as shown for comparison.

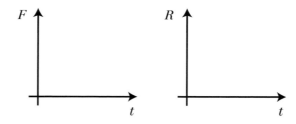

RECIPROCAL GRAPHS $y = \dfrac{a}{x}$

A reciprocal function is in the form $y = \dfrac{a}{x}$ where a is a constant.

The graph of a reciprocal function produces a curve called a hyperbola.

ACTIVITY 2

SKILLS

ANALYSIS

Draw the graph of $y = \dfrac{1}{x}$, where $x \neq 0$, for $-3 \leq x \leq 3$

Dividing a number by 0 gives no value, so it is necessary to investigate the behaviour of y as x gets closer to zero from both sides. Use this table of values.

x	-3	-2	-1	$-\dfrac{1}{2}$	$-\dfrac{1}{4}$	$\dfrac{1}{4}$	$\dfrac{1}{2}$	1	2	3
y										

The graph of a **reciprocal** function has two parts, both of which are smooth curves.

KEY POINT

- The graph of $y = \dfrac{a}{x}$ is a **hyperbola**.

  The curve approaches, but never touches, the axes. The axes are called **asymptotes** to the curve.

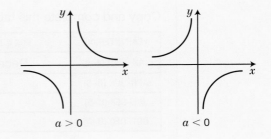

Other reciprocal graphs which involve division by an x term include, for example,

$y = 1 + \dfrac{4}{x}$, $y = 2x - \dfrac{1}{x}$, $y = \dfrac{12}{x-1}$ and $y = 3x + \dfrac{1}{x^2}$

EXAMPLE 3

SKILLS

ANALYSIS

a Draw the graph of $y = \dfrac{10}{x} + x - 5$ for $1 \leq x \leq 6$ by first creating a table of values.

b Use the graph to estimate the smallest value of y, and the value of x where this occurs.

c State the values of x for which $y = 2.5$

a

x	1	2	3	4	5	6
$\dfrac{10}{x}$	10	5	$3\dfrac{1}{3}$	$2\dfrac{1}{2}$	2	$1\dfrac{2}{3}$
x	1	2	3	4	5	6
-5	-5	-5	-5	-5	-5	-5
y	6	2	$1\dfrac{1}{3}$	$1\dfrac{1}{2}$	2	$2\dfrac{2}{3}$

b The graph shows that the minimum value of $y \simeq 1.3$ when $x \simeq 3.2$

c When $y = 2.5$, $x \simeq 1.7$ or 5.8

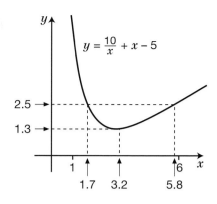

$$y = \frac{10}{x} + x - 5$$

2.5

1.3

1   1.7   3.2   5.8   6 x

EXERCISE 2 Draw the graphs between the stated x values after creating a suitable table of values.

1 ▶ $y = \dfrac{4}{x}$ $-4 \leq x \leq 4$ **3 ▶** $y = \dfrac{10}{x}$ $-5 \leq x \leq 5$

2 ▶ $y = -\dfrac{4}{x}$ $-4 \leq x \leq 4$ **4 ▶** $y = -\dfrac{8}{x}$ $-4 \leq x \leq 4$

5 ▶ An insect colony decreases after the spread of a virus.

Its population y after t months is given by the equation

$y = \dfrac{2000}{t}$ valid for $1 \leq t \leq 6$

a Copy and complete this table of values for $y = \dfrac{2000}{t}$

t (months)	1	2	3	4	5	6
y	2000					

b Draw a graph of y against t and name the shape of graph.

c Use your graph to estimate when the population is decreased by 70% from its size after 1 month.

d How long does it take for the population to decrease from 1500 to 500?

6 ▶ A water tank gets a leak. The volume v, in m³, at time t hours after the leak occurs is given by the equation $v = \dfrac{1000}{t}$ valid for $1 \leq t \leq 20$

a Copy and complete this table of values for $v = \dfrac{1000}{t}$

t (hours)	1	5		15	
v (m³)			100		50

b Draw the graph of $v = \dfrac{1000}{t}$ for $1 \leq t \leq 20$

c Use the graph to estimate when the volume of water is reduced by 750 m³ from its value after 1 hour.

d How much water has been lost between 8 hours and 16 hours?

7 ▶ Edna calculates that the temperature of a cup of tea is t degrees Celsius, m minutes after it has been made, as given by the equation $t = \dfrac{k}{m}$ where k is a constant and the equation is valid for $5 \le m \le 10$

a Use the figures in this table to find the value of k, and hence copy and complete the table.

m (minutes)	5	6	7	8	9	10
t (°C)						40

b Draw the graph of $t = \dfrac{k}{m}$ for $5 \le m \le 10$

c Use your graph to estimate the temperature of a cup of tea after 450 seconds.

d After how long is the temperature of the cup of tea 60 °C?

e Edna only drinks tea at a temperature of between 50 °C and 75 °C. Use your graph to find between what times Edna will drink her cup of tea.

EXERCISE 2*

Draw the graphs between the stated x values after creating a suitable table of values. State what values of x make y undefined.

1 ▶ $y = 1 + \dfrac{4}{x}$ 　 $-4 \le x \le 4$

3 ▶ $y = x^2 + \dfrac{2}{x}$ 　 $-4 \le x \le 4$

2 ▶ $y = \dfrac{6}{x - 2}$ 　 $-3 \le x \le 6$

4 ▶ $y = \dfrac{8}{x} + x - 4$ 　 $1 \le x \le 6$

5 ▶ Jacqui is training for an athletics competition at school. She experiments with various angles of projection x for putting the shot, and finds that the horizontal distance d, in metres, is given by

$d = 100 - x - \dfrac{2000}{x}$ valid for

$30° \le x \le 60°$

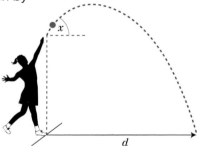

a Copy and complete the table of values for $d = 100 - x - \dfrac{2000}{x}$

x (°)	30	35	40	45	50	55	60
100			100				
$-x$			−40				
$-\dfrac{2000}{x}$			−50				
d (m)			10				

b Draw the graph of $d = 100 - x - \dfrac{2000}{x}$ for $30° \le x \le 60°$

c Use your graph to estimate the greatest distance that Jacqui can get, and the angle necessary to achieve it.

d What values can x take if Jacqui is to put (throw) the shot at least 9 m?

6 ▶ A snowboard company called Zoom hires out x hundred boards per week. The amount received R and the costs C, both measured in $1000s, are given by

$$R = \frac{6x}{x+1} \text{ and } C = x + 1 \text{ both valid for } 0 \le x \le 5$$

a Create a table of values for $0 \le x \le 5$ and then draw graphs of R against x and C against x on a single set of axes.

b Zoom's profit P in $1000s per week is given by $P = R - C$. Use your graphs to estimate how many boards must be hired out per week for Zoom to make a profit.

c What is the greatest weekly profit that Zoom can make, and the number of boards that must be hired out per week for this to be achieved?

7 ▶ The Dimox paint factory wants to store $50 \, \text{m}^3$ of paint in a closed cylindrical tank. To reduce costs, it wants to use the minimum possible surface area (including the top and bottom).

a If the total surface area of the tank is $A \, \text{m}^2$, and the radius is $r \, \text{m}$, show that the height $h \, \text{m}$ of the tank is given by $h = \dfrac{50}{\pi r^2}$

Hence, show that $A = 2\pi r^2 + \dfrac{100}{r}$

b Construct a suitable table of values for $1 \le r \le 5$ and draw a graph of A against r.

c Use your graph to estimate the value of r that produces the smallest surface area, and this value of A.

ACTIVITY 3

SKILLS

PROBLEM
SOLVING

Find the shape of the reciprocal curve $y = \dfrac{k}{x^2}$ for $k > 0$ and $k < 0$.

HINT
Consider the graphs of $y = \dfrac{10}{x^2}$ and $y = -\dfrac{10}{x^2}$ for $-2 \le x \le 2$.

EXERCISE 3

REVISION

Draw the graphs of these equations between the stated x values by first creating suitable tables of values.

1 ▶ $y = x^3 + x - 3$ $-3 \le x \le 3$ 2 ▶ $y = x^3 + x^2 + 3$ $-3 \le x \le 3$

3 ▶ The profit P in $ millions earned by Pixel Internet after t years is given by the equation $P = t^3 - 6t - 6$ for $0 \le t \le 6$

a Draw a graph of P against t in the given range by first creating a suitable table of values.

b Use your graph to estimate when Pixel Internet first made a profit.

c How long will it take for a profit of $100 000 000 to be made?

4 ▶ Nick goes on a diet, which claims that his weight w kg after t weeks between weeks 30 and 40 will be given by $w = \dfrac{k}{t}$ where k is a constant whole number.

a Use the data in this table to find the value of k, and hence copy and complete the table, giving your answers **correct to 2 significant figures**.

t (weeks)	30	32	34	36	38	40
w (kg)						70

b Draw a graph of $w = \dfrac{k}{t}$ for $t = 30$ to $t = 40$

c Use your graph to estimate when Nick's weight should reach 80 kg.

d Why is there only a limited range of t values for which the equation works?

5 ▶ The temperature of Kim's cup of coffee, $t\,°C$, m minutes after it has been poured into her mug, is given by the equation

$t = \dfrac{425}{m}$ valid for $5 \le m \le 10$

a Copy and complete this table and use it to draw the graph of t against m for $5 \le m \le 10$.

m	5	6	7	8	9	10
t	85			53.1		

b Kim likes her coffee between the temperatures of 50 °C and 70 °C. Use your graph to find at what times Kim should drink her coffee.

EXERCISE 3*

REVISION

Draw the graphs of these equations between the stated x values by first creating suitable tables of values.

1 ▶ $y = 2x^3 - x^2 - 3x$ $\quad -3 \le x \le 3$ $\qquad$ **2 ▶** $y = 3x(x + 2)^2 - 5$ $\quad -3 \le x \le 2$

3 ▶ State whether each graph is linear, quadratic, cubic or reciprocal.

A
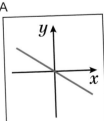

B

C
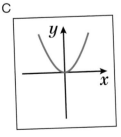

D
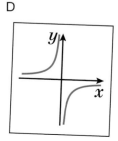

E
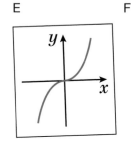

F
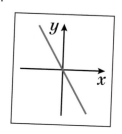

4 ▶ A pig farmer wants to enclose 600 m² of land for a rectangular pig pen. Three of the sides will consist of fencing, while the remaining side will consist of an existing stone wall. Two sides of the fence will be of length x metres.

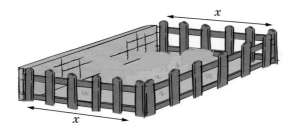

a Write the length of the third side in terms of x.

b Show that the total length L in metres of the fence is given by $L = 2x + \dfrac{600}{x}$

c Construct a suitable table of values for $5 \leq x \leq 40$. Use this table to draw a graph of L against x.

d Use your graph to estimate the minimum fence length possible, and the value of x for which it occurs.

e What values can x take, if the fence length should not exceed 75 m?

5 ▶ The graph of $y = x^3 - 5x^2 + ax + b$, where a and b are constants, passes through the points P(0, 5) and Q(1, 6).

a Show that the values of a and b are both 5 by making sensible substitutions using the co-ordinates of P and Q.

b Copy and complete this table of values for $y = x^3 - 5x^2 + 5x + 5$ for $-1 \leq x \leq 4$

x	−1	0	1	2	3	4
y		5	6			9

c Draw the graph of $y = x^3 - 5x^2 + 5x + 5$ for $-1 \leq x \leq 4$

d The curve represents a cross-section of a hillside. The top of the curve (R) represents the top of a hill and the bottom of the curve (S) represents the bottom of a valley. State the co-ordinates of R and S.

EXAM PRACTICE: GRAPHS 6

1   **a** Draw the graph of $y = x^3 + 3x^2 - x - 3$ for $-5 \le x \le 3$ by first copying and completing the table of values.

x	−5	−4	−3	−2	−1	0	1	2	3
y	−48		0		0		0		48

[4]

   b State the x values where this graph cuts the x-axis.   [2]

2   The flow $Q\,\text{m}^3/\text{s}$ of a small river t hours after midnight is observed after a storm, and is given by the equation $Q = t^3 - 8t^2 + 14t + 10$ for $0 \le t \le 5$

   a Draw a graph of Q against t by first copying and completing the table of values for the stated values of t.

t (hours after midnight)	0	1	2	3	4	5
Q (m³/s)	10		14		2	

[4]

   b Use your graph to estimate the maximum flow and the time when this occurs.   [3]
   c Between what times does the river flood, if this occurs when the flow exceeds $10\,\text{m}^3/\text{s}$?   [2]

3   The petrol consumption (y kilometres per litre) of a car is related to its speed (x kilometres per hour) by the formula $y = 70 - \dfrac{x}{3} - \dfrac{2400}{x}$

   a Copy and complete the table of values for $y = 70 - \dfrac{x}{3} - \dfrac{2400}{x}$

x	60	70	80	90	100	110	120	130
70	70	70	70	70	70	70	70	70
$-\dfrac{x}{3}$		−23.3	−26.7			−36.7		−43.3
$-\dfrac{2400}{x}$		−34.3		−26.7		−21.8		−18.5
y		12.4				11.5		8.2

[3]

   b Draw the graph of $y = 70 - \dfrac{x}{3} - \dfrac{2400}{x}$ for values of x from 60 to 130.

Draw the x-axis for $60 \le x \le 130$ using a **scale** of 2 cm to 10 km/h and the y-axis for $7 \le y \le 14$ using a scale of 4 cm to 1 km per litre.   [3]

   c Use your graph to estimate the petrol consumption at 115 km/h.   [2]

   d Use your graph to estimate the most economical speed at which to travel to conserve petrol.   [2]

[Total 25 marks]

CHAPTER SUMMARY: GRAPHS 6

CUBIC GRAPHS $y = ax^3 + bx^2 + cx + d$

A cubic function is one in which the highest power of x is x^3.

All cubic functions can be written in the form $y = ax^3 + bx^2 + cx + d$ where a, b, c and d are constants.

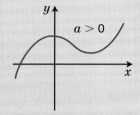
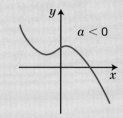

RECIPROCAL GRAPHS $y = \dfrac{a}{x}$ AND $y = \dfrac{a}{x^2}$

A reciprocal function is in the form $\dfrac{a}{x}$ or $y = \dfrac{a}{x^2}$ where a is a number.

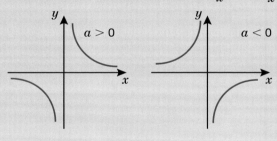

$$y = \frac{a}{x} \qquad\qquad\qquad y = \frac{a}{x^2}$$

Graphs of type $y = Ax + \dfrac{B}{x^2} + \dfrac{C}{x^3}$ (A, B and C are constants):

Draw the graph of $y = 2x + \dfrac{3}{x^2} - \dfrac{4}{x^3}$ for $-2 \leq x \leq 2$

Create a suitable table of values, plot the points and draw a smooth curve.

x	-2	-1	0	1	2
$2x$	-4	-2	0	2	4
$\dfrac{3}{x^2}$	0.75	3		3	0.75
$-\dfrac{4}{x^3}$	0.5	4		-4	-0.5
y	-2.75	5		1	4.25

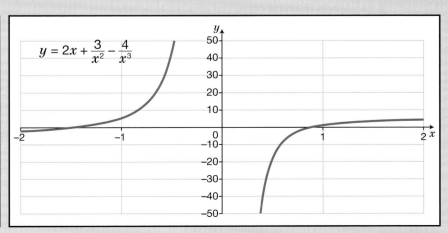

$$y = 2x + \frac{3}{x^2} - \frac{4}{x^3}$$

SHAPE AND SPACE 7

Very early civilisations realised there was a connection between the circumference of a circle and its diameter and used the poor approximation $C = 3d$ or $\pi \approx 3$. The ancient Greeks used a clever idea to work out much better approximations. If a regular polygon with perimeter P_1 is drawn inside a circle and another with perimeter P_2 outside a circle, then $P_1 < C < P_2$. By using 96-sided polygons, Archimedes (c. 225 BC) worked out $3.1408 < \pi < 3.1428$

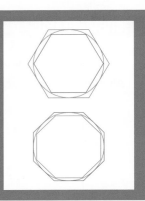

LEARNING OBJECTIVES

- Calculate the area and circumference of a circle
- Calculate the perimeter and area of semicircles and quarter circles
- Calculate arc lengths, angles and areas of sectors of circles

- Calculate the volume and surface area of a prism, pyramid, cone and sphere
- Use links between scale factors for length, area and volume
- Solve problems involving the areas and volumes of similar shapes

BASIC PRINCIPLES

- The **perimeter** of a shape is the distance all the way round the shape.

  - Circle
    The perimeter of a circle is called the **circumference**.
    If C is the circumference, A the area and r the **radius**, then
    $C = 2\pi r$
    $A = \pi r^2$

  - Semicircle
    A semicircle is half a circle cut along a **diameter**.
    The perimeter is half the circumference of the circle plus the diameter, so $P = \pi r + 2r$

    The area is half of the area of the circle, so $A = \dfrac{\pi r^2}{2}$

  - Quadrant
    A quadrant is quarter of a circle.
    The perimeter is a quarter of the circumference of the circle plus twice the radius, so $P = \dfrac{\pi r}{2} + 2r$

    The area is a quarter of the area of the circle, so $A = \dfrac{\pi r^2}{4}$

- Find the perimeter of rectangles and triangles.
- Use Pythagoras' theorem.
- Recognise **similar** shapes.
- Use the **ratio** of corresponding sides to work out **scale factors**.
- Find missing lengths on similar shapes.

CIRCLES

EXAMPLE 1

The circumference of a circle is 10 cm. Find the radius.

Using $C = 2\pi r$

$\quad 10 = 2\pi r$ (Make r the **subject** of the equation)

$\quad r = \dfrac{10}{2\pi}$

$\quad\quad = 1.59$ cm to 3 s.f.

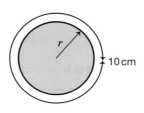

10 cm

EXAMPLE 2

The area of a circle is 24 cm². Find the radius.

Using $A = \pi r^2$

$\quad 24 = \pi r^2$ (Make r the subject of the equation)

$\quad r^2 = \dfrac{24}{\pi}$

$\quad r = \sqrt{\dfrac{24}{\pi}}$

$\quad\quad = 2.76$ cm to 3 s.f.

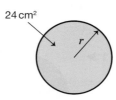

24 cm²

EXAMPLE 3

Find the perimeter and area of the shape shown.

The radius of the quadrant BCD is 3 cm, so BC = 3 cm.

Perimeter = AB + BC + **arc** CD + DE + EA

$\quad\quad = 4 + 3 + \dfrac{2 \times \pi \times 3}{4} + 4 + 3$

$\quad\quad = 18.7$ cm (to 3 s.f.)

Area = area of quadrant BCD + area of rectangle ABDE

$\quad\quad = \dfrac{9\pi}{4} + 12$

$\quad\quad = 19.1$ cm² (to 3 s.f.)

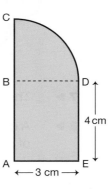

C

B - - - - - - D

A   E

4 cm

←— 3 cm —→

EXERCISE 1

Find the perimeter and area of each of the following shapes, giving answers to 3 s.f.
All dimensions are in cm. All arcs are parts of circles.

1 ▶

←— 8 —→

3 ▶
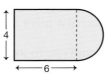
4
←— 6 —→

5 ▶
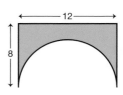
←——— 12 ———→
8

2 ▶
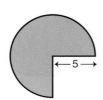
←— 5 —→

4 ▶

←——— 10 ———→
6

6 ▶
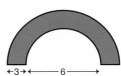
←3→ ←—— 6 ——→

7 ▶ Find the radius and area of a circle with circumference of 6 cm.

8 ▶ Find the radius and circumference of a circle with area 14 cm².

9 ▶ The minute hand of a clock is 80 mm long. How many metres does the end of the hand travel in 12 hours?

10 ▶ A car wheel has a diameter of 48 cm. How many revolutions (one circular movement around an axis) does the wheel make on a journey of 12 km? Give your answer to 3 s.f.

EXERCISE 1*

Find the perimeter and area of each of the following shapes, giving answers to 3 s.f.
All dimensions are in cm. All arcs are parts of circles.

1 ▶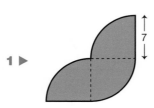

3 ▶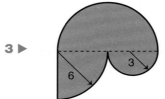

5 ▶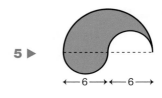

2 ▶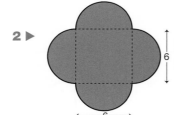

4 ▶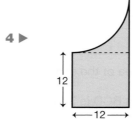

6 ▶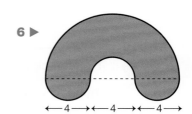

7 ▶ The area of a quadrant of a circle is 8 cm². Find the radius and perimeter.

8 ▶ A cow is tied by a rope to one corner of a 20 m square field.

The cow can move inside half the area of the field. How long is the rope?

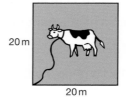

9 ▶ A new circular coin has just been made which has a circumference in cm that is numerically the same value as the area in cm². What is the radius of the coin?

10 ▶ The radius of the Earth is 6380 km.

    a How far does a point on the equator travel in 24 hours?

    b Find the speed of a point on the equator in m/s.

Q11 HINT
You do not need to know the radius of the Earth.

11 ▶ A hot-air balloon travels round the Earth 1 km above the surface, following the equator. How much further does it travel than the distance around the equator?

12 ▶ The shape shown consists of a square and a semicircle. The perimeter is 22 cm.

Find the radius of the semicircle and the area of the shape.

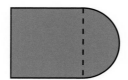

ARCS

An arc is part of the circumference of a circle.

The arc shown is the fraction $\frac{x}{360}$ of the whole circumference.

So the arc length is $\frac{x}{360} \times 2\pi r$

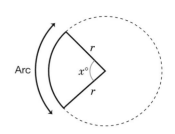

EXAMPLE 4

Find the perimeter of the shape shown.

Using Arc length $= \frac{x}{360} \times 2\pi r$

Arc length $= \frac{80}{360} \times 2\pi \times 4 = 5.585$ cm

Perimeter $= 5.585 + 4 + 4 = 13.6$ cm to 3 s.f.

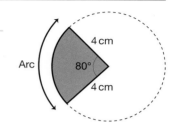

EXAMPLE 5

Find the angle marked x.

Using Arc length $= \frac{x}{360} \times 2\pi r$

$12 = \frac{x}{360} \times 2\pi \times 9$ (Make x the subject of the equation)

$x = \frac{12 \times 360}{2\pi \times 9}$

$= 76.4°$ to 3 s.f.

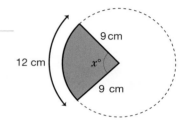

EXAMPLE 6

Find the radius r.

Using Arc length $= \frac{x}{360} \times 2\pi r$

$20 = \frac{50}{360} \times 2\pi r$ (Make r the subject of the equation)

$r = \frac{20 \times 360}{50 \times 2\pi}$

$= 22.9$ cm to 3 s.f.

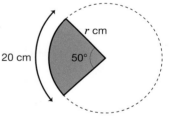

KEY POINT

■ Arc length $= \frac{x}{360} \times 2\pi r$

EXERCISE 2

In Questions 1–4, find the perimeter of the shape. Give answers to 3 s.f.

1 ▶

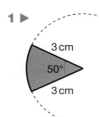

2 ▶

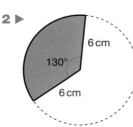

3 ▶

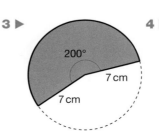

4 ▶

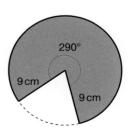

In Questions 5 and 6, find the angle marked x

5 ▶

5 cm

3 cm $x°$

5 cm

6 ▶

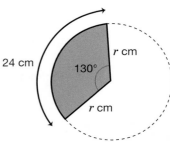

12 cm   6 cm

$x°$

6 cm

In Questions 7 and 8, find the radius r.

7 ▶

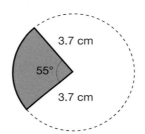

r cm

10 cm   40°

r cm

8 ▶

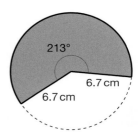

r cm

24 cm   130°

r cm

EXERCISE 2* ▷

In Questions 1 and 2, find the perimeter of the shape. Give answers to 3 s.f.

9th

11th

1 ▶

3.7 cm

55°

3.7 cm

2 ▶

213°

6.7 cm

6.7 cm

In Questions 3 and 4, find the angle marked x.

3 ▶

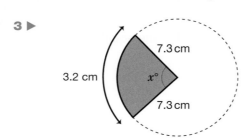

7.3 cm

3.2 cm $x°$

7.3 cm

4 ▶

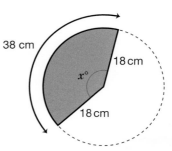

38 cm   18 cm

$x°$

18 cm

In Questions 5 and 6, find the radius r.

5 ▶

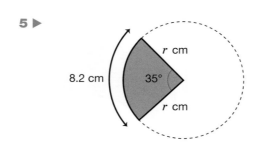

r cm

8.2 cm   35°

r cm

6 ▶

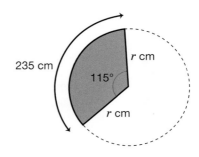

r cm

235 cm   115°

r cm

7 ▶ The minute hand of a clock is 9 cm long. How far does the end travel in 35 minutes?

8 ▶ Find the perimeter of the shape to 3 s.f.

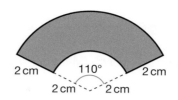

9 ▶ The perimeter of the shape is 28 cm.
Find the value of r.

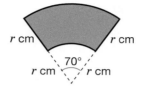

SECTORS

A **sector** of a circle is a region whose perimeter is an arc and two radii.

The sector shown is the fraction $\frac{x}{360}$ of the whole circle.

So the sector area is $\frac{x}{360} \times \pi r^2$

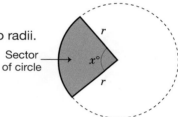

Sector of circle

EXAMPLE 7

Find the area of the sector shown.

Using Sector area $= \frac{x}{360} \times \pi r^2$

$A = \frac{65}{360} \times \pi \times 7^2$

$\quad = 27.8 \text{ cm}^2$ to 3 s.f.

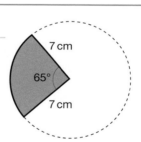

EXAMPLE 8

Find the angle marked x.

Using Sector area $= \frac{x}{360} \times \pi r^2$

$12 = \frac{x}{360} \times \pi \times 5^2$ (Make x the subject of the equation)

$x = \frac{12 \times 360}{\pi \times 5^2}$

$\quad = 55.0°$ to 3 s.f.

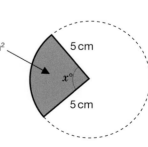

Area = 12 cm²

EXAMPLE 9

Find the radius of the sector shown.

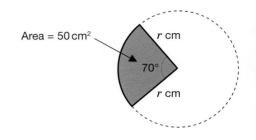

Area = 50 cm²

Using Sector area = $\dfrac{x}{360} \times \pi r^2$

$50 = \dfrac{70}{360} \times \pi r^2$ (Make r the subject of the equation)

$r^2 = \dfrac{50 \times 360}{70 \times \pi}$

$r = \sqrt{\dfrac{50 \times 360}{70 \times \pi}}$

   = 9.05 cm (to 3 s.f.)

KEY POINT

■   Sector area = $\dfrac{x}{360} \times \pi r^2$

EXERCISE 3

In Questions 1–4, find the area of the shape. Give answers to 3 s.f.

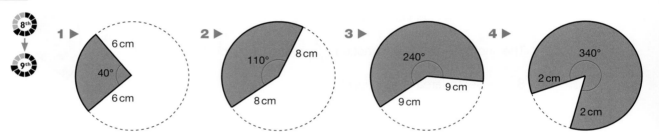

1 ▶   6 cm   40°   6 cm

2 ▶   110°   8 cm   8 cm

3 ▶   240°   9 cm   9 cm

4 ▶   340°   2 cm   2 cm

In Questions 5 and 6, find the angle marked x.

5 ▶   3 cm   6 cm² $x°$ 3 cm

6 ▶   72 cm² $x°$ 8 cm   8 cm

In Questions 7 and 8, find the radius r.

7 ▶ r cm   12 cm²   40° r cm

8 ▶   82 cm² r cm   130° r cm

EXERCISE 3*

In Questions 1 and 2, find the area of the shape. Give answers to 3 s.f.

1 ▶

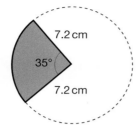

7.2 cm
35°
7.2 cm

2 ▶

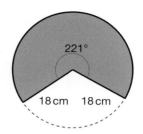

221°
18 cm   18 cm

In Questions 3 and 4, find the angle marked x.

3 ▶

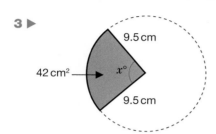

9.5 cm
42 cm² → $x°$
9.5 cm

4 ▶

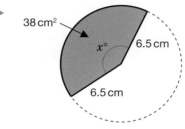

38 cm² →
$x°$ 6.5 cm
6.5 cm

In Questions 5 and 6, find the radius r.

5 ▶

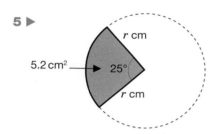

r cm
5.2 cm² → 25°
r cm

6 ▶

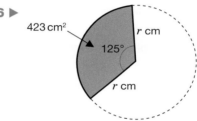

423 cm² →
125° r cm
r cm

7 ▶ Find the area of the shape to 3 s.f.

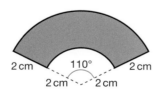

2 cm   110°   2 cm
2 cm ⌣ 2 cm

8 ▶ The area of the shape is 54 cm². Find the value of r.

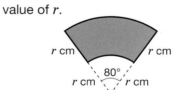

r cm r cm
80°
r cm r cm

9 ▶ Find the shaded area.

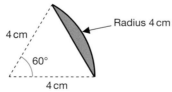

Radius 4 cm
4 cm
60°
4 cm

10 ▶ Three circular pencils, each with a diameter of 1 cm, are held together by an elastic band. What is the (stretched) length of the band?

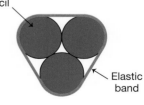

Pencil

Elastic band

11 ▶ Three drinks mats, each with a diameter of 8 cm, are placed on a table as shown. Find the blue shaded area.

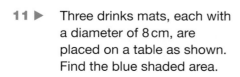

SOLIDS

SURFACE AREA AND VOLUME OF A PRISM

ACTIVITY 1

SKILLS

REASONING

Curved surface area of cylinder

Take a rectangular piece of paper and make it into a hollow (empty) cylinder.

   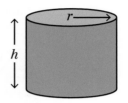

Use your paper to explain why the curved surface area of a cylinder = $2\pi rh$

KEY POINTS

■ Any solid with parallel sides that has a constant
cross-section is called a **prism**.

■ Volume of a prism = area of cross-section × length

■ A **cuboid** is a prism with a rectangular cross-section.

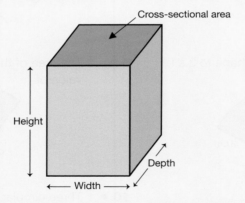

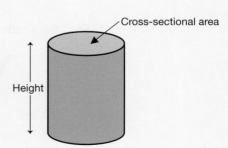

■ Volume of a cuboid = width × depth × height

■ A cylinder is a prism with a circular cross-section.

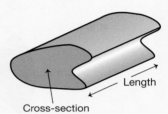

■ If the height is h and the radius r, then volume of a cylinder = $\pi r^2 h$
■ Curved surface area of a cylinder = $2\pi rh$

EXAMPLE 10

Calculate the volume and surface area of the prism shown.

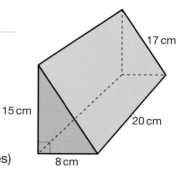

The cross-section is a **right-angled triangle**.

Area of cross-section $= \frac{1}{2} \times 8 \times 15 = 60 \, \text{cm}^2$

So

Volume = area of cross-section × length

Volume $= 60 \times 20 = 1200 \, \text{cm}^3$

Surface area = **sum** of the areas of the 5 faces (2 triangles + 3 rectangles)

$$= 2(\tfrac{1}{2} \times 8 \times 15) + (17 \times 20) + (8 \times 20) + (15 \times 20)$$
$$= 920 \, \text{cm}^2$$

EXAMPLE 11

Calculate the volume and total surface area of an unopened cola can that is a cylinder with diameter 6 cm and height 11 cm.

Using $V = \pi r^2 h$ with $r = 3$ and $h = 11$

$V = \pi \times 3^2 \times 11$

$\quad = 311 \, \text{cm}^3$ to 3 s.f.

Total surface area = area of 2 circles + curved surface area

$A = 2 \times \pi r^2 + 2\pi rh$

$\quad = 2 \times \pi \times 3^2 + 2 \times \pi \times 3 \times 11$

$\quad = 264 \, \text{cm}^2$ to 3 s.f.

EXERCISE 4

1 ▶   Find the volume of the prism shown.

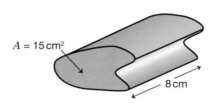

$A = 15 \, \text{cm}^2$

8 cm

2 ▶   Find the volume and surface area of this shape.

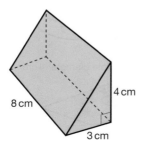

4 cm

8 cm

3 cm

3 ▶ Find the volume and surface area of this can of drink.

16 cm

← 6 cm →

4 ▶ A swimming pool has the dimensions shown.
Find the volume in m³.

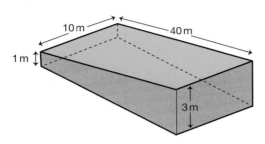

10 m
40 m
1 m
3 m

5 ▶ A water container has the dimensions shown.
Find the volume in m³.

1 m
2 m
Semicircle

6 ▶ A 500 cm³ jar of olive oil is
a cylinder 8 cm in diameter.
How tall is it?

OLIVE
OIL

EXERCISE 4*

6th

8th

1 ▶ The diagram shows a metal piece of a conservatory roof.
Find the volume of the piece in cm³.

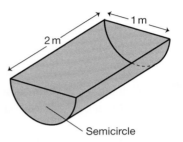

$A = 16\,cm^2$
3 m

2 ▶ The diagram shows some steps.
Find the volume in cm³ and the surface area in cm².

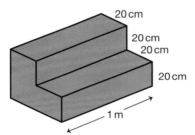

20 cm
20 cm
20 cm
20 cm
1 m

3 ▶ The diagram shows a can of food with semicircular ends.
Find the volume and surface area.

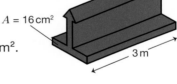

8 cm
3 cm
6 cm

4 ▶ The diagram shows a sweet.
Find the volume in cm³ and the
surface area in cm².

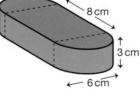

2 cm
diameter
Hole 1 cm
diameter
5 mm

5 ▶ A cylindrical hole is cut
into a brick as shown.
Find the volume and surface
area of the brick.

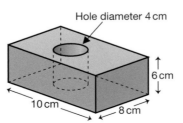

Hole diameter 4 cm
6 cm
10 cm
8 cm

6 ▶ A roll of sticky tape has the dimensions shown.
If the tape is 25 m long, how thick is the tape?

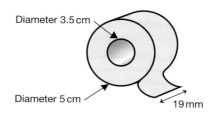

Diameter 3.5 cm

Diameter 5 cm

19 mm

VOLUME AND SURFACE AREA OF A PYRAMID, CONE AND SPHERE

KEY POINTS

■ Volume of a pyramid $= \frac{1}{3} \times$ area of base $\times$ vertical height

■ A cone is a pyramid with a circular base.

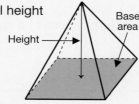

Height

Base area

■ Volume of a cone $= \frac{1}{3} \times$ area of base $\times$ vertical height

$$= \frac{1}{3} \times \pi r^2 \times h, \text{ where } r = \text{radius and } h = \text{vertical height}$$

■ Curved surface area of a cone $= \pi r l$, where l is the slant (sloped) height

■ For a sphere of radius r,

■ Volume $= \frac{4}{3} \pi r^3$

■ Surface area $= 4\pi r^2$

EXAMPLE 12

Find the volume of the rectangular-based pyramid shown.

Using $V = \frac{1}{3} \times$ area of base $\times$ vertical height

$$V = \frac{1}{3} \times 8 \times 10 \times 12$$

$$= 320 \text{ cm}^3$$

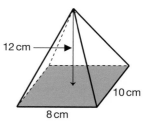

12 cm

10 cm

8 cm

EXAMPLE 13

Find the total surface area of the cone shown.

Use Pythagoras' Theorem to work out l.
$l^2 = 5^2 + 12^2$
$l^2 = 169$
$l = 13$

Curved surface area of a cone $= \pi r l$
$$= \pi \times 5 \times 13$$
$$= 204 \text{ cm}^2$$

The base is a circle with area $\pi r^2 = \pi \times 5^2$
$$= 78.5 \text{ cm}^2$$

Total surface area = curved surface area + area of base
$$= 204 + 78.5$$
$$= 283 \text{ cm}^2 \text{ to 3 s.f.}$$

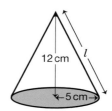

12 cm

l

5 cm

EXAMPLE 14

A table tennis ball has a volume of 33 cm³. Find the radius and surface area.

Using Volume of a sphere $= \frac{4}{3}\pi r^3$

$$33 = \frac{4}{3}\pi r^3 \qquad \text{(Make } r \text{ the subject of the equation)}$$

$$r^3 = \frac{33 \times 3}{4 \times \pi}$$

$$r = \sqrt[3]{\frac{33 \times 3}{4 \times \pi}}$$

$$= 1.99 \, \text{cm}$$

Using Surface area of a sphere $= 4\pi r^2$

$$A = 4 \times \pi \times 1.99^2$$

$$= 49.8 \, \text{cm}^2 \text{ to 3 s.f.}$$

EXERCISE 5

8th → 10th

1 ▶ The glass pyramid at the Louvre in Paris, France has a square base of side 35 m and vertical height 21.5 m.

Calculate its volume in m³.

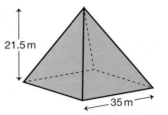

2 ▶ A traffic cone has the dimensions shown.

Find the volume in cm³ and the curved surface area in cm².

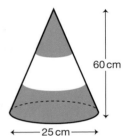

3 ▶ A hanging flower basket is a **hemisphere** with diameter 30 cm.

Find the volume and the external curved surface area of the basket.

4 ▶ A scoop (a rounded spoon) for ground coffee is a hollow hemisphere with diameter 4 cm.

When the scoop is full, the coffee forms a cone on top of the scoop.

Find the volume of coffee.

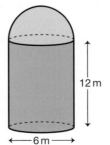

5 ▶ A grain tower is a cylinder with a hemisphere on top, with the dimensions shown.

Find the volume and total surface area of the tower.

6 ▶ The volume of a cricket ball is 180 cm³.

Find the radius and surface area.

7 ▶ A flat roof is a rectangle measuring 6 m by 8 m.
The rain from the roof flows into a cylindrical water container with radius 50 cm.

By how much does the water level in the container rise if 1 cm of rain falls?
(Assume the container does not overflow.)

8 ▶ A fuel tanker (carrier) is pumping fuel into an aircraft's fuel tank. The tanker is a cylinder 2 m in diameter and 3 m long.

The aircraft's tank is a cuboid 5 m × 4 m × 1 m high. Before pumping, the fuel tanker is full and the aircraft's fuel tank is empty.

How deep is the fuel in the aircraft's tank after pumping is complete?

EXERCISE 5*

1 ▶ A crystal consists of two square-based pyramids as shown.
Calculate the volume of the crystal.

5 mm
5 mm

2 ▶ An ice-cream cone is full of ice cream as shown.

What is the volume of ice cream?

Hemisphere diameter 5 cm
10 cm

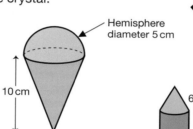

6 cm
22 cm
10 cm

3 ▶ A water bottle is a cylinder with a cone at one end and a hemisphere at the other.

Find the volume and surface area.

4 ▶ A monument in South America is in the shape of a truncated (shortened by having its top cut off) pyramid.

Find the volume of the monument.

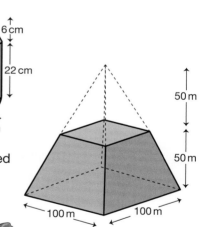

50 m
50 m
100 m
100 m

5 ▶ A vase is a truncated cone.

Find the volume of the vase.

12 cm diameter

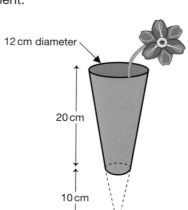

20 cm
10 cm

6 ▶ The volume of the Earth is 1.09×10^{12} km³.

Find the surface area of the Earth assuming that it is a sphere.

7 ▶ A spherical stone ball of diameter 10 cm is dropped into a cylindrical barrel of water and sinks to the bottom. The ball is completely covered by water.

By how much does the water rise in the barrel?

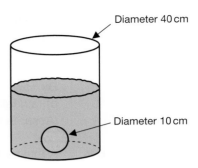

Diameter 40 cm

Diameter 10 cm

8 ▶ The sphere and the cone shown have the same volume. Calculate the height of the cone.

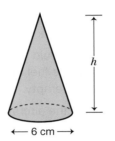

h

← 6 cm →

← 6 cm →

9 ▶ A spherical drop of oil with diameter 3 mm falls onto a water surface and produces a circular oil film of radius 10 cm.

Calculate the thickness of the oil film.

SIMILAR SHAPES

AREAS OF SIMILAR SHAPES

When a shape doubles in size, then the area does NOT double, but increases by a factor of four.

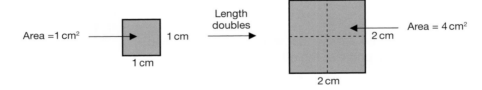

Length doubles

Area =1 cm²   1 cm

1 cm

Area = 4 cm²

2 cm

2 cm

The linear scale factor is 2, and the area scale factor is 4.

If the shape triples in size, then the area increases by a factor of nine.

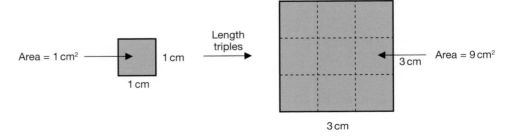

Length triples

Area = 1 cm²   1 cm

1 cm

Area = 9 cm²

3 cm

3 cm

Note: The two shapes must be similar. Two similar shapes that are the same size are **congruent**.

If a shape increases by a linear scale factor of k, then the area scale factor is k^2.

This applies even if the shape is irregular.

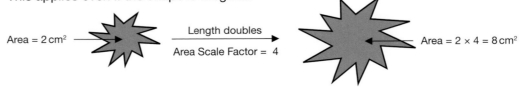

Area = 2 cm²

Length doubles

Area Scale Factor = 4

Area = 2 × 4 = 8 cm²

EXAMPLE 15

The two shapes shown are similar.
The area of shape A is 10 cm².
Find the area of shape B.

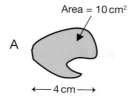

The linear scale factor $k = \dfrac{8}{4} = 2$

(Divide the length of shape B by the length of shape A.)

The area scale factor $k^2 = 2^2 = 4$

So the area of shape B $= 10 \times 4 = 40$ cm²

EXAMPLE 16

The two shapes shown are similar.
The area of shape A is 18 cm².
Find the area of shape B.

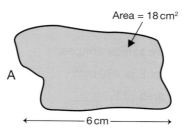

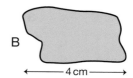

The linear scale factor $k = \dfrac{4}{6} = \dfrac{2}{3}$

(Divide the length of shape B by the length of shape A.)

The area scale factor $k^2 = \left(\dfrac{2}{3}\right)^2 = \dfrac{4}{9}$

So the area of shape B $= 18 \times \dfrac{4}{9} = 8$ cm²

EXAMPLE 17

The two triangles are similar, with dimensions
and areas as shown.

What is the value of x?

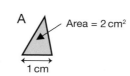

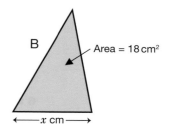

The area scale factor $k^2 = \dfrac{18}{2} = 9$

(Divide the area of shape B by the area of shape A.)

The linear scale factor $k = \sqrt{9} = 3$

So $x = 1 \times 3 = 3$ cm

KEY POINT

■ When a shape is enlarged by linear scale factor k, the area of the shape is enlarged
by scale factor k^2.

1 ▶ A and B are similar shapes.

The area of A is 4 cm².

Find the area of B.

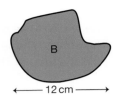

←6 cm→

← 12 cm →

2 ▶

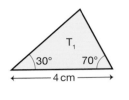
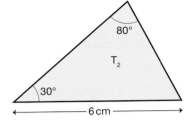

T₁
30° 70°
← 4 cm →

80°
T₂
30°
← 6 cm →

a Why are the two triangles similar?

b If the area of T₁ is 3.8 cm², find the area of T₂.

3 ▶ E and F are similar shapes.

The area of E is 480 cm².

Find the area of F.

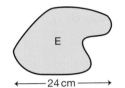

E

F

←16 cm→

—— 24 cm ——

4 ▶ I and J are similar shapes.

The area of I is 150 cm².

Find the area of J.

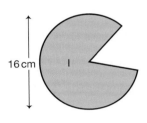
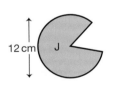

16 cm I

12 cm J

5 ▶ The shapes M and N are similar.

The area of M is 8 cm² and the area of N is 32 cm².

Find x.

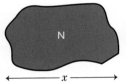

M

N

←3 cm→

←——— x ———→

6 ▶ Q and R are similar shapes.

The area of Q is 5 cm² and the area of R is 11.25 cm².

Find x.

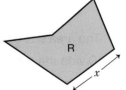

Q

R

2 cm

x

7 ▶ U and V are similar shapes.

The area of U is 48 cm² and the area of V is 12 cm².

Find x.

U

V

← 6 cm →

←x→

8 ▶ A and B are similar pentagons.

The area of A is 160 cm² and the area of B is 90 cm².

Find x.

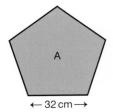

A

B

← 32 cm →

←x→

EXERCISE 6*

1 ▶ The two stars are similar in shape.

The area of the smaller star is 300 cm².

Find the area of the larger star.

2 ▶ The two shapes are similar.

The area of the larger shape is 125 cm².

Find the area of the smaller shape.

3 ▶ The two shapes are similar.

Find x.

4 ▶ The two leaves are similar in shape.

Find x.

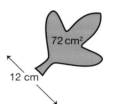

5 ▶ A model aeroplane is made to a scale of $\frac{1}{20}$ of the size of the real plane. The area of the wings of the real plane are 4×10^5 cm². Find the area of the wings of the model in cm².

6 ▶ An oil slick increases in length by 20%. Assuming the shape is similar to the original shape, what is the percentage increase in area?

7 ▶ Meera washes some cloths in hot water and they shrink by 10%. What is the percentage reduction in area?

8 ▶ Calculate the shaded area A.

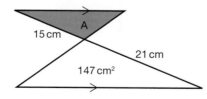

9 ▶ Calculate the shaded area B.

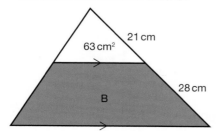

VOLUMES OF SIMILAR SHAPES

When a solid doubles in size, the volume does NOT double, but increases by a factor of eight.

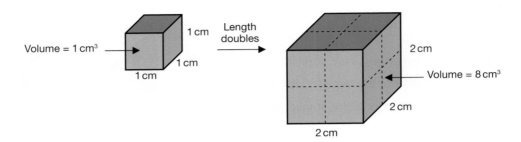

The linear scale factor is 2, and the volume scale factor is 8.

If the solid triples in size, then the volume increases by a factor of 27.

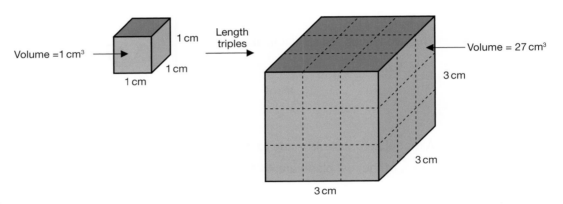

If a solid increases by a linear scale factor of k, then the volume scale factor is k^3.

This applies even if the solid is irregular.

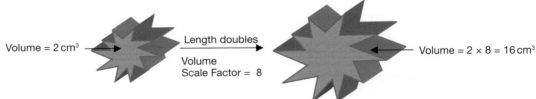

Note: The two solids must be similar.

EXAMPLE 18

The two solids shown are similar.
The volume of solid A is 20 cm³.
Find the volume of solid B.

The linear scale factor $k = \dfrac{8}{4} = 2$

(Divide the length of solid B by the
length of solid A.)

The volume scale factor $k^3 = 2^3 = 8$

So the volume of solid B = 20 × 8 = 160 cm³

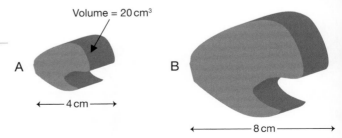

EXAMPLE 19

The two cylinders shown are similar.
The volume of cylinder C is 54 cm³.
Find the volume of cylinder D.

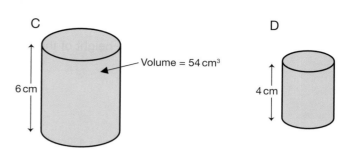

C

Volume = 54 cm³

6 cm

D

4 cm

The linear scale factor $k = \dfrac{4}{6} = \dfrac{2}{3}$

(Divide the height of cylinder D by the height of cylinder C.)

The volume scale factor $k^3 = \left(\dfrac{2}{3}\right)^3 = \dfrac{8}{27}$

So the volume of cylinder D = $54 \times \dfrac{8}{27} = 16$ cm³

EXAMPLE 20

The two prisms are similar, with dimensions and volumes as shown.
What is the value of x?

The volume scale factor $k^3 = \dfrac{54}{2} = 27$

(Divide the volume of solid D by the volume of solid C.)

The linear scale factor $k = \sqrt[3]{27} = 3$

So $x = 1 \times 3 = 3$ cm

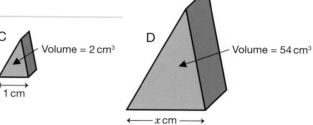

C

Volume = 2 cm³

1 cm

D

Volume = 54 cm³

← x cm →

KEY POINT

■ When a shape is enlarged by linear scale factor k, the volume of the shape is enlarged by scale factor k^3.

EXERCISE 7

1 ▶ The cones shown are similar.

Find the volume of the larger cone.

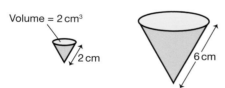

Volume = 2 cm³

2 cm

6 cm

2 ▶ The statues shown are similar.

Find the volume of the larger statue.

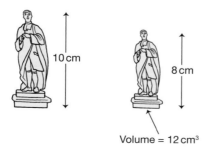

10 cm

8 cm

Volume = 12 cm³

3 ▶ The two bottles are similar.

Find the volume of the smaller bottle.

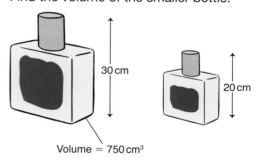

30 cm

20 cm

Volume = 750 cm³

4 ▶ The two smartphones are similar.

Find the volume of the smaller phone.

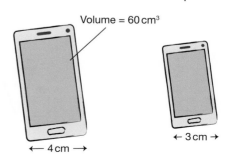

Volume = 60 cm³

← 4 cm →

← 3 cm →

5 ▶ The two glasses are similar.

Find the height of the larger glass.

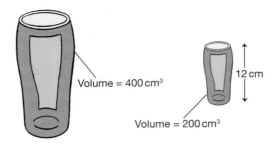

Volume = 400 cm³

12 cm

Volume = 200 cm³

6 ▶ The two pencils are similar.

Find the diameter of the larger pencil.

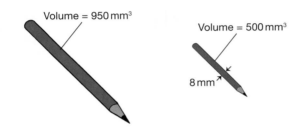

Volume = 950 mm³

Volume = 500 mm³

8 mm

7 ▶ The two eggs are similar.

Find the height of the smaller egg.

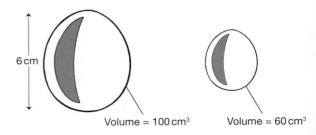

6 cm

Volume = 100 cm³

Volume = 60 cm³

8 ▶ The two candles are similar. The larger candle has a volume of 160 cm³ and a surface area of 200 cm². The smaller candle has a volume of 20 cm³. Find the surface area of the smaller candle.

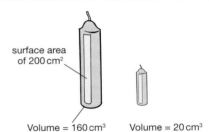

surface area of 200 cm²

Volume = 160 cm³   Volume = 20 cm³

EXERCISE 7*

1 ▶ X and Y are similar shapes.

The volume of Y is 50 cm³.

Find the volume of X.

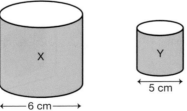

X

Y

5 cm

6 cm

2 ▶ X and Y are similar shapes.

The volume of Y is 128 cm³.

Find the volume of X.

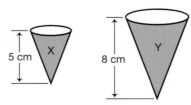

5 cm   X

8 cm   Y

3 ▶ The two candlesticks are similar.

Find the height of the larger candlestick.

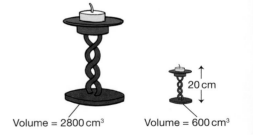

20 cm

Volume = 2800 cm³   Volume = 600 cm³

4 ▶ The two bottles of shampoo are similar.

Find the height of the smaller bottle.

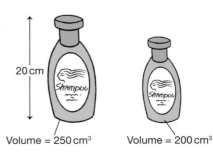

20 cm

Shampoo

Shampoo

Volume = 250 cm³   Volume = 200 cm³

5 ▶ The manufacturers of a chocolate bar decide to produce a similar bar by increasing all dimensions by 20%. What would be a fair percentage increase in price?

6 ▶ Two wedding cakes are made from the same mixture and have similar shapes.

The larger cake costs $135 and is 30 cm in diameter.

Find the cost of the smaller cake, which has a diameter of 20 cm.

7 ▶ A supermarket stocks similar small and large cans of beans. The areas of their labels are 63 cm² and 112 cm² respectively.

   a The weight of the large can is 640 g. What is the weight of the small can?

   b The height of the small can is 12 cm. What is the height of the large can?

8 ▶ Two solid statues are similar in shape and made of the same material.

One is 1 m high and weighs 64 kg. The other weighs 1 kg

   a What is the height of the smaller statue?

   b If 3 g of gold is required to cover the smaller statue, how much is needed for the larger one?

9 ▶ A solid sphere weighs 10 g.

   a What will be the weight of another sphere made from the same material but having three times the diameter?

   b The surface area of the 10 g sphere is 20 cm². What is the surface area of the larger sphere?

10 ▶ Suppose that an adult hedgehog is an exact enlargement of a baby hedgehog on a scale of 3 : 2, and that the baby hedgehog has 2000 quills with a total length of 15 m and a skin area of 360 cm².

   a How many quills would the grown-up hedgehog have?

   b What would be their total length?

   c What would be the grown-up hedgehog's skin area?

   d If the grown-up hedgehog weighed 810 g, what would the baby hedgehog weigh?

ACTIVITY 2

Imagine a strange animal consisting of a spherical body supported by one leg. The body weighs 100 kg, the cross-sectional area of the leg is 100 cm² and the height is 1 m.

How much weight in kg does each square cm of leg support?

A similar animal is 2 m tall. For this animal:

How many kg does it weigh?

What is the cross-sectional area of its leg in cm²?

How much weight in kg does each square cm of leg support?

If the leg can support a maximum of 4 kg/cm², what is the maximum height, in metres, of a similar animal?

This helps explain why animals cannot get bigger and bigger as there is a limit to the load that bone can support. Calculations like this have led archaeologists to suspect that the largest dinosaurs lived in shallow lakes to support their weight.

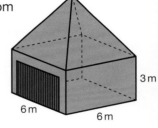

REVISION

1 ▶ Find the area and perimeter of the shape shown.

4 cm

←2 cm→

2 ▶ Find the area and perimeter of the shape shown.

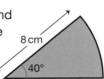

8 cm

40°

3 ▶ Find the volume and surface area of this prism.

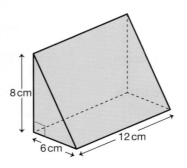

8 cm

12 cm

6 cm

4 ▶ The diagram shows a garage.

The height of the top of the pyramid roof is 5 m from ground level.

Find the volume of the garage.

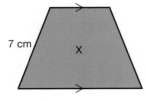

3 m

6 m

6 m

5 ▶ X and Y are similar shapes. The area of Y is 50 cm². Find the area of X.

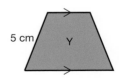

7 cm   X

5 cm   Y

6 ▶ Two similar buckets have depths of 30 cm and 20 cm.

The smaller bucket holds 8 litres of water. Find the capacity of the larger bucket.

REVISION

1 ▶　Find the area and perimeter of the shaded shape.

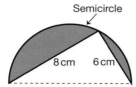

Semicircle

8 cm　6 cm

2 ▶　Find the area and perimeter of the shaded shape.

50°

3 cm　2 cm

3 ▶　Find the volume and surface area of this prism.

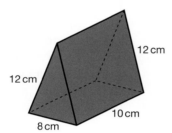

12 cm

12 cm

10 cm

8 cm

4 ▶　After burning, a hemispherical depression is left in a candle as shown

Find the volume and surface area of this partially burnt candle.

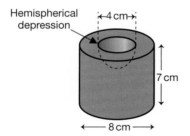

Hemispherical depression

4 cm

7 cm

8 cm

5 ▶　X and Y are similar shapes. The area of X is 81 cm²; the area of Y is 49 cm². Find x.

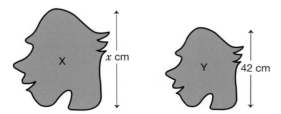

X

x cm

Y

42 cm

6 ▶　'McEaters' sells drinks in three similar cups, small, medium and large.

The height of the small cup is 10 cm and the height of the large cup is 15 cm.

a　A small drink costs $2. What is a fair price for a large drink?

b　A medium drink costs $3.47. What is the height of a medium drink?

EXAM PRACTICE: SHAPE AND SPACE 7

1 Find the perimeter and area of the shape shown. Give your answers to 3 s.f.

[4]

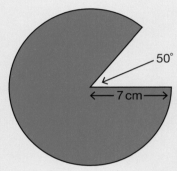

8 cm

←— 12 cm —→

2 The radius of a circle is 7 cm. Find the area and perimeter of the sector shown. Give your answers to 3 s.f.

[4]

50°

←— 7 cm —→

3 The diagram shows a hollow concrete pipe. Find the volume in m³.

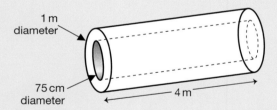

1 m
diameter

75 cm
diameter

←— 4 m —→

[5]

4 The two chocolates are similar. The height of the smaller chocolate is 2 cm, the height of the larger chocolate is 3 cm.

a If the surface area of the smaller chocolate is 20 cm², what is the surface area of the larger chocolate in cm²?

b If the volume of the larger chocolate is 24 cm³, what is the volume of the smaller chocolate in cm³ to 3 s.f.?

[6]

5 A drinks can is 15 cm tall and has a volume of 425 cm³. A similar can contains 217 cm³ and has a surface area of 230 cm².

a What is the height of the smaller can in cm to 3 s.f?

b What is the surface area of the larger can in cm² to 3 s.f?

[6]

[Total 25 marks]

CHAPTER SUMMARY: SHAPE AND SPACE 7

CIRCLES

The perimeter of a circle is called the circumference.

Circumference,
$C = \pi d$ or $C = 2\pi r$
Area, $A = \pi r^2$

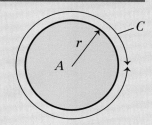

Arc length $= \dfrac{x}{360} \times 2\pi r$

Sector area $= \dfrac{x}{360} \times \pi r^2$

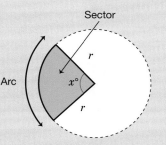

SOLIDS

For a prism,

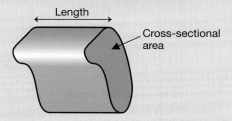

Length

Cross-sectional area

Volume = area of cross-section × length

Surface area = total area of all its faces

A cylinder is a prism with a circular cross-section.

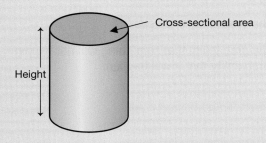

Cross-sectional area

Height

Volume = $\pi r^2 h$

Curved surface area = $2\pi rh$

For a pyramid,

Volume $= \dfrac{1}{3} \times$ area of base × vertical height

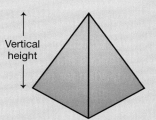

Vertical height

For a cone,

Volume $= \dfrac{1}{3} \pi r^2 h$
Curved surface area $= \pi rl$

For a sphere,

Volume $= \dfrac{4}{3} \pi r^3$ Surface area $= 4\pi r^2$

SIMILAR SHAPES

When a shape is enlarged by linear scale factor k, the area of the shape is enlarged by scale factor k^2.

When a shape is enlarged by linear scale factor k, the volume of the shape is enlarged by scale factor k^3.

When the linear scale factor is k:

- Lengths are multiplied by k
- Area is multiplied by k^2
- Volume is multiplied by k^3

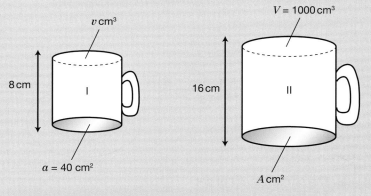

$v\,\text{cm}^3$

$V = 1000\,\text{cm}^3$

8 cm

16 cm

I

II

$a = 40\,\text{cm}^2$

$A\,\text{cm}^2$

The linear scale factor (k) from I to II is $8k = 16 \Rightarrow k = 2$

The area scale factor (k^2) is $2^2 = 4$
$\Rightarrow A = 40 \times 4 = 160\,\text{cm}^2$

The volume scale factor (k^3) is $2^3 = 8$
$\Rightarrow v \times 8 = 1000 \Rightarrow v = 125\,\text{cm}^3$

SETS 3

In many sports, drug-testing is routinely carried out in order to discover which athletes are using performance-enhancing drugs – substances that can improve the performance of athletes but are forbidden for use in sport. If tests reveal quantities over a critical value then it is assumed that the athlete has been cheating. However it is possible to fail the test even if none of these drugs have been taken. The appropriate figure to use to decide whether the athlete has been cheating is a *conditional* probability, meaning the probability the athlete has cheated *given* they have failed the test, written as P(cheated | failed the test).

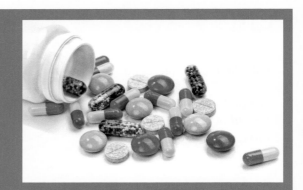

LEARNING OBJECTIVE

- Use Venn diagrams to calculate probability

BASIC PRINCIPLES

- A **set** is a collection of objects, described by a list or a rule. A = {1, 3, 5}

- The number of elements of set A is given by n(A). n(A) = 3

- The **universal set** contains all the elements being discussed in a particular problem. $\mathscr{E}$

$\mathscr{E}$

- B is a **subset** of A if every member of B is a member of A.
  B ⊂ A

- The **complement** of set A is the set of all elements not in A.
  A′

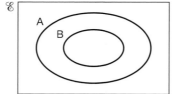

- The **intersection** of A and B is the set of elements which are in both A and B.
  A ∩ B

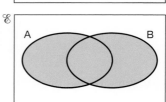

- The **union** of A and B is the set of elements which are in A or B or both.
  A ∪ B

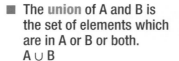

- Use **Venn diagrams** to represent two or three sets.

- Use algebra to solve problems involving sets.

- Calculate the probability of an event.

- Calculate the probability that something will not happen given the probability that it will happen.

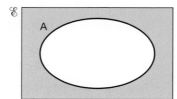

PROBABILITY

Venn diagrams can be very useful in solving certain probability questions. Venn diagrams can only be used if all the outcomes are equally likely.

The probability of the event A happening is given by $\dfrac{n(A)}{n(\mathcal{E})}$

EXAMPLE 1

SKILLS

PROBLEM
SOLVING

A fair six-sided die is thrown. What is the probability of throwing

a a **prime number**
b a number greater than 2
c a prime number or a number greater than 2
d a prime number that is greater than 2?

As all the outcomes are equally likely a Venn diagram can be used.

$n(\mathcal{E}) = 6$

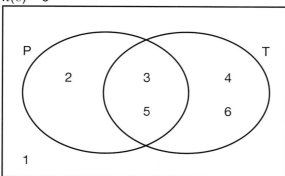

The Venn diagram shows all the outcomes and the subsets
P = {prime numbers}, T = {numbers greater than 2}.

There are six possible outcomes so $n(\mathcal{E}) = 6$

a The probability of a prime number is $\dfrac{n(P)}{n(\mathcal{E})} = \dfrac{3}{6} = \dfrac{1}{2}$

b The probability of a number greater than 2 is $\dfrac{n(T)}{n(\mathcal{E})} = \dfrac{4}{6} = \dfrac{2}{3}$

c A prime number or a number greater than 2 is the set P ∪ T, so the probability is $\dfrac{n(P \cup T)}{n(\mathcal{E})} = \dfrac{5}{6}$

d A prime number that is greater than 2 is the set P ∩ T, so the probability is $\dfrac{n(P \cap T)}{n(\mathcal{E})} = \dfrac{2}{6} = \dfrac{1}{3}$

Note: You cannot add the probabilities from parts **a** and **b** to obtain the answer to part **c**. The Venn diagram shows that if you do this you count the numbers 3 and 5 twice.

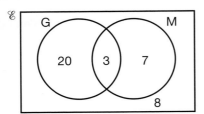

The Venn diagram shows the number of students studying German (G) and Mandarin (M).

SKILLS

PROBLEM
SOLVING

$\mathscr{E}$

G 20 3 7 M
8

A student is picked at **random**. Work out
a P(G ∩ M)
b P(G′)
c P(G ∪ M)

a Total number of students: 20 + 3 + 7 + 8 = 38

$$P(G \cap M) = \frac{\text{number of students in } G \cap M}{\text{total number of students}} = \frac{3}{38}$$

b $P(G') = \frac{7 + 8}{38} = \frac{\text{number of students in } G'}{\text{total number of students}} = \frac{15}{38}$

c $P(G \cup M) = \frac{20 + 3 + 7}{38} = \frac{30}{38}$

KEY POINTS

- Venn diagrams can be used to work out probabilities if all the outcomes are equally likely.
  The probability of event A, written P(A), is $\frac{n(A)}{n(\mathscr{E})}$

- Venn diagrams are very useful in problems that ask for the probability of A and B or the probability of A or B.
  - The probability of A and B, written P(A ∩ B), is $\frac{n(A \cap B)}{n(\mathscr{E})}$
  - The probability of A or B (or both), written P(A ∪ B), is $\frac{n(A \cup B)}{n(\mathscr{E})}$

EXERCISE 1

1 ▶ A group of people complete this survey about watching TV.

Tick all that apply.
☐ I watch TV shows as they are shown.
☐ I watch TV shows on catch-up websites.
☐ I record TV shows to watch later.
☐ I don't watch any TV shows.

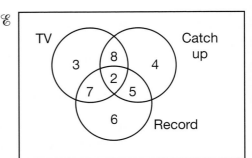

The Venn diagram shows the results.

What is the probability that a person selected at random
a only watches TV shows as they are shown
b never records TV shows to watch later
c records shows, uses websites and watches shows as they are shown?

2 ▶ In a class of 27 students: 12 have black hair, 17 have brown eyes, 9 have black hair and brown eyes.

a Draw a Venn diagram to show this information.

b What is the probability that a student picked at random from this class

   i has black hair and brown eyes
   ii has black hair but not brown eyes
   iii doesn't have black hair or brown eyes?

3 ▶ Dan asked the 30 students in his class if they were studying French (F) or Spanish (S). 15 were studying both and a total of 21 were studying French. They were all studying at least one language.

a Draw a Venn diagram to show Dan's data.

b Work out

   i P(S)
   ii P(F ∩ S)
   iii P(F ∪ S)
   iv P(F′ ∩ S)

4 ▶ 50 people attended a charity lunch. 30 chose a baguette only, 14 chose soup only and 2 chose neither baguette nor soup.

a Draw a Venn diagram to show this information.

b Work out   **i** P(B ∩ S)   **ii** P(B′ ∪ S′)

5 ▶ One day Keshni records the number of customers to her jewellery shop and the numbers of rings (R), bracelets (B) and necklaces (N) sold. The Venn diagram shows the results.

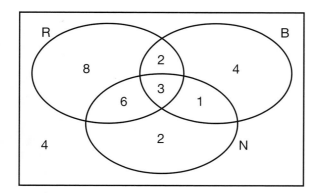

a How many customers came that day?

b How many customers bought both a ring and a necklace?

c Work out the probability that a customer chosen at random

   i bought a bracelet
   ii bought a ring and a necklace but not a bracelet
   iii bought a bracelet and a necklace.

EXERCISE 1*

1 ▶ Charlie asks the 30 students in his class if they passed their English (E) and Mathematics (M) tests.

21 students passed both their English and Mathematics tests. 2 students didn't pass either test. 25 students passed their Mathematics test.

  a Draw a Venn diagram to show Charlie's data.

  b Work out

    i $P(E)$ **ii** $P(E \cap M)$ **iii** $P(E \cup M)$ **iv** $P(E' \cap M)$

  c What does $P(E' \cap M)$ mean in the context of the question?

2 ▶ Mike is a stamp collector. The Venn diagram shows information about his stamp collection.

$\mathscr{E} = \{$Mike's full collection of 720 stamps$\}$
$C = \{$stamps from the 20th century$\}$
$B = \{$British stamps$\}$

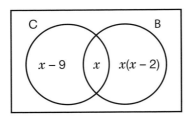

A stamp is chosen at random. It is from the 20th century. Work out the probability that it is British.

3 ▶ A fair 12-sided dice numbered 1 to 12 is rolled.
$X = \{$number rolled is prime$\}$, $Y = \{$number rolled is a **multiple** of 2$\}$.

  a Draw a Venn diagram to show X and Y.

  Work out

  b **i** $P(X \cap Y)$ **ii** $P(Y')$ **iii** $P(X' \cup Y)$

4 ▶ Ahmed asked 28 friends if they had texted (T) or emailed (E) that day. The number who had only texted was the same as the number who had done both. The number who had only emailed was two less than the number who had done both. The number who had neither texted nor emailed was twice the number who had done both.

  a Work out $P(E' \cup T)$.   **b** Work out $P(E \cap T')$.

5 ▶ 123 year 11 students chose the courses they wanted at the school prom dinner. 75 chose a starter, 106 chose a main course and 81 chose a dessert. 67 chose a main course and dessert, 64 chose a starter and a main course while 49 chose a starter and dessert. Everybody ordered at least one course. Find the probability that a student chosen at random chose all three courses.

CONDITIONAL PROBABILITY USING VENN DIAGRAMS

Sometimes additional information is given which makes the calculated probabilities **conditional** on an event having happened.

A mathematics teacher sets two tests for 50 students. 40 students pass the first test, 38 students pass the second test and 30 students pass both tests. Find the probability that a student selected at random

a passed both tests

b passed the second test given that the student passed the first test.

Let F = {students who passed the first test}
Let S = {students who passed the second test}

The information is shown in a Venn diagram.

The numbers must add up to 50, so 2 students failed both tests.

a 30 out of 50 students passed both tests so the probability is $\frac{30}{50} = 0.6$

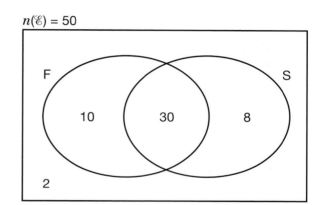

$n(\mathscr{E}) = 50$

b As the student passed the first test the student must be in the set F. The relevant part of the Venn diagram is shown.

40 students passed the first test. Of these 40 students, 30 also passed the second test so the probability is $\frac{30}{40} = 0.75$

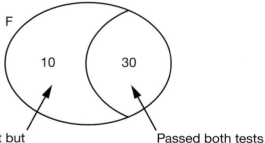

Passed first test but failed second test

Passed both tests

When you are given further information, then you are selecting from a subset rather than from the universal set. This subset becomes the new universal set for that part of the question.

The notation P(A|B) means 'the probability of A given B has occurred' or more simply 'the probability of A given B'.

EXAMPLE 4

SKILLS

PROBLEM
SOLVING

The Venn diagram shows the results of a survey of shopping habits.

F = {people who bought food online}

C = {people who bought clothes online}

Work out P(F|C) for this survey.

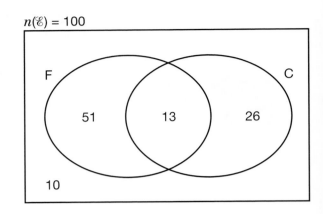

$n(\mathcal{E}) = 100$

P(F|C) means the probability a person buys food online given that they buy clothes online.

The subset to select from is C

$n(C) = 13 + 26 = 39$

$\Rightarrow P(F|C) = \dfrac{13}{39} = \dfrac{1}{3}$

KEY POINTS

- Conditional probability means selecting from a subset of the Venn diagram.
- P(A|B) means 'the probability of A *given* B'.

EXERCISE 2

1 ▶ In a swimming club, 15 people swim front crawl, 12 people swim breaststroke while 7 people swim both front crawl and breaststroke.

    a Draw a Venn diagram to show this data.

What is the probability that a person chosen at random

    b swims front crawl and breaststroke

    c swims front crawl only

    d swims breaststroke given that they swim front crawl?

2 ▶ Levi surveyed 140 students in his year group to find out if they sent text messages or emails last week.

79 students sent text messages and emails. 126 students sent text messages. All of the students had done at least one of these.

    a Draw a Venn diagram to show Levi's data.

A student is chosen at random.

    b Work out the probability that the student sent an email last week.

    c Given that the student sent a text message, work out the probability that they sent an email.

3 ▶ There are 1000 pairs of shoes in a shoe shop. There are 50 pairs of trainers (T) and 100 pairs of pink shoes (P). There are 860 pairs of shoes that are neither trainers nor pink.

    a Draw a Venn diagram to show this data.

    b If a pair of shoes is selected at random work out

        i $P(T \cap P)$ **ii** $P(T|P)$

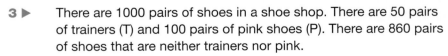

4 ▶ Lucy carried out a survey of 150 students to find out how many students play an instrument (I) and how many play for a school sports team (S).

63 students play for a school sports team. 27 students play an instrument and play for a school sports team. 72 students neither play an instrument nor play for a school sports team.

a Draw a Venn diagram to show Lucy's data.

b Work out the probability that a student plays an instrument.

c Work out the probability that a student plays an instrument given that they play for a school sports team.

5 ▶ The Venn diagram shows people's choice of chocolate (C), strawberry (S) and vanilla (V) ice-cream flavours for the 'three scoops' dessert.

a How many people had all three flavours?

b How many people chose ice-cream for dessert?

c Work out
   i $P(C \cap S \cap V)$ **ii** $P(S \cap V)$ **iii** $P(V|S)$

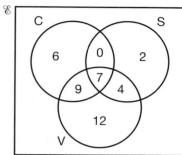

EXERCISE 2*

1 ▶ The Venn diagram shows customers' choice of cheese (C), tuna (T) and egg (E) fillings for a sandwich in a café.

a How many people chose all three fillings?

A customer is chosen at random.

b Work out
   i $P(T \cap E)$ **ii** $P(C \cap T \cap E)$ **iii** $P(E|C)$

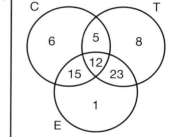

2 ▶ Caitlin did a survey of the pets that her neighbours own: cats (C), dogs (D) and fish (F).

The Venn diagram shows her results.

a How many people took part in the survey?

One of the pet owners is chosen at random.

b Work out
   i $P(C)$ **ii** $P(C|D)$ **iii** $P(F|C')$

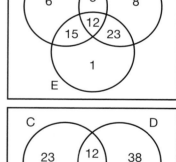

3 ▶ There are 30 dogs at the animal rescue centre. Each dog is at least one colour from black, brown and white.

3 of the dogs are black, brown and white. 15 of the dogs are black and white. 5 of the dogs are brown and white. 8 of the dogs are black and brown. 22 of the dogs are black. 15 of the dogs are brown.

a Draw a Venn diagram to show this information.

One of the dogs is chosen at random.

b Work out the probability that this dog is brown but not white.

c Given that the dog is black, work out the probability that this dog is also brown.

4 ▶ 150 people were asked which of the countries France, the Netherlands and Spain they had visited.

80 people had been to France, 52 to the Netherlands and 63 to Spain. 21 people had been to France and the Netherlands. 28 people had been to France and Spain. 25 people had been to the Netherlands and Spain. 17 people had visited none of these countries.

a Draw a Venn diagram to represent this information.

b Work out the probability that a person, picked from this group at random, had visited only two of the three countries.

c Given that a person had visited Spain work out the probability that they had also visited France.

5 ▶ At a party 45 children chose from three flavours of ice-cream, strawberry (S), chocolate (C) or mint (M). 18 chose strawberry, 24 chose chocolate, 14 chose mint and 8 chose nothing. 10 children chose strawberry and chocolate, 7 chose chocolate and mint while 5 chose mint and strawberry. A child is picked at random.

Work out   **a** $P(S \cap C \cap M)$ **b** $P(C|M)$ **c** $P(C|M')$

ACTIVITY 1

SKILLS

PROBLEM
SOLVING

In a group of 1000 athletes it is known that 5% have taken a performance-enhancing drug. 98% of those who have taken the drug will test positive under a new test, but 2% of those who have not taken the drug will also test positive.

Let   D = {athletes who have taken the drug}

N = {athletes who have not taken the drug}

P = {athletes who test positive}

Copy and complete the Venn diagram to show this data.

$n(\mathscr{E}) = 1000$

D   P   N

What is the probability that an athlete has not taken the drug given that their test result is positive? Comment on your answer.

EXERCISE 3 ▷ **REVISION**

1 ▶ People were asked if they had

■ one or more brothers (B)

■ one or more sisters (S).

The Venn diagram shows the results.

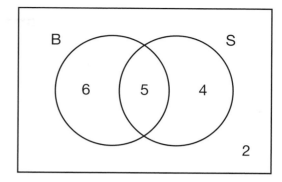

How many people

a had brothers but no sisters

b had brothers and sisters

c had no brothers or sisters

d were asked?

e What is the probability that a person picked from this group at random has one or more sisters?

f What is the probability that a person picked from this group at random has sisters but no brothers?

2 ▶ $\mathscr{E}$ is the set of students in a film club. A is the set of students who like action films while H is the set of students who like horror films.

$n(\mathscr{E}) = 63$, $n(A) = 37$, $n(H) = 28$ and $n(A \cup H)' = 5$

Work out

a $P(A \cap H')$ **b** $P(A' \cup H')$

3 ▶ Lily carried out a survey of 92 students to find out how many had a holiday at home (H) and how many had a holiday abroad (A) this year.

35 students had a holiday at home. 11 students had a holiday at home and a holiday abroad. 19 students did not have a holiday at all.

a Draw a Venn diagram to show Lily's data.

A student is picked at random from this group.

Work out the probability that they

b went on holiday abroad

c went on holiday abroad, given that they had a holiday at home.

4 ▶ The Venn diagram shows the numbers of students who take Mathematics (M), English (E) and History (H).

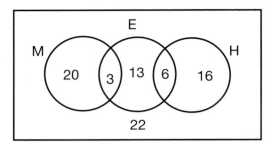

Work out   **a** P(M ∪ E)   **b** P(H|E)

5 ▶ The Venn diagram shows the instruments played by members of an orchestra: violin (V), flute (F) and piano (P).

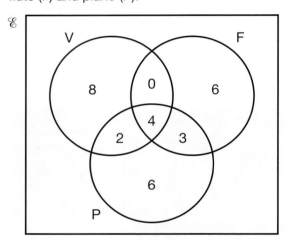

a How many people play all three instruments?
b How many people are in the orchestra?
c Work out

    i P(V ∩ P ∩ F)   **ii** P(V ∩ F)   **iii** P(V|P)

EXERCISE 3*

REVISION

1 ▶ A teacher asks the 26 students in her class if they like to sing or to play an instrument. 13 students like to sing, 16 like to play an instrument and 2 like neither.

A student is selected at random. Work out the probability that this student

a likes to sing and play an instrument

b likes to sing or play an instrument but not both.

Mike now joins the class, and the probability that a student only likes to sing is now $\frac{1}{3}$
c What does Mike like to do?

2 ▶ A 10-sided fair die is thrown. S = {number is square}, F = {number is a **factor** of 6} and N = {number is prime}.

Work out

a P(S ∩ F)   **b** P(S ∪ N)   **c** P(F′ ∪ N)   **d** P(S ∩ F ∩ N)

3 ▶ 90% of teenagers receive money every month from their parents and 40% do jobs around the house. 5% don't receive money every month or do jobs around the house.

Work out the probability that a teenager selected at random

a receives money every month and does jobs around the house

b receives money every month given that they do jobs around the house.

4 ▶ There are 100 students studying Mathematics.

All 100 study at least one of three courses: Pure, Mechanics and Statistics. 18 study all three. 24 study Pure and Mechanics. 31 study Pure and Statistics. 22 study Mechanics and Statistics. 57 study Pure. 37 study Mechanics.

a Draw a Venn diagram to show this information.

One of the students is chosen at random.

b Work out the probability that this student studies Pure but not Mechanics.

c Given that the student studies Statistics, work out the probability that this student also studies Mechanics.

5 ▶ A gym offers three different classes: Total Spin (T), Bootcamp (B) and Zumba (Z). 70 members of the gym were asked which of these classes they had attended.

24 people had attended Total Spin. 28 people had attended Bootcamp. 30 people had attended Zumba. 10 people had attended Total Spin and Bootcamp. 12 people had attended Bootcamp and Zumba. 7 people had attended Total Spin and Zumba. 13 people had attended none of these classes.

a Draw a Venn diagram to represent this information.

A person is picked at random.

b Work out

   i $P(T \cup B \cup Z)$ **ii** $P(Z|B)$ **iii** $P(B|Z')$

EXAM PRACTICE: SETS 3

1 Maisie asks the 24 students in her class if they passed their Science (S) and Mathematics (M) tests.

18 students passed Science and Mathematics. 3 students failed both tests. 20 students passed Mathematics.

a Draw a Venn diagram to show Maisie's results.

b Work out

   i P(S) ii P(S ∩ M) iii P(S ∪ M) iv P(S' ∩ M) **[5]**

2 The manager of a coffee shop records customers' purchases one day. Of the 210 customers surveyed, 180 bought a drink and 90 bought a snack. All the customers bought something.

What is the probability that a customer picked at random

a bought a drink and a snack

b bought a drink or a snack but not both? **[5]**

3 75% of teenagers own a smartphone and 35% own a tablet. 10% own neither of these. What is the probability that a teenager selected at random

a has a smartphone or a tablet but not both

b has a tablet given that they have a smartphone? **[5]**

4 At a school, the students can play football (F), hockey (H) or tennis (T).

Nahal carried out a survey to find out which sport students in his year played.

He recorded his results in a Venn diagram.

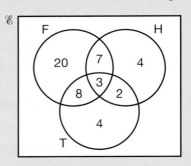

a How many students took part in the survey?

One of the students is chosen at random.

b Work out

   i P(H)   ii P(F ∩ H ∩ T)   iii P(T|H)   **[5]**

5 There are 40 people in a dance class. All 40 can do at least one of the dances: waltz (W), jive (J) or tango (T). 30 people can waltz, 21 can jive and 14 can tango. 13 people can waltz and jive, 6 can jive and tango while 11 can tango and waltz. A person is selected at random. Work out

a P(W ∩ J ∩ T)   b P(T|J)   **[5]**

[Total 25 marks]

CHAPTER SUMMARY: SETS 3

PROBABILITY

Venn diagrams can be used to work out probabilities if all the outcomes are equally likely.

The probability of event A, written P(A), is $\dfrac{n(A)}{n(\mathscr{E})}$

Venn diagrams are very useful in problems that ask for the probability of A and B or the probability of A or B.

The probability of A and B, written P(A ∩ B), is $\dfrac{n(A \cap B)}{n(\mathscr{E})}$

The probability of A or B (or both), written P(A ∪ B), is $\dfrac{n(A \cup B)}{n(\mathscr{E})}$

CONDITIONAL PROBABILITY

Conditional probability means selecting from a subset of the Venn diagram.

The notation P(A|B) means 'the probability of A given B'.

The Venn diagram shows the numbers of students with a smartphone (S) or a laptop (L).

$n(\mathscr{E})$ = 36

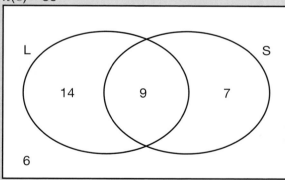

If a student is selected at random then:

P(L ∩ S) means the probability of having a laptop and a smartphone = $\dfrac{9}{36} = \dfrac{1}{4}$

P(L ∪ S) means the probability of having a laptop or a smartphone (or both) = $\dfrac{14 + 9 + 7}{36} = \dfrac{5}{6}$

P(L|S) means the probability of having a laptop given they have a smartphone = $\dfrac{9}{9 + 7} = \dfrac{9}{16}$

P(S|L) means the probability of having a smartphone given they have a laptop = $\dfrac{9}{14 + 9} = \dfrac{9}{23}$

UNIT 8

Eight is the largest cube (2^3) in the Fibonacci sequence.

A polygon with 8 sides is called an octagon.

There are eight planets in our Solar System (Mercury, Venus, Earth, Mars, Jupiter, Saturn, Uranus and Neptune).

The infinity symbol ∞ is often described as a sideways eight.

NUMBER 8

Pressure is an example of a compound measure. The pressure exerted on the ground by a person increases as the area of the shoe in contact with the ground decreases. So the pressure exerted by a stiletto heel is immense. In fact, the wearing of stiletto heels has been banned at some archaeological sites due to the damage they cause.

LEARNING OBJECTIVES

- Convert between metric units of area
- Convert between metric units of volume
- Calculate rates
- Convert between metric speed measures
- Solve problems involving compound measures, including density and pressure

BASIC PRINCIPLES

- Rearrange formulae
- You will be expected to remember these conversions:

    - 1000 metres (m) = 1 kilometre (km)

    - 1000 millimetres (mm) = 1 metre (m)

    - 100 centimetres (cm) = 1 metre (m)

    - 10 millimetres (mm) = 1 centimetre (cm)

    - 1 millilitre (ml) = 1 cm^3

    - 1 litre = 1000 ml = 1000 cm^3

    - 1 kilogram (kg) = 1000 grams (g)

    - 1 tonne (t) = 1000 kilograms (kg)

CONVERTING BETWEEN UNITS OF LENGTH

EXAMPLE 1

SKILLS

PROBLEM SOLVING

The **circumference** of the Earth is approximately 40 075 km.

Change 40 075 km to mm.

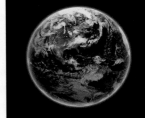

40 075 km	$= 40\,075 \times 1000$ m	(as 1 km = 1000 m)
	$= 40\,075 \times 1000 \times 1000$ mm	(as 1 m = 1000 mm)
	$= 4.0075 \times 10^{10}$ mm	

EXAMPLE 2

SKILLS

PROBLEM SOLVING

The thickness of paper is around 0.05 mm. Change 0.05 mm to km.

0.05 mm	$= \dfrac{0.05}{1000}$ m	(as 1000 mm = 1 m)
	$= \dfrac{0.05}{1000 \times 1000}$ km	(as 1000 m = 1 km)
	$= 5 \times 10^{-8}$ km	

EXERCISE 1

Give answers to 3 s.f. and in **standard form** where appropriate.

1 ▶ A blue whale is the largest living animal and is the heaviest known to have existed. The longest recorded blue whale measured 33 m long. Convert 33 m to km.

2 ▶ The distance light travels in one year is about 9.46×10^{12} km. Convert 9.46×10^{12} km to cm.

3 ▶ The River Nile, at 6.695×10^{8} cm, is the longest river in the world. Convert 6.695×10^{8} cm to km.

4 ▶ The length of the human intestines in an adult is about 820 cm. Convert this length to km.

5 ▶ A cough virus is about 9 nanometres in **diameter**. How many mm is this? (A nanometre = 10^{-9} metres)

6 ▶ The height of Mount Everest is about 8.848 km. Convert 8.848 km into micrometres. (A micrometre = 10^{-6} metres)

EXERCISE 1*

Give answers to 3 s.f. and in standard form where appropriate.

1 ▶ The distance to the nearest star is about 3.99×10^{13} km. Convert 3.99×10^{13} km into mm.

2 ▶ A human hair is about 50 micrometres in diameter. How many km is this? (A micrometre = 10^{-6} metres)

3 ▶ The tallest human in recorded history was 8 feet 11 inches tall. Convert this height to metres. (1 foot = 12 inches and 1 inch = 2.54 cm)

4 ▶ At 6561 feet the Pronutro Zip line in Sun City, South Africa is the longest zip wire in the world. Convert 6561 feet into km. (1 foot = 12 inches and 1 inch = 2.54 cm)

5 ▶ The latest estimate of the length of the Great Wall of China is 13 171 miles.
Convert 13 171 miles to km.
(1 mile = 1760 yards, 1 yard = 36 inches and 1 inch = 2.54 cm)

6 ▶ The tallest skyscraper in the world, Burj Khalifa in Dubai, is 829.8 m in height.
How high is this in feet? (1 foot = 12 inches and 1 inch = 2.54 cm)

CONVERTING BETWEEN UNITS OF AREA

A diagram is useful, as is shown in the following examples.

EXAMPLE 3 A rectangle measures 1 m by 2 m. Find the area in mm².

1 m is 1000 mm

2 m is 2000 mm

So the diagram is as shown.

So the area is 1000 × 2000 mm²

$= 2\,000\,000$ mm²

$= 2 \times 10^6$ mm²

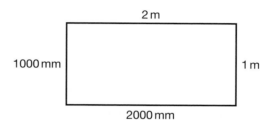

EXAMPLE 4 Change 30 000 cm² to m².

$1\,m^2 = 1\,m \times 1\,m$

$= 100\,cm \times 100\,cm$

$= 10\,000\,cm^2$

So $30\,000\,cm^2 = \dfrac{30\,000}{10\,000}\,m^2$

$= 3\,m^2$

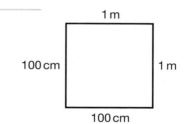

KEY POINTS
- A diagram can help convert areas.
- $1\,m^2 = 10^6\,mm^2$

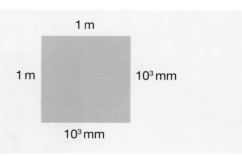

EXERCISE 2

Give answers to 3 s.f. and in standard form where appropriate.

1 ▶ The surface area of a coke can is about 290 cm².

    a What is the surface area in mm²?
    b What is the surface area in m²?

2 ▶ A Swedish artist spent $2\frac{1}{2}$ years and 100 tons of paint to create the largest painting in the world (8000 m²) done by a single artist. The painting is called 'Mother Earth' and depicts a woman holding a peace sign. Find the area in km².

3 ▶ A4 paper measures 297 mm by 210 mm. Find the area in m².

4 ▶ The area of a pixel in a typical computer display is about 55 000 μm². Find this area in mm². (1 μm = 1 micrometre = 10^{-6} metres)

5 ▶ The head of a pin has an area of about 2 mm². Find this area in m².

6 ▶ A singles tennis court is a rectangle 78 feet long by 27 feet wide. Find the area in m². (1 foot = 12 inches and 1 inch = 2.54 cm)

EXERCISE 2*

Give answers to 3 s.f. and in standard form where appropriate.

1 ▶ The surface area of the Earth is about 5.1×10^8 km². 70% of this area is covered by water. How many mm² of dry land is there?

2 ▶ The surface area of a red blood cell is about 100 μm². Find this area in km². (1 μm = 1 micrometre = 10^{-6} metres)

3 ▶ The length of a football pitch must be between 100 yards and 130 yards and the width must be between 50 yards and 100 yards. Find the maximum and minimum area in hectares. (1 hectare = 10^4 m² and 1 yard = 914.4 mm)

4 ▶ One estimate gives 3×10^{12} for the number of trees in the world, each with a **mean** leaf area of 2800 square feet. If the world population is 7×10^9, how many cm² of leaf area are there per person?

5 ▶ Estimates of the total surface area of lungs vary from 50 to 75 square metres. Convert these estimates into square feet. (1 foot = 12 inches and 1 inch = 2.54 cm)

6 ▶ The largest TV screen in the world measures 370 inches along the diagonal. That's approximately the size of two elephants trunk to tail with another two elephants trunk to tail on top of them. The **ratio** of the lengths of the sides is 16 : 9. Find the area of the screen in mm². (1 inch = 2.54 cm)

CONVERTING BETWEEN UNITS OF VOLUME

Diagrams are very useful.

EXAMPLE 5

A **cuboid** measures 1 m by 2 m by 3 m.
Find the volume in mm³.

1 m is 1000 mm
2 m is 2000 mm
3 m is 3000 mm
So the volume is $1000 \times 2000 \times 3000 = 6 \times 10^9 \, mm^3$

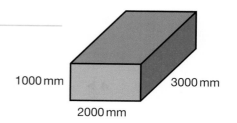
1000 mm   3000 mm
2000 mm

EXAMPLE 6

Change $10^7 \, cm^3$ to m³.

$1\,m^3 = 1\,m \times 1\,m \times 1\,m$
$\quad = 100\,cm \times 100\,cm \times 100\,cm$
$\quad = 10^6\,cm^3$
So $10^7\,cm^3 = \dfrac{10^7}{10^6}\,m^3$
$\quad = 10\,m^3$

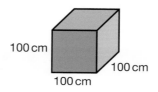
100 cm
100 cm
100 cm

KEY POINTS

- A diagram can help convert volumes.
- $1\,m^3 = 10^6\,cm^3$

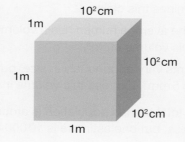
$10^2\,cm$
1m
$10^2\,cm$
1m
$10^2\,cm$
1m

EXERCISE 3

Give answers to 3 s.f. and in standard form where appropriate.

1 ▶ Roughly 0.2 km³ of material was excavated during the construction of the Panama Canal. How many m³ is this?

2 ▶ The world's largest mango had a volume of around 850 cm³. How many mm³ is this?

3 ▶ The approximate amount of rock ejected during the 1980 eruption of Mount St Helens was $1.2 \times 10^9 \, m^3$. Express this volume in km³.

4 ▶ A basketball has a volume of 8014 cm³. Find this volume in m³.

5 ▶ A medium grain of sand has a volume of about $6.2 \times 10^{-11} \, m^3$.

   a Find the volume in mm³.
   b How many grains of sand are there in 1 cm³?

6 ▶ The volume of crude oil that can be carried on the super tanker 'Knock Nevis' is $6.5 \times 10^5 \, m^3$. How many litres is this?

EXERCISE 3*

Give answers to 3 s.f. and in standard form where appropriate.

1 ▶ The Mediterranean Sea has a volume of around $3.75 \times 10^6 \, \text{km}^3$.
Find this volume in litres.

2 ▶ The volume of the Earth is approximately $1.08 \times 10^{27} \, \text{cm}^3$. Express this in km^3.

3 ▶   **a** A grain of sugar has a volume of about $0.016 \, \text{mm}^3$. Express this volume in cm^3.
     b A bag of sugar is a cuboid measuring 11 cm by 7.5 cm by 6 cm.
       Estimate how many grains of sugar are in the bag.

4 ▶ It is estimated that there are 1.5×10^9 cows on the planet, each producing 180 litres of methane each day. Find how many cubic metres of methane are produced each day.

5 ▶ A typical virus has a volume of around $5 \times 10^{-21} \, \text{m}^3$. A teaspoon has a capacity of 5 ml. How many viruses are needed to fill the teaspoon?

6 ▶ A British gallon has a volume of 277.42 cubic inches. How many litres are there in one gallon? (1 inch = 2.54 cm)

ACTIVITY 1

SKILLS

PROBLEM
SOLVING

It is sometimes claimed that the total volume of ants on the planet exceeds the total volume of human beings. This activity examines this claim.

a The average human has a volume of $7.1 \times 10^{-2} \, \text{m}^3$. Express this volume in cm^3.

b Ants vary enormously in size, but a reasonable average figure is $5 \, \text{mm}^3$. Express this volume in cm^3.

The total human population is around 7.4×10^9. The total ant population is impossible to assess, but one estimate is 10 000 trillion ants, that is 10^{16}.

c According to these figures, which has the greater volume, ants or humans?

COMPOUND MEASURES

A compound measure is made up of two or more different measurements and is often a measure of a **rate** of change.

Examples of a compound measure are speed, density and pressure.

Speed is a compound measure because $\text{speed} = \dfrac{\text{distance}}{\text{time}}$

Density is a compound measure because $\text{density} = \dfrac{\text{mass}}{\text{volume}}$

Pressure is a compound measure because $\text{pressure} = \dfrac{\text{force}}{\text{area}}$

EXAMPLE 7

A car is traveling at 36 km/hr.
What is its speed in m/s?

36 km/hr = 36 × 1000 m/hr = 36 000 m/hr
(to convert km/hr to m/hr × 1000)

36 000 m/hr = 36 000 ÷ 60 m/min = 600 m/min
(to convert m/hr to m/min ÷ 60)

600 m/min = 600 ÷ 60 m/s = 10 m/s
(to convert m/min to m/s ÷ 60)

Alternatively

distance = 36 km = 36 × 1000 m

time = 1 hour = 60 × 60 seconds

$$\text{speed} = \frac{\text{distance}}{\text{time}} = \frac{36 \times 1000}{60 \times 60} = 10 \text{ m/s}$$

Density is the **mass** of a substance contained in a certain volume. It is usually measured in grams per cubic centimetre (g/cm³). If a substance has a density of 4 g/cm³ then 1 cm³ has a mass of 4 g, 2 cm³ has a mass of 8 g and so on.

EXAMPLE 8

The diagram shows a block of wood in the shape of a cuboid.

The density of wood is 0.6 g/cm³.

Work out the mass of the block of wood.

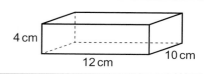

4 cm   12 cm   10 cm

The formula to calculate density is

$$\text{density} = \frac{\text{mass}}{\text{volume}}$$

The density is given and the mass needs to be found.

So work out the volume in cm³:

Volume of block = $l \times w \times h$

= 12 × 10 × 4 = 480 cm³

Substitute into the formula:

$$0.6 = \frac{\text{mass}}{480}$$

Multiply both sides by 480:

$$0.6 \times 480 = \frac{\text{mass}}{480} \times 480$$

Mass = 288 g

Pressure is the force applied over a certain area. It is usually measured in newtons per cm² (N/cm²) or newtons per m² (N/m²). If a force of 12 N is applied to an area of 4 cm² then the pressure is 12 ÷ 4 = 3 N/cm².

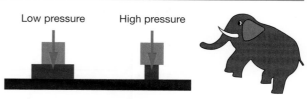

Low pressure   High pressure

Large area   Small area

EXAMPLE 9

The piston of a bicycle pump has an area of 7 cm².
A force of 189 N is applied to the piston. What is the pressure generated?

$$\text{pressure} = \frac{\text{force}}{\text{area}}$$

$$\text{pressure} = 189 \div 7 = 27 \text{ N/cm}^2$$

KEY POINTS

- $\blacksquare$ $\text{speed} = \dfrac{\text{distance}}{\text{time}}$ usually measured in m/s or km/hr

- $\blacksquare$ $\text{density} = \dfrac{\text{mass}}{\text{volume}}$ usually measured in g/cm³ or kg/m³

- $\blacksquare$ $\text{pressure} = \dfrac{\text{force}}{\text{area}}$ usually measured in N/cm² or N/m²

EXERCISE 4

1 ▶   A hiker walks 8.1 km in 1.5 hours.
     a What is the speed of the hiker in km/hr?
     b What is this speed in m/s?

2 ▶   A commercial aeroplane has a cruising speed of 250 m/s.
     What is this speed in km/h?

3 ▶   A Formula 1 racing car has a top speed of 350 km/h.
     A peregrine falcon is the fastest bird with a speed of 108 m/s.
     Which is faster? Explain your answer.

4 ▶   1 m³ of silver has a mass of 10.5 tonnes.
     Find the volume of a silver cup of mass 420 g.

5 ▶   An aluminium kettle of volume 80 cm³ has a density of 2.6 g/cm³. Find its mass.

6 ▶   Gold has a density of 19 g/cm³.
     a Find the mass in kg of a gold ingot of dimensions 20 cm by 6 cm by 3 cm.
     b Find the volume of a gold ingot that has a mass of 11.97 kg.

7 ▶   The mass of this plastic cuboid is 2208 g.

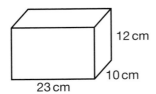

12 cm
10 cm
23 cm

     Work out the density of the plastic in grams per cm³.

8 ▶   A force of 45 N is applied to an area of 26 000 cm². Work out the pressure in N/m².

9 ▶   A force applied to an area of 4.5 m² produces a pressure of 20 N/m². Work out the force in N.

10 ▶   Copy and complete the table.
     Give your answers to 3 **significant figures**.

FORCE	AREA	PRESSURE
60 N	2.6 m²	__ N/m²
__ N	4.8 m²	15.2 N/m²
100 N	__ m²	12 N/m²

11 ▶ On average, a human consumes 1 m³ of fluid a year.
Calculate a person's average daily fluid consumption in litres per day.

12 ▶ **a** Reno cycles at 4.5 m/s for 40 minutes. How many km does he travel?
b Hannah runs 100 m in 12 seconds. What is her speed in km/hr?

EXERCISE 4*

1 ▶ A Boeing 737 uses 15 400 litres of fuel in 5.5 hours.
Given that 1000 litres = 1 m³, find the consumption in cubic metres per minute.

2 ▶ **a** The bullet train in Japan travels at 89 m/s for 0.4 hours. How many km does it travel?
b A drag car covers a 1000 feet course in 3.7 secs. What is the car's average speed in km/hr? (1 m = 3.28 feet).

3 ▶ The greatest recorded speed of Usain Bolt is 12.3 m/s. The greatest speed of a great white shark is 40 km/h. Which is faster? Explain your answer.

4 ▶ In 1999, the shadow of the total eclipse travelled at 1700 miles/hour.
Given that 1 km = 0.6215 miles, change this to metres per second.

5 ▶ Karl travels 35 miles in 45 minutes then 65 km in $1\frac{1}{2}$ hours.
What is his average speed for the total journey in km/h? (5 miles = 8 kilometres)

6 ▶ Mahogany has a density of 0.75 g/cm³.

a Find the mass, in kg, of a mahogany plank measuring 20 mm by 9 cm by 2.5 m.
b Find the volume in m³ of a piece of mahogany that weighs 120 kg.

7 ▶ The mass of air in a classroom of dimensions 4 m by 5 m by 6 m is 120 kg. Find the density of air in g/cm³.

8 ▶ Each year, about 4.8 billion aluminium cans are dumped in landfill sites in the UK.
One aluminium can weighs 20 g and the density of aluminium is 2.7 g/cm³. What is the volume, in cubic metres to 1 s.f, of the cans dumped in landfill sites?

9 ▶ The density of juice is 1.1 grams per cm³.
The density of water is 1 gram per cm³.
270 cm³ of drink is made by mixing 40 cm³ of juice with 230 cm³ of water.
Work out the density of the drink.

10 ▶ A cylindrical bottle of water has a flat, circular base with a diameter of 0.1 m.
The bottle is on a table and exerts a force of 12 N on the table.
Work out the pressure in N/cm².

11 ▶ The pressure between a car tyre and the road is 99 960 N/m². The car tyres have a combined area of 0.12 m² in contact with the road.
What is the force exerted by the car on the road?

12 ▶ Jamie sits on a chair with four identical legs.
Each chair leg has a flat square base measuring 2 cm by 2 cm.
Jamie has a weight of 750 N and the chair has a weight of 50 N.

a Work out the pressure on the floor in N/cm², when only the four chair legs are in contact with the floor.
b The area of Jamie's trainers is 0.04 m². Work out the pressure on the floor when Jamie is standing up. Give your answer in N/m².
c Does Jamie exert a greater pressure on the floor when he is standing up or sitting on the chair?

ACTIVITY 2

It has been claimed that the pressure exerted by a stiletto heel is greater than that exerted by an elephant. This activity will examine this claim.

a The area of one stiletto heel can be taken as 1 cm². An average 50 kg woman will exert a force of 490 N on the ground. Assuming the woman is resting on two stiletto heels, calculate the pressure in N/cm² she exerts on the floor.

b An average 4 tonne elephant will exert a force of 39 200 N on the ground. Assume it is standing on all four feet equally, and that each foot has an area of 324 square inches. Calculate the pressure in N/cm² that the elephant exerts on the floor. (1 inch = 2.54 cm)

c Find the ratio 'pressure exerted by stiletto heel : pressure exerted by elephant' in the form $n : 1$. Comment on the claim.

EXERCISE 5

REVISION

1 ▶ Olivia's earrings are 35 mm long. How long are they in km?

2 ▶ In winter the Antarctic icepack covers an area of about 1.4×10^9 km². Convert this area into cm².

3 ▶ A world record wedding dress train was 2750 m long by 6 m wide. Find the area in km².

4 ▶ A hectare is 10^4 m². How many hectares are there in 1 km²?

5 ▶ A cargo container has a volume of 67.5 m³. What is the volume in mm³?

6 ▶ An Olympic swimming pool has a volume of 2500 m³.

a How many litres of water does it hold?
b Express the volume in km³.

7 ▶ Water is leaking from a water butt at a rate of 4.5 litres per hour.

a Work out how much water leaks from the water butt in i 20 mins ii 50 mins.
b Initially there are 180 litres of water in the water butt. Work out how long it takes for all the water to leak from the water butt.

8 ▶ A car travels 320 km and uses 20 litres of petrol.

    a Work out the average rate of petrol usage. State the units with your answer.
    b Estimate the amount of petrol that would be used when the car has travelled 65 km.

9 ▶ Kelly flies for 45 minutes at an average speed of 15 m/s in a hang glider.
How far has Kelly flown in kilometres?

10 ▶ Archie skis cross-country for a distance of 7.5 km. It takes him 40 minutes.
What is his average skiing speed in m/s?

11 ▶ Paul travels 45 miles in 1 hour and 20 minutes and then 120 km in $2\frac{1}{2}$ hours.

What is his average speed for the total journey in km/hr to the nearest km?

(5 miles = 8 kilometres)

12 ▶ A cubic block of wood has side length 1 m and mass 690 kg.
What is the density of the wood in g/cm³?

13 ▶ The cross-sectional area of the plastic **cylinder** is 300 cm².
Its length is 100 cm.
The plastic has a density of 2.34 g/cm³.
What is the mass of the cylinder?

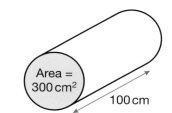

14 ▶ Air has density 0.0012 g/cm³. The mass of air in a room is 54 kg.
What is the volume of the room?

15 ▶ A force of 54 N is applied to an area of 21 600 cm². Work out the pressure in N/m².

16 ▶ A force applied to an area of 0.36 m² produces a pressure of 50 N/m².
Work out the force in N.

EXERCISE 5*   **REVISION**

1 ▶ In 1991 Mike Powell jumped a distance of 29 feet $4\frac{1}{4}$ inches in the long jump. How far is
that in km? (1 foot = 12 inches and 1 inch = 2.54 cm)

2 ▶ The area of a dot printed on a computer printer using 300 dots per inch resolution is about
7000 µm². Find this area in cm². (1 µm = 10^{-6} m)

3 ▶ A smartphone measures 2.5 inches by 4.75 inches. Find the area in m².
(1 inch = 2.54 cm)

4 ▶ A picometre is 10^{-12} m. How many cubic picometres are there in 1 km³?

5 ▶ A small grain of rice has a volume of about $2 \times 10^{-8} \, m^3$.

    a Find the volume in cm^3.
    b The yearly volume of rice produced is about $0.8 \, km^3$.

    How many grains of rice are there in $0.8 \, km^3$?

6 ▶ The British pint is $5.68 \times 10^{-4} \, m^3$. How many litres is this?

7 ▶ Water flows from a hosepipe at a rate of 1200 litres per hour.

    a Work out how much water flows from the hosepipe in
      i 10 minutes   **ii** 35 minutes.

Two identical hosepipes are used at the same time to fill a garden pool.

They each have the same flow rate as in part **a**.

The pool has a capacity of 216 000 litres. Initially the pool is empty.

    b Work out how many days it takes to fill the pool.

8 ▶ Dave drives his motorbike for 180 miles and uses 25 litres of petrol.

    a Work out the average rate of petrol usage. State the units with your answer.
    b Estimate the amount of petrol Dave would use if he travels 275 miles on the motorbike.

    Give your answer to an appropriate **degree of accuracy**.

9 ▶ The speed limit on a motorway is 110 km/hr. What is this in m/s?

10 ▶ Paul swims 750 metres in 25 minutes. What is his average speed in km/h?

11 ▶ **a** A bike travels at b m/s. Write an expression for this speed in km/h.
     b A man runs at c km/h. Write an expression for this speed in m/s.

12 ▶ $1 \, m^3$ of water from the North Sea has mass 1025 kg.

$1 \, cm^3$ of water from the Dead Sea has mass 1.24 g.

In which sea is the water more dense? Justify your answer.

13 ▶ An alloy is made from a mix of copper and tin.

The density of copper is 8.94 grams per cm^3.

The density of tin is 7.3 grams per cm^3.

$500 \, cm^3$ of the alloy is made from $360 \, cm^3$ of copper with $140 \, cm^3$ of tin.

Work out the density of the alloy to 3 s.f. in g/cm^3.

14 ▶ A plastic has density y g/cm^3. Write an expression for its density in kg/m^3.

15 ▶ A cylindrical plant pot has a circular base with diameter 0.3 m.

The plant pot exerts a force of 60 N on the ground.

Work out the pressure in N/cm^2.

Give your answer to 3 significant figures.

16 ▶ A cylinder with a movable piston in an engine contains hot gas.

The pressure of the gas is $350 \, N/m^2$.

The area of the piston is $0.05 \, m^2$.

What is the force exerted by the piston? What force in Newtons is exerted?

EXAM PRACTICE: NUMBER 8

1 A stamp measures 20 mm by 24 mm. Find the area of the stamp in km². **[2]**

2 A micrometre is 10^{-6} m. How many cubic micrometres are there in 1 cm³? **[3]**

3 The speed of light is 3×10^8 m/s. A light year is the distance light travels in a year.

a How many km are there in a light year?

b A region of space is in the shape of a cube with side length of 1 light year. Work out the volume of the region. **[3]**

4 The average human in the developed world consumes about 900 kg of food per year. What is this in grams per minute? **[2]**

5 a A swallow flies for 40 minutes at an average speed of 11 m/s.

How far does the swallow fly in kilometres?

b A bee flies 45 m in 8 secs. What is its speed in km/hr? **[6]**

6 a The hub of a car wheel has a density of 2.5 g/cm³.

Find its volume if it weighs 4750 g.

b A cubic metre of concrete weighs 2.4 tonnes. Find its density in g/cm³. **[6]**

7 Copy and complete the table.

Give your answers to 3 significant figures.

FORCE	AREA	PRESSURE
40 N	3.2 m²	__N/m²
__N	6.4 m²	16.5 N/m²
2000 N	___m²	250 N/m²

[3]

[Total 25 marks]

CHAPTER SUMMARY: NUMBER 8

CONVERTING BETWEEN UNITS OF LENGTH

Convert 5 mm to km.

$5 \text{ mm} = 5 \div 1000 \text{ m} = 5 \div 1000 \div 1000 \text{ km} = 5 \times 10^{-6} \text{ km}$

CONVERTING BETWEEN UNITS OF AREA

A diagram can help convert areas.

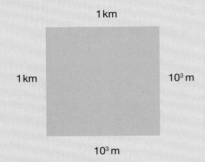

1 km

1 km 10^3 m

10^3 m

Convert 1 km² to m².

The diagram shows that 1 km² equals
$10^3 \times 10^3 \text{ m}^2 = 10^6 \text{ m}^2$

CONVERTING BETWEEN UNITS OF VOLUME

A diagram can help convert volumes.

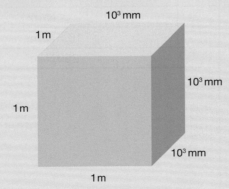

10^3 mm

1 m

10^3 mm

1 m

10^3 mm

1 m

Convert 1 m³ to mm³.

The diagram shows that 1 m³ equals
$10^3 \times 10^3 \times 10^3 \text{ m}^3 = 10^9 \text{ m}^3$

COMPOUND MEASURES

$$\text{speed} = \frac{\text{distance}}{\text{time}}$$

The maximum speed of a cheetah is 120 km/hr. What is this speed in m/s?

distance = 120 km = 120 × 1000 m

time = 1 hour = 60 × 60 seconds

$$\text{speed} = \frac{\text{distance}}{\text{time}} = \frac{120 \times 1000}{60 \times 60} = 33\frac{1}{3} \text{ m/s}$$

$$\text{density} = \frac{\text{mass}}{\text{volume}}$$

100 cm³ of lead weighs 1.134 kg.
What is its density in g/cm³?

mass = 1.134 kg = 1134 g

volume = 100 cm³

$$\text{density} = \frac{\text{mass}}{\text{volume}} = \frac{1134}{100} = 11.34 \text{ g/cm}^3$$

$$\text{pressure} = \frac{\text{force}}{\text{area}}$$

A pressure of 12 N/cm² acts on an area of 8 cm². What force is exerted?

Rearrange $\text{pressure} = \dfrac{\text{force}}{\text{area}}$ to give

force = pressure × area

Substituting the values ⇒ force = 12 × 8 = 96 N

ALGEBRA 8

The 'Enigma' machine is a piece of spy hardware from World War II. 'Enigma' performed a mathematical function by which input text was converted to coded text (the output). Decoding the coded text required the inverse function to be found. Security on the internet today depends on mathematicians finding functions that make it easy to generate coded information but at the same time are very difficult to decode.

LEARNING OBJECTIVES

- Determine whether a mapping is a function
- Use function notation
- Find the range of a function
- Understand which values may need to be excluded from a domain
- Find composite functions
- Find inverse functions

BASIC PRINCIPLES

- Use function machines to find the output for different inputs (number functions, and those containing a variable such as x).
- Use function machines to find inputs for different outputs.
- Write an expression to describe a function (given as a function machine).
- Find the function of a function machine given the outputs for a set of inputs.
- Draw mapping diagrams for simple functions (to show, e.g., the inputs/outputs for a function machine).
- Substitute positive and negative integers and fractions into expressions, including those involving small powers.
- Solve simple linear equations.
- Solve quadratic equations by factorising.
- Rearrange simple formulae.
- Recognise that $\sqrt{x}$ means the positive square root of x.
- Identify inverse operations.
- Draw linear and quadratic graphs.

FUNCTIONS

IDENTIFYING FUNCTIONS

The keys on your calculator are divided into two main classes, operation keys and function keys. The operation keys, such as ➕ and ✖️, operate on two numbers to give a single result.

The function keys, such as x^2 and $x!$, operate on one number to give a single result.

With a function key it is possible for two different inputs to give the same output, for example $2^2 = 4$ and $(-2)^2 = 4$

Relationships that are 'one to one' and 'many to one' are functions. A mapping diagram makes it easy to decide if it is a function or not. If only one arrow leaves each member of the **set** then the relationship is a function.

EXAMPLE 1

SKILLS

INTERPRETATION

State, giving a reason, whether each mapping shows a function or not.

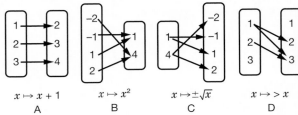

$x \mapsto x + 1$ $x \mapsto x^2$ $x \mapsto \pm\sqrt{x}$ $x \mapsto > x$
     A   B   C   D

A is a function as each element of the first set maps to exactly one element of the second set. It is called a 'one to one' function.

B is a function as each element of the first set maps to exactly one element of the second set. Since the elements in the second set have more than one element from the first set mapped to them, it is called a 'many to one' function.

C is not a function as each element of the first set maps to more than one element of the second set.

D is not a function as at least one element of the first set maps to more than one element of the second set.

A mapping diagram can be used if the size of the sets in the diagram is small. If the sets in the diagram are infinite (for example the set of all **real numbers**) then the vertical line test on a graph is used.

EXAMPLE 2

SKILLS

INTERPRETATION

Use the vertical line test to decide if the graph shows a function or not, giving a reason. If it is a function, state if it is 'one to one' or 'many to one'.

Wherever the red vertical line is placed on the graph, it will only **intersect** at one point, showing, for example, that $1.5 \mapsto 2.5$, so this is a function. It is an example of a 'one to one' function.

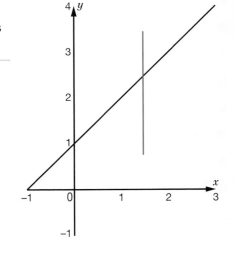

The vertical lines only intersect the graph on the right at one point wherever they are placed, so this is also a function.

The two blue lines intersect the graph at the same y value, showing that $1 \mapsto 2$ and $2 \mapsto 2$ so this is an example of a 'many to one' function.

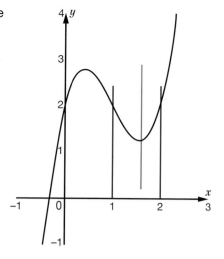

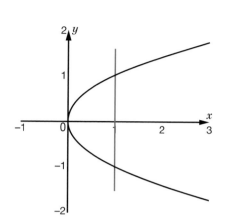

The red vertical line intersects the graph on the left at two points, showing, for example, that $1 \mapsto 1$ and $1 \mapsto -1$, so this is not a function.

KEY POINTS

- Relationships that are 'one to one' and 'many to one' are functions.
- If only one arrow leaves each member of the set then the relationship is a function.
- If a vertical line placed anywhere on a graph intersects the graph at only one point then the graph shows a function.

EXERCISE 1

In Questions 1–4 state, giving a reason, whether the mapping diagram shows a function or not.

1 ▶ 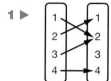   2 ▶ 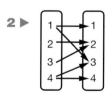   3 ▶ 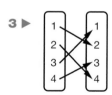   4 ▶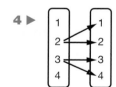

In Questions 5–12 use the vertical line test to decide if the graph shows a function or not, giving a reason. If it is a function, state if it is 'one to one' or 'many to one'.

5 ▶ 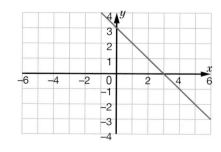   6 ▶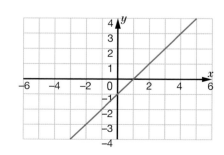

7 ▶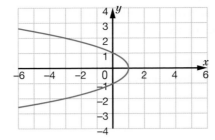

8 ▶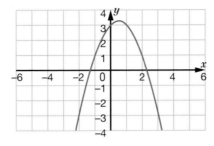

9 ▶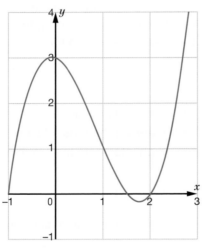

10 ▶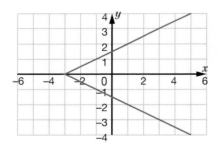

11 ▶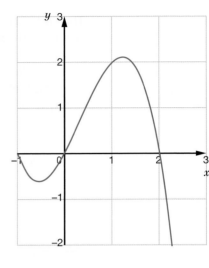

12 ▶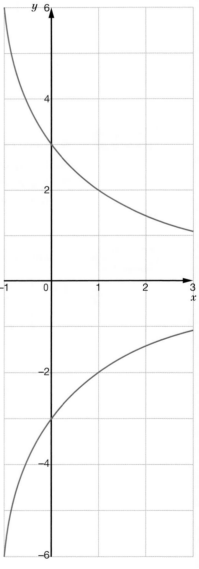

In Questions 1–4 complete the mapping diagram for the sets shown.
State, giving a reason, whether each mapping diagram shows a function or not.

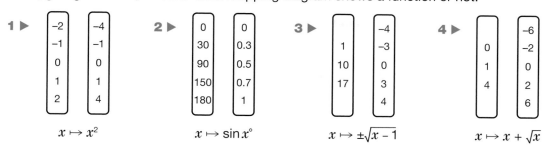

1 ▶
−2	−4
−1	−1
0	0
1	1
2	4

$x \mapsto x^2$

2 ▶
0	0
30	0.3
90	0.5
150	0.7
180	1

$x \mapsto \sin x°$

3 ▶
1	−4
10	−3
17	0
	3
	4

$x \mapsto \pm\sqrt{x-1}$

4 ▶
0	−6
1	−2
4	0
	2
	6

$x \mapsto x + \sqrt{x}$

In Questions 5–12 use the vertical line test to decide if the graph shows a function or not, giving a reason. If it is a function, state if it is 'one to one' or 'many to one'.

5 ▶

6 ▶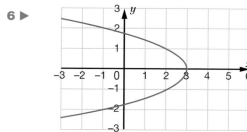

7 ▶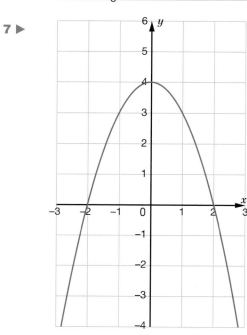

8 ▶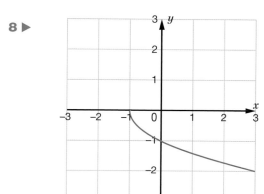

9 ▶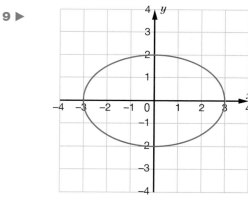

10 ▶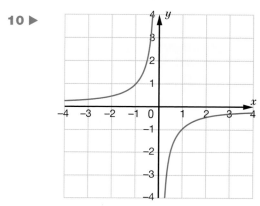

11 ▶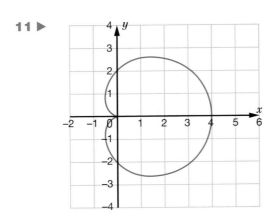

12 ▶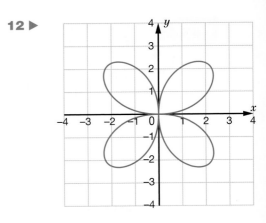

FUNCTION NOTATION

A function is a set of rules for turning one number into another. Functions are very useful, for example, they are often used in computer spreadsheets. A function is a mathematical computer, an imaginary box that turns an input number into an output number.

Number in → | Function | → Number out

If the function doubles every input number then

2 → | Double input | → 4

A letter can be used to represent the rule. If we call the doubling function f then

5 → | f | → 10

f has operated on 5 to give 10, so we write $f(5) = 10$

If x is input then $2x$ is output, so we write $f(x) = 2x$ or $f : x \mapsto 2x$

x → | f | → $2x$

EXAMPLE 3

SKILLS

INTERPRETATION

a If $h : x \mapsto 3x - 2$ find

   i $h(2)$
   ii $h(y)$

b If $g(x) = \sqrt{5 - x}$, find $g(-4)$

a i $h(2) = 3 \times 2 - 2 = 4$
   ii $h(y) = 3y - 2$

b $g(-4) = \sqrt{5 - (-4)} = \sqrt{9} = 3$

EXAMPLE 4

SKILLS

INTERPRETATION

a   If f(x) = 2 + 3x and f(x) = 8, find x

b   If g(x) = $\dfrac{1}{x - 3}$ and g(x) = $\dfrac{1}{2}$, find x

a   2 + 3x = 8 $\Rightarrow$ 3x = 6 $\Rightarrow$ x = 2

b $\dfrac{1}{x - 3} = \dfrac{1}{2}$ $\Rightarrow$ x − 3 = 2 $\Rightarrow$ x = 5

KEY POINT

■ A function is a rule for turning one number into another.

EXERCISE 2

In Questions 1–4, f$:x \mapsto 3x - 2$, g(x) = $2x^2$, h$:x \mapsto x^2 + 2x$, k(x) = $\dfrac{18}{x}$

1 ▶   Calculate   **a** f(2) **b** g(0) **c** k(6)

2 ▶   Calculate   **a** f(3) + g(−2) **b** h(1) − f(0)   **c** f(1) − h(−1) + k(1)

3 ▶   Calculate   **a** f(4) × k(2)   **b** g(3) ÷ k(3)

4 ▶   Calculate   f(−1) + h(2) × k(9) − g(−1)

5 ▶   If g(x) = $\sqrt{x + 1}$, calculate   **a** g(3) **b** g(−1) **c** g(99)

6 ▶   If f$:x \mapsto \dfrac{1}{1 + 2x}$, calculate   **a** f(2) **b** f(−1) **c** f(a)

7 ▶   If h(x) = $2 - \dfrac{1}{x}$, calculate   **a** h(2) **b** h(−2) **c** h(y)

8 ▶   If f(x) = 2x + 2 and f(x) = 8, find x

9 ▶   If p(x) = $\dfrac{1}{x + 1}$ and p(x) = $\dfrac{1}{4}$, find x

10 ▶   If g(x) = $\sqrt{5x + 1}$ and g(x) = 4, find x

EXERCISE 2*

In Questions 1–4, f$:x \mapsto 3 - 2x$, g(x) = $x^2 + 4x$, p$:x \mapsto x^3 - 2x^2$, q(x) = $\dfrac{12}{x - 1}$

1 ▶   Calculate   **a** f(−3) **b** p(−1)   **c** q(5)

2 ▶   Calculate   **a** g(2) + p(2) **b** f(4) − q(3)   **c** q(2) − g(3) + p(−2)

3 ▶   Calculate   **a** g(−2) × p(1) **b** q(7) ÷ f(2)

4 ▶   Calculate   f(−2) − g(4) ÷ q(13) × p(2)

5 ▶   If g(x) = $\dfrac{1}{3 - 2x}$, calculate   **a** g(3) **b** g(−1) **c** g(99)

6 ▶   If f$:x \mapsto \sqrt{x^2 + 2x}$, calculate   **a** f(2) **b** f(−2) **c** f(2a)

7 ▶   If h(x) = $\dfrac{3x + 2}{x - 4}$, calculate   **a** h(2) **b** h(−2) **c** h(3y)

8 ▶   If f(x) = $\dfrac{2}{3x + 1}$ and f(x) = $\dfrac{1}{2}$, find x

9 ▶   If p(x) = $x^2 - x - 4$ and p(x) = 2, find x

10 ▶   If g(x) = $\dfrac{2x + 17}{5 - x}$ and g(x) = 4, find x

EXAMPLE 5

SKILLS

INTERPRETATION

Given $f(x) = 7x + 5$ and $h(x) = 6 + 2x$,
find the value of x such that $f(x) = h(x)$.

$7x + 5 = 6 + 2x \quad \Rightarrow \quad 5x = 1 \quad \Rightarrow \quad x = \dfrac{1}{5}$

A function operates on all of the input. If the function is double and add one then if $4x$ is input, $8x + 1$ is output.

$4x \longrightarrow$ | Double and add 1 | $\xrightarrow{8x + 1}$

EXAMPLE 6

SKILLS

INTERPRETATION

If $f(x) = 3x + 1$, find

a $f(2x)$ **b** $2f(x)$ **c** $f(x - 1)$ **d** $f(-x)$

a $f(2x) = 3(2x) + 1 = 6x + 1$
b $2f(x) = 2(3x + 1) = 6x + 2$
c $f(x - 1) = 3(x - 1) + 1 = 3x - 2$
d $f(-x) = 3(-x) + 1 = 1 - 3x$

KEY POINT

■ A function operates on all of its input.

EXERCISE 3

1 ▶ If $f(x) = 2x + 1$, find   **a** $f(-x)$ **b** $f(x + 2)$ **c** $f(x) + 2$

2 ▶ If $f(x) = 4x - 3$, find   **a** $f(x + 1)$ **b** $f(2x)$ **c** $2f(x)$

3 ▶ If $f(x) = 3 - x$, find   **a** $f(-x)$ **b** $f(-3x)$ **c** $-3f(x)$

4 ▶ If $f(x) = x^2 + 2x$, find   **a** $f(x - 1)$ **b** $f(x - 1) + 1$ **c** $1 - f(x - 1)$

5 ▶ If $f(x) = x^2 - x$, find   **a** $f(3x)$ **b** $3f(x)$ **c** $f(-x)$

6 ▶ Given $f(x) = 4 + 3x$ and $g(x) = 8 - x$, find the value of x such that $f(x) = g(3x)$

7 ▶ Given $f(x) = x^2 + 2x$ and $g(x) = x^2 + 6x + 3$
find the values of x such that $2f(x) = g(x)$

8 ▶ Given $p(x) = \dfrac{1}{3 - x}$ and $q(x) = \dfrac{1}{x + 4}$
find the value of x such that $2p(x) = q(2x)$

EXERCISE 3*

1 ▶ If $f(x) = 2 - x$, find   **a** $f(-x)$ **b** $f(-2x)$ **c** $-2f(x)$

2 ▶ If $f(x) = x^2 + 1$, find   **a** $f(x + 2)$ **b** $f(x) + 2$ **c** $f(-x)$

3 ▶ If $f(x) = 2x^2 - x$, find   **a** $f(-x)$ **b** $-f(x)$ **c** $f(2x)$

4 ▶ If $f(x) = 3 - x^2$, find   **a** $f(3x)$ **b** $3f(x)$ **c** $3f(-x)$

5 ▶ If $f(x) = \dfrac{1}{x^2}$, find   **a** $\dfrac{1}{f(x)}$ **b** $f\left(\dfrac{1}{x}\right)$ **c** $\dfrac{1}{f\left(\dfrac{1}{x}\right)}$

6 ▶ Given $f(x) = x^2 + x$ and $h(x) = 2x^2 + 2x - 18$
find the values of x such that $f(x) = h(-x)$

7 ▶ Given $p(x) = (x + 3)^2$ and $q(x) = x - 3$
find the values of x such that $p(2x) = q(-x) + 2$

8 ▶ Given $g(x) = \sqrt{2x + 1}$ and $h(x) = x - 1$
find the values of x such that $g(x + 1) = h(x) + 1$

DOMAIN AND RANGE

Here the only numbers the function can use are {1, 2, 4, 7}.

This set is called the **domain** of the function.

The set {3, 4, 6, 9} produced by the function is called the **range** of the function.

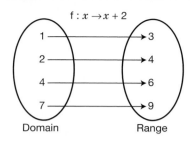

EXAMPLE 7

Find the range of the function $f: x \mapsto 2x + 1$ if the domain is $\{-1, 0, 1, 2\}$.

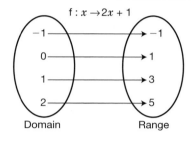

The diagram shows that the range is {−1, 1, 3, 5}.

EXAMPLE 8

Find the range of the function $g(x) = x + 2$ if the domain is $\{x: x \geq -2, x$ is an integer$\}$.

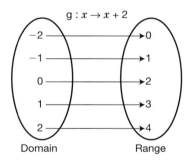

The diagram shows some of the domain. As the function changes an integer into another integer, the range is $\{y: y \geq 0, y$ is an integer$\}$.

EXAMPLE 9

If $f(x) = 2x + 5$ has a domain of $-2 \leq x \leq 2$, find the range of $f(x)$.

Range of $f(x)$ is $1 \leq f(x) \leq 9$.

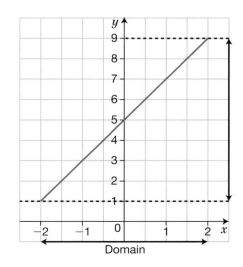

Range of f(x) is
$1 \leq f(x) \leq 9$

EXAMPLE 10

Find the range of

a $h : x \mapsto x^2$ b $f : x \mapsto x^2 + 2$

if the domain of both functions is all real numbers.

---

a Since $x^2 \geq 0 \Rightarrow$ the range is $\{y : y \geq 0, y$ a real number$\}$
b Since $x^2 \geq 0$, then $x^2 + 2 \geq 2 \Rightarrow$ the range is $\{y : y \geq 2, y$ a real number$\}$

The graph of the function gives a useful picture of the domain and range. The domain corresponds to the x-axis, and the range to the y-axis. The graphs of the functions used in Example 10 are shown below.

The first graph ($y = x^2$) shows that all the y values are greater than or equal to zero, meaning that the range is $\{y : y \geq 0, y$ a real number$\}$.

The second graph ($y = x^2 + 2$) shows that all the y values are greater than or equal to 2, meaning that the range is $\{y : y \geq 2, y$ a real number$\}$.

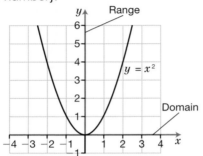

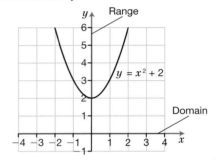

KEY POINTS

■ The graph of the function gives a useful picture of the domain and range.
■ The domain corresponds to the x-axis, and the range to the y-axis.

EXERCISE 4

For each function, find the range for the given domains.

	FUNCTION	a	b
1 ▶	x	$0 \leq x \leq 2$	$-2 \leq x \leq 2$
2 ▶	$2x$	$0 \leq x \leq 5$	$-5 \leq x \leq 5$
3 ▶	$x + 1$	$0 \leq x \leq 3$	$-3 \leq x \leq 3$
4 ▶	$2x - 3$	$0 \leq x \leq 4$	$-4 \leq x \leq 4$
5 ▶	$x - 1$	$0 \leq x \leq 1$	$-1 \leq x \leq 1$
6 ▶	$3x + 11$	$0 \leq x \leq 3$	$-3 \leq x \leq 3$
7 ▶	$4 - x$	$0 \leq x \leq 4$	$-4 \leq x \leq 4$
8 ▶	$5 - x$	$-1 \leq x \leq 1$	$-10 \leq x \leq 10$
9 ▶	$\frac{1}{2}x + \frac{1}{2}$	$-2 \leq x \leq 2$	$-10 \leq x \leq 10$
10 ▶	$3(x + 5)$	$-1 \leq x \leq 1$	$-10 \leq x \leq 10$

For each function, find the range for the given domains.

	FUNCTION	a	b
1 ▶	$2 - 3x$	$\{-2, -1, 0, 1\}$	All real numbers
2 ▶	$5 + 4x$	$\{-1, 0, 1, 2\}$	All real numbers
3 ▶	$x^2 + 2x$	$\{-2, 0, 2, 4\}$	$\{x: x \geq 0, x$ a real number$\}$
4 ▶	$x^2 - x$	$\{-2, -1, 0, 1\}$	$\{x: x \geq 4, x$ a real number$\}$
5 ▶	$(x - 1)^2 + 2$	$\{-4, -2, 0, 2\}$	All real numbers
6 ▶	$(x + 1)^2 - 2$	$\{-4, -2, 0, 2\}$	All real numbers
7 ▶	$x^3 + x$	$\{-2, 0, 2, 4\}$	$\{x: x \geq 1, x$ a real number$\}$
8 ▶	$(x - 1)^3$	$\{-4, -2, 0, 2\}$	$\{x: x \geq 0, x$ a real number$\}$
9 ▶	$\dfrac{1}{x + 1}$	$\{0, 1, 2, 3\}$	$\{x: x \geq 0, x$ a real number$\}$
10 ▶	$x + \dfrac{12}{x}$	$\{2, 4, 6, 12\}$	$\{x: x \geq 6, x$ a real number$\}$

VALUES EXCLUDED FROM THE DOMAIN

Sometimes there are values which cannot be used for the domain as they lead to impossible operations, usually division by zero or the square root of a negative number.

State which values (if any) must be excluded from the domain of these functions.

a $f(x) = \dfrac{1}{x}$ b $g(x) = \dfrac{1}{x - 2}$

a   Division by zero is not allowed, so $x = 0$ must be excluded from the domain of f (see graph).

b   Division by zero is not allowed, so $x - 2 \neq 0$ which means $x = 2$ must be excluded from the domain of g (see graph).

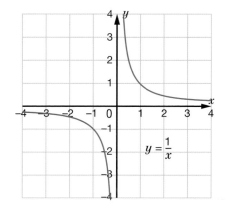

$y = \dfrac{1}{x}$

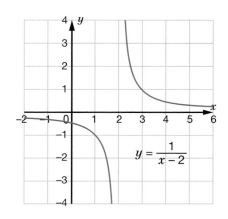

$y = \dfrac{1}{x - 2}$

EXAMPLE 12

State which values (if any) must be excluded from the domain of these functions.

a $f(x) = \sqrt{x}$ **b** $g(x) = 1 + \sqrt{x - 2}$

a The square root of a negative number is not allowed, though it is possible to square root zero, so $x < 0$ must be excluded from the domain of f (see graph).

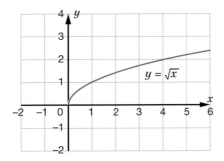

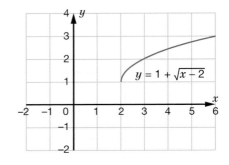

b If $x - 2 < 0$ then x must be excluded from the domain. $x - 2 < 0 \Rightarrow x < 2$, so $x < 2$ must be excluded from the domain of g (see graph).

KEY POINTS

■ Some numbers cannot be used for the domain as they lead to impossible operations.

■ These operations are usually division by zero or the square root of a negative number.

EXERCISE 5

State which values (if any) must be excluded from the domain of these functions.

1 ▶ $f : x \to \dfrac{1}{x + 1}$

2 ▶ $g : x \to \dfrac{1}{x - 1}$

3 ▶ $h : x \to \sqrt{x - 2}$

4 ▶ $f : x \to \sqrt{2 - x}$

5 ▶ $g(x) = x - \dfrac{1}{x}$

6 ▶ $h(x) = x + \dfrac{1}{x^2}$

7 ▶ $p(x) = x^2 + 3$

8 ▶ $q(x) = 5x - 1$

9 ▶ $r : x \to \sqrt{x^2 - 4}$

10 ▶ $s(x) = \sqrt{9 - x^2}$

EXERCISE 5*

State which values (if any) must be excluded from the domain of these functions.

1 ▶ $h(x) = \dfrac{5}{2x - 1}$

2 ▶ $g(x) = \dfrac{3}{4x - 3}$

3 ▶ $f : x \to \sqrt{9 - x}$

4 ▶ $h : x \to \sqrt{x + 4}$

5 ▶ $p(x) = \dfrac{1}{(x + 1)^2}$

6 ▶ $q(x) = \dfrac{1}{(1 - x)^2}$

7 ▶ $r : x \to \dfrac{1}{x^2 - 1}$

8 ▶ $s : x \to \dfrac{1}{x^2 + 1}$

9 ▶ $f(x) = \dfrac{1}{\sqrt{x + 2}}$

10 ▶ $g(x) = \dfrac{1}{\sqrt{2 - x}}$

COMPOSITE FUNCTIONS

When one function is followed by another, the result is a composite function.

If $f : x \mapsto 2x$ and $g : x \mapsto x + 3$, then

If the order of these functions is changed, then the output is different:

If x is input then

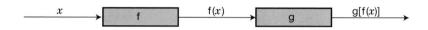

$g[f(x)]$ is usually written without the square brackets as $gf(x)$.

$gf(x)$ means do f first, followed by g.

Note that the domain of g is the range of f.

In the same way, $fg(x)$ means do g first, followed by f.

EXAMPLE 13

SKILLS

INTERPRETATION

If $f(x) = x^2$ and $g(x) = x + 2$, find

a $fg(3)$ **b** $gf(3)$ **c** $fg(x)$ **d** $gf(x)$ **e** $gg(x)$

---

a $g(3) = 5$, so $fg(3) = f(5) = 25$

b $f(3) = 9$, so $gf(3) = g(9) = 11$

c $g(x) = x + 2$, so $fg(x) = f(x + 2) = (x + 2)^2$

d $f(x) = x^2$, so $gf(x) = g(x^2) = x^2 + 2$

e $gg(x) = g(x + 2) = x + 4$

KEY POINTS

- $fg(x)$ and $gf(x)$ are composite functions.
- To work out $fg(x)$, first work out $g(x)$ and then substitute the answer into $f(x)$.
- To work out $gf(x)$, first work out $f(x)$ and then substitute the answer into $g(x)$.

EXERCISE 6

1 ▶ Find $fg(3)$ and $gf(3)$ if $f(x) = x + 5$ and $g(x) = x - 2$

2 ▶ Find $fg(1)$ and $gf(1)$ if $f(x) = x^2$ and $g(x) = x + 2$

3 ▶ Find $fg(4)$ and $gf(4)$ if $f(x) = \dfrac{1}{x}$ and $g(x) = \dfrac{1}{x + 1}$

For Questions 4–7, find

a fg(x)   **b** gf(x)   **c** ff(x)   **d** gg(x) when

4 ▶ $f(x) = x - 4$ $g(x) = x + 3$

5 ▶ $f(x) = 2x$ $g(x) = x + 2$

6 ▶ $f(x) = x^2$ $g(x) = x + 2$

7 ▶ $f(x) = x - 6$ $g(x) = x + 6$

8 ▶ $f(x) = \dfrac{x}{2}$ and $g(x) = x + 1$

       Find x if   **a** fg(x) = 4   **b** gf(x) = 4

EXERCISE 6*

1 ▶   Find fg(−3) and gf(−3) if $f(x) = 2x + 3$ and $g(x) = 5 - x$

2 ▶   Find fg(2) and gf(2) if $f(x) = x^2 + 1$ and $g(x) = (x + 1)^2$

3 ▶   Find fg(−3) and gf(−3) if $f(x) = x + \dfrac{2}{x}$ and $g(x) = \dfrac{2}{x - 1}$

For Questions 4–7, find

a fg(x)   **b** gf(x)   **c** ff(x)   **d** gg(x) when

4 ▶ $f(x) = \dfrac{x - 4}{2}$ $g(x) = 2x$

5 ▶ $f(x) = 2x^2$ $g(x) = x - 2$

6 ▶ $f(x) = \dfrac{1}{x - 2}$ $g(x) = x + 2$

7 ▶ $f(x) = 4x$ $g(x) = \sqrt{\left(\dfrac{1}{4}x + 4\right)}$

8 ▶ $f(x) = 1 + \dfrac{x}{2}$ and $g(x) = 4x + 1$

       Find x if   **a** fg(x) = 4   **b** gf(x) = 4

9 ▶ $f(x) = 1 + x^2$ and $g(x) = \dfrac{1}{x - 5}$

       What is the domain of   **a** fg(x)   **b** gf(x)?

10 ▶ $f(x) = \sqrt{2x + 4}$ and $g(x) = 4x + 2$

       What is the domain of   **a** fg(x)   **b** gf(x)?

INVERSE FUNCTION

The inverse function undoes whatever has been done by the function.

If the function is travel 1 km North, then the inverse function is travel 1 km South.

If the function is 'add one' then the inverse function is 'subtract one'.

These functions are $f : x \mapsto x + 1$ ('add one') and $g : x - 1$ ('subtract one').

If f is followed by g then whatever number is input is also the output.

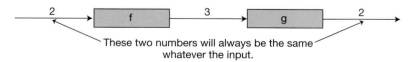

These two numbers will always be the same whatever the input.

If x is the input then x is also the output.

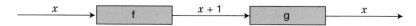

The function g is called the inverse of the function f.
The inverse of f is the function that undoes whatever f has done.
The notation f^{-1} is used for the inverse of f.

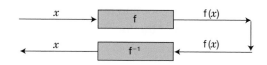

Note that graphically $f^{-1}(x)$ is a reflection of $f(x)$ in $y = x$

FINDING THE INVERSE FUNCTION

If the inverse function is not obvious, the following steps will find the inverse function.

Step 1 Write the function as $y = \ldots$

Step 2 Change any x to y and any y to x.

Step 3 Make y the **subject** giving the inverse function and use the correct $f^{-1}(x)$ notation.

EXAMPLE 14 Find the inverse of $f(x) = 2x - 5$

Step 1 $y = 2x - 5$

Step 2 $x = 2y - 5$

Step 3 $x = 2y - 5 \Rightarrow 2y = x + 5 \Rightarrow y = \frac{1}{2}(x + 5) \Rightarrow f^{-1}(x) = \frac{1}{2}(x + 5)$

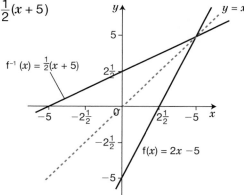

The graph shows that $f^{-1}(x) = \frac{1}{2}(x + 5)$ is a reflection
of $f(x) = 2x - 5$ in the line $y = x$

EXAMPLE 15 Find the inverse of $g(x) = 2 + \dfrac{3}{x}$

Step 1 $y = 2 + \dfrac{3}{x}$

Step 2 $x = 2 + \dfrac{3}{y}$

Step 3 $x - 2 = \dfrac{3}{y} \Rightarrow y = \dfrac{3}{x - 2} \Rightarrow g^{-1}(x) = \dfrac{3}{x - 2}$

KEY POINT

- To find the inverse function:
  - Step 1   Write the function as $y = \ldots$
  - Step 2   Change any x to y and any y to x.
  - Step 3   Make y the subject giving the inverse function and use the correct $f^{-1}(x)$ notation.
- The graph of the inverse function is the reflection of the function in the line $y = x$.

ACTIVITY 1

SKILLS

INTERPRETATION

If $f : x \mapsto x + 1$ and $g : x - 1$, show that f is the inverse of g.

If $f(x) = 2x$, show that $g(x) = \frac{x}{2}$ is the inverse of f.

Is f also the inverse of g?

If $f(x) = 4 - x$, show that f is the inverse of f.

(Functions like this are called 'self inverse'.)

EXERCISE 7

For Questions 1–8 find the inverse of the function given.

1 ▶ $f : x \mapsto 6x + 4$

2 ▶ $f(x) = \dfrac{(x + 3)}{2}$

3 ▶ $f : x \mapsto 12 - 5x$

4 ▶ $f(x) = 3(x - 6)$

5 ▶ $f : x \mapsto 6(x + 5)$

6 ▶ $g(x) = \dfrac{1}{3x + 4}$

7 ▶ $p(x) = 4 - \dfrac{3}{x}$

8 ▶ $f : x \mapsto = x^2 + 7$

9 ▶   If $f(x) = 2x - 5$, find   **a** $f^{-1}(3)$ **b** $f^{-1}(0)$ **c** $f^{-1}(-3)$

10 ▶ $f(x) = 2x + 5$. Solve the equation $f(x) = f^{-1}(x)$

EXERCISE 7*

For Questions 1–8 find the inverse of the function given.

1 ▶ $f : x \mapsto 3(x + 4) - 5$

2 ▶ $f : x \mapsto 12 - 4x$

3 ▶ $f(x) = 8(4 - 3x)$

4 ▶ $f(x) = \dfrac{3}{4 - 2x}$

5 ▶ $f : x \mapsto 4 - \dfrac{7}{x}$

6 ▶ $g(x) = \sqrt{x^2 + 7}$

7 ▶ $p : x \mapsto 2x^2 + 16$

8 ▶ $r(x) = \dfrac{2x + 3}{4 - x}$

9 ▶   If $f(x) = x^2 - 5$, find   **a** $f^{-1}(11)$ **b** $f^{-1}(44)$ **c** $f^{-1}(-5)$

10 ▶ $f(x) = 2(4x - 7)$. Solve the equation $f(x) = f^{-1}(x)$

11 ▶ $f(x) = 3 - \dfrac{2}{x}$. Solve the equation $f(x) = f^{-1}(x)$

12 ▶ $f(x) = \dfrac{3}{x + 2}$. Solve the equation $f(x) = f^{-1}(x)$

ACTIVITY 2

SKILLS

PROBLEM
SOLVING

ANCIENT CODING

One of the first people that we know of who used codes was Julius Caesar, who invented his own. He would take a message and move each letter three places down the alphabet, so that a ↦ d, b ↦ e, c ↦ f and so on. At the end of the alphabet he imagined the alphabet starting again so w ↦ z, x ↦ a, y ↦ b and z ↦ c. Spaces were ignored.

a Copy and complete this table where the first row is the alphabet and the second row is the coded letter.

A	B	C	D	E	F	G	H	I	J	K	L	M	N	O	P	Q	R	S	T	U	V	W	X	Y	Z
D	E	F																				Z	A	B	C

b Decode 'lkdwhpdwkv'.

c Code your own message and ask someone in your class to decode it.

Julius Caesar was using a function to code his message and the inverse function to decode it. The system is simple and easy to break, even if each letter is moved more than three places down the alphabet.

d Code another message, moving the letters by more than three places.

e Ask someone in your class to decode your message. Do not tell them how many places you moved the letters.

Modern computers make it easy to deal with codes like this. If you can program a spreadsheet, try using one to code and decode messages, moving the letters more than three places.

(The spreadsheet function =CODE(letter) returns a number for the letter while the function =CHAR(number) is the inverse function giving the letter corresponding to the number.)

The ideal function to use is a 'trapdoor function' which is one that is simple to use but with an inverse that is very difficult to find unless you know the key. Modern functions use the problem of factorising the **product** of two huge **prime** numbers, often over forty digits long. Multiplying the primes is straightforward, but even with the fastest modern computers the factorising can take hundreds of years!

EXERCISE 8

REVISION

1 ▶ Decide, giving a reason, if the graph shows a function or not.

a

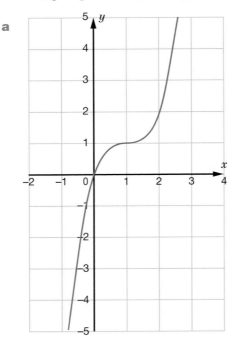

b

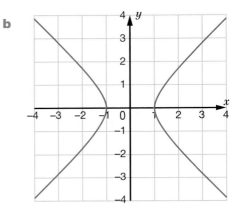

2 ▶ If $g : x \mapsto 3x + 7$, calculate   **a** $g(2)$ **b** $g(-3)$ **c** $g(0)$

3 ▶ If $g(x) = 3 - 4x$, calculate x if   **a** $g(x) = 5$ **b** $g(x) = -2$

4 ▶ If $f : x \mapsto 5x - 2$, find   **a** $f(x) + 1$ **b** $f(x + 1)$

5 ▶ Given $f(x) = 3x - 4$ and $g(x) = 2(x + 3)$, find the value of x such that $f(x) = g(x)$

6 ▶ State which values of x must be excluded from the domain of

   a $f(x) = \dfrac{1}{x - 1}$ **c** $h(x) = \sqrt{x + 1}$

   b $g : x \to \dfrac{3}{2x - 1}$ **d** $p : x \to \sqrt{3x - 6}$

7 ▶ Find the range of these functions if the domain is all real numbers.

   a $f(x) = 2x + 1$ **c** $h(x) = (x + 1)^2$

   b $g(x) = x^2 + 1$ **d** $f(x) = x^3$

8 ▶ If $f(x) = x^2 + 1$ and $g(x) = \dfrac{1}{x}$

   a Find   **i** $fg(x)$ **ii** $gf(x)$

   b Find and **simplify** $gg(x)$

9 ▶ Find the inverse of

   a $p : x \to 4(2x + 3)$ **c** $r : x \to \dfrac{1}{x + 3}$

   b $q(x) = 7 - x$ **d** $s(x) = x^2 + 4$

10 ▶ $f(x) = 4x - 3, \quad g(x) = \dfrac{x + 3}{4}$

   a Find the function $fg(x)$

   b Hence describe the relationship between the functions f and g.

EXERCISE 8*   REVISION

1 ▶   Decide, giving a reason, if the graph shows a function or not.

a

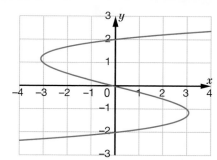

b

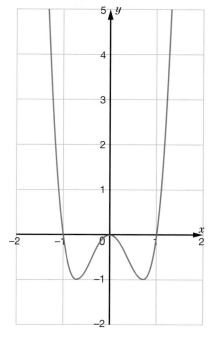

2 ▶   If $h : x \mapsto \sqrt{x + 9}$, calculate   **a** h(7)   **b** h(0)   **c** h(−9)

3 ▶   If $g(x) = x^2 - x$, calculate x if   **a** g(x) = 6   **b** g(x + 1) = 12

4 ▶   If $f : x \mapsto 5 - 2x$, find   **a** f(x) − 1   **b** f(x − 1)

5 ▶   Given $f(x) = (x + 3)^2$ and $g(x) = 2x^2 + 8x + 1$,
find the values of x such that f(x) = g(x)

6 ▶   State which values of x must be excluded from the domain of
   a $f(x) = \dfrac{5}{4 - 3x}$ **b** $g : x \rightarrow \dfrac{7}{(x + 1)^2}$ **c** $h(x) = \sqrt{2 + 5x}$ **d** $p : x \rightarrow \sqrt{x^2 - 9}$

7 ▶   Find the range of these functions if the domain is all real numbers.
   a $f(x) = 2x^2 + 3$ **b** $g(x) = (x - 2)^2$ **c** $h(x) = \sqrt{x + 2}$ **d** $f(x) = x^3 - 1$

8 ▶   If $f(x) = x^3$ and $g(x) = \dfrac{1}{x - 8}$
   a Find   **i** fg(x)   **ii** gf(x)
   b What values should be excluded from the domain of   **i** fg(x)   **ii** gf(x)?
   c Find and simplify gg(x)

9 ▶   Find the inverse of
   a $p : x \rightarrow 4(1 - 2x)$ **c** $r : x \rightarrow \sqrt{2x - 3}$
   b $q(x) = 2 - \dfrac{3}{4 - x}$ **d** $s(x) = (x - 2)^2$

10 ▶ $p(x) = \dfrac{1}{x - 2},$ $q(x) = \dfrac{1}{x} + 2$
   a Find the function pq(x)
   b Hence describe the relationship between the functions p and q.
   c Write down the exact value of $pq\sqrt{7}$.

EXAM PRACTICE: ALGEBRA 8

1 Decide, giving a reason, if the graph shows a function or not. **[3]**

a

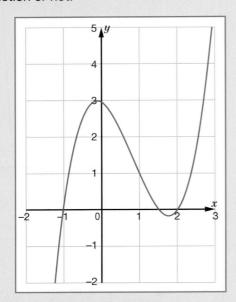

b

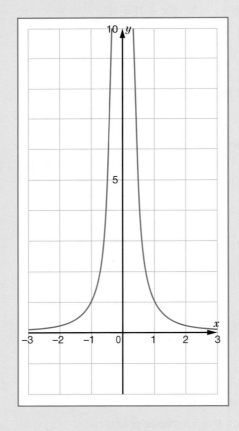

c

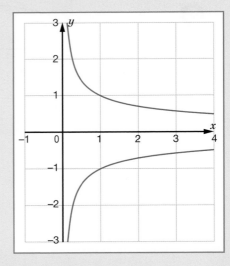

2 If $f : x \mapsto x^2 + x$ find **a** $f(2x)$ **b** $2f(x)$ **c** $f(-x)$ **[3]**

3 If the domain of $f : x \mapsto x^2$ is all real numbers, what is the range? **[2]**

4 State which values (if any) must be excluded from the domain of these functions.

a $f : x \mapsto \sqrt{x - 4}$

b $f(x) = x^2$

c $f(x) = \dfrac{2}{x - 3}$ **[3]**

5 If $f(x) = 2x - 3$ and $g(x) = x + 1$, find $gf(2)$ **[2]**

6 If $f(x) = (x + 2)^2$ and $g(x) = 2x$, find

a $fg(x)$

b $gg(x)$ **[4]**

7 Find the inverse of

a $f(x) = \dfrac{x}{2} - 3$

b $f(x) = \dfrac{2}{5 - x}$ **[4]**

8 $f(x) = 3(x + 2)$ and $g(x) = 3x - 1$
If $f^{-1}(a) + g^{-1}(a) = 1$, find a. **[4]**

[Total 25 marks]

CHAPTER SUMMARY: ALGEBRA 8

FUNCTIONS

Relationships that are 'one to one' and 'many to one' are functions. If only one arrow leaves the members of the set then the relationship is a function.

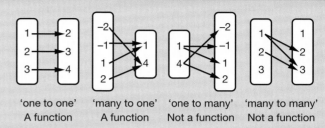

'one to one'   'many to one'   'one to many'   'many to many'
A function   A function   Not a function   Not a function

If a vertical line placed anywhere on a graph intersects the graph at only one point then the graph shows a function.

A function operates on all of the input. If the function is treble (three times the original) and add four ($f(x) = 3x + 4$) then if $2x$ is input, $6x + 4$ is output

so $f(2x) = 6x + 4$

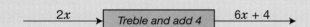

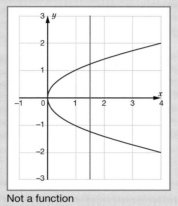

Not a function

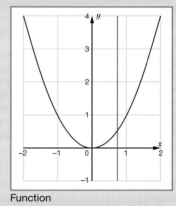

Function

DOMAIN AND RANGE

The graph of the function gives a useful picture of the domain and range.
The domain corresponds to the x-axis, and the range to the y-axis.
Some numbers cannot be used for the domain as they lead to impossible operations, usually division by zero or the square root of a negative number.

The domain of $f(x) = \dfrac{1}{x}$ is all real numbers except zero as division by zero is not possible.

COMPOSITE FUNCTIONS

$gf(x)$ means do f first, followed by g.
$fg(x)$ means do g first, followed by f.
If $f(x) = x^2$ and $g(x) = x + 1$ then $gf(x) = x^2 + 1$ and $fg(x) = (x + 1)^2$

INVERSE FUNCTION

To find the inverse function:
Step 1   Write the function as $y = \ldots$
Step 2   Change any x to y and any y to x.
Step 3   Make y the subject giving the inverse function and use the
        correct $f^{-1}(x)$ notation.
The graph of the inverse function is the reflection of the function in the line $y = x$

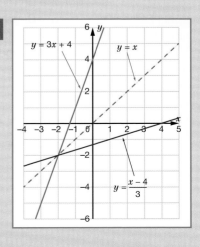

Find the inverse of $f(x) = 3x + 4$
Step 1 $y = 3x + 4$
Step 2 $x = 3y + 4$
Step 3 $y = \dfrac{x - 4}{3}$ $\Rightarrow f^{-1}(x) = \dfrac{x - 4}{3}$

GRAPHS 7

Every time you plot a graph you are using the Cartesian co-ordinate system named after René Descartes (1596–1650). The idea for the co-ordinate system came to him when he was ill. Lying in bed watching a fly buzzing around, he realised that he could describe the fly's position using three numbers: how far along one wall, how far across the adjacent wall and how far up from the floor. For a graph on a sheet of paper, only two numbers are needed.

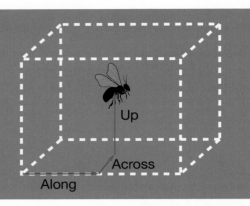

LEARNING OBJECTIVES

- Use graphs to solve quadratic equations
- Use graphs to solve cubic equations
- Use a graphical method to solve simultaneous equations with one linear equation and one non-linear equation

BASIC PRINCIPLES

- Plot graphs of linear, quadratic, cubic and reciprocal functions using a table of values.
- Use graphs to solve quadratic equations of the form $ax^2 + bx + c = 0$
- Solve a pair of linear simultaneous equations graphically (recognising that the solution is the point of intersection).

USING GRAPHS TO SOLVE QUADRATIC EQUATIONS

An accurately drawn graph can be used to solve equations that may be difficult to solve by other methods.

The graph of $y = x^2$ is easy to draw and can be used to solve many quadratic equations.

EXAMPLE 1

SKILLS

PROBLEM SOLVING

Here is the graph of $y = x^2$. By drawing a suitable straight line on the graph, solve the equation $x^2 - x - 3 = 0$, giving answers **correct to** 1 d.p.

Rearrange the equation so that one side is x^2.

$x^2 - x - 3 = 0$

$x^2 = x + 3$

Draw the line $y = x + 3$

Find where $y = x^2$ **intersects** $y = x + 3$

The graph shows the solutions are $x = -1.3$ or $x = 2.3$

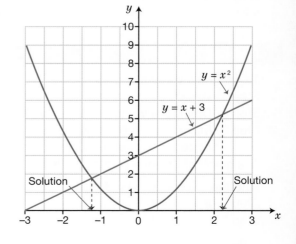

EXAMPLE 2

SKILLS

PROBLEM
SOLVING

Here is the graph of $y = x^2$. By drawing a suitable straight line on the graph, solve the equation $2x^2 + x - 8 = 0$, giving answers correct to 1 d.p.

Rearrange the equation so that one side is x^2.

$$2x^2 + x - 8 = 0$$

$$x^2 = 4 - \frac{1}{2}x$$

Draw the line $y = 4 - \frac{1}{2}x$

Find where $y = x^2$ intersects $y = 4 - \frac{1}{2}x$

The graph shows the solutions are $x = -2.3$ or $x = 1.8$

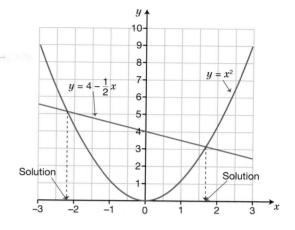

KEY POINTS

- The graph of $y = x^2$ can be used to solve quadratic equations of the form $ax^2 + bx = c = 0$
- Rearrange the equation so that $x^2 = f(x)$, where $f(x)$ is a linear **function**.
- Draw $y = f(x)$ and find the x co-ordinates of the intersection points of the curve $y = x^2$ and the line $y = f(x)$

EXERCISE 1

Draw an accurate graph of $y = x^2$ for $-4 \le x \le 4$. Use your graph to solve these equations.

1 ▶ $x^2 - 5 = 0$ 3 ▶ $x^2 + 2x - 7 = 0$ 5 ▶ $2x^2 - x - 20 = 0$

2 ▶ $x^2 - x - 2 = 0$ 4 ▶ $x^2 - 4x + 2 = 0$ 6 ▶ $3x^2 + x - 1 = 0$

EXERCISE 1*

Draw an accurate graph of $y = x^2$ for $-4 \le x \le 4$. Use your graph to solve these equations.

1 ▶ $x^2 - x - 3 = 0$ 3 ▶ $x^2 - 4x + 4 = 0$ 5 ▶ $3x^2 - x - 27 = 0$

2 ▶ $x^2 + 3x + 1 = 0$ 4 ▶ $2x^2 + x - 12 = 0$ 6 ▶ $4x^2 + 3x - 6 = 0$

EXAMPLE 3

SKILLS

PROBLEM
SOLVING

Here is the graph of $y = x^2 - 5x + 5$ for $0 \le x \le 5$

By drawing suitable straight lines on the graph, solve these equations, giving answers to 1 d.p.

a $0 = x^2 - 5x + 5$
b $0 = x^2 - 5x + 3$
c $0 = x^2 - 4x + 4$

a   Find where $y = x^2 - 5x + 5$ intersects $y = 0$ (the x-axis).

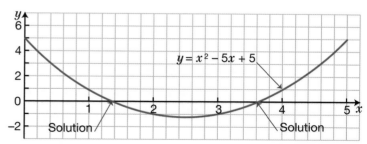

The graph shows the solutions are $x = 1.4$ and $x = 3.6$ to 1 d.p.

b Rearrange the equation so that one side is
$x^2 - 5x + 5$

$0 = x^2 - 5x + 3$ (Add 2 to both sides)

$2 = x^2 - 5x + 5$

Find where $y = x^2 - 5x + 5$ intersects $y = 2$

The graph shows the solutions are
$x = 0.7$ and $x = 4.3$ to 1 d.p.

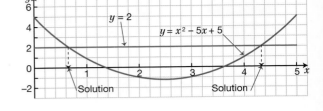

c Rearrange the equation so that one side is
$x^2 - 5x + 5$

$0 = x^2 - 4x + 4$ (Add 1 to both sides)

$1 = x^2 - 4x + 5$ (Subtract x from both sides)

$1 - x = x^2 - 5x + 5$

Find where $y = x^2 - 5x + 5$ intersects $y = 1 - x$

The graph shows the solution is $x = 2$ to 1.d.p.

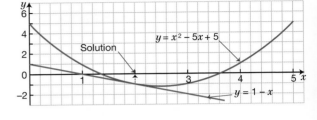

Note: If the line does not cut the graph, there will be no real solutions.

KEY POINT

■ The graph of one quadratic equation can be used to solve other quadratic equations with suitable rearrangement.

EXERCISE 2

1 ▶ Draw the graph of $y = x^2 - 3x$ for $-1 \leq x \leq 5$

Use your graph to solve these equations.

    a $x^2 - 3x = 0$ **c** $x^2 - 3x = -1$ **e** $x^2 - 3x - 3 = 0$

    b $x^2 - 3x = 2$ **d** $x^2 - 3x = x + 1$ **f** $x^2 - 5x + 1 = 0$

2 ▶ Draw the graph of $y = x^2 - 4x + 3$ for $-1 \leq x \leq 5$

Use your graph to solve these equations.

    a $x^2 - 4x + 3 = 0$ **c** $x^2 - 5x + 3 = 0$

    b $x^2 - 4x - 2 = 0$ **d** $x^2 - 3x - 2 = 0$

3 ▶ Find the equations solved by the intersection of these pairs of graphs.

    a $y = 2x^2 - x + 2$, $y = 3 - 3x$

    b $y = 4 - 3x - x^2$, $y = 2x - 1$

4 ▶ Using a graph of $y = 3x^2 + 4x - 2$, find the equations of the lines that should be drawn to solve these equations.

    a $3x^2 + 2x - 4 = 0$

    b $3x^2 + 3x - 2 = 0$

    c $3x^2 + 7x + 1 = 0$

5 ▶ Romeo is throwing a rose up to Juliet's balcony.
The balcony is 2 m away from him and 3.5 m above him.
The equation of the path of the rose is $y = 4x - x^2$, where
the origin is at Romeo's feet.

   a Find by a graphical method where the rose lands.

   b The balcony has a 1 m high wall. Does the rose pass
over the wall?

6 ▶ A cat is sitting on a 2 m high fence when it sees a mouse
1.5 m away from the foot of the fence The cat leaps along
the path $y = -0.6x - x^2$, where the origin is where the cat
was sitting and x is measured in metres. Find, by a
graphical method, whether the cat lands on the mouse.

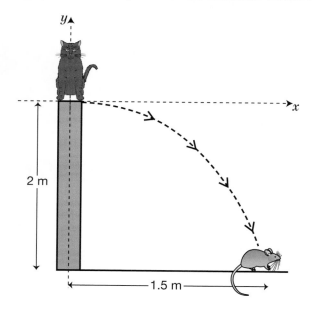

EXERCISE 2*

1 ▶ Draw the graph of $y = 5x - x^2$ for $-1 \le x \le 6$

Use your graph to solve these equations.

   a $5x - x^2 = 0$ **b** $5x - x^2 = 3$ **c** $5x - x^2 = x + 1$ **d** $x^2 - 6x + 4 = 0$

2 ▶ Draw the graph of $y = 2x^2 + 3x - 1$ for $-3 \le x \le 2$

Use your graph to solve these equations.

   a $2x^2 + 3x - 1 = 0$ **b** $2x^2 + 3x - 4 = 0$ **c** $2x^2 + 5x + 1 = 0$

3 ▶ Find the equations solved by the intersection of these pairs of graphs.

   a $y = 6x^2 - 4x + 3$, $y = 3x + 5$ **b** $y = 7 + 2x - 5x^2$, $y = 3 - 5x$

4 ▶ Using a graph of $y = 5x^2 - 9x - 6$, find the equations of the lines that should be drawn
to solve these equations.

   a $5x^2 - 10x - 8 = 0$ **b** $5x^2 - 7x - 5 = 0$

5 ▶ Jason is serving in tennis. He hits the ball from a height of 2.5 m and the path of the ball is given by $y = -0.05x - 0.005x^2$, where the origin is the point where he hits the ball.

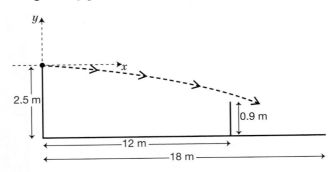

   a The net is 0.9 m high and is 12 m away. Does the ball pass over the net?

   b For the serve to be allowed it must land between the net and the service line, which is 18 m away. Is the serve allowed?

6 ▶ A food parcel is dropped by a low-flying aeroplane flying over sloping ground. The path of the food parcel is given by $y = 40 - 0.005x^2$ and the slope of the ground is given by $y = 0.2x$. Use a graphical method to find the co-ordinates of the point where the food parcel will land. (Use $0 \le x \le 100$)

USING GRAPHS TO SOLVE OTHER EQUATIONS

EXAMPLE 4

SKILLS

PROBLEM
SOLVING

Here is the graph of $y = x^3$

By drawing suitable straight lines on the graph, solve these equations, giving the answers to 1 d.p.

   a $x^3 + 2x - 4 = 0$

   b $x^3 - 3x + 1 = 0$

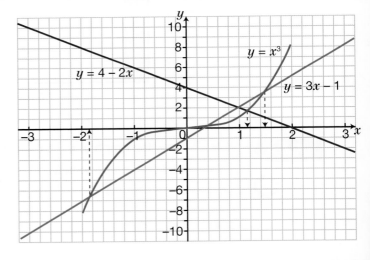

a Rearrange the equation so that one side is x^3

$x^3 + 2x - 4 = 0$ (Add 4 to both sides)

$x^3 + 2x = 4$ (Subtract $2x$ from both sides)

$x^3 = 4 - 2x$

Find where $y = x^3$ and $y = 4 - 2x$ intersect.

The graph shows that there is only one solution.

The graph shows the solution is $x = 1.2$ to 1 d.p.

b Rearrange the equation so that one side is x^3

$x^3 - 3x + 1 = 0$ (Subtract 1 from both sides)

$x^3 - 3x = -1$ (Add $3x$ to both sides)

$x^3 = 3x - 1$

Find where $y = x^3$ and $y = 3x - 1$ intersect.

The graph shows that there are three solutions.

The graph shows the solutions are $x = -1.9$, $x = 0.4$ or $x = 1.5$ to 1 d.p.

EXERCISE 3

1 ▶　**a** Draw the graph of $y = x^3$ for $-3 \le x \le 3$

　　b Use your graph to solve these equations.
　　　i $x^3 - 3x = 0$　　　**ii** $x^3 - 3x - 1 = 0$　　　**iii** $x^3 - 2x + 1 = 0$

2 ▶　**a** Copy and complete this table of values for $y = x^3 - 5x + 1$, giving values to 1 d.p.

x	–3	–2.5	–2	–1.5	–1	–0.5	0	0.5	1	1.5	2	2.5	3
y		–2.1		5.1		3.4		–1.4		–3.1		4.1	

　　b Draw the graph of $y = x^3 - 5x + 1$ for $-3 \le x \le 3$

　　c Use your graph to solve these equations.
　　　i $x^3 - 5x + 1 = 0$　　**ii** $x^3 - 5x - 2 = 0$　　　**iii** $x^3 - 7x - 1 = 0$

3 ▶　**a** Copy and complete this table of values for $y = \dfrac{6}{x}$

x	–3	–2.5	–2	–1.5	–1	–0.5	0.5	1	1.5	2	2.5	3
y		–2.4		–4		–12		6				2

　　b Draw the graph of $y = \dfrac{6}{x}$ for $-3 \le x \le 3$ where $x \ne 0$

　　c Use your graph to solve these equations.

　　　i $\dfrac{6}{x} - 5 = 0$　　　　**ii** $\dfrac{6}{x} - 2x - 1 = 0$

4 ▶　The graph of $y = x^3 + 3x - 4$ has been drawn. What lines should be drawn on this graph to solve the following equations?

　　a $x^3 + 3x + 1 = 0$　　　　**b** $x^3 + x - 4 = 0$　　　　**c** $x^3 - 3x + 4 = 0$

5 ▶　The graph of $y = \dfrac{4}{x} + x^2$ has been drawn. What lines should be drawn on this graph to solve the following equations?

　　a $\dfrac{4}{x} + x^2 - 6 = 0$　　**b** $\dfrac{4}{x} + x^2 + 2x - 7 = 0$　　**c** $\dfrac{4}{x} + x + 1 = 0$

EXERCISE 3*

1 ▶　**a** Draw the graph of $y = 3x^2 - x^3 - 1$ for $-2 \le x \le 3$

　　b Use your graph to solve these equations.
　　　i $3x^2 - x^3 - 1 = 0$　　**ii** $3x^2 - x^3 - 4 = 0$　　　**iii** $3x^2 - x^3 - 4 + x = 0$

2 ▶　**a** Draw the graph of $y = x^4 - 4x^2 + 2$ for $-3 \le x \le 3$

　　b Use your graph to solve these equations.
　　　i $x^4 - 4x^2 + 1 = 0$　　**ii** $x^4 - 4x^2 - 2x + 3 = 0$　　**iii** $2x^4 - 8x^2 + x + 2 = 0$

3 ▶　**a** Draw the graph of $y = \dfrac{12}{x^2}$ for $-5 \le x \le 5$ where $x \ne 0$

　　b Use your graph to solve these equations.

　　　i $\dfrac{12}{x^2} - x - 2 = 0$　　**ii** $\dfrac{12}{x^2} + x - 5 = 0$　　　**iii** $12 - x^3 + x^2 = 0$

4 ▶　The graph of $y = 3x^3 + 6x^2 - 5x + 3$ has been drawn. What lines should be drawn on this graph to solve the following equations?

　　a $3x^3 + 6x^2 - 1 = 0$　　**b** $3x^3 + 6x^2 - 2x + 5 = 0$　　**c** $x^3 + 2x^2 - 2x + 1 = 0$

5 ▶   The graph of $y = x^2 + \dfrac{16}{x}$ has been drawn. What lines should be drawn on this graph to solve the following equations?

a $x^3 - x^2 + 16 = 0$ b $x^3 - 3x^2 - 8x + 16 = 0$

USING GRAPHS TO SOLVE NON-LINEAR SIMULTANEOUS EQUATIONS

You can use a graphical method to solve a pair of **simultaneous equations** where one equation is linear and the other is non-linear.

ACTIVITY 1

Mary is watering her garden with a hose. Her little brother, Peter, is annoying her so she tries to spray him with water.

The path of the water jet is given by

$y = 2x - \dfrac{1}{4}x^2$

The slope of the garden is given by

$y = \dfrac{1}{4}x - 1$

Peter is standing at (8, 1)

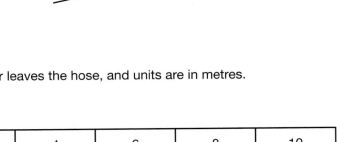

The origin is the point where the water leaves the hose, and units are in metres.

Copy and complete these tables.

x	0	2	4	6	8	10
$2x$			8			
$-\dfrac{1}{4}x^2$				–9		
$y = 2x - \dfrac{1}{4}x^2$		3				

x	0	4	8
$\dfrac{1}{4}x$			2
$y = \dfrac{1}{4}x - 1$	–1		

On one set of axes, draw the two graphs representing the path of the water and the slope of the garden.

Does the water hit Peter? Give a reason for your answer.

Mary changes the angle of the hose so that the path of the water is given by $y = x - 0.1x^2$

Draw in the new path. Does the water hit Peter this time?

In Activity 1, the simultaneous equations $y = 2x - \frac{1}{4}x^2$ and $y = \frac{1}{4}x - 1$ were solved graphically by drawing both graphs on the same axes and finding the x co-ordinates of the points of intersection.

Some non-linear simultaneous equations can be solved algebraically and this is the preferred method as it gives accurate solutions. When this is impossible then graphical methods must be used.

EXAMPLE 5

Solve the simultaneous equations $y = x^2 - 5$ and $y = x + 1$ graphically.

SKILLS

PROBLEM SOLVING

Construct a tables of values and draw both graphs on one set of axes.

x	−3	−2	−1	0	1	2	3
$x^2 - 5$	4	−1	−4	−5	−4	−1	4

x	−3	0	3
$x + 1$	−2	1	4

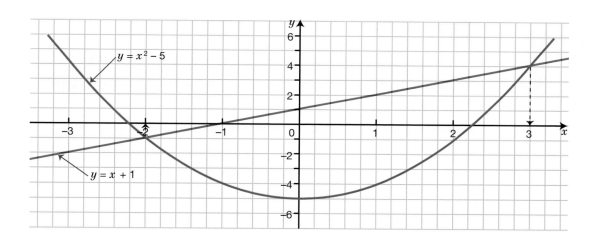

The co-ordinates of the intersection points are (−2, −1) and (3, 4) so the solutions are

$x = -2, y = -1$ or $x = 3, y = 4$

KEY POINT

■ To solve simultaneous equations graphically, draw both graphs on one set of axes.
The co-ordinates of the intersection points are the solutions of the simultaneous equations.

EXERCISE 4

Solve the simultaneous equations graphically, drawing graphs from $-4 \leq x \leq 4$

1 ▶ $y = 4 - x^2, y = 1 + 2x$

2 ▶ $y = x^2 + 2x - 1, 1 + 3x - y = 0$

3 ▶ $y = x^2 - 4x + 6, y + 2 = 2x$

4 ▶ $x^2 + y = 4, y = 1 - \dfrac{x}{4}$

5 ▶ $y = \dfrac{4}{x}, y + 1 = x$

6 ▶ $y = x^3 + 2x^2, y - 1 = \dfrac{1}{2}x$

EXERCISE 4*

Solve the simultaneous equations graphically.

1 ▶ $y = x^2 - x - 5, y = 1 - 2x$

2 ▶ $y = 2x^2 - 2x - 4, y = 6 - x$

3 ▶ $y = 10x^2 + 3x - 4, y = 2x - 2$

4 ▶ $(x + 1)^2 + y = 6, y = x + 3$

5 ▶ $y = x^3 - 4x^2 + 5, y = 3 - 2x$

6 ▶ $y = \dfrac{10}{x} + 4, y = 5x + 2$

EXERCISE 5

REVISION

1 ▶ An emergency rocket is launched out to sea from the top of a 50 m high cliff.

Taking the origin at the top of the cliff, the path of the rocket is given by

$y = x - 0.01x^2$

Use a graphical method to find where the rocket lands in the sea.

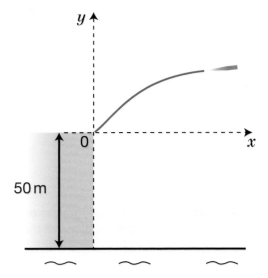

2 ▶ Draw the graph of $y = x^2 - 2x - 1$ for $-2 \leq x \leq 4$. Use the graph to solve these equations.

a $x^2 - 2x - 1 = 0$

b $x^2 - 2x - 4 = 0$

c $x^2 - x - 3 = 0$

3 ▶ The graph of $y = 3x^2 - x + 1$ has been drawn. What lines should be drawn to solve the following equations?

a $3x^2 - x - 2 = 0$

b $3x^2 + x - 4 = 0$

4 ▶ a Find the equation that is solved by finding the intersection of the graph of $y = 2x^2 - x + 2$ with the graph of $y = 2x + 3$

b Find the equation of the line that should be drawn on the graph of $y = 2x^2 - x + 2$ to solve the equation $2x^2 - 4x = 0$

5 ▶ The graph of $y = 2x^3 + 3x - 5$ has been drawn. What lines should be drawn on this graph to solve the following equations?

a $2x^3 + 3x - 9 = 0$

b $2x^3 - 2x - 5 = 0$

c $2x^3 + 6x - 7 = 0$

6 ▶ Solve the simultaneous equations $y = 1 + 3x - x^2$ and $y = 3 - x$ graphically. Plot your graphs for $-1 \le x \le 4$ and give your answers to 1 d.p.

EXERCISE 5*

REVISION

1 ▶ Draw the graph of $y = 5 + 3x - 2x^2$ for $-2 \le x \le 4$
Use the graph to solve these equations.

a $2 + 3x - 2x^2 = 0$

b $7 + x - 2x^2 = 0$

c $2 + 2x - x^2 = 0$

2 ▶ The graph of $y = 4x^2 + 2x - 4$ has been drawn. What lines should be drawn to solve the following equations?

a $4x^2 - x - 3 = 0$

b $2x^2 + 3x - 5 = 0$

3 ▶ The graph of $y = 6x^3 - 3x^2 + 12x - 18$ has been drawn. What lines should be drawn to solve the following equations?

a $6x^3 - 3x^2 - 18 = 0$

b $6x^3 - 3x^2 + 16x - 38 = 0$

c $2x^3 - x^2 + x - 1 = 0$

4 ▶ a Find the equation that is solved by the intersection of the graph of $y = 2x^3 - 6x^2 - 5x + 7$ with the graph of $y = 2 + 3x - 5x^2$

b Find the equation of the line that should be drawn on the graph of $y = 2x^3 - 6x^2 - 5x + 7$ to solve the equation $2x^3 - 5x + 5 = 0$

5 ▶ Solve the simultaneous equations $y = x^3$ and $y = 4 - 4x^2$ graphically.

6 ▶ The area of a rectangle is $30\,cm^2$ and the **perimeter** is $24\,cm$. If x is the length of the rectangle and y is the width, form two equations for x and y and solve them graphically to find the dimensions of the rectangle.

EXAM PRACTICE: GRAPHS 7

1 **a** Draw the graph of $y = x^2 - 2x$ for $-2 \leq x \leq 4$, by copying and completing the table below.

x	−2	−1	0	1	2	3	4
y	8		0				8

b By drawing suitable lines on your graph, solve

i $x^2 - 2x = 1 - x$

ii $x^2 - 4x + 2 = 0$ [8]

2 If the graph of $y = 3x^2 - 3x + 5$ has been drawn, find the equations of the lines that should be drawn to solve these equations.

a $3x^2 - 4x - 1 = 0$

b $3x^2 - 2x - 2 = 0$

c $3x^2 + x - 3 = 0$ [6]

3 If the graph of $y = 5x^3 - x^2 + 4x + 1$ has been drawn, find the equations of the lines that should be drawn to solve these equations.

a $5x^3 - x^2 + 1 = 0$

b $5x^3 - x^2 + 6x - 3 = 0$ [4]

4 **a** Draw the graph of $y = 4 + 2x - x^2$ for $-2 \leq x \leq 4$, by copying and completing the table below.

x	−2	−1	0	1	2	3	4
y			4				−4

b Use this graph to solve the simultaneous equations
$y = 4 + 3x - x^2$ and $x + 2y = 6$, giving your answers to 1 d.p. [7]

[Total 25 marks]

CHAPTER SUMMARY: GRAPHS 7

USING GRAPHS TO SOLVE QUADRATIC EQUATIONS

The graph of $y = x^2$ can be used to solve quadratic equations of the form $ax^2 + bx + c = 0$

Rearrange the equation so that $x^2 = f(x)$, where $f(x)$ is a linear function.

Draw $y = f(x)$ and find the x co-ordinates of the intersection points of the curve $y = x^2$ and the line $y = f(x)$

To solve $x^2 + 2x - 2 = 0$, rearrange the equation so that one side is x^2

$x^2 = 2 - 2x$

Draw the line $y = 2 - 2x$ and find where it intersects $y = x^2$

The graph shows the solutions are $x \approx -2.7$ or $x \approx 0.7$

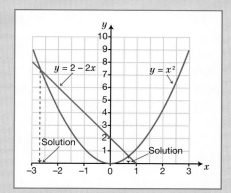

USING GRAPHS TO SOLVE OTHER EQUATIONS

The graph of one quadratic equation can be used to solve other quadratic equations with suitable rearrangement.

If the graph of $y = x^2 - 3x - 4$ has been drawn, then the x co-ordinates of the intersection with $y = x - 1$ will solve
$x^2 - 3x - 4 = x - 1$ or $x^2 - 4x - 3 = 0$

The graph show that the solutions are $x \approx -0.6$ and $x \approx 4.6$

The graph of one cubic equation can be used to solve other cubic equations with suitable rearrangement.

If the graph of $y = x^3 - 2x^2 + 4x - 3$ has been drawn, then the x co-ordinates of the intersection with $y = 2x - 5$ will solve $x^3 - 2x^2 + 4x - 3 = 2x - 5$ or
$x^3 - 2x^2 + 2x + 2 = 0$

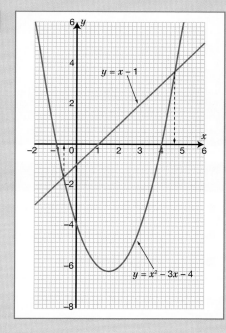

USING GRAPHS TO SOLVE NON-LINEAR SIMULTANEOUS EQUATIONS

To solve simultaneous equations graphically, draw both graphs on one set of axes. The co-ordinates of the intersection points are the solutions of the simultaneous equations.

To solve $y = x^3 + 1$ and $y = \dfrac{1}{x}$ simultaneously draw both graphs.

The graphs show the solutions are approximately $(-1.2, -0.8)$ and $(0.7, 1.4)$

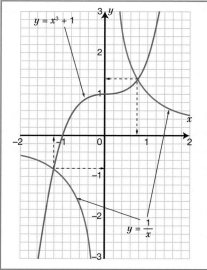

SHAPE AND SPACE 8

Vectors are used to make calculations more direct, sometimes involving motion and forces. Present-day analysis by algebra was first explained by Josiah Willard Gibbs (1839–1903). A modern application is when air traffic controllers describe aircraft motion along vector routes.

LEARNING OBJECTIVES

- Understand and use vector notation
- Calculate the magnitude of a vector
- Calculate using vectors and represent the solutions graphically
- Use the scalar multiple of a vector

- Calculate the resultant of two or more vectors
- Solve geometric problems in two dimensions using vector methods
- Apply vector methods for simple geometric proofs

BASIC PRINCIPLES

- Understand and use of **column vectors**.
- Know what a **resultant vector** is.
- A knowledge of **bearings**.

- Understand and use of Pythagoras' theorem.
- **Sketch** 2D shapes.
- Knowledge of the properties of **quadrilaterals** and **polygons**.

VECTORS AND VECTOR NOTATION

A **vector** has both **magnitude** (size) and direction and can be represented by an arrow.
A **scalar** has only size.

Vectors: Force, **velocity**, 10 km on a 060° bearing, **acceleration**…

Scalars: Temperature, time, area, length…

In this book, vectors are written as bold letters (such as **a**, **p** and **x**) or capitals covered by an arrow (such as $\overrightarrow{AB}$, $\overrightarrow{PQ}$ and $\overrightarrow{XY}$). In other books, you might find vectors written as bold italic letters (**a**, **p**, **x**). When hand-writing vectors, they are written with a wavy or straight underline (a͟, p͟, x͟ or a̲, p̲, x̲).

On co-ordinate axes, a vector can be described by a column vector, which can be used to find the magnitude and angle of the vector.

The magnitude of a vector is its length. So, the magnitude of the vector $\begin{pmatrix} x \\ y \end{pmatrix}$ is $\sqrt{x^2 + y^2}$

The magnitude of the vector **a** is often written as |a|.

EXAMPLE 1

SKILLS

PROBLEM SOLVING

a Express vector **s** as a column vector.
b Find the magnitude of vector **s**.
c Calculate the size of angle x.

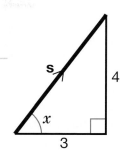

a $\mathbf{s} = \begin{pmatrix} 3 \\ 4 \end{pmatrix}$

b $|\mathbf{s}| = \sqrt{3^2 + 4^2} = \sqrt{25} = 5$

c $\tan x = \dfrac{4}{3} \Rightarrow x = 53.1°$ (to 3 s.f.)

EXAMPLE 2

SKILLS

PROBLEM SOLVING

a Express vector $\overrightarrow{PQ}$ as a column vector.
b Find the magnitude of vector $\overrightarrow{PQ}$.
c Calculate the size of angle y.

a $\overrightarrow{PQ} = \begin{pmatrix} 6 \\ -3 \end{pmatrix}$

b Length of $\overrightarrow{PQ} = \sqrt{6^2 + (-3)^2} = \sqrt{45} = 6.71$ (to 3 s.f.)

c $\tan y = \dfrac{3}{6} \Rightarrow y = 26.6°$ (to 3 s.f.)

KEY POINT

■ The magnitude of the vector $\begin{pmatrix} x \\ y \end{pmatrix}$ is its length: $\sqrt{x^2 + y^2}$

ACTIVITY 1

SKILLS

PROBLEM SOLVING

Franz and Nina are playing golf. Their shots to the hole (H) are shown as vectors.

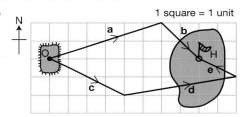

Copy and complete this table by using the grid on the right.

VECTOR	COLUMN VECTOR	MAGNITUDE (TO 3 s.f.)	BEARING
a	$\begin{pmatrix} 6 \\ 2 \end{pmatrix}$	6.32	072°
b			
c			
d			
e			

Write down the vector $\overrightarrow{OH}$ and state if there is a connection between $\overrightarrow{OH}$ and the vectors **a**, **b** and vectors **c**, **d**, **e**.

ADDITION OF VECTORS

The result of adding a set of vectors is the vector representing their total effect. This is the resultant of the vectors.

To add a number of vectors, they are placed end to end so that the next vector starts where the last one finished. The resultant vector joins the start of the first vector to the end of the last one.

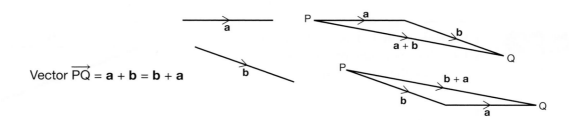

Vector $\overrightarrow{PQ} = \mathbf{a} + \mathbf{b} = \mathbf{b} + \mathbf{a}$

PARALLEL AND EQUIVALENT VECTORS

Two vectors are parallel if they have the same direction but not necessarily equal length.

For example, these vectors **a** and **b** are parallel.

$\mathbf{a} = \begin{pmatrix} 3 \\ 2 \end{pmatrix}$ $\mathbf{b} = \begin{pmatrix} 6 \\ 4 \end{pmatrix}$

Two vectors are equivalent if they have the same direction and length.

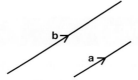

MULTIPLICATION OF A VECTOR BY A SCALAR

When a vector is multiplied by a scalar, its length is multiplied by this number; but its direction is unchanged, unless the scalar is negative, in which case the direction is reversed.

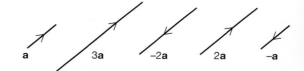

- $\mathbf{a} = 2\mathbf{b} \Rightarrow \mathbf{a}$ is parallel to **b** and **a** is twice as long as **b**.
- $\mathbf{a} = k\mathbf{b} \Rightarrow \mathbf{a}$ is parallel to **b** and **a** is k times as long as **b**.

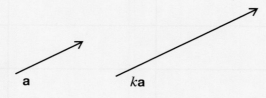

ADDITION, SUBTRACTION AND MULTIPLICATION OF VECTORS

Vectors can be added, subtracted and multiplied using their components.

EXAMPLE 3

SKILLS

ANALYSIS

$$\mathbf{s} = \begin{pmatrix} 1 \\ 2 \end{pmatrix}, \mathbf{t} = \begin{pmatrix} 3 \\ 0 \end{pmatrix} \text{ and } \mathbf{u} = \begin{pmatrix} -2 \\ 5 \end{pmatrix}$$

a Express in column vectors

$$\mathbf{p} = \mathbf{s} + \mathbf{t} + \mathbf{u}, \mathbf{q} = \mathbf{s} - 2\mathbf{t} - \mathbf{u} \text{ and } \mathbf{r} = 3\mathbf{s} + \mathbf{t} - 2\mathbf{u}$$

b Sketch the resultants **p**, **q** and **r** accurately.

c Find their magnitudes.

a CALCULATION	**b** SKETCH	**c** MAGNITUDE
$\mathbf{p} = \mathbf{s} + \mathbf{t} + \mathbf{u}$ $= \begin{pmatrix} 1 \\ 2 \end{pmatrix} + \begin{pmatrix} 3 \\ 0 \end{pmatrix} + \begin{pmatrix} -2 \\ 5 \end{pmatrix}$ $= \begin{pmatrix} 2 \\ 7 \end{pmatrix}$		Length of **p** $= \sqrt{2^2 + 7^2}$ $= \sqrt{53}$ $= 7.3 \text{ to 1 d.p.}$
$\mathbf{q} = \mathbf{s} - 2\mathbf{t} - \mathbf{u}$ $= \begin{pmatrix} 1 \\ 2 \end{pmatrix} - 2\begin{pmatrix} 3 \\ 0 \end{pmatrix} - \begin{pmatrix} -2 \\ 5 \end{pmatrix}$ $= \begin{pmatrix} 1 \\ 2 \end{pmatrix} + \begin{pmatrix} -6 \\ 0 \end{pmatrix} + \begin{pmatrix} 2 \\ -5 \end{pmatrix}$ $= \begin{pmatrix} -3 \\ -3 \end{pmatrix}$		Length of **q** $= \sqrt{(-3)^2 + (-3)^2}$ $= \sqrt{18}$ $= 4.2 \text{ to 1 d.p.}$
$\mathbf{r} = 3\mathbf{s} + \mathbf{t} - 2\mathbf{u}$ $= 3\begin{pmatrix} 1 \\ 2 \end{pmatrix} + \begin{pmatrix} 3 \\ 0 \end{pmatrix} - 2\begin{pmatrix} -2 \\ 5 \end{pmatrix}$ $= \begin{pmatrix} 3 \\ 6 \end{pmatrix} + \begin{pmatrix} 3 \\ 0 \end{pmatrix} + \begin{pmatrix} 4 \\ -10 \end{pmatrix}$ $= \begin{pmatrix} 10 \\ -4 \end{pmatrix}$		Length of **r** $= \sqrt{10^2 + (-4)^2}$ $= \sqrt{116}$ $= 10.8 \text{ to 1 d.p.}$

EXERCISE 1

1 ▶ Given that $\mathbf{p} = \begin{pmatrix} 2 \\ 3 \end{pmatrix}$ and $\mathbf{q} = \begin{pmatrix} 4 \\ 5 \end{pmatrix}$

simplify and express $\mathbf{p} + \mathbf{q}$, $\mathbf{p} - \mathbf{q}$ and $2\mathbf{p} + 3\mathbf{q}$ as column vectors.

2 ▶ Given that $\mathbf{u} = \begin{pmatrix} 1 \\ 2 \end{pmatrix}$, $\mathbf{v} = \begin{pmatrix} -4 \\ 3 \end{pmatrix}$ and $\mathbf{w} = \begin{pmatrix} 2 \\ -5 \end{pmatrix}$

simplify and express $\mathbf{u} + \mathbf{v} + \mathbf{w}$, $\mathbf{u} + 2\mathbf{v} - 3\mathbf{w}$ and $3\mathbf{u} - 2\mathbf{v} - \mathbf{w}$ as column vectors.

3 ▶ Given that $\mathbf{p} = \begin{pmatrix} 1 \\ 2 \end{pmatrix}$ and $\mathbf{q} = \begin{pmatrix} 3 \\ 4 \end{pmatrix}$

simplify and express $\mathbf{p} + \mathbf{q}$, $\mathbf{p} - \mathbf{q}$ and $2\mathbf{p} + 5\mathbf{q}$ as column vectors.

4 ▶ Given that $\mathbf{s} = \begin{pmatrix} 1 \\ -3 \end{pmatrix}$, $\mathbf{t} = \begin{pmatrix} 2 \\ 3 \end{pmatrix}$ and $\mathbf{u} = \begin{pmatrix} 4 \\ -5 \end{pmatrix}$

simplify and express $\mathbf{s} + \mathbf{t} + \mathbf{u}$, $2\mathbf{s} - \mathbf{t} + 2\mathbf{u}$ and $2\mathbf{u} - 3\mathbf{s}$ as column vectors.

5 ▶ Two vectors are defined as $\mathbf{v} = \begin{pmatrix} 3 \\ 1 \end{pmatrix}$ and $\mathbf{w} = \begin{pmatrix} 1 \\ 4 \end{pmatrix}$

Express $\mathbf{v} + \mathbf{w}$, $2\mathbf{v} - \mathbf{w}$ and $\mathbf{v} - 2\mathbf{w}$ as column vectors, find the magnitude and draw the resultant vector triangle for each vector.

6 ▶ Two vectors are defined as $\mathbf{p} = \begin{pmatrix} 2 \\ -1 \end{pmatrix}$ and $\mathbf{q} = \begin{pmatrix} 3 \\ 5 \end{pmatrix}$

Express $\mathbf{p} + \mathbf{q}$, $3\mathbf{p} + \mathbf{q}$ and $\mathbf{p} - 3\mathbf{q}$ as column vectors, find the magnitude and draw the resultant vector triangle for each vector.

7 ▶ The points A, B, C and D are the **vertices** of a quadrilateral where A has co-ordinates (3, 2).

$\overrightarrow{AB} = \begin{pmatrix} 2 \\ 3 \end{pmatrix}$, $\overrightarrow{BC} = \begin{pmatrix} 3 \\ 1 \end{pmatrix}$ and $\overrightarrow{CD} = \begin{pmatrix} -2 \\ -3 \end{pmatrix}$

a Draw quadrilateral ABCD on squared paper.

b Write $\overrightarrow{AD}$ as a column vector.

c What type of quadrilateral is ABCD?

d What do you notice about $\overrightarrow{BC}$ and $\overrightarrow{AD}$?

8 ▶ The points A, B, C and D are the vertices of a rectangle.

A has co-ordinates (2, −1), $\overrightarrow{AB} = \begin{pmatrix} 4 \\ 0 \end{pmatrix}$ and $\overrightarrow{AD} = \begin{pmatrix} 0 \\ -3 \end{pmatrix}$

a Draw rectangle ABCD on squared paper.

b Write as a column vector   **i** $\overrightarrow{CB}$ **ii** $\overrightarrow{BC}$

What do you notice?

c What do you notice about   **i** $\overrightarrow{AB}$ and $\overrightarrow{DC}$ **ii** $\overrightarrow{AD}$ and $\overrightarrow{BC}$?

9 ▶ In quadrilateral ABCD, $\overrightarrow{AB} = \begin{pmatrix} 2 \\ 3 \end{pmatrix}$, $\overrightarrow{BC} = \begin{pmatrix} 1 \\ -3 \end{pmatrix}$, $\overrightarrow{CD} = \begin{pmatrix} -2 \\ -3 \end{pmatrix}$ and $\overrightarrow{DA} = \begin{pmatrix} -1 \\ 3 \end{pmatrix}$

What type of quadrilateral is ABCD?

10 ▶ P is the point (5, 6). $\overrightarrow{PQ} = \begin{pmatrix} 3 \\ 1 \end{pmatrix}$

a Find the co-ordinates of Q.

b R is the point (7, 4). Express $\overrightarrow{PR}$ as a column vector.

c $\overrightarrow{RT} = \begin{pmatrix} 3 \\ -5 \end{pmatrix}$. Calculate the length of PT.

Give your answer to 3 **significant figures**.

EXERCISE 1*

1 ▶ Given that $\mathbf{p} = \begin{pmatrix} 2 \\ 1 \end{pmatrix}$ and $\mathbf{q} = \begin{pmatrix} 3 \\ -1 \end{pmatrix}$

find the magnitude and bearing of the vectors $\mathbf{p} + \mathbf{q}$, $\mathbf{p} - \mathbf{q}$ and $2\mathbf{p} - 3\mathbf{q}$

2 ▶ Given that $\mathbf{r} = \begin{pmatrix} 1 \\ -3 \end{pmatrix}$ and $\mathbf{s} = \begin{pmatrix} 4 \\ 1 \end{pmatrix}$

find the magnitude and bearing of the vectors $2(\mathbf{r} + \mathbf{s})$, $3(\mathbf{r} - 2\mathbf{s})$ and $(4\mathbf{r} - 6\mathbf{s})\sin 30°$.

3 ▶ Given that $\mathbf{t} + \mathbf{u} = \begin{pmatrix} 1 \\ 1 \end{pmatrix}$, where $\mathbf{t} = \begin{pmatrix} m \\ 3 \end{pmatrix}$, $\mathbf{u} = \begin{pmatrix} 2 \\ n \end{pmatrix}$ and m and n are constants, find the values of m and n.

4 ▶ Given that $2\mathbf{v} - 3\mathbf{w} = \begin{pmatrix} 2 \\ -3 \end{pmatrix}$, where $\mathbf{v} = \begin{pmatrix} 5 \\ -m \end{pmatrix}$, $\mathbf{w} = \begin{pmatrix} n \\ 4 \end{pmatrix}$ and m and n are constants, find the values of m and n.

5 ▶ Chloe, Leo and Max enter an orienteering competition. They have to find their way in the countryside using a map and compass. Each person decides to take a different route, described using these column vectors, where the units are in km:

$$\mathbf{s} = \begin{pmatrix} 1 \\ 1 \end{pmatrix} \qquad \mathbf{t} = \begin{pmatrix} 2 \\ 3 \end{pmatrix}$$

They all start from the same point P, and take 3 hours to complete their routes.

Chloe: $\mathbf{s} + 2\mathbf{t}$　　　　　Leo: $2\mathbf{s} + \mathbf{t}$　　　　　Max: $5\mathbf{s} - \mathbf{t}$

a Express each journey as a column vector.

b Find the length of each journey, and therefore calculate the average speed of each person in km/hr.

6 ▶ Use the information in Question 5 to answer this question.
Chloe, Leo and Max were all aiming to be at their first marker position Q, which is at

position vector $\begin{pmatrix} 1 \\ 5 \end{pmatrix}$ from point P.

a Find how far each person was from Q after their journeys.

b Calculate the bearing of Q from each person after their journeys.

7 ▶ These vectors represent journeys made by yachts in km.
Express each one in column vector form.

a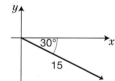

b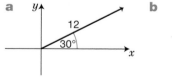

8 ▶ These vectors represent journeys completed by crows in km.
Express each one in column vector form.

a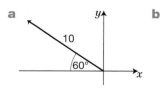

b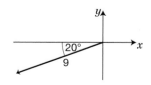

9 ▶ $\mathbf{s} = \begin{pmatrix} 2 \\ 3 \end{pmatrix}$ and $\mathbf{t} = \begin{pmatrix} -6 \\ 1 \end{pmatrix}$

a If $m\mathbf{s} + n\mathbf{t} = \begin{pmatrix} -16 \\ 6 \end{pmatrix}$, solve this vector equation to find the constants m and n.

b If $p\mathbf{s} + \mathbf{t} = \begin{pmatrix} 0 \\ q \end{pmatrix}$, solve this vector equation to find the constants p and q.

10 ▶ The centre spot O of a hockey pitch is the origin of the co-ordinate axes on the right, with the x-axis positioned across the pitch, and the y-axis positioned up the centre. All units are in metres.

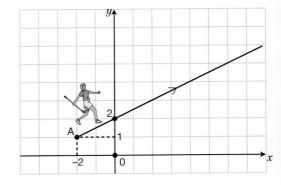

At $t = 0$, Anne is at A. When $t = 1$, Anne is at $(0, 2)$. Anne's position after t seconds is shown by the vector starting at point A. She runs at a constant speed.

a Explain why after t seconds Anne's position vector $\mathbf{r}$ is given by $\mathbf{r} = \begin{pmatrix} -2 \\ 1 \end{pmatrix} + t\begin{pmatrix} 2 \\ 1 \end{pmatrix}$

b Find Anne's speed in m/s.

At $t = 2$, Fleur, who is positioned on the centre spot, O, hits the ball towards Anne's path so that Anne receives it 5 s after setting off.

The position vector, $\mathbf{s}$, of the ball hit by Fleur is given by $\mathbf{s} = (t - 2)\begin{pmatrix} a \\ b \end{pmatrix}$

c Find the value of constants a and b and therefore the speed of the ball in m/s.

VECTOR GEOMETRY

Vector methods can be used to solve geometric problems in two dimensions and produce simple geometric proofs.

EXAMPLE 4

SKILLS

ANALYSIS

Given vectors **a**, **b** and **c** as shown, draw the vector **r** where $\mathbf{r} = \mathbf{a} + \mathbf{b} - \mathbf{c}$

Here is the resultant of $\mathbf{a} + \mathbf{b} - \mathbf{c} = \mathbf{r}$

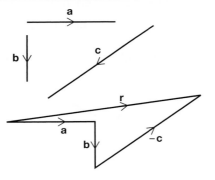

EXAMPLE 5

SKILLS

ANALYSIS

ABCDEF is a regular hexagon with centre O.
$\overrightarrow{AB} = \mathbf{x}$ and $\overrightarrow{BC} = \mathbf{y}$. Express the vectors $\overrightarrow{ED}$, $\overrightarrow{DE}$, $\overrightarrow{FE}$, $\overrightarrow{AC}$, $\overrightarrow{FA}$ and $\overrightarrow{AE}$ in terms of $\mathbf{x}$ and $\mathbf{y}$.

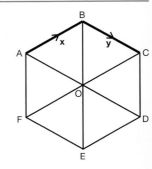

$\overrightarrow{ED} = \mathbf{x}$ 　　　 $\overrightarrow{DE} = -\mathbf{x}$ 　　 $\overrightarrow{FE} = \mathbf{y}$

$\overrightarrow{AC} = \mathbf{x} + \mathbf{y}$ 　 $\overrightarrow{FA} = \mathbf{x} - \mathbf{y}$ 　 $\overrightarrow{AE} = 2\mathbf{y} - \mathbf{x}$

EXERCISE 2   Use this diagram to answer Questions 1–4.

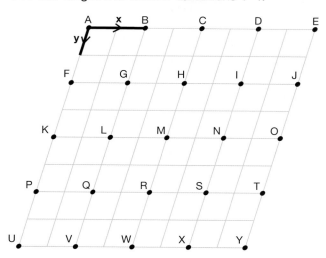

For Questions 1 and 2, express each vector in terms of **x** and **y**.

1 ▶   a $\overrightarrow{XY}$ b $\overrightarrow{EO}$ c $\overrightarrow{WC}$ d $\overrightarrow{TP}$ **2 ▶**   a $\overrightarrow{KC}$ b $\overrightarrow{VC}$ c $\overrightarrow{CU}$ d $\overrightarrow{AS}$

For Questions 3 and 4, find the vector formed when the vectors given are added to point H.
Write the vector as capital letters (e.g. $\overrightarrow{HO}$).

3 ▶   a 2**x**   b **x** + 2**y**   c 2**y** − **x**   d 2**x** + 2**y**

4 ▶   a 4**y** + 2**x**   b 4**y** − 2**x**   c **x** − 2**y**   d 2**x** + 6**y**

In Questions 5 and 6, express each vector in terms of **x** and **y**.

5 ▶   ABCD is a rectangle.

Find
a $\overrightarrow{DC}$ b $\overrightarrow{DB}$
c $\overrightarrow{BC}$ d $\overrightarrow{AC}$

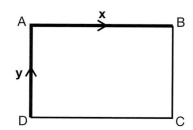

6 ▶   ABCD is a **trapezium**.

Find
a $\overrightarrow{AC}$ b $\overrightarrow{DB}$
c $\overrightarrow{BC}$ d $\overrightarrow{CB}$

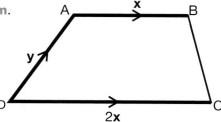

7 ▶   In the diagram $\overrightarrow{BA}$ = **a** and $\overrightarrow{AP}$ = **b**
P is the **mid-point** of AC.

Write down in terms of **a** and/or **b**
a $\overrightarrow{AC}$ b $\overrightarrow{BP}$ c $\overrightarrow{BC}$

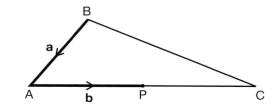

8 ▶ ABCD is a rectangle.

M is the mid-point of DC.

$\overrightarrow{AB} = \textbf{q}$ and $\overrightarrow{DA} = \textbf{p}$

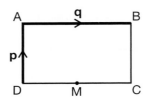

Write in terms of **p** and **q**

a $\overrightarrow{DC}$ **b** $\overrightarrow{DM}$ **c** $\overrightarrow{AM}$ **d** $\overrightarrow{BM}$

9 ▶ Here are five vectors.

$\overrightarrow{AB} = 2\textbf{a} - 8\textbf{b}$ $\overrightarrow{CD} = \textbf{a} + 4\textbf{b}$ $\overrightarrow{EF} = 4\textbf{a} - 16\textbf{b}$

$\overrightarrow{GH} = -2\textbf{a} + 8\textbf{b}$ $\overrightarrow{IJ} = \textbf{a} - 7\textbf{b}$

a Three of these vectors are parallel.

Which three?

b Simplify   **i** $5\textbf{p} - 6\textbf{q} + 2\textbf{p} - 7\textbf{q}$ **ii** $3(\textbf{a} - 2\textbf{b}) + \frac{1}{2}(3\textbf{a} + 4\textbf{b})$

10 ▶ In **parallelogram** ABCD, $\overrightarrow{AB} = \textbf{a}$ and $\overrightarrow{BC} = \textbf{b}$

M is the mid-point of AD.

Write in terms of **a** and **b**

a $\overrightarrow{AM}$ **b** $\overrightarrow{BM}$ **c** $\overrightarrow{CM}$

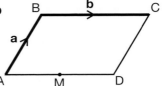

In Questions 1–4, express each vector in terms of **x** and **y**.

1 ▶ ABCD is a parallelogram.

Find

a $\overrightarrow{DC}$ **b** $\overrightarrow{AC}$ **c** $\overrightarrow{BD}$ **d** $\overrightarrow{AE}$

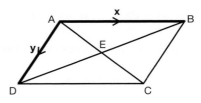

2 ▶ ABCD is a **rhombus**.

Find

a $\overrightarrow{BD}$ **b** $\overrightarrow{BE}$ **c** $\overrightarrow{AC}$ **d** $\overrightarrow{AE}$

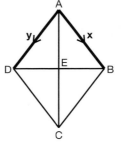

3 ▶ ABCDEF is an irregular hexagon.

Find

a $\overrightarrow{AB}$ **b** $\overrightarrow{AD}$ **c** $\overrightarrow{CF}$ **d** $\overrightarrow{CA}$

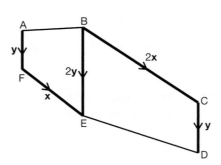

4 ▶ ABC is an **equilateral triangle**,
with $\overrightarrow{AC} = 2\mathbf{x}$, $\overrightarrow{AB} = 2\mathbf{y}$

Find

a $\overrightarrow{PQ}$ **b** $\overrightarrow{PC}$ **c** $\overrightarrow{QB}$ **d** $\overrightarrow{BC}$

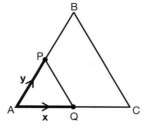

5 ▶ ABCD is a square.

M is the mid-point of AB.

$\overrightarrow{DC} = \mathbf{r}$ and $\overrightarrow{DA} = \mathbf{s}$

Write in terms of **r** and **s**

a $\overrightarrow{AB}$ **b** $\overrightarrow{BC}$ **c** $\overrightarrow{AM}$ **d** $\overrightarrow{DM}$

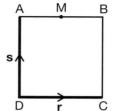

In Questions 6–10, express each vector in terms of **x** and **y**, where $\overrightarrow{OA} = \mathbf{x}$ and $\overrightarrow{OB} = \mathbf{y}$

6 ▶ M is the mid-point of AB.

Find

a $\overrightarrow{AB}$ **b** $\overrightarrow{AM}$ **c** $\overrightarrow{OM}$

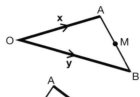

7 ▶ The **ratio** of AM : MB = 1 : 2.

Find

a $\overrightarrow{AB}$ **b** $\overrightarrow{AM}$ **c** $\overrightarrow{OM}$

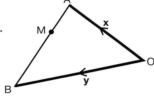

8 ▶ The ratio of OA : AD = 1 : 2 and B is the mid-point of OC.

a Find $\overrightarrow{AB}$, $\overrightarrow{OD}$ and $\overrightarrow{DC}$.

b M is the mid-point of CD. Find $\overrightarrow{OM}$.

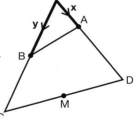

9 ▶ OABCDE is a regular hexagon.

Find $\overrightarrow{AB}$, $\overrightarrow{BC}$, $\overrightarrow{AD}$ and $\overrightarrow{BD}$.

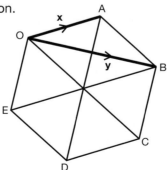

10 ▶ ABP is an equilateral triangle.

OAB is an **isosceles triangle**, where OA = AB.

a Find $\overrightarrow{OP}$, $\overrightarrow{AB}$ and $\overrightarrow{BP}$.

b M is the mid-point of BP. Find $\overrightarrow{OM}$.

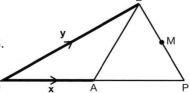

PARALLEL VECTORS AND COLLINEAR POINTS

Simple vector geometry can be used to prove that lines are parallel and points lie on the same straight line (**collinear**).

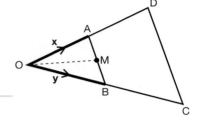

EXAMPLE 6

SKILLS

ANALYSIS

In △OAB, the mid-point of AB is M.

$\overrightarrow{OA} = \textbf{x}, \overrightarrow{OB} = \textbf{y}, \overrightarrow{OD} = 2\textbf{x}$ and $\overrightarrow{OC} = 2\textbf{y}$

a Express $\overrightarrow{AB}$, $\overrightarrow{OM}$ and $\overrightarrow{DC}$ in terms of **x** and **y**.

b What do these results show about AB and DC?

a $\begin{aligned} \overrightarrow{AB} &= \overrightarrow{AO} + \overrightarrow{OB} \\ &= -\textbf{x} + \textbf{y} \\ &= \textbf{y} - \textbf{x} \end{aligned}$

$\begin{aligned} \overrightarrow{OM} &= \overrightarrow{OA} + \overrightarrow{AM} \\ &= \overrightarrow{OA} + \tfrac{1}{2}\overrightarrow{AB} \\ &= \textbf{x} + \tfrac{1}{2}(\textbf{y} - \textbf{x}) \\ &= \textbf{x} + \tfrac{1}{2}\textbf{y} - \tfrac{1}{2}\textbf{x} \\ &= \tfrac{1}{2}\textbf{x} + \tfrac{1}{2}\textbf{y} \\ &= \tfrac{1}{2}(\textbf{x} + \textbf{y}) \end{aligned}$

$\begin{aligned} \overrightarrow{DC} &= \overrightarrow{DO} + \overrightarrow{OC} \\ &= -2\textbf{x} + 2\textbf{y} \\ &= 2\textbf{y} - 2\textbf{x} \\ &= 2(\textbf{y} - \textbf{x}) \end{aligned}$

b $\overrightarrow{DC} = 2\overrightarrow{AB}$, so AB is parallel to DC and the length of DC is twice the length of AB.

EXERCISE 3

8th

11th

1 ▶ $\overrightarrow{AC} = \textbf{a}$ and $\overrightarrow{CM} = \textbf{b}$.
M is the mid-point of CB.

Write down in terms of **a** and/or **b**

a $\overrightarrow{CB}$ **b** $\overrightarrow{MA}$ **c** $\overrightarrow{AB}$

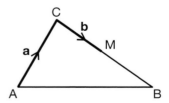

2 ▶ The points P, Q, R and S have co-ordinates
(1, 3), (7, 5), (−12, −15) and (24, −3) respectively.
O is the origin.

a Write down the position vectors $\overrightarrow{OP}$ and $\overrightarrow{OQ}$.
b Write down as a column vector   **i** $\overrightarrow{PQ}$ **ii** $\overrightarrow{RS}$
c What do these results show about the lines PQ and RS?

3 ▶ $\mathbf{a} = \begin{pmatrix} -1 \\ 3 \end{pmatrix}$ and $\mathbf{b} = \begin{pmatrix} 4 \\ -2 \end{pmatrix}$

Find a vector **c** such that $\mathbf{a} + \mathbf{c}$ is parallel to $\mathbf{a} - \mathbf{b}$

4 ▶   OABC is a quadrilateral in which $\overrightarrow{OA} = 2\mathbf{a}$, $\overrightarrow{OB} = 2\mathbf{a} + \mathbf{b}$ and $\overrightarrow{OC} = \frac{1}{2}\mathbf{b}$

a Find $\overrightarrow{AB}$ in terms of **a** and **b**.

What does this tell you about $\overrightarrow{AB}$ and $\overrightarrow{OC}$?

b Find $\overrightarrow{BC}$ in terms of **a** and **b**.

What does this tell you about $\overrightarrow{OA}$ and $\overrightarrow{BC}$?

c What type of quadrilateral is OABC?

5 ▶   The points A, B and C have co-ordinates (1, 5), (3, 12) and (5, 19) respectively.

a Find as column vectors

i $\overrightarrow{AB}$ ii $\overrightarrow{AC}$

b What do these results show you about the points A, B and C?

6 ▶   The point P has co-ordinates (3, 2).
The point Q has co-ordinates (7, 7).
The point R has co-ordinates (15, 17).
Show that points P, Q and R are collinear.

OABC is a quadrilateral in which $\overrightarrow{OA} = \mathbf{a}$, $\overrightarrow{OB} = \mathbf{a} + 2\mathbf{b}$ and $\overrightarrow{OC} = 4\mathbf{b}$

D is the point so that $\overrightarrow{BD} = \overrightarrow{OC}$ and X is the mid-point of BC.

a Find in terms of **a** and **b**

i $\overrightarrow{OD}$ ii $\overrightarrow{OX}$

b Explain what your answers to parts **a i** and **a ii** mean.

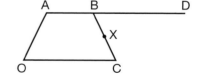

---

a i $\overrightarrow{OD} = \overrightarrow{OB} + \overrightarrow{BD}$

$= \mathbf{a} + 2\mathbf{b} + 4\mathbf{b}$

$= \mathbf{a} + 6\mathbf{b}$

ii $\overrightarrow{OX} = \overrightarrow{OC} + \overrightarrow{CX}$

$\overrightarrow{CX} = \frac{1}{2}\overrightarrow{CB}$

$\overrightarrow{CB} = \overrightarrow{CO} + \overrightarrow{OB}$

$= \overrightarrow{OB} - \overrightarrow{OC}$

$= \mathbf{a} + 2\mathbf{b} - 4\mathbf{b}$

$= \mathbf{a} - 2\mathbf{b}$

$\overrightarrow{CX} = \frac{1}{2}(\mathbf{a} - 2\mathbf{b}) = \frac{1}{2}\mathbf{a} - \mathbf{b}$

$\overrightarrow{OX} = 4\mathbf{b} + \frac{1}{2}\mathbf{a} - \mathbf{b} = \frac{1}{2}\mathbf{a} + 3\mathbf{b}$

b $\overrightarrow{OD} = \mathbf{a} + 6\mathbf{b} = 2(\frac{1}{2}\mathbf{a} + 3\mathbf{b})$

$\overrightarrow{OD} = 2\overrightarrow{OX}$

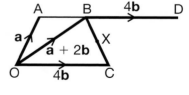

So OD and OX are parallel with a common point. This means that O, X and D lie on the same straight line. The length of OD is 2 times the length of OX. So X is the mid-point of OD.

- $\overrightarrow{PQ} = k\overrightarrow{QR}$ shows that the lines PQ and QR are parallel. Also they both pass through point Q so PQ and QR are part of the same straight line.

- P, Q and R are said to be collinear (they all lie on the same straight line).

EXERCISE 3*

1 ▶ ABCDEF is a regular hexagon with centre O.

$\overrightarrow{OA} = 6\mathbf{a}$, $\overrightarrow{OB} = 6\mathbf{b}$

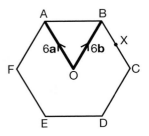

a Express in terms of **a** and/or **b**

  i $\overrightarrow{AB}$ **ii** $\overrightarrow{EF}$

X is the mid-point of BC.

b Express $\overrightarrow{EX}$ in terms of **a** and/or **b**.

Y is the point on AB extended, such that AB : BY = 3 : 2

c Prove that E, X and Y lie on the same straight line.

2 ▶ JKLM is a parallelogram.
The diagonals of the parallelogram **intersect** at O.

$\overrightarrow{OJ} = \mathbf{j}$ and $\overrightarrow{OK} = \mathbf{k}$

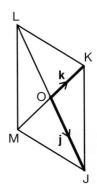

a Write an expression, in terms of **j** and **k**, for

  i $\overrightarrow{LJ}$ **ii** $\overrightarrow{KJ}$ **iii** $\overrightarrow{KL}$

X is the point such that $\overrightarrow{OX} = 2\mathbf{j} - \mathbf{k}$

b i Write down an expression, in terms of **j** and **k**, for $\overrightarrow{JX}$.

  ii Explain why J, K and X lie on the same straight line.

3 ▶ APB is a triangle. N is a point on AP.

$\overrightarrow{AB} = \mathbf{a}$, $\overrightarrow{AN} = 2\mathbf{b}$, $\overrightarrow{NP} = \mathbf{b}$

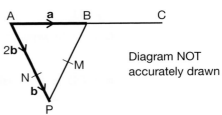
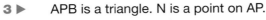

a Find the vector $\overrightarrow{PB}$, in terms of **a** and **b**.
B is the mid-point of AC.
M is the mid-point of PB.

b Show that NMC is a straight line.

Diagram NOT
accurately drawn

4 ▶ OPQ is a triangle. B is the mid-point of PQ.

$\overrightarrow{OA} = 3\mathbf{p}$, $\overrightarrow{AP} = \mathbf{p}$, $\overrightarrow{OQ} = 2\mathbf{q}$ and $\overrightarrow{QC} = \mathbf{q}$

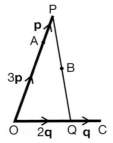
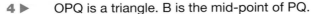

a Find, in terms of **p** and **q**, the vectors

  i $\overrightarrow{PQ}$ **ii** $\overrightarrow{AC}$ **iii** $\overrightarrow{BC}$

b Therefore explain why ABC is a straight line.
The length of AB is 3 cm.

c Find the length of AC.

5 ▶ ABCDEF is a regular hexagon, with centre O.

$\overrightarrow{OA} = \mathbf{a}$ and $\overrightarrow{OB} = \mathbf{b}$

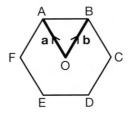

Diagram NOT
accurately drawn

a Write the vector $\overrightarrow{AB}$ in terms of **a** and **b**.
The line AB is extended to the point K so that
AB : BK = 1 : 2

b Write the vector $\overrightarrow{CK}$ in terms of **a** and **b**.
Give your answer in its simplest form.

6 ▶ OAYB is a quadrilateral.

$\overrightarrow{OA} = 3\mathbf{a}$

$\overrightarrow{OB} = 6\mathbf{b}$

a Express $\overrightarrow{AB}$ in terms of **a** and **b**.

X is the point on AB such that AX : XB = 1 : 2
and $\overrightarrow{BY} = 5\mathbf{a} - \mathbf{b}$.

b Prove that $\overrightarrow{OX} = \frac{2}{5}\overrightarrow{OY}$

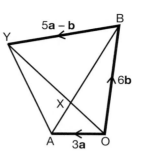

Diagram NOT
accurately drawn

ACTIVITY 2

A radar-tracking station O is positioned at the origin of x and y axes, where the x-axis points due East and the y-axis points due North.

All the units are in km.

A helicopter is noticed t minutes after midday with position vector **r**, where

$$\mathbf{r} = \begin{pmatrix} 12 \\ 5 \end{pmatrix} + t\begin{pmatrix} -3 \\ 4 \end{pmatrix}$$

Copy and complete this table and use it to plot the journey of the helicopter.

TIME	12:00 $t = 0$	12:01 $t = 1$	12:02 $t = 2$	12:03 $t = 3$	12:04 $t = 4$
r	$\begin{pmatrix} 12 \\ 5 \end{pmatrix}$				

Calculate the speed of the helicopter in km/hour and its bearing **correct to 1 decimal place**.

An international border is described by the line $y = 5x$.

Draw the border on your graph.

Estimate the time the helicopter crosses the border by looking carefully at your graph.

Considering the helicopter's position vector $\mathbf{r} = \begin{pmatrix} x \\ y \end{pmatrix}$, express x and y in terms of t and use these equations with $y = 5x$ to find the time when the border is crossed, correct to the nearest second.

EXERCISE 4   REVISION

1 ▶ Given that $\mathbf{p} = \begin{pmatrix} 3 \\ 4 \end{pmatrix}$ and $\mathbf{q} = \begin{pmatrix} -2 \\ 1 \end{pmatrix}$ simplify $\mathbf{p} + \mathbf{q}$, $\mathbf{p} - \mathbf{q}$ and $3\mathbf{p} - 2\mathbf{q}$ as column vectors and find the magnitude of each vector.

2 ▶ Given that $\mathbf{r} = \begin{pmatrix} 2 \\ -5 \end{pmatrix}$ and $\mathbf{s} = \begin{pmatrix} 3 \\ 4 \end{pmatrix}$ simplify $2\mathbf{r} + \mathbf{s}$, $2\mathbf{r} - \mathbf{s}$ and $\mathbf{s} - 2\mathbf{r}$ as column vectors and find the magnitude of each vector.

3 ▶ In terms of vectors **x** and **y**, find these vectors.

   a $\overrightarrow{AB}$ **b** $\overrightarrow{AC}$ **c** $\overrightarrow{CB}$

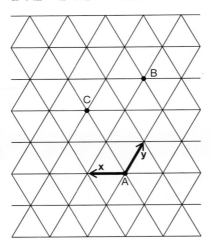

4 ▶ $\overrightarrow{OA} = $ **v** and $\overrightarrow{OB} = $ **w** and M is the mid-point of AB.
Find these vectors in terms of **v** and **w**.

   a $\overrightarrow{AB}$ **b** $\overrightarrow{AM}$ **c** $\overrightarrow{OM}$

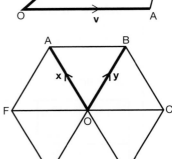

5 ▶ ABCDEF is a regular hexagon.

   $\overrightarrow{OA} = $ **x** and $\overrightarrow{OB} = $ **y**

   Find these vectors in terms of **x** and **y**.

   a $\overrightarrow{AB}$ **b** $\overrightarrow{FB}$ **c** $\overrightarrow{FD}$

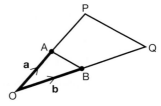

6 ▶ In the triangle OPQ, A and B are mid-points
of sides OP and OQ respectively,

   $\overrightarrow{OA} = $ **a** and $\overrightarrow{OB} = $ **b**

   a Find in terms of **a** and **b**: $\overrightarrow{OP}$, $\overrightarrow{OQ}$, $\overrightarrow{AB}$ and $\overrightarrow{PQ}$.

   b What can you conclude about AB and PQ?

7 ▶ $\mathbf{r} = \begin{pmatrix} 1 \\ 3 \end{pmatrix}$ and $\mathbf{s} = \begin{pmatrix} -2 \\ 5 \end{pmatrix}$

   a Calculate $2\mathbf{r} - \mathbf{s}$

   b Calculate $2(\mathbf{r} - \mathbf{s})$

   c Calculate the length of vector **s** in surd form.

   d $v\mathbf{r} + w\mathbf{s} = \begin{pmatrix} -3 \\ 13 \end{pmatrix}$

   What are the values of the constants v and w?

8 ▶ ABCDEF is a regular hexagon.

   $\overrightarrow{AB} = $ **a**, $\overrightarrow{BC} = $ **b** and $\overrightarrow{FC} = 2$**a**

   a Find in terms of **a** and **b**

   i $\overrightarrow{FE}$ **ii** $\overrightarrow{CE}$

   $\overrightarrow{CE} = \overrightarrow{EX}$

   b Prove that FX is parallel to CD.

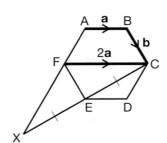

REVISION

1 ▶ Given that $2\mathbf{p} - 3\mathbf{q} = \begin{pmatrix} 5 \\ 15 \end{pmatrix}$, where $\mathbf{p} = \begin{pmatrix} 4 \\ m \end{pmatrix}$, $\mathbf{q} = \begin{pmatrix} n \\ -3 \end{pmatrix}$ and m and n are constants, find the values of m and n.

2 ▶ If $\mathbf{r} = \begin{pmatrix} 4 \\ -1 \end{pmatrix}$, $\mathbf{s} = \begin{pmatrix} 3 \\ 7 \end{pmatrix}$ and $m\mathbf{r} + n\mathbf{s} = \begin{pmatrix} 7 \\ 37 \end{pmatrix}$, find the constants m and n.

3 ▶ OXYZ is a parallelogram. M is the mid-point of XY.

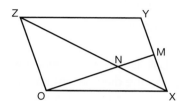

a Given that $\overrightarrow{OX} = \begin{pmatrix} 8 \\ 0 \end{pmatrix}$ and $\overrightarrow{OZ} = \begin{pmatrix} -2 \\ 6 \end{pmatrix}$, write down the vectors $\overrightarrow{XM}$ and $\overrightarrow{XZ}$.

b Given that $\overrightarrow{ON} = v\overrightarrow{OM}$, write down in terms of v the vector $\overrightarrow{ON}$.

c Given that $\overrightarrow{ON} = \overrightarrow{OX} + w\overrightarrow{XZ}$, find in terms of w the vector $\overrightarrow{ON}$.

d Solve two **simultaneous equations** to find v and w.

4 ▶ ABCD is a parallelogram in which $\overrightarrow{AB} = \mathbf{x}$ and $\overrightarrow{BC} = \mathbf{y}$
AE : ED = 1 : 2
a Express in terms of $\mathbf{x}$ and $\mathbf{y}$, $\overrightarrow{AC}$ and $\overrightarrow{BE}$.

b AC and BE intersect at F, such that $\overrightarrow{BF} = v\overrightarrow{BE}$
  i Express $\overrightarrow{BF}$ in terms of $\mathbf{x}$, $\mathbf{y}$ and v.
  ii Show that $\overrightarrow{AF} = (1 - v)\mathbf{x} + \frac{1}{3}v\mathbf{y}$
  iii Use this expression for $\overrightarrow{AF}$ to find the value of v.

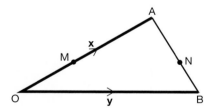

5 ▶ OM : MA = 2 : 3 and AN = $\frac{3}{5}$AB.

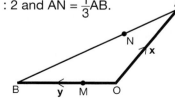

a Find $\overrightarrow{MA}$, $\overrightarrow{AB}$, $\overrightarrow{AN}$ and $\overrightarrow{MN}$.

b How are OB and MN related?

6 ▶ OM : MB = 1 : 2 and AN = $\frac{1}{3}$AB.

a Find $\overrightarrow{AB}$ and $\overrightarrow{MN}$.

b How are OA and MN related?

7 ▶ ABCDEF is a regular hexagon with centre O.

$\overrightarrow{DA} = 8\mathbf{a}$, $\overrightarrow{EB} = 8\mathbf{b}$

a Express in terms of **a** and/or **b**

   i $\overrightarrow{OA}$ **ii** $\overrightarrow{OB}$ **iii** $\overrightarrow{AB}$

X is the mid-point of ED.

b Express $\overrightarrow{AX}$ in terms of **a** and/or **b**.

Y is the point on AB extended, such that $\overrightarrow{AY} = 2\overrightarrow{AB}$

c Show that D, C and Y lie on a straight line.

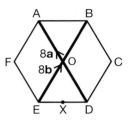

8 ▶ Amila and Winnie are playing basketball. During the game, their position vectors on the court are defined relative to the axes on the diagram. At time t seconds after the game starts their position vectors are given by **a** and **w** respectively:

$$\mathbf{a} = \begin{pmatrix} 2 \\ -1 \end{pmatrix} + t\begin{pmatrix} 1 \\ 2 \end{pmatrix}$$

$$\mathbf{w} = \begin{pmatrix} -3 \\ 4 \end{pmatrix} + t\begin{pmatrix} 3 \\ 1 \end{pmatrix}$$

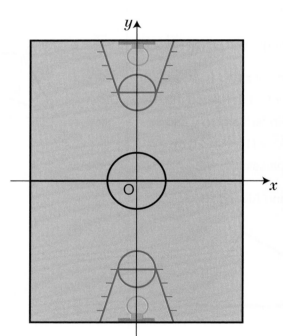

a Find the position vectors for Amila and Winnie after

   i 1 s   **ii** 2 s

b Write down the vector from Amilia to Winnie after 2 s and use it to find how far apart they are at this moment. Leave your answer in **surd** form.

c Calculate the speeds of the two girls in surd form.

EXAM PRACTICE: SHAPE AND SPACE 8

1 $\mathbf{p} = \begin{pmatrix} 2 \\ -2 \end{pmatrix}$　　$\mathbf{q} = \begin{pmatrix} 6 \\ 2 \end{pmatrix}$　　$\mathbf{r} = \begin{pmatrix} -1 \\ 5 \end{pmatrix}$

$\mathbf{p} + 3\mathbf{r} = 2\mathbf{q} - \mathbf{s}$

  a Work out **s** as a column vector.　　**[3]**

  b Calculate the magnitude of **s**. Give your answer
    to 1 d.p.　　**[2]**

2 The position of a point is determined by its
position vector $\begin{pmatrix} x \\ y \end{pmatrix}$ relative to the origin O.
A and B have position vectors

$\overrightarrow{OA} = \begin{pmatrix} 20 \\ 15 \end{pmatrix}$ and $\overrightarrow{OB} = \begin{pmatrix} 30 \\ 40 \end{pmatrix}$

  a Write down $\overrightarrow{AB}$.　　**[2]**

  b Calculate the magnitude of the vector $\overrightarrow{AB}$.
    Give your answer to 1 d.p.　　**[2]**

  c Calculate the angle that vector $\overrightarrow{AB}$ makes
    with the x-axis.　　**[2]**

  d X is the mid-point of AB. Write $\overrightarrow{OX}$ as a column
    vector.　　**[2]**

3 OACB is a parallelogram. M is the mid-point of OB
and N is the mid-point of OA.

$\overrightarrow{OM} = \mathbf{m}$

$\overrightarrow{ON} = \mathbf{n}$

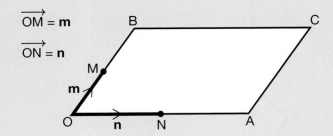

Express these vectors in terms of **m** and **n**.
  a $\overrightarrow{OA}$　　**[2]**
  b $\overrightarrow{OB}$　　**[2]**
  c $\overrightarrow{AB}$　　**[2]**
  d $\overrightarrow{NM}$　　**[2]**
  e What can you conclude about $\overrightarrow{NM}$ and $\overrightarrow{AB}$?　　**[1]**

4 WXYZ is a trapezium, with XY parallel to WZ.

$\overrightarrow{XZ} = 5\mathbf{a} - 2\mathbf{b}$, $\overrightarrow{ZY} = 3\mathbf{a} + 4\mathbf{b}$ and $\overrightarrow{WZ} = 6\mathbf{a} + k\mathbf{b}$
where k is an unknown constant.

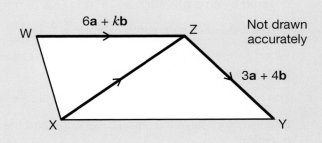

Work out the value of k.

Show **working** to justify your answer.　　**[3]**

[Total 25 marks]

CHAPTER SUMMARY: SHAPE AND SPACE 8

VECTORS AND VECTOR NOTATION

Vectors have magnitude and direction and can be written as bold letters such as **v** and **u**; with capitals covered by an arrow such as $\overrightarrow{OP}$ and $\overrightarrow{OQ}$; or on co-ordinate axes as column vectors: $\begin{pmatrix} 3 \\ -1 \end{pmatrix}$, $\begin{pmatrix} 0 \\ 6 \end{pmatrix}$

When handwriting the vector, underline the letter: a͟, p͟, x͟. |**a**| is the magnitude of vector **a**.

The magnitude of a vector $\begin{pmatrix} x \\ y \end{pmatrix}$ is its length: $\sqrt{x^2 + y^2}$

Equal vectors have the same magnitude and the same direction.

MULTIPLICATION OF A VECTOR BY A SCALAR

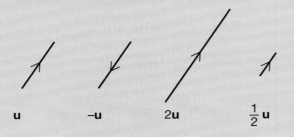

u   **–u**   **2u** $\frac{1}{2}$**u**

u = 2**w** ⇒ **u** is parallel to **w**, **u** is twice as long as **w**.
u = k**w** ⇒ **u** is parallel to **w**, **u** is k times as long as **w**.

$\overrightarrow{PQ} = k\overrightarrow{QR}$ shows that the lines PQ and QR are parallel.

Also they both pass through point Q so PQ and QR are part of the same straight line.

P, Q and R are said to be collinear (they all lie on the same straight line).

VECTOR GEOMETRY

$$\overrightarrow{AC} = \overrightarrow{AB} + \overrightarrow{BC} = 2\mathbf{a} + \mathbf{b}$$
$$\Rightarrow \overrightarrow{AD} = \overrightarrow{AC} + \overrightarrow{CD}$$
$$= 2\mathbf{a} + \mathbf{b} + (-3\mathbf{a}) = \mathbf{b} - \mathbf{a}$$

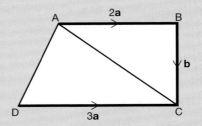

AB is parallel to DC ⇒ ABCD is a trapezium.

Ratio of AB : DC = 2 : 3 ⇒ 2DC = 3AB

$$\overrightarrow{OD} = \begin{pmatrix} 3 \\ 0 \end{pmatrix}, \overrightarrow{DC} = \begin{pmatrix} -2 \\ 4 \end{pmatrix}, \overrightarrow{CA} = \begin{pmatrix} -2 \\ -2 \end{pmatrix}$$

$$\Rightarrow \overrightarrow{OA} = \overrightarrow{OD} + \overrightarrow{DC} + \overrightarrow{CA}$$
$$= \begin{pmatrix} 3 \\ 0 \end{pmatrix} + \begin{pmatrix} -2 \\ 4 \end{pmatrix} + \begin{pmatrix} -2 \\ -2 \end{pmatrix}$$
$$= \begin{pmatrix} -1 \\ 2 \end{pmatrix}$$

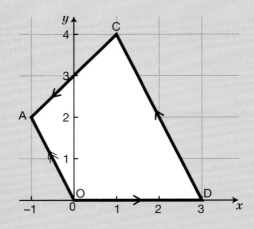

Length (magnitude) of
$$\overrightarrow{OA} = \sqrt{(-1)^2 + 2^2}$$
$$= \sqrt{5}$$

HANDLING DATA 5

Jakob Bernoulli's (1655–1705) revolutionary work "The Art of Conjecturing" (1713) contained many of his most famous concepts: permutations (arrangements) and combinations, treatment of mathematical predictability; and the subject of probability. Bernoulli had been writing the book on and off for over twenty years, but was never satisfied with it and therefore never published it. The book was finally published in 1713, 8 years after his death, after the work was given to a publisher by Bernoulli's nephew Nicholas.

LEARNING OBJECTIVES

- Add probabilities for mutually exclusive events
- Find the probability of independent events
- Draw and use tree diagrams to calculate the probability of independent events
- Decide whether two events are independent or dependent
- Draw and use tree diagrams to calculate conditional probability
- Use two-way tables to calculate conditional probability

BASIC PRINCIPLES

- For equally likely outcomes, probability $= \dfrac{\text{number of successful outcomes}}{\text{total number of possible outcomes}}$
- P(A) means the probability of event A occurring.
- P(A′) means the probability of event A not occurring.
- $0 \leq P(A) \leq 1$
- P(A) + P(A′) = 1, so P(A′) = 1 − P(A)
- P(A|B) means the probability of A occurring given that B has already happened.

LAWS OF PROBABILITY

EXAMPLE 1

SKILLS

REASONING

Calculate the probability that a **prime** number is not obtained when a fair 10-sided spinner that is numbered from 1 to 10 is spun.

Let A be the event that a prime number is obtained.

P(A′) = 1 − P(A)

$= 1 - \dfrac{4}{10}$ (There are 4 prime numbers: 2, 3, 5, 7)

$= \dfrac{6}{10} = \dfrac{3}{5}$

INDEPENDENT EVENTS

If two events have no influence on each other, they are **independent events**.

If it snows in Moscow, it would be unlikely that this event would have any influence on your teacher winning the lottery on the same day. These events are said to be independent.

MUTUALLY EXCLUSIVE EVENTS

Two events are **mutually exclusive** when they cannot happen at the same time. For example, a number rolled on a die cannot be both odd and even.

COMBINED EVENTS

MULTIPLICATION ('AND') RULE

EXAMPLE 2

Two dice are rolled together. One is a fair die numbered 1 to 6. On the other, each **face** is of a different colour: red, yellow, blue, green, white and purple

What is the probability that the dice will show an odd number and a purple face?

All possible outcomes are shown in this sample space diagram.

	R	Y	B	G	W	P
1	•	•	•	•	•	⊙
2	•	•	•	•	•	•
3	•	•	•	•	•	⊙
4	•	•	•	•	•	•
5	•	•	•	•	•	⊙
6	•	•	•	•	•	•

Let event A be that the dice will show an odd number and a purple face.

By inspection of the sample space diagram, $P(A) = \frac{3}{36} = \frac{1}{12}$

Alternatively:

If event O is that an odd number is thrown: $P(O) = \frac{1}{2}$

If event P is that a purple colour is thrown: $P(P) = \frac{1}{6}$

So, $P(A) = P(O \text{ and } P) = P(O) \times P(P) = \frac{1}{2} \times \frac{1}{6} = \frac{1}{12}$

KEY POINT

■ For two independent events A and B, $P(A \text{ and } B) = P(A) \times P(B)$

ADDITION ('OR') RULE

EXAMPLE 3

A card is **randomly** selected from a pack of 52 playing cards.

Find the probability that either an Ace or a Queen is selected.

Event A is an Ace is picked out: $P(A) = \dfrac{4}{52}$

Event Q is a Queen is picked out: $P(Q) = \dfrac{4}{52}$

Probability of A or Q: $P(A \text{ or } Q) = P(A) + P(Q)$

$$= \frac{4}{52} + \frac{4}{52}$$
$$= \frac{8}{52}$$
$$= \frac{2}{13}$$

4 Aces

4 Queens

KEY POINT

■ For mutually exclusive events A and B, $P(A \text{ or } B) = P(A) + P(B)$

The above rule makes sense, as adding fractions produces a larger fraction and the condition of one or other event happening suggests a greater chance.

INDEPENDENT EVENTS AND TREE DIAGRAMS

A **tree diagram** shows two or more events and their probabilities. The probabilities are written on each branch of the diagram.

EXAMPLE 4

SKILLS

PROBLEM
SOLVING

This fair five-sided spinner is spun twice.
a Draw a tree diagram to show the probabilities.
b What is the probability of both spins landing on red?
c What is the probability of landing on one red and one blue?

a

1st spin   **2nd spin**

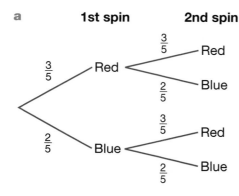

b Go along the branches for Red, Red. The 1st and 2nd spins are independent so:

$$P(R, R) = \frac{3}{5} \times \frac{3}{5} = \frac{9}{25}$$

c Go along the branches for Red, Blue and Blue, Red:

$$P(R, B) = \frac{3}{5} \times \frac{2}{5} = \frac{6}{25}$$

$$P(B, R) = \frac{2}{5} \times \frac{3}{5} = \frac{6}{25}$$

The outcomes Red, Blue and Blue, Red are mutually exclusive so:

$$P(R, B \text{ or } B, R) = \frac{6}{25} + \frac{6}{25} = \frac{12}{25}$$

THE 'AT LEAST' SITUATION

EXAMPLE 5

SKILLS

REASONING

A box contains three letter tiles:
A tile is chosen at random from the box and then replaced.
A second tile is now chosen at random from the box and then returned.

a Draw a tree diagram to show the probabilities.

b Work out the probability that at least one A is chosen.

a **First selection**   **Second selection**

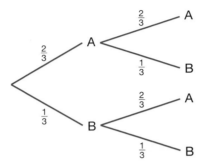

b Let the number of letter A's taken out be X.
At least one A means that $X \geq 1$

Method 1

List all the possibilities.

$P(X \geq 1) = P(X = 1) + P(X = 2) = P(AB) + P(BA) + P(AA)$

$$= \frac{2}{3} \times \frac{1}{3} + \frac{1}{3} \times \frac{2}{3} + \frac{2}{3} \times \frac{2}{3}$$

$$= \frac{2}{9} + \frac{2}{9} + \frac{4}{9} = \frac{8}{9}$$

Method 2

Use the rule $P(E) + P(E') = 1$

$P(X \geq 1) + P(X = 0) = 1$

$P(X \geq 1) = 1 - P(X = 0)$

$$= 1 - P(BB)$$

$$= 1 - \frac{1}{3} \times \frac{1}{3}$$

$$= 1 - \frac{1}{9} = \frac{8}{9}$$

Note: Both methods need to be understood and applied.

EXERCISE 1

Use tree diagrams to solve these problems.

1 ▶ A fair six-sided die is thrown twice.
Calculate the probability of obtaining these scores.

a Two sixes
b No sixes
c A six and not a six, in that order
d A six and not a six, in any order

2 ▶ A box contains two red and three green beads.
One bead is randomly chosen and replaced before another is chosen.
Calculate the probability of obtaining these combinations of beads.

a Two red beads
b Two green beads
c A red bead and a green bead, in that order
d A red bead and a green bead, in any order

3 ▶ A chest of drawers contains four yellow ties and six blue ties.
One is randomly selected and replaced before another is chosen.
Calculate the probability of obtaining these ties.

   a Two yellow ties
   b Two blue ties
   c A yellow tie and a blue tie, in that order
   d A yellow tie and a blue tie, in any order

4 ▶ A spice shelf contains three jars of chilli and four jars of mint.
One is randomly selected and replaced before another is chosen.

   a Calculate the probability of selecting two jars of chilli.
   b What is the probability of selecting a jar of chilli and a jar of mint?

5 ▶ In a game of basketball the probability of scoring from a free shot is $\frac{2}{3}$
A player has two consecutive free shots.

   a Calculate the probability that he scores two baskets.
   b What is the probability that he scores one basket?
   c What is the probability that he scores no baskets?

6 ▶ Each evening Dina either reads a book or watches television.

The probability that she watches television is $\frac{3}{4}$, and if she does this the probability that
she will fall asleep is $\frac{4}{7}$. If she reads a book the probability that she will fall asleep is $\frac{3}{7}$

   a Calculate the probability that she does not fall asleep.
   b What is the probability that she does fall asleep?

7 ▶ Two cards are picked at random from a pack of 52 playing cards. The cards are replaced
after each selection.
Calculate the probability that these cards are picked.

   a Two kings
   b A red card and a black card
   c A picture card and an odd number card
   d A heart and a diamond

8 ▶ 80 people with similar symptoms were tested for a virus using a new trial medical test.
19 of the people tested showed a positive result.
The virus only developed in 11 of the people who tested positive.
A total of 67 people did not develop the virus at all.

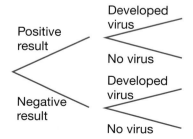

   a Copy and complete the tree diagram (frequency tree).
   b Work out the probability that a person develops the virus.

EXERCISE 1*

1 ▶ On a 'hook-a-duck' game at a fair you win a prize if you pick a duck with an 'X' on its base. Aaron picks a duck at random, replaces it and then picks another one.

a Copy and complete the tree diagram to show the probabilities.

b What is the probability of
  i winning two prizes
  ii winning nothing
  iii winning one prize
  iv winning at least one prize?

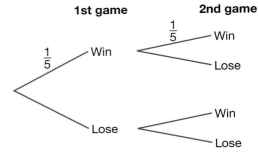

2 ▶ Megan has two bags of discs.
In bag A there are 3 red and 5 green discs.
In bag B there are 1 red and 5 green discs.
A disc is chosen at random from each bag.

a Copy and complete the tree diagram to show the probabilities.

b Work out the probability of choosing
  i two discs that are the same colour
  ii one red and one green disc
  iii no red discs
  iv at least one red disc.

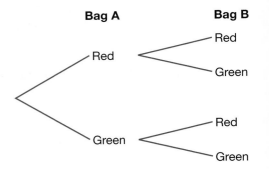

3 ▶ In a game, a set of cards is numbered 1–10.
If you pick a prime number you win a prize.
James plays the game twice.

a Copy and complete the tree diagram to show the probabilities.

b What is the probability of winning
  i two prizes
  ii nothing
  iii one prize
  iv at least one prize?

 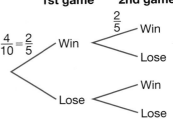

4 ▶ Marganita has two bags of sweets.
Bag A contains 7 caramels and 3 mints.
Bag B contains 4 caramels and 6 mints.
She picks a sweet at random from each bag.

a Copy and complete the tree diagram to show the probabilities.

b Work out the probability of picking
  i two sweets the same
  ii one caramel and one mint
  iii no mints
  iv at least one caramel.

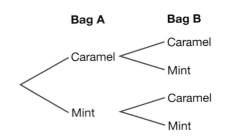

5 ▶ Terry plays two games on his phone. He has a 1 in 10 chance of winning the first game and a 1 in 5 chance of winning the second.

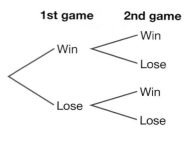

1st game   2nd game

a Copy and complete the tree diagram to show the probabilities.

b Work out the probability that
   i he wins both games
   ii he wins the first game but not the second.

6 ▶ An archer shoots his arrows at a target. The probability that he scores a bullseye (a shot in the middle of the target) in any one attempt is $\frac{1}{3}$. If he shoots twice, calculate these probabilities:

a P(two bullseyes)
b P(no bullseyes)
c P(at least one bullseye)

7 ▶ A netball shooter has two free shots. The probability that she scores (or misses) with each shot can be found from this table.

	FIRST SHOT	SECOND SHOT
SCORES	$\frac{2}{3}$	
MISSES		$\frac{3}{7}$

a Copy and complete the table and then draw a tree diagram to show the probabilities.

b For the two shots, calculate
   i P(both shots miss)
   ii P(one shot scores)
   iii P(at least one shot scores)

8 ▶ The spinner is spun twice. Use tree diagrams to help you to calculate these probabilities.

a P(two even numbers)
b P(an even number and an odd number)
c P(a black number and a white number)
d P(a white even number and a black odd number)

CONDITIONAL PROBABILITY

The probability of an event based on the occurrence of a previous event is called a **conditional probability**.

Two events are **dependent** if one event depends upon the outcome of another event. For example, removing a King from a pack of playing cards reduces the chance of choosing another King.

A conditional probability is the probability of a dependent event.
Tree diagrams can be used to solve problems involving dependent events.

EXAMPLE 6

SKILLS

INTERPRETATION

Two cards are randomly selected one after the other from a pack of 52 playing cards and **not** replaced.

Calculate the probability that the second card is a King given that the first card is

a a King

b not a King.

a P(second is a King given the first card is a King) = $\frac{3}{51}$
(3 kings left in a pack of 51)

b P(second is a King given the first card is not a King) = $\frac{4}{51}$
(4 kings left in a pack of 51)

EXAMPLE 7

SKILLS

REASONING

A group of Border Collie puppies contains four females (F) and two males (M). A vet randomly removes one from their basket and it is **not** replaced before another one is chosen.

a Draw a tree diagram to show all the possible outcomes.

b What is the probability that the vet removes

   i two males **ii** one male and one female?

a   **First puppy**   **Second puppy**

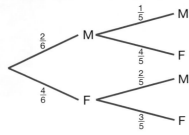

b i When the second puppy is taken, there are only five left in the basket.

Let event A be that two males are chosen.

P(A) = P(M₁ and M₂) (M₁ means the first puppy is a male,
                        M₂ means the second puppy is a male)

    = P(M₁) × P(M₂) (Given by tree diagram route MM)

    = $\frac{2}{6} \times \frac{1}{5} = \frac{2}{30} = \frac{1}{15}$

ii Let event B be that a male and a female are chosen.

  P(B) = P(M₁ and F₂) or P(F₁ and M₂)

  (Given by tree diagram routes MF and FM)

      = $\frac{2}{6} \times \frac{4}{5} + \frac{4}{6} \times \frac{2}{5}$

      = $\frac{8}{30} + \frac{8}{30} = \frac{16}{30} = \frac{8}{15}$

ACTIVITY 1

The Titanic hit an iceberg on April 15th 1912.

Many people lost their lives in this famous disaster.

The two-way table shows the passenger list deaths by cabin class on the ship.

CABIN CLASS	RESCUED	DEATHS	TOTAL
1ST	203	122	325
2ND	118	167	285
3RD	178	528	706
SHIP'S CREW	212	673	885
TOTAL	**711**	**1490**	**2201**

An estimate for the probability that a passenger did not survive given that they were in a

$$\text{2nd class cabin} = \frac{\text{number of fatalities in 2nd class}}{\text{total number of passengers in 2nd class}} = \frac{167}{285} \approx 0.59$$

Work out an estimate for the probability that a passenger was rescued given that they were in

i 1st   **ii** 2nd   **iii** 3rd class cabins.

Was it more likely for a person not to be rescued if they were a passenger or if they were a crew member?

1 ▶ State whether each of these are independent or dependent events.

    a Picking a card from a pack of 52 playing cards, replacing it and then picking another one.
    b Flipping a coin and rolling a dice.
    c Picking two marbles from a bag, one after the other.
    d Picking a plastic piece from a bag, then rolling a dice.
    e Picking a student from a class, then picking another student.

2 ▶ High school students in a school have to study one subject from each of two lists.
The two-way table shows their choices.

		LIST A			
		HISTORY	GEOGRAPHY	FRENCH	TOTAL
LIST B	BUSINESS	12	7	8	27
	ICT	8	17	12	37
	ART	3	6	7	16
	TOTAL	23	30	27	80

Work out the probability that a student chosen at random
a studies history
b studies business
c studies ICT, given the student studies history
d studies French, given that the student studies art.

3 ▶ Marion has a bag containing 6 chocolates and 6 mints.

She chooses a sweet at random and eats it.

She then chooses another sweet at random.

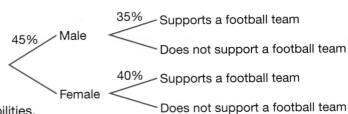

1st sweet   2nd sweet

a Copy and complete the tree diagram to show all the probabilities.

b Work out the probability that the two sweets consist of

   i two chocolates
   ii a chocolate and a mint
   iii two mints
   iv at least one chocolate.

4 ▶ In a survey, 45% of the people asked were male. 35% of the men and 40% of the women supported a football team.

45% Male
35% Supports a football team
Does not support a football team

Female
40% Supports a football team
Does not support a football team

a Copy and complete the tree diagram to show all the probabilities.

b One person is chosen at random.
Find the probability that this is a woman who does not support a football team.

5 ▶ A train is either late or on time.

The probability it is late is 0.85. If the train is late, the probability Mr Murphy is late for work is 0.7.
If the train is on time, the probability he is late is 0.1

Work out the probability that Mr Murphy arrives at work on time.

6 ▶ There are 3 red, 4 blue and 5 green marbles in a bag. Husni picks a marble then John picks a marble.

1st marble   2nd marble

$\frac{3}{12} = \frac{1}{4}$ Red
$\frac{2}{11}$ Red
Blue
Green

Blue
Red
Blue
Green

Green
Red
Blue
Green

a Copy and complete the tree diagram to show all the probabilities.

b What is the probability that they both pick the same colour?

7 ▶ There are 3 tins of red paint and 11 tins of green paint.

James and Melek both take a tin at random.

Work out the probability that they do not pick the same colour.

8 ▶ Jon has 8 packets of soup in his cupboard, but all the labels are missing.

He knows that there are 5 packets of tomato soup and 3 packets of mushroom soup.

He opens three packets at random.

Work out the probability that

a all three packets are the same variety of soup
b he opens more packets of mushroom soup than tomato soup.

EXAMPLE 8

SKILLS

REASONING

A fruit basket contains two oranges (O) and three apples (A).

A fruit is selected at random and not replaced before another is randomly selected.

Calculate the probability of choosing at least one apple.

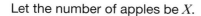

Let the number of apples be X.

At least one apple means that $X \geq 1$

First consider the tree diagram showing all the possible outcomes.

First selection   Second selection

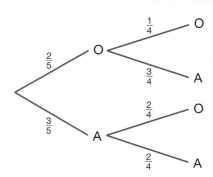

Method 1

List all the possibilities.

$P(X \geq 1) = P(X = 1) + P(X = 2) = P(OA) + P(AO) + P(AA)$

$$= \frac{2}{5} \times \frac{3}{4} + \frac{3}{5} \times \frac{2}{4} + \frac{3}{5} \times \frac{2}{4}$$

$$= \frac{6}{20} + \frac{6}{20} + \frac{6}{20} = \frac{18}{20} = \frac{9}{10}$$

Method 2

Use the rule: $P(E) + P(E') = 1$

$$P(X \geq 1) + P(X = 0) = 1$$

$$P(X \geq 1) = 1 - P(X = 0)$$

$$= 1 - P(OO)$$

$$= 1 - \frac{2}{5} \times \frac{1}{4}$$

$$= 1 - \frac{1}{10} = \frac{9}{10}$$

Note: Both methods need to be understood and applied.

EXERCISE 2*

1 ▶ A box contains two black stones and four white stones.

One is randomly selected and *not* replaced before another is randomly selected.

  a Draw a tree diagram to show all the possible outcomes.

  b Calculate the probability of selecting
    i a black stone and a white stone, in that order
    ii a black stone and a white stone, in any order
    iii at least one black stone.

2 ▶ A bag contains three orange discs and four purple discs. One is randomly selected and replaced *together with a disc of the colour not picked out* (orange or purple). Another disc is then randomly selected.

  a Draw a tree diagram to show all the possible outcomes.

  b Calculate the probability of selecting
    i two orange discs
    ii one disc of each colour
    iii at least one purple disc.

3 ▶ There are only black and white marbles in bag A and bag B. A marble is randomly chosen from bag A and is then placed in bag B. A marble is then randomly chosen from bag B.

Bag A   Bag B

  a Draw a tree diagram to show all the possible outcomes.
  b Calculate the probability of choosing
    i a black marble
    ii a white marble.

4 ▶ Helga oversleeps on one day in 5, and when this happens she breaks her shoelace 2 out of 3 times. When she does not oversleep, she breaks her shoelace only 1 out of 6 times. If she breaks her shoelace, she is late for school.

  a Calculate the probability that Helga is late for school.
  b What is the probability that she is not late for school?
  c In 30 school days, how many times would you expect Helga to be late?

5 ▶ Dan has to attend a meeting .
The probabilities of dry weather (D), rain (R) or snow (S) are
$P(D) = 0.5$, $P(R) = 0.3$, $P(S) = 0.2$
If it is dry the probability that Dan will arrive in time for the meeting is 0.8
If it rains the probability that he will arrive in time for the meeting is 0.4
If it snows the probability that he will arrive in time for the meeting is 0.15
Is he more likely to arrive in time for the meeting or be late?

6 ▶ Andy and Roger play a tennis match.
The probability that Andy wins the first set is 0.55
When Andy wins a set, the probability that he wins the next set is 0.65
When Roger wins a set, the probability that he wins the next set is 0.6
The first person to win two sets wins the match.

  a Draw a tree diagram to show all the possible outcomes.
  b Calculate the probability that Andy wins the match by 2 sets to 1.
  c Calculate the probability that Roger wins the match.

7 ▶ A box contains two red sweets and three green sweets. A sweet is picked at random and eaten. If a red sweet is picked first, then two extra red sweets are placed in the box. If a green sweet is picked first, then three extra green sweets are placed in the box. If two picks are made, calculate

  a P(two red sweets)
  b P(a red sweet and a green sweet)
  c P(at least one green sweet).

8 ▶ Suzi has just taken up golf, and she buys a golf bag containing five different clubs. Unfortunately, she does not know when to use each club, and so chooses them randomly for each shot. The probabilities for each shot that Suzi makes are shown in this table.

	GOOD SHOT	BAD SHOT
RIGHT CLUB	$\frac{2}{3}$	
WRONG CLUB		$\frac{3}{4}$

  a Copy and complete the table.
  b Use the table to calculate the probability that for the first two shots
    i both are good
    ii at least one is bad.

ACTIVITY 2

'The Book of Odds' is a collection of the probabilities that reflect life events in the USA. A few of these are shown below.

EVENT	PROBABILITY EXPRESSED AS 1 : n
Killed by a meteorite	1 : 700 000
Being made the next American President	1 : 10 000 000
Becoming a movie star	1 : 1 500 000
Becoming an astronaut	1 : 12 100 000
Going to hospital with a pogo-stick injury	1 : 115 300
Having a genius child	1 : 1 000 000
Winning an Olympic Gold Medal	1 : 662 000
Death of a left-handed person mis-using a right-handed product	1 : 7 000 000
Being wrongfully convicted of a crime	1 : 3703
Being made a saint	1 : 20 000 000

Select three of these USA 'life-events' and give a simple explanation of how these figures might have been calculated.

EXERCISE 3

REVISION

For these questions, use tree diagrams where appropriate.

1 ▶ A box contains two Maths books and three French books. A book is removed and replaced before another is taken.

    a Draw a tree diagram to show all the possible outcomes.
    b Find the probability of removing
      i two French books
      ii a Maths book and a French book.

2 ▶ A football penalty taker has a $\frac{3}{4}$ probability of scoring a goal.

    a If she takes two penalties, find the probability that she scores no goals.
    b What is the probability that she scores one goal from two penalties?

3 ▶ The probability that a biased coin shows tails is $\frac{2}{5}$

When the coin is tossed two times, find the probability of getting

    a two heads
    b exactly one tail.

Head   Tail

4 ▶ Assume that the weather is always either sunny or rainy in Italy.

If the weather is sunny one day, the probability that it is sunny the day after is $\frac{1}{5}$

If it rains one day, the probability that it rains the next day is $\frac{3}{4}$

   a If it is sunny on Sunday, calculate the probability that it rains on Monday.
   b It is sunny on Sunday. What is the probability that it is sunny on Tuesday?

5 ▶ The letters of the word HYPOTHETICAL appear on plastic squares that are placed in a bag and mixed up. A square is randomly selected and replaced before another is taken. Calculate

   a P(two H)
   b P(one T)
   c P(a vowel).

6 ▶ All female chaffinches have the same patterns of laying eggs.

The probability that any female chaffinch will lay a certain number of eggs is given in the table.

NUMBER OF EGGS	0	1	2	3	4 or more
PROBABILITY	0.1	0.3	0.3	0.2	x

   a Calculate the value of x.
   b Calculate the probability that a female chaffinch will lay fewer than 3 eggs.
   c Calculate the probability that two female chaffinches will lay a total of 2 eggs.

EXERCISE 3*

REVISION

For these questions, use tree diagrams where appropriate.

1 ▶ A chocolate box contains four milk chocolates and five coconut sweets. Gina loves milk chocolates, and hates coconut sweets. She takes one chocolate randomly, and *does not replace it* before picking out another one at random.

   a Calculate the probability that Gina is very happy.
   b What is the probability that Gina is very unhappy?
   c What is the probability that she has at least one milk chocolate?

2 ▶ The probability that Mr Glum remembers his wife's birthday and buys her a present is $\frac{1}{3}$

The probability that he does not lose the present on the way home is $\frac{2}{3}$

The probability that Mrs Glum likes the present is $\frac{1}{5}$

   a Calculate the probability that Mrs Glum receives a birthday present.
   b What is the probability that Mrs Glum receives a birthday present but dislikes it?
   c What is the probability that she is happy on her birthday?
   d What is the probability that she is not happy on her birthday?

3 ▶ A sleepwalker (a person that walks in their sleep) gets out of bed and is five steps away from his bedroom door. He is equally likely to take a step forward as he is to take one backwards.

    a Calculate the probability that after five steps he is at his bedroom door.

    b What is the probability that after five steps he is only one step away from his bedroom door?

    c What is the probability that, after taking five steps, he is closer to his bedroom door than to his bed?

4 ▶ There are 25 balls in a bag.
Some of the balls are red.
All the other balls are blue.
Kate picks two balls at random and does not replace them.
The probability that she will pick two red balls is 0.07
Calculate the probability that the two balls she picks will be of different colours.

5 ▶ The diagram shows an archery target that is divided into three regions by concentric circles with radii that are in the **ratio** $1 : 2 : 3$

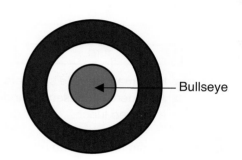

— Bullseye

    a Find the ratio of the areas of the three regions in the form $a : b : c$, where a, b and c are **integers**.

The probability that an arrow will hit the target is $\frac{4}{5}$
If it does hit the target, the probability of it hitting any region is proportional to the area of that same region.

    b Calculate the probability that an arrow will hit the bullseye.

    c Two arrows are shot. Calculate the probability of these events.

      i They both hit the bullseye.

      ii The first hits the bullseye and the second does not.

      iii At least one hits the bullseye.

6 ▶ The two-way table shows the number of deaths and serious injuries caused by road traffic accidents in Great Britain in 2013.

		SPEED LIMIT			
		20 mph	**30 mph**	**40 mph**	**Total**
	FATAL	6	520	155	681
TYPE OF INJURY	**SERIOUS**	420	11 582	1662	13 664
	TOTAL	426	12 102	1817	14 345

Work out an estimate for the probability

a that the accident is fatal given that the speed limit is 30 mph

b that the accident happens at 20 mph given that the accident is serious

c that the accident is serious given that the speed limit is 40 mph.

Give your answers to 2 **decimal places**.

EXAM PRACTICE: HANDLING DATA 5

1 The probability that Richard beats John at badminton is 0.7
The probability that Richard beats John at squash is 0.6
These events are independent.
Calculate the probability that, in a week when they play one game of badminton and one game of squash

  a Richard wins both games   **[2]**
  b Richard wins one game and loses the other.   **[2]**

2 Mohammed carried out a survey on the students in his year group to see who had school dinners and who had a packed lunch. The two-way table shows his results.

	SCHOOL DINNER	PACKED LUNCH	TOTAL
MALE	27	46	73
FEMALE	36	41	77
TOTAL	63	87	150

  a How many female students are there?   **[2]**
  b A student is chosen at random. Given that the student is female, work out the probability that the student has a packed lunch.   **[2]**

3 A certain teacher who often forgets things has a newspaper delivered to his home each morning, and he tries to remember to carry it from home to school, and back home again, each day. However, in practice he often forgets, and the probability that he remembers to take his newspaper on either journey is $\frac{2}{3}$

Also, the probability that he loses it on any journey (if he sets off with it) is $\frac{1}{5}$

Work out the probabilities, on a particular day, that

  a he takes his newspaper with him to school and loses it on the way   **[2]**
  b he arrives at school with his newspaper   **[2]**
  c he leaves the school with his newspaper   **[2]**
  d he arrives at home with his newspaper   **[2]**
  e he has that day's newspaper at home in the evening.   **[2]**

4 Alison, Brenda, Claire and Donna are the runners in a race.
The probabilities of Alison, Brenda, Claire and Donna winning the race are show below.

ALISON	BRENDA	CLAIRE	DONNA
0.31	0.28	0.24	0.17

  a Calculate the probability that either Alison or Claire will win the race.   **[2]**

Hannah and Tracy play each other in a tennis match.
The probability of Hannah winning the tennis match is 0.47

  b Copy and complete the probability tree diagram.   **[2]**

  c Calculate the probability that Brenda will win the race and Tracy will win the tennis match.   **[3]**

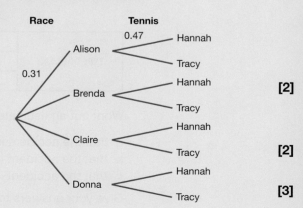

[Total 25 marks]

CHAPTER SUMMARY: HANDLING DATA 5

LAWS OF PROBABILITY

INDEPENDENT EVENTS

If two events have no effect on each other, they are independent.

For example, the fact that it rains in Hong Kong on Tuesday is not going to have an effect on whether Manchester United win on Saturday.

For two independent events A and B, the probability of both events occurring is

$$P(A \text{ and } B) = P(A) \times P(B)$$

MUTUALLY EXCLUSIVE EVENTS

A light cannot be both on and off at the same time. Such events are said to be mutually exclusive.

For two mutually exclusive events A and B, the probability of event A or event B occurring is

$$P(A \text{ or } B) = P(A) + P(B)$$

COMBINED EVENTS

An ordinary die is thrown twice.

Let event A be that a 6 is thrown. Then A′ is the event that a 6 is not thrown.

P(throwing **only one** 6)

$$= (P(A) \times P(A')) + (P(A') \times P(A))$$

$$= \left(\frac{1}{6} \times \frac{5}{6}\right) + \left(\frac{5}{6} \times \frac{1}{6}\right)$$

$$= \frac{10}{36}$$

$$= \frac{5}{18}$$

P(throwing **at least one** 6)

$$= P(AA) + P(AA') + P(A'A)$$

$$= 1 - P(A'A')$$

$$= 1 - \frac{5}{6} \times \frac{5}{6}$$

$$= 1 - \frac{25}{36}$$

$$= \frac{11}{36}$$

CONDITIONAL PROBABILITY

The probability of an event based on the occurrence of a previous event is called a conditional probability.

The sample space has been changed by a previous event.

If a Queen is picked from a pack of 52 playing cards, the probability that the next card picked is a Queen

$$= \frac{\text{number of Queens after one is picked}}{\text{total cards left in the pack after one is picked}} = \frac{3}{51}$$

UNIT 9

$9 = 3^{2^1}$ which makes it an exponential factorial number.

The decimal representation of π, starting at the 762nd decimal place, is a sequence of six 9s.

A polygon with nine sides is called a nonagon.

The Nine Muses in Greek mythology were daughters of Zeus and Mnemosyne.

NUMBER 9

In the UK, characters feature on banknotes to celebrate individuals who have shaped British thought, values and society. In 2007 a picture of Adam Smith (1723–1790), a Scottish philosopher and political economist, was put on the back of a £20 note. Smith, best known for his book *The Wealth of Nations*, is often called the 'father of modern economics', and his work still influences economic policies today.

LEARNING OBJECTIVES

- Decide which product or service is better value for money

- Carry out calculations involving money

- Solve real-life problems involving percentages and money

- Convert between currencies

BASIC PRINCIPLES

- Global financial processes can be complex. The ones in this section involve the simple day-to-day concepts of comparative costs, salaries and taxes, sales tax and foreign currency.

- The mathematical processes involved in this section have all been met before.

- The key skills all involve percentages.

- To calculate x **as a percentage of** y: $\frac{x}{y} \times 100$

- To calculate x **per cent of** y: 1% of $y = \frac{y}{100}$

  so $x\%$ of $y = x \times \frac{y}{100} = y \times \left(\frac{x}{100} \right)$

- 5% of a quantity can be found by multiplying by $\frac{5}{100}$ or 0.05

- 95% of a quantity can be found by multiplying by $\frac{95}{100}$ or 0.95

- $1\% = \frac{1}{100} = 0.01$ $10\% = \frac{10}{100} = \frac{1}{10} = 0.1$

- $50\% = \frac{50}{100} = \frac{1}{2} = 0.5$ $75\% = \frac{75}{100} = \frac{3}{4} = 0.75$

- Percentage change $= \dfrac{\text{value of change}}{\text{original value}} \times 100$

- Per annum (p.a.) is frequently used and means 'per year'.

PERCENTAGE INCREASE AND DECREASE

To increase a quantity by $R\%$, multiply it by $1 + \dfrac{R}{100}$

To decrease a quantity by $R\%$, multiply it by $1 - \dfrac{R}{100}$

PERCENTAGE CHANGE	MULTIPLY BY
+ 5%	1.05
+ 95%	1.95
− 5%	0.95
− 95%	0.05

COMPARATIVE COSTS

To help shoppers to compare the prices of packaged items, shopkeepers often show the cost per 100 grams or cost per litre. Other units will enable a consumer to compare the value of items they might wish to buy.

EXAMPLE 1

SKILLS

PROBLEM
SOLVING

A 100 g jar of Brazilian coffee costs €4.50, and a 250 g jar of the same coffee costs €12.

Which is the better value?

100 g jar:

$\text{cost/g} = \dfrac{450}{100}\,\text{cents/g}$

$\qquad = 4.50\,\text{cents/g}$

250 g jar:

$\text{cost/g} = \dfrac{1200}{250}\,\text{cents/g}$

$\qquad = 4.80\,\text{cents/g}$

The 100 g jar of coffee is better value.

EXERCISE 1

Where appropriate, give your answers to 3 significant figures.

1 ▶ Which tin of paint is the better value,
tin X, 1 litre for $4.50, or
tin Y, 2.5 litres for $10.50?

2 ▶ Which bag of rice is the better value,
bag P, 1 kg for $3.25, or
bag Q, 5 kg for $16.50?

3 ▶ Which roll of the same type of curtain material is the better value,
roll A, 10 m for $65, or
roll B, 16 m for $100?

4 ▶ Which bottle of cranberry juice is the better value,
bottle I, 330 ml for $1.18, or
bottle II, 990 ml for $3.20?

5 ▶ Which is the better value in tinned pineapples,
tin A, 450 g at $2.70, or
tin B, 240 g at $1.36?

EXERCISE 1*

Where appropriate give your answers to 3 **significant figures**.

1 ▶ A garden centre sells various sizes of monkey-puzzle trees.

A 1.5 m tree costs $45, while a 2 m tree is $55 and a 3.5 m
tree is sold at $110.

By calculating the cost for each tree in dollars per metre,
list the trees in order of value for money.

2 ▶ Mrs Becker wants to cover a kitchen floor with tiles. She considers three types of
square tile:

marble, 20 cm × 20 cm for $3.00 each, or
slate, 25 cm × 25 cm for $4.50 each, or
limestone, 0.3 m × 0.3 m for $6.50 each.

For each type of tile, calculate the cost in dollars per square metre and therefore find
the cheapest choice for Mrs Becker.

3 ▶ Which on-line movie rental company gives the best value?

Mega-Movie: 1 DVD at $7 for a week
Films R Us: 2 DVDs at $12 for five days

Q4 HINT
Find the value of
$ per ml for each.

4 ▶ Which bottle of 'Feline' perfume is the best value?

Small:	80 ml	$100
Medium:	160 ml	$180
Large:	250 ml	$280

5 ▶ Calculate the total cost of three companies' mobile phone
charges for 10 hours of calls and state which telephone
company is the best value over this period of time.

TELEPHONE COMPANY	MOBILE PHONE COST ($)	CALL COST PER MIN ($)
Yellow	75	0.20
Lime	50	0.25
Rainbow	15	0.32

TAXATION

Governments collect money from their citizens through tax to pay for public services such as education, health care, military defence and transport.

SALES TAX

This is the tax paid on spending. It is included in the price paid for most articles. In many countries some articles are free from sales tax, for example, children's clothing.

EXAMPLE 2

SKILLS

PROBLEM SOLVING

Rita buys a tennis racket for $54 plus sales tax at 15%.
Calculate the selling price.

Selling price = $54 × 1.15 = $62.10

EXAMPLE 3

SKILLS

PROBLEM SOLVING

Liam buys a computer game for $36 including a sales tax of 20%.
Calculate the sales tax.

Let the price excluding sales tax be p.

$p \times 1.20 = 36$

$p = \dfrac{36}{1.20} = 30$

Sales tax = $36 − $30 = $6

EXERCISE 2

1 ▶ A computer costs $1200 plus sales tax at 15%. Calculate the selling price.

2 ▶ The price of a food mixer is advertised as £320 excluding sales tax at 15%. Calculate the selling price.

3 ▶ Mika is a builder who charges €24 000 to build a house extension excluding sales tax at 17.5%.

What is Mika's final bill?

4 ▶ Frida buys 15 books for her local library at €7.50 each excluding sales tax at 12%.

What is Frida's total bill for these books?

5 ▶ Tomiwa buys a house for £750 000 excluding a sales tax of 15%.

What is Tomiwa's bill for the house?

EXERCISE 2*

1 ▶ Ian buys a vintage car for £36 800 inclusive of sales tax at 15%.
Calculate the amount of sales tax paid.

2 ▶ Saskia buys a new coat for €460 inclusive of sales tax at 15%.
Calculate the cost of the coat excluding sales tax.

3 ▶ Herbert has some trees removed from his garden. He is told that the bill excluding sales tax at 17.5% is $750. He is then charged $885.
Herbert thinks he has been overcharged. Is he correct?

4 ▶ Ruth hires a plumber who charges £40 per hour plus sales tax at 15%.
She is charged £368 for her total bill.
How many hours did Ruth hire the plumber for?

5 ▶ Zac buys a watch for $y including sales tax at 15%.
Show that the sales tax paid by Zac is $\dfrac{3y}{23}$

SALARIES AND INCOME TAX

A salary is a fixed annual sum of money usually paid each month, but the salary is normally stated per annum (p.a.).

Income tax is paid on money earned and most governments believe that richer people should pay more tax than poorer people. As a result of this the amount of tax falls into different rates or 'tax bands'.

The country of Kalculus has these tax rates.

TAX RATES	SALARY P.A
20%	$0–$50 000
40%	> $50 000

The examples below explain how these tax rates are used to find the tax owed.
Any salary over $50 000 p.a. pays 40% on the amount over $50 000 paid p.a.

EXAMPLE 4

SKILLS

PROBLEM SOLVING

Lauren earns $30 000 p.a. as a waitress in the country of Kalculus.
What is her monthly income tax bill?

Tax rate on $30 000 is 20%.

Annual tax = $30 000 × 0.20 = $6000

Monthly tax bill = $\dfrac{\$6000}{12} = \$500$

EXAMPLE 5

SKILLS

PROBLEM SOLVING

Frankie earns $1 000 000 p.a. as a professional footballer in the country of Kalculus.
What is his monthly income tax bill?

The first $50 000 is taxed at 20%, the remainder is taxed at 40%.

Annual tax at 20%:	$50 000 × 0.20	= $10 000	p.a.
Annual tax at 40%:	($1 000 000 − $50 000) × 0.40	= $380 000	p.a.
Total tax p.a.	$10 000 + $380 000	= $390 000	p.a.

Monthly income tax bill = $\dfrac{\$390\,000}{12}$ = $32 500

EXERCISE 3

Assume that all the taxation questions in Exercise 3 and Exercise 3* are based in the country of Kalculus, so the tax rates in the previous table are applied.

1 ▶ Emma works in a cake shop earning $40 000 p.a.
What is her annual income tax bill?

2 ▶ Dominic earns $50 000 p.a. as a computer technician.
What is his annual income tax bill?

3 ▶ Claudia earns $120 000 p.a. as a bank manager.
What is her monthly income tax bill?

4 ▶ Mimi earns $250 000 p.a. as a brain surgeon.
How much will she earn per month after income tax has been deducted?

5 ▶ Mr and Mrs Gregor earn $140 000 p.a. and $70 000 p.a. respectively.
What is their monthly income tax bill?

EXERCISE 3*

1 ▶ Bruno earns $750 000 p.a. as an accountant.
Calculate what percentage of his salary he pays in annual income tax.

2 ▶ Shakira earns $60 000 p.a. as a librarian.
Calculate what percentage of her salary she pays in annual income tax.

3 ▶ Najid paid $10 000 income tax in a year.
What was his salary that year?

4 ▶ Sofya paid $1 000 000 income tax in a year.
What was her salary that year?

5 ▶ Fedor earns x p.a. where $x > \$50\,000$. His monthly income tax bill is y.
Show that $y = \dfrac{x - 25000}{30}$

FOREIGN CURRENCY

Different currencies are exchanged and converted around the world, so an agreed rate of conversion is needed.

The table below shows some examples of values compared to $1 (USA).

COUNTRY OR CONTINENT	CURRENCY	EXCHANGE RATE TO $1 (US dollar)
Australia	dollar	$1.46
China	yuan	元6.57
Europe	euro	€0.92
India	rupee	₹67.79
Nigeria	naira	₦199.25
Russia	ruble	₽77.75
South Africa	rand	R16.78
UK	pound	£0.70

EXAMPLE 6

How many euros will $150 (US dollars) buy?

SKILLS

PROBLEM SOLVING

Using the conversion rates in the table above:
$1 = €0.92
$150 = 150 × €0.92 = €138

EXERCISE 4

Use the exchange rates in the table above.

1 ▶ Convert 250 US dollars into
   a UK pounds   b euros   c Russian rubles.

2 ▶ Convert 75 US dollars into
   a South African rand   b Nigerian naira   c Chinese yuan.

3 ▶ Convert these into US dollars.
   a £1500   b ₹1500   c €1500

4 ▶ How many US dollars is an Australian dollar millionaire worth?

5 ▶ An American house is advertised in the UK for £192 000 and in France for €250 000.
   Which price is cheaper in US dollars and by how much?

EXAMPLE 7   How many UK pounds will 元150 buy?

SKILLS   Using the conversion rates in the table on page 271:

PROBLEM
SOLVING $1 = 元6.57   (Divide both sides by 6.57)

$0.15 = 元1

元150 = 150 × $0.15 = $22.50

$22.50 = 22.50 × £0.70 = £15.75

EXERCISE 4*   Use the exchange rates in the table on page 271.

1 ▶   Convert £250 into
    a US dollars   **b** euros   **c** rupees.

2 ▶   Convert 元1000 into
    a rand   **b** rubles   **c** Australian dollars.

3 ▶   How many UK pounds is a Russian ruble millionaire worth?

4 ▶   Deon has £2400 and plans to travel through Nigeria and South Africa. He wants to convert all of his money to the currencies of Nigeria (naira) and South Africa (rand) in a **ratio** of 1:2. How much of each currency will he have?

5 ▶   Shakina has €3600 and plans to travels through China, Australia and America. She wants to convert all of her money to the currencies of China (yuan), Australia (Australian dollars) and America (US dollars) in the ratio of 1:2:3.
How much of each currency will she have?

EXERCISE 5   REVISION

1 ▶   The price of two brands of loose-leaf tea are given as:
    Green:$1.90/500g   Mint:$0.76/200g
    Which brand is the better value?

2 ▶   Rita buys a watch for $2115 inclusive of sales tax at 17.5%.
    Calculate how much sales tax she paid.

3 ▶   Mimi earns $90 000 p.a. while her brother, Willem, earns $70 000 p.a.
    Calculate the difference in their monthly income tax bills.
    Use the tax rates for Kalculus on page 269.

4 ▶ $1 (US dollars) = £0.70 (UK pounds) = €0.92 (European euros)

    Convert
    a £120 into $ **b** €120 into £.

5 ▶ $1 (US dollars) = 元6.57 (Chinese yuan)
    Teresa invests 元1000 for 5 years with a **compound interest** of 5%.
    Calculate her profit in US dollars.

EXERCISE 5*

REVISION

1 ▶ Three car models are tested for their fuel efficiency by driving them at 80 km/h.

CAR MODEL	DISTANCE TRAVELLED (km)	FUEL USED (litres)
Fizz	50	2.2
Tyrol	100	4.5
Wessex	300	12.9

Rank the cars for fuel efficiency.

2 ▶ Lakshand buys a large tin of paint. The price of the paint, £x, is reduced by 15%. Sales tax of 15% is then added so the final cost is £46.92. Find the value of x.

3 ▶ Roberto is a professional basketball player who earns $100 000 per week.
Calculate how much income tax Roberto pays per day.
(Use the tax rates for Kalculus on page 269.)

4 ▶ Martha wants to buy a new Tornado Camper Van paying in US dollars.
She is able to purchase it for £45 000 in the UK or €60 000 in Spain.
If $1 (US dollars) = £0.70 (UK pounds) = €0.92 (European euros), where should Martha buy the van?
Show clear reasons for your answer.

5 ▶ Mr Dupois earns $$x$ p.a. and Mrs Dupois earns $$y$ p.a. in the country of Kalculus, where $x < 50 000$ and $y > 50 000$.
The Dupois family monthly income tax bill is $$z$.
Show that $z = \dfrac{x + 2y - 50000}{60}$
(Use the tax rates for Kalculus on page 269.)

EXAM PRACTICE: NUMBER 9

1 Which car has the better fuel efficiency?
Pluto: 11 litres/200 km
Jupiter: 14.1 litres/300 km   **[4]**

2 Hilda buys a coffee maker for $437 inclusive of sales tax at 15%.
Calculate the original price excluding sales tax. **[3]**

3 The country of Kalculus has these tax rates:

TAX RATES	SALARY PER ANNUM (P.A)
20%	$0–$50 000
40%	> $50 000

Mr Hildenberg earns $120 000 p.a. as a dentist.
Mrs Hildenberg earns $150 000 p.a. as an engineer.
How much does the Hildenberg family pay in tax per week?   **[8]**

4 $1 (US dollars) = £0.70 (UK pounds) = €0.92 (European euros)

Convert
a $25 into £   **b** £25 into $ **c** €25 into $ **[6]**

5 Simeon wants to buy a French house for €165 000 by using UK pounds.
The exchange rate for £ (UK pounds) to € (European euros) changes dramatically over a few months.

January 1st: £1 = €1.25
April 1st: £1 = €1.65

How much will Simeon save in UK pounds if he makes the purchase on April 1st rather than January 1st?   **[4]**

[Total 25 marks]

CHAPTER SUMMARY: NUMBER 9

COMPARATIVE COSTS

Shopkeepers show the cost per 100 grams or cost per litre.

This enables a consumer to compare the value of items they might wish to buy.

Cars can be compared for efficiency of fuel by finding miles per litre of fuel used by the engine at a fixed speed, often 80 km/h.

TAXATION

Governments collect money from their citizens through tax to pay for public services such as education, health care, defence and transport.

SALES TAX

Sales tax is paid on spending and is a fixed rate, often 15% or 20%. It is included in the price paid for most articles. In many countries some articles are free from sales tax, for example, children's clothing.

SALARIES AND INCOME TAX

A salary is a fixed annual sum on payment for a job of that is normally stated per annum (p.a.).

Income tax is paid on money earned. Most governments believe that richer people should pay more tax than poorer people. Therefore the amount of tax falls into different tax rates or 'tax bands'.

FOREIGN CURRENCY

A person visiting a foreign country for a holiday or on a business trip may need to buy foreign currency. Different currencies are exchanged and converted around the world, so an agreed rate of conversion is needed.

These rates are constantly changing and are shown in comparison to the unit rate of currency in the country where the money is being exchanged, for example, £1, $1 or €1.

	Selling Rate (£)	Country	Currency	Buying Rate (£)	Selling Rate (£)	Country	Currency	Buying Rate
47987	1.5025		Kroner Norway	9.38	8.88		Lita Lithuania	4.12
569869	1.485		Kroner Sweden	11.4	10.78		Lat Latvia	0.545
149425	1.1125		Kroner Iceland	299	184		Ruble Russia	47.46
168224	1.09		Pound Egypt	8.84	7.8		Dollars Barbados	3.15
1.724138	1.626		Dollars HongKong	12.23	11.6		Leu Rumania	4.87
1.7487	1.61		Dirham UAE	5.77	5.43		Lev Bulgaria	2.31
1.599999	1.51		Baht Thailand	51.42	47.45		Sheqel Israel	5.93
1.6252	1.49		Dollar Singapore	2.2	2.0875		Dinar Jordan	1.1279
148.5884	140.6		Ringgit Malaysia	5.15	4.74		Riyal Qatar	5.77
1.663894	1.574		Koruna Czech	29.94	27.7		Rial Oman	0.8101
2.22	2.115		Forint Hungary	313.61	287		Riyal Saudi Arabia	5.94
11.59	10.8		Zloty Poland	4.53	4.21		Real Brazil	2.85
2.38	2.26		Kroons Estonia	18.47	16.88		Peso Argentina	5.89
	8.23		Kuna Croatia	8.84	7.85		Peso Chile	860.5

Commission 0%   Commission 0%

ALGEBRA 9

One of the most famous theorems in mathematics is Fermat's Last Theorem which states that $x^n + y^n = z^n$ has no non-zero integer solutions when $n > 2$. Fermat wrote in the margin of his notebook in 1637 'I have discovered a truly remarkable proof which this margin is too small to contain.' Encouraged by this statement, mathematicians struggled for 358 years to prove this theorem before a proof was published in 1995 by Andrew Wiles. The proof itself is over 150 pages long and took seven years to complete.

Pierre de Fermat 1601–65 ▲   Andrew Wiles 1953– ▲

LEARNING OBJECTIVES

■ Solve simultaneous equations with one equation being quadratic

■ Solve simultaneous equations with one equation being a circle

■ Prove a result using algebra

BASIC PRINCIPLES

■ Solve **quadratic equations** (using **factorisation** or the quadratic formula).

■ Solve simultaneous equations (by substitution, elimination or graphically).

■ **Expand** brackets.

■ Expand the **product** of two linear expressions.

■ Form and **simplify** expressions.

■ Factorise expressions.

■ Complete the square for a quadratic expression.

SOLVING TWO SIMULTANEOUS EQUATIONS – ONE LINEAR AND ONE NON-LINEAR

ACTIVITY 1

SKILLS

ANALYSIS

Use the graph to solve the simultaneous equations
$x + 2y = 10$ and $x^2 + y^2 = 25$

What is the connection between the line $3y = 4x - 25$ and the circle $x^2 + y^2 = 25$?

Are there any real solutions to the simultaneous equations
$3y = 18 - x$ and $x^2 + y^2 = 25$?

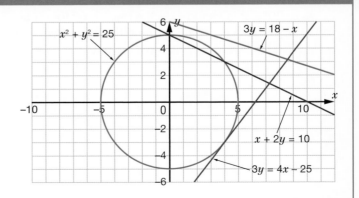

KEY POINTS

■ When solving simultaneous equations where one equation is linear and the other is non-linear:
  - ■ If there is one solution, the line is a **tangent** to the curve.
  - ■ If there is no solution, the line does not **intersect** the curve.

Drawing graphs is one way of solving **simultaneous equations** where one equation is linear and the other quadratic. Sometimes they can be solved algebraically.

EXAMPLE 1

SKILLS

ANALYSIS

Solve the simultaneous equations $y = x + 6$ and $y = 2x^2$ algebraically and show the result graphically.

$y = x + 6$ (1)

$y = 2x^2$ (2)

Substitute (2) into (1):

$2x^2 = x + 6$ (Rearrange)

$2x^2 - x - 6 = 0$ (Factorise)

$(2x + 3)(x - 2) = 0$ (Solve)

So either $(2x + 3) = 0$ or $(x - 2) = 0$

$x = -1\frac{1}{2}$ or $x = 2$

Substitute $x = -1\frac{1}{2}$ into (1) to give $y = 4\frac{1}{2}$

Substitute $x = 2$ into (1) to give $y = 8$

So the solutions are $x = -1\frac{1}{2}$, $y = 4\frac{1}{2}$ and $x = 2$, $y = 8$

The graphs of the equations are shown below.

The solutions correspond to the intersection points $(-1\frac{1}{2}, 4\frac{1}{2})$ and $(2, 8)$.

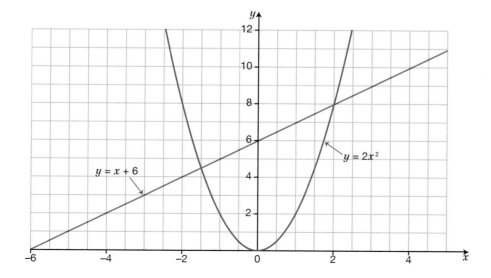

KEY POINTS

■ If the two equations are of the form $y = f(x)$ and $y = g(x)$:

  ■ Solve the equation $f(x) = g(x)$ to find x.
  ■ When x has been found, find y using the easier of the original equations.
  ■ Write out your solutions in the correct pairs.

EXERCISE 1

Solve the simultaneous equations.

1 ▶ $y = x + 6$, $y = x^2$

2 ▶ $y = 2x + 3$, $y = x^2$

3 ▶ $y = 3x + 4$, $y = x^2$

4 ▶ $y = 2x + 8$, $y = x^2$

5 ▶ $y = x + 1$, $y = x^2 - 2x + 3$

6 ▶ $y = x - 1$, $y = x^2 + 2x - 7$

7 ▶ $y = x + 1$, $y = \dfrac{2}{x}$

8 ▶ $y = 1 + \dfrac{2}{x}$, $y = x$

EXERCISE 1*

Solve the simultaneous equations, giving your answers **correct to** 3 s.f. where appropriate.

1 ▶ $y = 2x - 1$, $y = x^2 + 4x - 6$

2 ▶ $y = 3x + 1$, $y = x^2 - x + 2$

3 ▶ $y = 4x + 2$, $y = x^2 + x - 5$

4 ▶ $y = 1 - 3x$, $y = x^2 - 7x + 3$

5 ▶ $y = x + 2$, $y = \dfrac{8}{x}$

6 ▶ $y = 1 + \dfrac{2}{x}$, $y = \dfrac{3}{x^2}$

7 ▶ $y = 3\sqrt{x}$, $y = x + 1$

8 ▶ $y = 1 + \dfrac{15}{x^4}$, $y = \dfrac{8}{x^2}$

Example 2 shows how to solve algebraically the pair of simultaneous equations from Activity 1.

EXAMPLE 2

SKILLS

ANALYSIS

Solve the simultaneous equations $x + 2y = 10$ and $x^2 + y^2 = 25$

$x + 2y = 10$ .... (1)

$x^2 + y^2 = 25$ .... (2)

Make x the **subject** of equation (1) (the linear equation):

$x = 10 - 2y$ .... (3)

Substitute (3) into (2) (the non-linear equation):

$(10 - 2y)^2 + y^2 = 25$ (Expand brackets)

$100 - 40y + 4y^2 + y^2 = 25$ (Simplify)

$5y^2 - 40y + 75 = 0$ (Divide both sides by 5)

$y^2 - 8y + 15 = 0$ (Solve by factorising)

$(y - 3)(y - 5) = 0$

$y = 3$ or $y = 5$

Substitute $y = 3$ into (1) to give $x = 4$

Substitute $y = 5$ into (1) to give $x = 0$

So the solutions are $x = 0$, $y = 5$ and $x = 4$, $y = 3$

EXAMPLE 3

SKILLS

ANALYSIS

Solve the simultaneous equations $x + y = 4$ and $x^2 + 2xy = 2$

$x + y = 4$ (1)

$x^2 + 2xy = 2$ (2)

Substituting for y from (1) into (2) will make the working easier.

Make y the subject of (1):

$y = 4 - x$ (3)

Substitute (3) into (2):

$x^2 + 2x(4 - x) = 2$ (Expand brackets)

$x^2 + 8x - 2x^2 = 2$ (Simplify)

$x^2 - 8x + 2 = 0$ (Solve using the quadratic formula)

$$x = \frac{8 \pm \sqrt{(-8)^2 - 4 \times 1 \times 2}}{2 \times 1}$$

$x = 0.258$ or $x = 7.74$ (to 3 s.f.)

Substituting $x = 0.258$ into (1) gives $y = 3.74$ (to 3 s.f.)

Substituting $x = 7.74$ into (1) gives $y = -3.74$ (to 3 s.f.)

So the solutions are
$x = 0.258$ and $y = 3.74$ or $x = 7.74$ and $y = -3.74$ (to 3 s.f.)

EXERCISE 2

For Questions 1–10, solve the simultaneous equations, giving your answers correct to 3 **significant figures** where appropriate.

1 ▶ $2x + y = 1$, $x^2 + y^2 = 2$ 6 ▶ $x + y = 3$, $x^2 - 2y^2 = 4$

2 ▶ $x + 2y = 2$, $x^2 + y^2 = 1$ 7 ▶ $x - 2y = 3$, $x^2 + 2y^2 = 3$

3 ▶ $x - 2y = 1$, $xy = 3$ 8 ▶ $2x + y = 2$, $4x^2 + y^2 = 2$

4 ▶ $3x + y = 4$, $xy = -4$ 9 ▶ $x - y = 2$, $x^2 + xy - 3y^2 = 5$

5 ▶ $x + y = 2$, $3x^2 - y^2 = 1$ 10 ▶ $x + y = 4$, $2x^2 - 3xy + y^2 = 4$

11 ▶ The rim (outer edge) of a bicycle wheel has a **radius** of 30 cm, and the inner hub (central part) has a radius of 3 cm. The spokes (bars connected to the centre) are tangents to the inner hub. The diagram shows just one spoke. The x and y axes are positioned with the origin at the centre of the wheel as shown. The equation of the rim is $x^2 + y^2 = 900$.

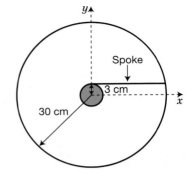

  a Write down the equation of the spoke.
  b Solve the equations simultaneously.
  c What is the length of the spoke?

12 ▶ The shape of the **cross-section** of a vase is given by $y^2 - 24y - 32x + 208 = 0$, with units in cm. The vase is 20 cm high. Find the radius of the top of the vase.

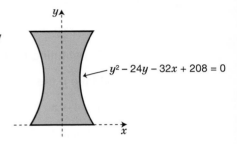

$y^2 - 24y - 32x + 208 = 0$

EXERCISE 2*

In Questions 1–3, solve the simultaneous equations, giving your answers to 3 significant figures where appropriate.

1 ▶ $2x + y = 2,\ 3x^2 - y^2 = 3$

2 ▶ $y - x = 4,\ 2x^2 + xy + y^2 = 8$

3 ▶ $x + y = 1,\ \dfrac{x}{y} + \dfrac{y}{x} = 2.5$

4 ▶ Find the points of intersection of the circle $x^2 - 6x + y^2 + 4y = 12$ and the line $4y = 3x - 42$. What is the connection between the line and the circle?

5 ▶   **a** Find the intersection points A and B of the line $4y + 3x = 22$ and the circle $(x - 2)^2 + (y - 4)^2 = 25$
  b Find the distance AB.

6 ▶ The design for some new glasses frames is shown in the diagram. They consist of two circles, both 2 cm radius, which are held 2 cm apart by a curved bridge piece and a straight length of wire AB.

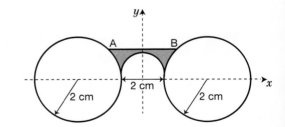

Axes are set up as shown, and AB is 1.5 cm above the x-axis.
The equation of the left-hand circle is $(x + 3)^2 + y^2 = 4$

  a Write down the equation of the line AB.
  b Find the co-ordinates of A and B and the length of the wire AB.

7 ▶ A rocket is launched from the surface of the Earth. The surface of the Earth can be modelled by the equation $x^2 + y^2 = 6400^2$ where the units are in km.

The path of the rocket can be modelled by the equation $y = 8000 - \dfrac{x^2}{2500}$

Find the co-ordinates of where the rocket takes off and where it lands.

8 ▶ Tracy is designing a desk decoration which is part of a
sphere 8 cm in **diameter**.
The decoration is 7 cm high.
The diagram shows a cross-section of the decoration.
The equation of the circle is $x^2 + y^2 = 16$

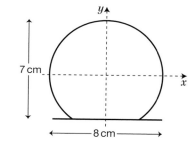

    a Write down the equation of the base.
    b Solve these simultaneous equations and find the
    diameter of the base.

9 ▶ Maria is watering her garden with a hose. Her little brother,
Peter, is annoying her and she tries to spray him with water.

The path of the water jet is given by $y = 2x - \frac{1}{4}x^2$ and the

slope of the garden is given by $y = \frac{1}{4}x - 1$. Peter is standing

at (8, 1). The origin is the point where the water leaves the
hose and units are in metres. Solve the simultaneous
equations to find where the water hits the ground. Does
Peter get wet?

10 ▶ During a football match José kicks a football onto the roof of the stadium. The path of

the football is given by $y = 2.5x - \frac{x^2}{15}$. The equation of the roof of the stadium is given by

$y = \frac{x}{2} + 10$ for $20 \leq x \leq 35$. All units are in metres. Solve the simultaneous equations and

find where the football lands on the roof.

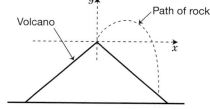

11 ▶ A volcano shaped like a cone ejects a rock from the top.
The path of the rock is given by $y = 2x - x^2$ and the side

of the volcano is given by $y = -\frac{2x}{3}$ where the units

are in km and the origin is at the top.
Solve the simultaneous equations and find where the
rock lands.

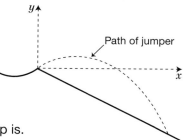

12 ▶ In a ski-jump the path of the skiier is given by

$y = 0.6x - \frac{x^2}{40}$

while the landing slope is given by $y = -\frac{x}{2}$. The origin is at

the take-off point of the skier, and all units are in metres.
Solve the simultaneous equations and find how long the jump is.

PROOF

Beal's Conjecture

There was once a billionaire called Beal,
Pursued his Conjecture with zeal,
If someone could crack it
They'd make a right packet:
But only genius will clinch'em the deal.

Limerick by Rebecca Siddall (an IGCSE student)

There is as yet no proof of Beal's Conjecture, made by Andrew Beal in 1993, which is
'If $a^x + b^y = c^z$ then a, b and c must have a common **prime factor**.'
(All letters are positive **integers** with x, y and $z > 2$.)

Andrew Beal himself has offered a reward of $\$10^6$ for a proof or **counter-example** to his conjecture.

To prove that a statement is true you must show that it is true in all cases. However, to prove that a statement is not true all you need to do is find a counter-example. A counter-example is an example that does not fit the statement.

EXAMPLE 4

SKILLS

ANALYSIS

Bailey says that substituting integers for n in the expression $n^2 + n + 1$ always produces **prime numbers**. Give a counter-example to prove that Bailey's statement is not true.

Substitute different integers for n.

When $n = 1$, $n^2 + n + 1 = 3$ which is prime.

When $n = 2$, $n^2 + n + 1 = 7$ which is prime.

When $n = 3$, $n^2 + n + 1 = 13$ which is prime.

When $n = 4$, $n^2 + n + 1 = 21$ which is not prime.

Disproving statements by considering a lot of cases and trying to find a counter-example can take a long time, and will not work if the statement turns out to be true. For example, the formula $2n^2 + 29$ will produce prime numbers up to $n = 28$ and most people would have given up long before the solution is found!

EXAMPLE 5

SKILLS

ANALYSIS

Find a counter-example to prove that $2n^2 + 29$ is not always prime.

Substitute $n = 29$ because $2n^2 + 29$ will then factorise.

When $n = 29$, $2 \times 29^2 + 29 = 1711$

However $2 \times 29^2 + 29 = 29(2 \times 29 + 1) = 29 \times 59$

So $1711 = 29 \times 59$ showing that 1711 is not prime.

EXERCISE 3

Give a counter-example to prove that these statements are not true.

1 ▶   The **sum** of two odd numbers is always odd.

2 ▶   A **quadrilateral** with sides of equal length is always a square.

3 ▶   All prime numbers are odd.

4 ▶   The difference between two numbers is always less than their sum.

5 ▶ $(x + 2)^2 = x^2 + 4$

6 ▶ $n^2 + n + 41$ always produces prime numbers.

7 ▶   The sum of two square numbers is always even.

8 ▶   The square of a number is always greater than the number itself.

EXERCISE 3*

Give a counter-example to prove that these statements are not true.

1 ▶ The difference between two square numbers is always odd.

2 ▶ The sum of two cubed numbers is always odd.

3 ▶ The product of two numbers is always greater than their sum.

4 ▶ The cube of a number is always greater than its square.

5 ▶ If $a < b$ then $ka < kb$

6 ▶ If $a < b$ then $a^2 < b^2$

7 ▶ $(x + y)^2 = x^2 + y^2$

8 ▶ $n^4 + 29n^2 + 101$ always produces prime numbers.

To prove that a statement is true, we must show that it is true in all cases.

EXAMPLE 6

Prove that the difference between the squares of any two **consecutive** integers is equal to the sum of these integers.

Let n be any integer. Then the next integer is $n + 1$

The difference between the squares of these two consecutive integers is

$(n + 1)^2 - n^2 = n^2 + 2n + 1 - n^2 = 2n + 1$

The sum of these two consecutive integers is $n + (n + 1) = 2n + 1$

So the difference between the squares of any two consecutive integers is equal to the sum of these integers.

KEY POINT

- When n is an integer, consecutive integers can be written in the form
  $..., n - 1, n, n + 1, n + 2, ...$

EXERCISE 4

1 ▶ Prove that the sum of any three consecutive integers is divisible by three.

2 ▶ The **median** of three consecutive integers is n.

   a Write down expressions in terms of n for the three integers.

   b Prove that the **mean** of the three integers is also n.

3 ▶ Prove that the sum of the squares of two consecutive integers is not divisible by two.

4 ▶ If n is an integer, prove that $(3n + 1)^2 - (3n - 1)^2$ is a **multiple** of 12.

5 ▶ a, b and c are three consecutive numbers. Prove that $c^2 - a^2 = 4b$

EXERCISE 4*

Q2 HINT
Triangle
numbers
can be
written as
$\dfrac{n(n+1)}{2}$

1 ▶   Prove that the sum of any four consecutive integers is not divisible by four.

2 ▶   Prove that eight times a triangle number is one less than a **perfect square**.

3 ▶   If a, b, c and n are any integers, prove that there is a value of n for which $an^2 + bn + c$ is not prime.

4 ▶   The nth rectangle number is given by $R_n = n(n + 1)$. Prove that $(R_n + R_{n+1})$ is twice a perfect square.

5 ▶   **Multiply out** $(n - 1)n(n + 1)$. Hence show $n^3 - n$ is divisible by   **a** 2   **b** 3   **c** 6

Q5 HINT
Consider the two cases when n is even and when n is odd.

ACTIVITY 2

SKILLS

ANALYSIS

1 ▶   In this 4×4 grid, a 2×2 square is highlighted.

1	2	3	4
5	6	7	8
9	10	11	12
13	14	15	16

   a   Work out $(11 + 12) - (7 + 8)$

   b   Choose another 2×2 square and work out (sum of the two larger numbers) − (sum of the two smaller numbers)

   c   Use algebra to prove that you will get the same result for all the 2×2 squares in the grid.

   d   Prove a similar result for all 2×2 squares in a 5×5 grid.

   e   Prove a similar result for all 2×2 squares in an $m \times m$ grid ($m \geq 2$).

2 ▶   In this 4×4 grid, a 2×2 square is highlighted.

1	2	3	4
5	6	7	8
9	10	11	12
13	14	15	16

   a   Prove that for any 2×2 square in this 4×4 grid, the products of the diagonals have a difference of 4.

   b   Prove a similar result for all 2×2 squares in an $m \times m$ grid ($m \geq 2$).

   c   Prove a similar result for all 2×2 squares in a rectangular grid with p rows and m columns ($m \geq 2$, $p \geq 2$).

EVEN AND ODD NUMBER PROOFS

An even number is divisible by two, so an even number can always be written as $2 \times$ another number or $2n$ where n is an integer.

An odd number is one more than an even number so $2n + 1$ is always odd where n is an integer.

EXAMPLE 7

SKILLS

ANALYSIS

Prove that the product of any two odd numbers is always odd.

Let the two odd numbers be $2n + 1$ and $2m + 1$

Then $(2n + 1)(2m + 1) = 4nm + 2n + 2m + 1$

$$= 2(2nm + n + m) + 1$$

But $2nm + n + m$ is an integer. Let $2nm + n + m = p$

Then $(2n + 1)(2m + 1) = 2p + 1$ which is odd.

KEY POINTS

■ When n is an integer
   ■ Any even number can be written in the form $2n$.
   ■ Consecutive even numbers can be written in the form $2n$, $2n + 2$, $2n + 4$, …
   ■ Any odd number can be written in the form $2n + 1$.
   ■ Consecutive odd numbers can be written in the form $2n + 1$, $2n + 3$, $2n + 5$, …

EXERCISE 5

1 ▶ Prove that the sum of any two consecutive odd numbers is a multiple of 4.

2 ▶ Prove that the difference of any two odd numbers is even.

3 ▶ Prove that an odd number squared is always odd.

4 ▶ Prove that the product of any even number and any odd number is even.

5 ▶ Given that $2(x - n) = x + 5$ where n is an integer, prove that x must be an odd number.

6 ▶ Roxy makes a square pattern with 25 plastic pieces. She removes two columns of plastic pieces and then replaces one piece. She notices that the remaining pieces can now be rearranged into a perfect square. Prove that that this will always be possible whatever sized square Roxy starts with.

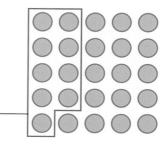

Counters removed

Q7 HINT

The number abc
$= 100a + 10b + c$

7 ▶ Prove that a three-digit number ending in 5 is always divisible by 5.

8 ▶ A triangle with sides of length 3, 4 and 5 cm is a **right-angled triangle**. Prove that this is the only right-angled triangle with sides that are consecutive integers.

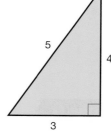

EXERCISE 5*

1 ▶ Prove that the sum of three consecutive even numbers is a multiple of 6.

2 ▶ Prove that an odd number cubed is always odd.

3 ▶ Prove that the difference between the squares of two consecutive odd numbers is equal to four times the integer between them.

4 ▶ Given that $4(x + n) = 3x + 10$ where n is an integer, prove that x must be an even number.

Q5 HINT

Consider the two cases, n odd or n even.

5 ▶ The sum $1 + 2 + 3 \ldots + n = \dfrac{n(n + 1)}{2}$. Prove that $\dfrac{n(n + 1)}{2}$ is always an integer.

Q6 HINT
The number abc
$= 100a + 10b + c$

Q8 HINT
The number abc
$= 100a + 10b + c$

6 ▶ Prove for a three-digit number that if the sum of the digits is divisible by three, then the number is divisible by 3.

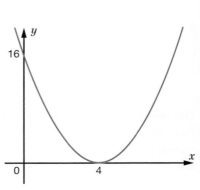

7 ▶ $a*b$ is defined as $a*b = 2a + b$. For example $4*3 = 2 \times 4 + 3 = 11$
Prove that, if a and b are integers and b is odd, then $a*b$ is odd.

8 ▶ abc is a three-digit number such that $a + c = b$. Prove that abc is divisible by 11. Find and prove a similar rule for a four-digit number to be divisible by 11.

ACTIVITY 3

SKILLS

ANALYSIS

Here are two 'proofs'. Try to find the mistakes.

1 ▶   Let $a = b$
$\Rightarrow a^2 = ab$
$\Rightarrow a^2 - b^2 = ab - b^2$
$\Rightarrow (a + b)(a - b) = b(a - b)$
$\Rightarrow a + b = b$
Substituting $a = b = 1$ gives $2 = 1$!

2 ▶   Assume a and b are positive and that $a > b$.
$a > b$
$\Rightarrow ab > b^2$
$\Rightarrow ab - a^2 > b^2 - a^2$
$\Rightarrow a(b - a) > (b + a)(b - a)$
$\Rightarrow a > b + a$
Substituting $a = 2, b = 1$ leads to $2 > 3$!

PROOFS USING COMPLETING THE SQUARE

EXAMPLE 8

SKILLS

ANALYSIS

Prove that $x^2 - 8x + 16 \geq 0$ for any value of x.
Hence **sketch** the graph of $y = x^2 - 8x + 16$

$x^2 - 8x + 16 = (x - 4)^2 - 16 + 16$ (by completing the square)
$= (x - 4)^2$

$(x - 4)^2 \geq 0$ for any value of x as any number squared is always positive, so $x^2 - 8x + 16 \geq 0$ for any value of x.

To sketch $y = (x - 4)^2$, note that when $x = 4$, $(x - 4)^2 = 0$ so the point (4, 0) lies on the graph.

This point must be the **minimum point** on the graph because $(x - 4)^2 \geq 0$. When $x = 0$, $y = 16$ (by substituting in $y = x^2 - 8x + 16$) so (0, 16) lies on the graph.

The graph is also a positive parobola so is 'U' shaped.

EXAMPLE 9

SKILLS

ANALYSIS

Prove that $8x - x^2 - 18 < 0$ for any value of x.

Hence find the largest value of $8x - x^2 - 18$ and sketch the graph of $y = 8x - x^2 - 18$

$$8x - x^2 - 18 = -(x^2 - 8x + 18)$$
$$= -((x - 4)^2 - 16 + 18) \quad \text{(by completing the square)}$$
$$= -((x - 4)^2 + 2)$$
$$= -(x - 4)^2 - 2$$

$(x - 4)^2 \geq 0$ for any value of $x \Rightarrow -(x - 4)^2 \leq 0$

so $-(x - 4)^2 - 2 < 0$ for any value of x.

When $x = 4$, $-(x - 4)^2 = 0$

so the largest value of $-(x - 4)^2 - 2$ is -2.

To sketch $y = -(x - 4)^2 - 2$, note that when $x = 4$, $y = -2$ and that this is the largest value of y so this is a **maximum point**.

Also when $x = 0$, $y = -18$ (substituting in $y = 8x - x^2 - 18$) so $(0, -18)$ lies on the graph. The graph is also a negative **parabola** so is '∩' shaped.

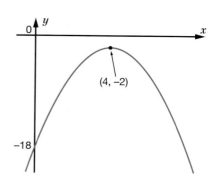

KEY POINTS

- $(x - a)^2 \geq 0$ and $(x + a)^2 \geq 0$ for all x.

- $(x - a)^2 = 0$ when $x = a$ and $(x + a)^2 = 0$ when $x = -a$.

- To prove a quadratic **function** is greater or less than zero, write it in completed square form.

- To find the co-ordinates of the **turning point** of a **quadratic graph**, write it in completed square form $y = a(x + b)^2 + c$. The turning point is then $(-b, c)$.

EXERCISE 6

1 ▶   a Prove that $x^2 + 4x + 4 \geq 0$ for any value of x.
       b Sketch the graph of $y = x^2 + 4x + 4$

2 ▶   a Prove that $2x - x^2 - 1 \leq 0$ for any value of x.
       b Sketch the graph of $y = 2x - x^2 - 1$

3 ▶   a Write $x^2 - 6x + 3$ in the form $(x + a)^2 + b$
       b What is the smallest value of $x^2 - 6x + 3$?
       c Sketch the graph of $y = x^2 - 6x + 3$

4 ▶   a Write $2x^2 + 4x - 3$ in the form $a(x + b)^2 + c$
       b Find the minimum point of $y = 2x^2 + 4x - 3$

5 ▶   a $x^2 + 10x + c \geq 0$ for all values of x. Find the value of c.
       b $x^2 - bx + 16 \geq 0$ for all values of x. Find the value of b.

EXERCISE 6*

1 ▶
  a Prove that $x^2 - 6x + 9 \geq 0$ for any value of x.
  b Sketch the graph of $y = x^2 - 6x + 9$

2 ▶
  a Prove that $14x - x^2 - 52 < 0$ for any value of x.
  b Find the largest value of $14x - x^2 - 52$
  c Sketch the graph of $y = 14x - x^2 - 52$

3 ▶ Prove that $2x^2 + 129 > 32x$ for all values of x.

4 ▶ Prove that the smallest value of $x^2 + 2bx + 4$ is $4 - b^2$

5 ▶
  a Prove that $x^2 + y^2 \geq 2xy$ for all values of x and y.
  b For what values of x and y does $x^2 + y^2 = 2xy$?

6 ▶ Amy and Reno are on a flat surface. Amy is 2 km due (directly) North of Reno. At midday, Amy starts walking slowly due East at 1 km/hr, while Reno starts walking slowly due North at 1 km per hour.

  a Find an expression for the square of the distance between them after t hours.
  b Prove that this distance has a minimum value of $\sqrt{2}$ km.
  c Find at what time this occurs.

Q6b HINT
Find the minimum of the square of the distance.

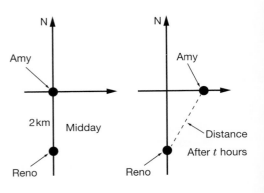

EXERCISE 7

REVISION

1 ▶ Solve these simultaneous equations.
  a $y = x^2$, $y = x + 12$
  b $y + x^2 = 6x$, $y = 2x - 5$

2 ▶ Solve these simultaneous equations, giving your answers correct to 2 d.p.
  a $y = x - 2$, $y = x^2 + 4x - 8$
  b $y = 1 - x$, $x^2 + y^2 = 4$

3 ▶ Robin Hood is designing a new bow in the shape of an **arc** of a circle of radius 2 m. Robin wants the distance between the string and the bow to be 30 cm. Use axes as shown in the diagram. The equation of the circle is $x^2 + y^2 = 4$

  a Write down the equation of AB.
  b Find the co-ordinates of A and B and the length of string that Robin needs.

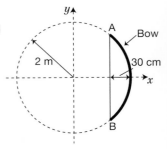

4 ▶ Give a counter-example to prove that the statement 'The sum of two square numbers is always odd' is not true.

5 ▶ Prove that the sum of four consecutive odd numbers is a multiple of 8.

6 ▶ Prove that the product of any two even numbers is divisible by 4.

7 ▶ a Prove that $x^2 + 8x + 16 \geq 0$ for all values of x.
     b Find the minimum point of the graph $y = x^2 + 8x + 16$
     c Sketch the graph $y = x^2 + 8x + 16$

EXERCISE 7*

REVISION

1 ▶ Solve these simultaneous equations.
     a $y = 2x^2$, $y = 5x + 3$
     b $xy = 4$, $y = 2x + 2$

2 ▶ Solve these simultaneous equations, giving your answers correct to 2 d.p.
     a $y = 2x - 1$, $y = 2x^2 + 7x - 5$
     b $x + y = 2$, $2x^2 - x + y^2 = 5$

3 ▶ The path of the comet Fermat is an ellipse whose equation relative to the Earth is
     $x^2 + 36y^2 = 324$, where the units are in AU
     (1 AU, called an astronomical unit, is the distance from the Earth to the Sun).

     The comet can be seen by eye at a distance of 3 AU from the Earth, and the equation
     of this circle is $(x - 17.5)^2 + y^2 = 9$

     Find the co-ordinates of the points where the comet can be seen from the Earth.

4 ▶ Give a counter-example to prove that the statement 'The difference between two cube
     numbers is always odd' is not true.

5 ▶ a, b and c are three consecutive numbers. Prove that the product of a and c is one less
     than b squared.

6 ▶ Prove that the product of two consecutive odd numbers is 1 less than a multiple of 4.

7 ▶ a Prove that $12x - x^2 - 40 < 0$
     b Sketch the graph $y = 12x - x^2 - 40$

EXAM PRACTICE: ALGEBRA 9

1 Solve these simultaneous equations, giving your answers to 3 s.f.

$y = x^2 - 4x + 3$ and $y = 2x - 3$ **[4]**

2 Solve these simultaneous equations: $2x - y = 4$ and $x^2 + y^2 = 16$ **[4]**

3 Give a counter-example to show that $(x + 4)^2 = x^2 + 16$ is false. **[2]**

4 Prove that if the difference of two numbers is 4, the difference of their squares is a multiple of 8. **[2]**

5 Prove that the product of two consecutive odd numbers is odd. **[4]**

6 a Prove that $x^2 - 10x + 25 \geq 0$ for all values of x. **[4]**

 b Find the minimum point of the graph $y = x^2 - 10x + 25$ **[3]**

 c Sketch the graph $y = x^2 - 10x + 25$ **[2]**

[Total 25 marks]

CHAPTER SUMMARY: ALGEBRA 9

SOLVING TWO SIMULTANEOUS EQUATIONS — ONE LINEAR AND ONE NON-LINEAR

Graphically this corresponds to the intersection of a line and a curve.
Always substitute the linear equation into the non-linear equation.
Solve the simultaneous equations $y = x^2 + 2$, $y = 3x$

$$y = x^2 + 2 \qquad (1)$$
$$y = 3x \qquad (2)$$

Substituting (2) into (1):

$$x^2 + 2 = 3x \qquad \text{(Rearrange)}$$
$$x^2 - 3x + 2 = 0 \qquad \text{(Factorise)}$$
$$(x - 1)(x - 2) = 0$$
$$x = 1 \text{ or } 2$$

Substituting into (2) gives the solutions as (1, 3) or (2, 6).

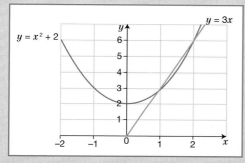

Solve the simultaneous equations $x^2 + y^2 = 13$, $x - y + 1 = 0$

$$x^2 + y^2 = 13 \quad (1)$$
$$x - y + 1 = 0 \quad (2)$$

The linear equation is equation (2).
Make y the subject of equation (2):

$y = x + 1 \Rightarrow y^2 = (x + 1)^2 \Rightarrow y^2 = x^2 + 2x + 1 \qquad (3)$

Substitute (3) into (1):

$$x^2 + x^2 + 2x + 1 = 13$$
$$2x^2 + 2x - 12 = 0 \qquad \text{(Divide by 2)}$$
$$x^2 + x - 6 = 0 \qquad \text{(Factorise)}$$
$$(x + 3)(x - 2) = 0$$
$$x = -3 \text{ or } 2$$

Substituting into (2) gives the solutions as (−3, −2) or (2, 3).

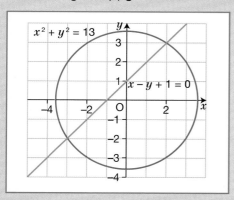

PROOF

To prove a statement is true you must show it is true in all cases.

To prove a statement is not true you need to find a counter-example – an example that does not fit the statement.

$n^3 + n^2 + 17$ does not always produce prime numbers as 17 is a **factor**.

If n is an integer
- $..., n - 1, n, n + 1, n + 2, ...$ are consecutive integers
- any even number can be written as $2n$
- consecutive even numbers can be written as $2n, 2n + 2, 2n + 6, ...$
- any odd number can be written as $2n + 1$
- consecutive odd numbers can be written as $2n + 1, 2n + 3, 2n + 5, ...$

Prove that the sum of two consecutive odd numbers is divisible by four.
$(2n + 1) + (2n + 3) = 2n + 1 + 2n + 3 = 4n + 4 = 4(n + 1)$

PROOFS USING COMPLETING THE SQUARE

- $(x - a)^2 \geq 0$ and $(x + a)^2 \geq 0$ for all x.
- $(x - a)^2 = 0$ when $x = a$ and $(x + a)^2 = 0$ when $x = -a$.
- To prove a quadratic function is greater or less than zero, write it in completed square form.
- To find the co-ordinates of the turning point of a quadratic graph, write it in completed square form $y = a(x + b)^2 + c$. The turning point is then $(-b, c)$.

Prove that $x^2 - 2x + 1 \geq 0$ for all x.

Hence find the turning point of $y = x^2 - 2x + 1$

$x^2 - 2x + 1 = (x - 1)^2 \geq 0$

$y = x^2 - 2x + 1 \Rightarrow y = (x - 1)^2 + 0$
so the turning point is $(- (-1), 0)$ which is $(1, 0)$.

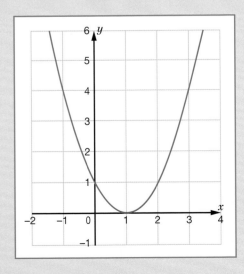

GRAPHS 8

Architects, engineers and designers all use different types of transformations of graphs to produce beautiful shapes. The photograph shows the Sydney Opera house with a transformed graph of $y = x^2$ superimposed. The graph of $y = x^2$ has been stretched, reflected and translated to fit. Graphs of higher powers of x would achieve a better fit.

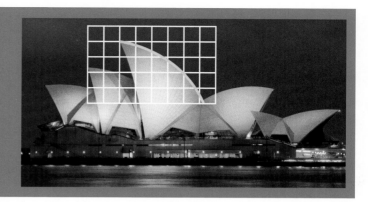

LEARNING OBJECTIVES

- Find the gradient of a tangent at a point
- Translate the graph of a function
- Understand the relationship between translating a graph and the change in its function form
- Reflect the graph of a function
- Understand the effect reflecting a curve in one of the axes has on its function form
- Stretch the graph of a function
- Understand the effect stretching a curve parallel to one of the axes has on its function form

BASIC PRINCIPLES

- Remember that a **tangent** is a straight line that touches a curve at one point only.
- Find the **gradient** of a line through two points.
- Find the gradient of a straight-line graph.
- Plot the graphs of linear and quadratic **functions** using a table of values.
- Interpret distance–time graphs.
- Interpret speed–time graphs.

- Identify transformations.
- Translate a shape using a **vector** and describe a **translation** using a **column vector**.
- Reflect a shape in the x- and y-axes and describe a reflection.
- Identify the image of a point after a reflection in the x-axis or a reflection in the y-axis.
- Use function notation.

GRADIENT OF A CURVE AT A POINT

Most graphs of real-life situations are curves rather than straight lines, but information on rates of change can still be found by drawing a tangent to the curve and using this to estimate the gradient of the curve at that point.

To find the gradient of a curve at a point, draw the tangent to the curve at the point. Do this by pivoting a ruler about the point until the angles between the ruler and the curve are as equal as possible.

The gradient of the tangent is the gradient of the curve at the point.

The gradient of the tangent is found by working out $\frac{\text{'rise'}}{\text{'run'}}$

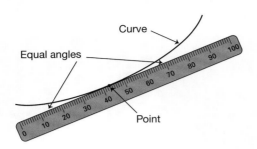

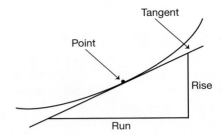

EXAMPLE 1

The graph of $y = x^2 - x$ is shown. Find the gradient of the graph at $x = 2$

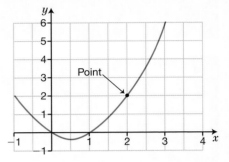

Use a ruler to draw a tangent to the curve at $x = 2$

Work out the rise and run.

Note: Be careful finding the rise and run when the **scales** on the axes are different as in this example.

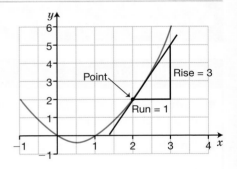

The gradient of the tangent is $\frac{\text{rise}}{\text{run}} = \frac{3}{1} = 3$

So the gradient of the curve at $x = 2$ is 3

Because the tangent is judged by eye, different people may get different answers for the gradient. The answers given are calculated using a different technique which you will learn in a later unit, so don't expect your answers to be exactly the same as those given.

KEY POINTS

- ■ To estimate the gradient of a curve at a point
    - ■ draw the best estimate of the tangent at the point
    - ■ find the gradient of this tangent.
- ■ Be careful finding the rise and run when the scales on the axes are different.

EXERCISE 1

1 ▶ By drawing suitable tangents on tracing paper, find the gradient of the graph at

a $x = 2$

b $x = 2\frac{1}{2}$

c $x = \frac{1}{2}$

d $x = 0$

e Where on the graph is the gradient equal to zero?

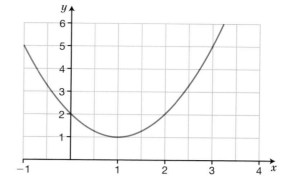

2 ▶ By drawing suitable tangents on tracing paper, find the gradient of the graph at

a $x = 0$

b $x = 2$

c $x = 0.4$

d $x = 2.5$

e Where on the graph is the gradient equal to 1?

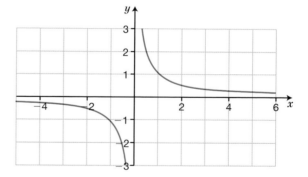

3 ▶ By drawing suitable tangents on tracing paper, find the gradient of the graph at

a $x = -2$

b $x = -1$

c $x = 1\frac{1}{2}$

d $x = \frac{1}{2}$

e Where on the graph is the gradient equal to −2?

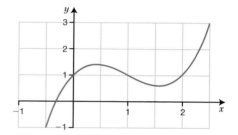

4 ▶ Plot the graph of $y = x(6 - x)$ for $0 \le x \le 6$
By drawing suitable tangents, find the gradient of the graph at

a $x = 1$

b $x = 2$

c $x = 5$

d Where on the curve is the gradient equal to zero?

EXERCISE 1*

1 ▶ By drawing suitable tangents on tracing paper, find the gradient of the graph at

a $x = \frac{1}{2}$

b $x = -\frac{1}{2}$

c $x = 2$

d $x = 3$

e Where on the graph is the gradient equal to zero?

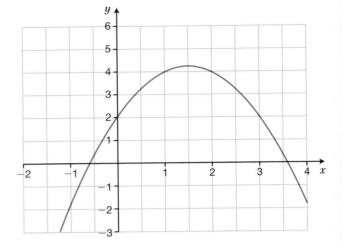

2 ▶ By drawing suitable tangents on tracing paper, find the gradient of the graph at

a $x = -1\frac{1}{2}$

b $x = -\frac{1}{2}$

c $x = \frac{1}{2}$

d $x = 1$

e Where on the graph is the gradient equal to zero?

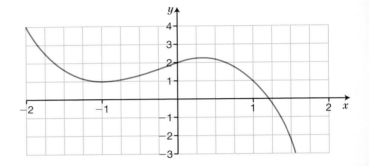

3 ▶ By drawing suitable tangents on tracing paper

a estimate the gradient at $x = -1$

b find another point with the same gradient as at $x = -1$

c find two points where the gradient is -2.

d Where on the graph is the gradient equal to zero?

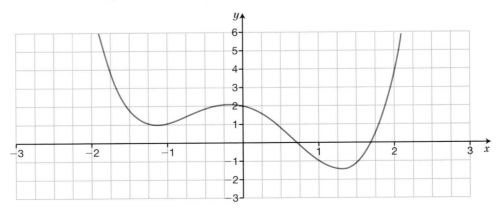

4 ▶   **a** Draw an accurate graph of $y = x^2$ for $-4 \leq x \leq 4$

     b Use your graph to complete the following table.

x-COORDINATE	−4	−3	−2	−1	0	1	2	3	4
GRADIENT	−8								8

     c Write down the connection between the x-coordinate and the gradient.

5 ▶   **a** Draw an accurate graph of $y = 2^x$ for $0 \leq x \leq 5$ by first copying and completing the table.

x	0	1	2	3	4	5
y						

     b Use this graph to estimate the gradient of the curve at $x = 1$ and $x = 3$

     c Where on the curve is the gradient equal to 12?

EXAMPLE 2   A dog is running in a straight line away from its owner.
Part of the distance–time graph describing the motion is shown.

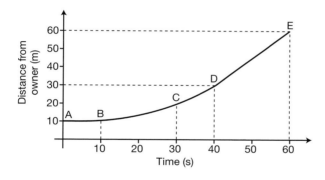

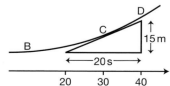

a Describe how the dog's speed varies.

b Estimate the dog's speed after 30 seconds.

Note: Remember that the gradient of a distance–time graph gives the speed.

---

a A to B: The gradient is zero, so the speed is zero. The dog is stationary for the first 10 seconds, 10 metres away from its owner.

     B to D: The gradient is gradually increasing, so the speed is gradually increasing. For the next 30 seconds the dog runs with increasing speed.

     D to E: The gradient is constant and equal to $\frac{30}{20}$ or 1.5, so the dog is running at a constant speed of 1.5 m/s.

b Draw a tangent at C and calculate the gradient of the tangent.

     The gradient is $\frac{15 \text{ m}}{20 \text{ s}} = 0.75$ m/s

     so the speed of the dog is approximately 0.75 m/s.

1 ▶ The graph shows part of the distance–time graph for a car caught in a traffic jam.

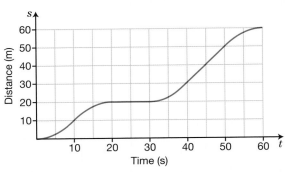

a By drawing suitable tangents on tracing paper, estimate the speed of the car when
i $t = 15$ s   **ii** $t = 25$ s   **iii** $t = 45$ s

b Describe how the car's speed changes over the 60 seconds.

2 ▶ The graph shows the speed–time graph for a young girl in a race.

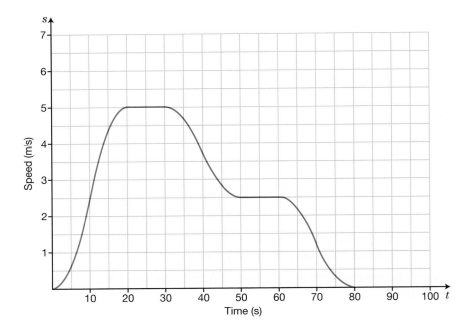

a By drawing suitable tangents on tracing paper, estimate the **acceleration** of the girl when
i $t = 15$ s   **ii** $t = 25$ s   **iii** $t = 70$ s

b Describe how the girl ran the race.

3 ▶ The temperature of a cup of coffee ($T°C$) after t minutes is given in the table.

t (min)	0	1	2	3	4	5	6
T (°C)	80	71	62	55	49	43	38

a Draw the temperature–time graph of this information.

b Use the graph to estimate the **rate** of change of temperature in °C/min when
i $t = 0$ **ii** $t = 3$ **iii** $t = 5$

4 ▶ The depth, d cm, of water in Ahmed's bath t minutes after he has pulled the plug out is given in the table.

t (min)	0	1	2	3	4	5	6
d (cm)	30	17	10	6	3.3	1.4	0

a Draw the graph showing depth against time.

b Use the graph to estimate the rate of change of depth in cm/min when
i $t = 0.5$ **ii** $t = 2.5$ **iii** $t = 5.5$

5 ▶ The **velocity**, v m/s, of a vintage aircraft t seconds after it starts to take off is given by $v = 0.025t^2$

a Draw a graph of v against t for $0 \le t \le 60$

b Use the graph to estimate the acceleration of the aircraft when
i $t = 10$ **ii** $t = 30$ **iii** $t = 50$

6 ▶ The distance, s m, fallen by a stone t seconds after being dropped down a well is given by $s = 5t^2$

a Draw a graph of s against t for $0 \le t \le 4$

b Use your graph to estimate the velocity of the stone when
i $t = 1$ **ii** $t = 2$ **iii** $t = 3$

EXAMPLE 3 ▶ The area of weed covering part of a pond doubles every 10 years. The area now covered is 100 m².

a Given that the area of weed, A m², after n years, is given by $A = 100 \times 2^{0.1n}$, draw the graph of A against n for $0 \le n \le 40$.

b By drawing suitable tangents, find the rate of growth of weed in m² per year after 10 years and after 30 years.

---

a

n (years)	0	10	20	30	40
A (m²)	100	200	400	800	1600

b Rate of growth at 10 years $= \dfrac{260 \text{ m}^2}{18 \text{ years}} \simeq 14 \text{ m}^2/\text{year}$

Rate of growth at 30 years $= \dfrac{800 \text{ m}^2}{14 \text{ years}} \simeq 57 \text{ m}^2/\text{year}$

The rate of growth is clearly increasing with time.

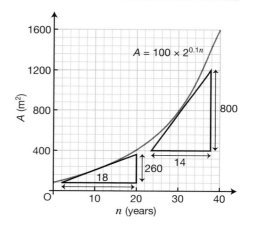

EXERCISE 2*

1 ▶ At time $t = 0$, ten bacteria are placed in a culture dish in a laboratory.
The number of bacteria, N, doubles every 10 minutes.

    a Copy and complete the table and use it to plot the graph of N against t for
$0 \leq t \leq 120$

t (min)	0	20	40	60	80	100	120
N	10		160				

    b By drawing suitable tangents, estimate the rate of change in bacteria per
minute when
      i $t = 0$ **ii** $t = 60$ **iii** $t = 100$

2 ▶ The sales, N, in a mobile phone network are increasing at a rate of 5% every month.
Present sales are two million.

    a Copy and complete the following table showing sales forecasts for the next nine months
and use it to plot the graph of N against t for $0 \leq t \leq 9$

t (months)	0	1	2	3	4	5	6	7	8	9
N (millions)	2		2.21							

    b By drawing suitable tangents, estimate the rate of increase of sales in numbers per
month when
      i $t = 0$ **ii** $t = 4$ **iii** $t = 8$

3 ▶ A party balloon of volume $2000\,cm^3$ loses 15% of its air every 10 minutes.

    a Copy and complete the table and use it to plot the graph of volume, V, against t for
$0 \leq t \leq 90$

t (min)	0	10	20	30	40	50	60	70	80	90
V (cm³)	2000		1445							

    b Use the graph to estimate the rate of change of the balloon's volume in cm^3/min when
      i $t = 10$ **ii** $t = 80$

    c When was the rate of change of the balloon's volume at its maximum value, and what
was this maximum value?

4 ▶ A radioactive isotope of **mass** 120 g decreases its mass, M, by 20% every 10 seconds.

    a Copy and complete the table and use it to plot the graph of M against t for
$0 \leq t \leq 90$

t (s)	0	10	20	30	40	50	60	70	80	90
M (g)	120				49.2					

    b Use the graph to estimate the rate of change of the isotope's mass in g/s when
      i $t = 20$ **ii** $t = 70$

    c When was the rate of change of the isotope's mass at its maximum value, and what was
this maximum value?

5 ▶ The graph shows the depth, y metres, of water at Brigstock Harbour t hours after 1200 hrs.

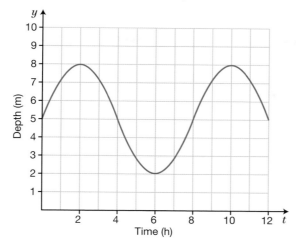

a Use the graph to estimate the rate of change of depth in m/h at
   i 1300 hrs   **ii** 1700 hrs   **iii** 2200 hrs

b At what time is the rate of change of depth a maximum, and what is this rate?

6 ▶ Steve performs a 'bungee-jump' from a platform above a river. His height, h metres, above the river t seconds after he jumps is shown on the graph.

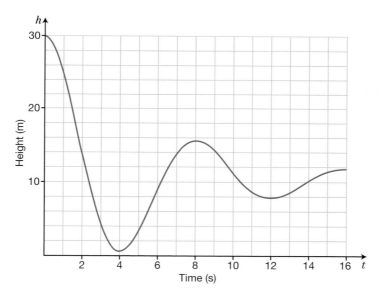

a Estimate Steve's velocity in m/s after
   i 1 s   **ii** 8 s   **iii** 14 s

b Estimate Steve's maximum velocity and the time it occurs.

TRANSLATING GRAPHS

EXAMPLE 4

SKILLS

ANALYSIS

The diagram shows the graphs of $y = x^2$, $y = x^2 - 1$, $y = x^2 - 2$ and $y = x^2 + 1$

Describe the transformation of the graph of

a $y = x^2$ to $y = x^2 - 1$

b $y = x^2$ to $y = x^2 - 2$

c $y = x^2$ to $y = x^2 + 1$

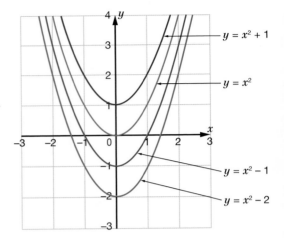

The transformation from $y = x^2$ to the other three graphs is a vertical translation.

a $y = x^2$ to $y = x^2 - 1$ is a translation of $\begin{pmatrix} 0 \\ -1 \end{pmatrix}$

b $y = x^2$ to $y = x^2 - 2$ is a translation of $\begin{pmatrix} 0 \\ -2 \end{pmatrix}$

c $y = x^2$ to $y = x^2 + 1$ is a translation of $\begin{pmatrix} 0 \\ 1 \end{pmatrix}$

EXAMPLE 5

SKILLS

ANALYSIS

Graph A is a translation of $y = f(x)$

a Describe the translation.

b Find the image of the **maximum point** (1, 2).

c Write down the equation of graph A.

d If $f(x) = -x^2 + 2x + 1$, find the equation of graph A.

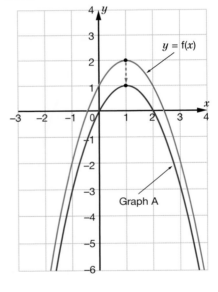

a A translation of $\begin{pmatrix} 0 \\ -1 \end{pmatrix}$

b $(1, 2) \rightarrow (1, 1)$

c $y = f(x) - 1$

d $y = (-x^2 + 2x + 1) - 1 = -x^2 + 2x$

EXAMPLE 6

SKILLS

ANALYSIS

The diagram shows the graphs of $y = x^2$, $y = (x - 1)^2$ and $y = (x + 1)^2$

Describe the transformation of the graph of

a $y = x^2$ to $y = (x - 1)^2$

b $y = x^2$ to $y = (x + 1)^2$

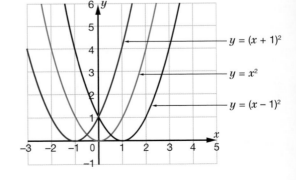

The transformation from $y = x^2$ to the other three graphs is a horizontal translation.

a $y = x^2$ to $y = (x - 1)^2$ is a translation of $\begin{pmatrix} 1 \\ 0 \end{pmatrix}$

b $y = x^2$ to $y = (x + 1)^2$ is a translation of $\begin{pmatrix} -1 \\ 0 \end{pmatrix}$

EXAMPLE 7

SKILLS

ANALYSIS

Graph A is a translation of $y = f(x)$

a Describe the translation.

b Find the image of the **minimum point** $(-2, 1)$.

c Write down the equation of graph A.

d If $f(x) = x^2 + 4x + 5$, find the equation of graph A.

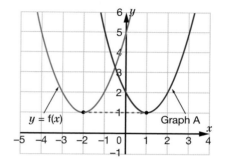

a A translation of $\begin{pmatrix} 3 \\ 0 \end{pmatrix}$

b $(1, 1)$

c $y = f(x - 3)$

d $y = (x - 3)^2 + 4(x - 3) + 5 = x^2 - 2x + 2$

KEY POINTS

■ The graph of $y = f(x) + a$ is a translation of the graph of $y = f(x)$ by $\begin{pmatrix} 0 \\ a \end{pmatrix}$

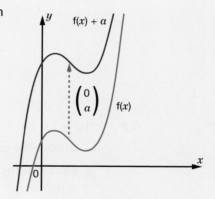

■ The graph of $y = f(x - a)$ is a translation of the graph of $y = f(x)$ by $\begin{pmatrix} a \\ 0 \end{pmatrix}$

■ The graph of $y = f(x + a)$ is a translation of the graph of $y = f(x)$ by $\begin{pmatrix} -a \\ 0 \end{pmatrix}$

■ Be very careful with the signs, they are the opposite to what most people expect.

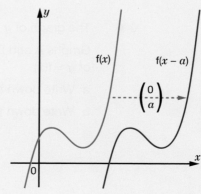

EXERCISE 3

1 ▶ $y = f(x)$ has a maximum point at $(0, 0)$.
Find the co-ordinates of the maximum point of

a $y = f(x) + 3$ **c** $y = f(x - 3)$

b $y = f(x) - 2$ **d** $y = f(x + 2)$

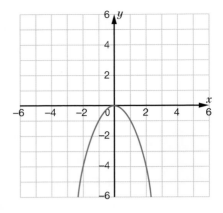

2 ▶ Write down the vector that translates $y = f(x)$ onto

a $y = f(x) - 7$ **c** $y = f(x) + 7$

b $y = f(x - 7)$ **d** $y = f(x + 7)$

3 ▶ The graph shows $f(x) = \dfrac{1}{x}$
Use tracing paper over the grid
to **sketch** these graphs.

a $f(x) = \dfrac{1}{x} - 2$ **c** $f(x) = \dfrac{1}{x} + 1$

b $f(x) = \dfrac{1}{x - 1}$ **d** $f(x) = \dfrac{1}{x + 2}$

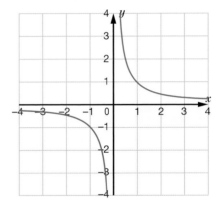

4 ▶ $f(x) = 1 - 2x$

a Sketch the graph of $y = f(x)$

b On the same grid sketch the graph of $y = f(x) + 3$

c Find the algebraic equation of $y = f(x) + 3$

5 ▶ $f(x) = x^2 + 1$

a Sketch the graph of $y = f(x)$

b On the same grid sketch the graph of $y = f(x - 1)$

c Find the algebraic equation of $y = f(x - 1)$

6 ▶ The graph of $y = f(x)$ is shown.

Graphs A and B are translations of the graph of $y = f(x)$

a Write down the equation of graph A.

b Write down the equation of graph B.

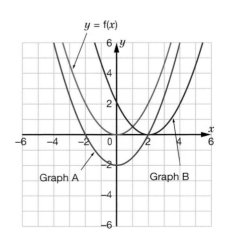

7 ▶ The graph of $y = x^2 + 5x - 1$ is translated by $\begin{pmatrix} 0 \\ 4 \end{pmatrix}$

Find the algebraic equation of the translated graph.

8 ▶ The graph of $y = 3 - x - x^2$ is translated by $\begin{pmatrix} -2 \\ 0 \end{pmatrix}$

Find the algebraic equation of the translated graph.

EXERCISE 3*

1 ▶ The graph shows $y = f(x)$. Sketch these graphs.

    a $y = f(x) + 2$ **c** $y = f(x + 1)$

    b $y = f(x - 2)$ **d** $y = f(x) - 1$

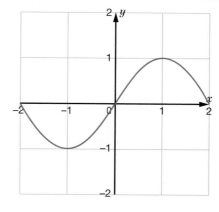

2 ▶ Write down the vector that translates $y = f(x)$ onto

    a $y = f(x) + 5$ **c** $y = f(x + 11)$

    b $y = f(x - 13)$ **d** $y = f(x) - 9$

3 ▶ $f(x) = \dfrac{1}{x} - 1$

    a Sketch the graph of $y = f(x)$

    b On the same grid sketch the graph of $y = f(x) + 2$

    c Find the algebraic equation of $y = f(x) + 2$

4 ▶ $f(x) = 4 - x^2$

    a Sketch the graph of $y = f(x)$

    b On the same grid sketch the graph of $y = f(x + 4)$

    c Find the algebraic equation of $y = f(x + 4)$

5 ▶ The graph of $y = f(x)$ is shown.

Graphs A and B are translations of the graph of $y = f(x)$

    a Write down the equation of graph A.

    b Write down the equation of graph B.

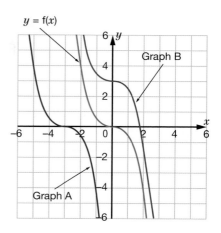

6 ▶ The graph of $y = 2x^3 + 5$ is translated by $\begin{pmatrix} 0 \\ -5 \end{pmatrix}$

Find the algebraic equation of the translated graph.

7 ▶ The graph of $y = 1 - 3x - x^2$ is translated by $\begin{pmatrix} 3 \\ 0 \end{pmatrix}$

Find the algebraic equation of the translated graph.

8 ▶ The graph of $y = 2 - x^2$ is translated by $\begin{pmatrix} 1 \\ 1 \end{pmatrix}$

a Sketch the graph of $y = 2 - x^2$

b Sketch the translated graph.

c Find the algebraic equation of the translated graph.

Q8c HINT
Translate by $\begin{pmatrix} 0 \\ 1 \end{pmatrix}$
and then
by $\begin{pmatrix} 1 \\ 0 \end{pmatrix}$

ACTIVITY 1

The diagrams show the graphs of
$y = \sin x$ and $y = \cos x$ for $-180° \le x \le 180°$

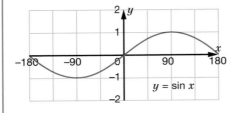 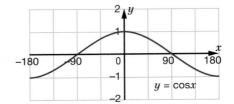

These are important graphs and you should memorise their shapes.

The reason for their distinctive shapes is explained in Unit 10.

■ Use your calculator to check some of the points, for example $\sin(-90°) = -1$
(Make sure your calculator is in **degree mode**.)

■ The sine graph can be translated to fit over the cosine graph. Find the value of a that **satisfies** $\sin(x + a) = \cos x$

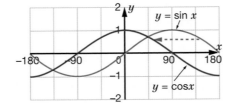

■ The cosine graph can be translated to fit over the sine graph. Find the value of a that satisfies $\cos(x + a) = \sin x$

REFLECTING GRAPHS

REFLECTION IN THE x-AXIS

The three graphs below show $y = f(x)$ and $y = -f(x)$ plotted on the same axes.
In each case the transformation is a reflection in the x-axis.

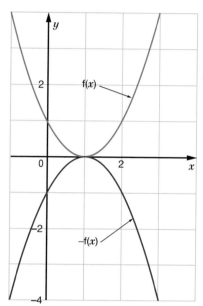
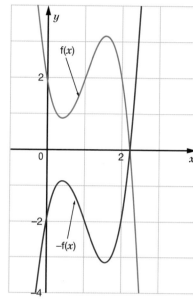
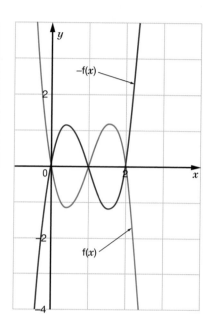

REFLECTION IN THE y-AXIS

The three graphs below show $y = f(x)$ and $y = f(-x)$ plotted on the same axes.
In each case the transformation is a reflection in the y-axis.

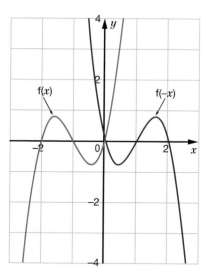
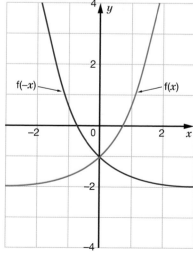
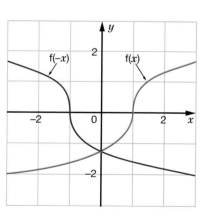

■ The graph of $y = -f(x)$ is a reflection of the graph of $y = f(x)$ in the x-axis.

■ The graph of $y = f(-x)$ is a reflection of the graph of $y = f(x)$ in the y-axis.

EXERCISE 4

1 ▶ The graph shows $y = f(x)$
Sketch these graphs.

   a $y = -f(x)$ **b** $y = f(-x)$

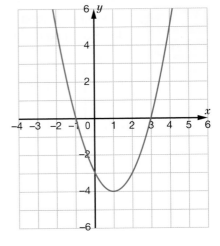

2 ▶ **a** Describe the transformation that maps the graph of $y = x + 1$ to the graph of
    i $y = -x - 1$ **ii** $y = -x + 1$

   b Sketch all three graphs on one set of axes.

3 ▶ The graph of $y = f(x)$ is shown.
Graphs A and B are reflections of the graph
of $y = f(x)$.

   a Write down the equation of graph A.

   b Write down the equation of graph B.

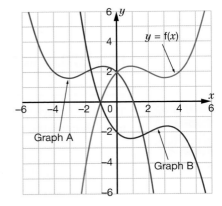

4 ▶ The graph shows $y = \sin x$ for $-180° \le x \le 180°$
Sketch

   a $y = -\sin x$ **b** $y = \sin(-x)$

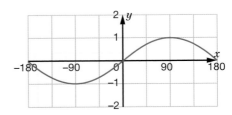

5 ▶ The graph of $y = f(x)$ passes through the points $(0, 4)$, $(2, 0)$ and $(4, -2)$. Find the
corresponding points that $y = f(-x)$ passes through.

EXERCISE 4*

1 ▶   **a** Describe the transformation that maps the graph of $y = 2 - x$ to the graph of
     i $y = 2 + x$ **ii** $y = -2 + x$

   b Sketch all three graphs on one set of axes.

2 ▶   The graph shows $y = x^2 - x + 1$

   a Sketch $y = x^2 + x + 1$

   b Sketch $y = x - x^2 - 1$

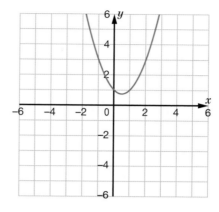

3 ▶   The graph of $y = f(x)$ is transformed to $y = -f(x)$. Find the image of the points

   a $(a, 0)$ **b** $(0, b)$ **c** (c, d)

4 ▶   **a** The graph of $y = x^3 - x^2 + x - 1$ is reflected in the y-axis. Find the equation of the
     reflected graph.

   b The graph of $y = x^3 - x^2 + x - 1$ is reflected in the x-axis. Find the equation of the
     reflected graph.

5 ▶ $f(x) = \cos x$

   a Sketch $y = f(x)$ for $-180° \leq x \leq 180°$ **c** What can you say about $y = f(x)$ and $y = f(-x)$?

   b Sketch $y = f(-x)$ for $-180° \leq x \leq 180°$ **d** Sketch $y = -f(x)$ for $-180° \leq x \leq 180°$

6 ▶   The graph shows $y = f(x)$

   a Sketch $y = f(-x)$ and then $y = -f(-x)$

   b What is the transformation from $y = f(x)$ to
     $y = -f(-x)$?

   c Does it matter if the reflections are performed
     in the opposite order
     i.e., $y = -f(x)$ then $y = -f(-x)$?

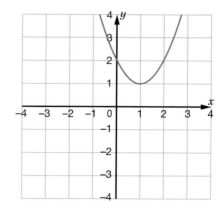

STRETCHING GRAPHS

STRETCHES PARALLEL TO THE y-AXIS

A graph can be stretched or compressed in the y direction by multiplying the function by a number.

EXAMPLE 10

SKILLS

ANALYSIS

$f(x) = x^2$

a Draw $y = f(x)$ and $y = 2f(x)$ on the same axes and describe the transformation.

b Draw $y = f(x)$ and $y = \frac{1}{2}f(x)$ on the same axes and describe the transformation.

a The arrows on the graph show that the transformation is a stretch of 2 in the y direction.

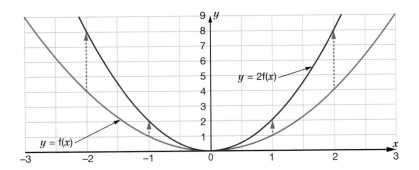

b The arrows on the graph show that the transformation is a stretch of $\frac{1}{2}$ in the y direction.

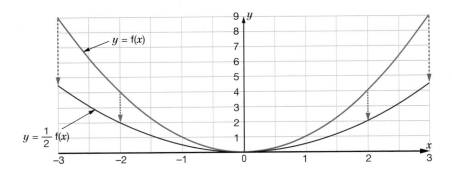

EXAMPLE 11

SKILLS

ANALYSIS

The diagram shows the graph of $y = f(x)$ which has a maximum point at (2, 4).

a Find the maximum point of $y = 4f(x)$

b Find the maximum point of $y = \frac{1}{4}f(x)$

---

a The graph of $y = 4f(x)$ is a stretch of the graph $y = f(x)$ in the y direction by a **scale factor** of 4, so the maximum point is (2, 16).

b The graph of $y = \frac{1}{4}f(x)$ is a stretch of the graph $y = f(x)$ in the y direction by a scale factor of $\frac{1}{4}$, so the maximum point is (2, 1).

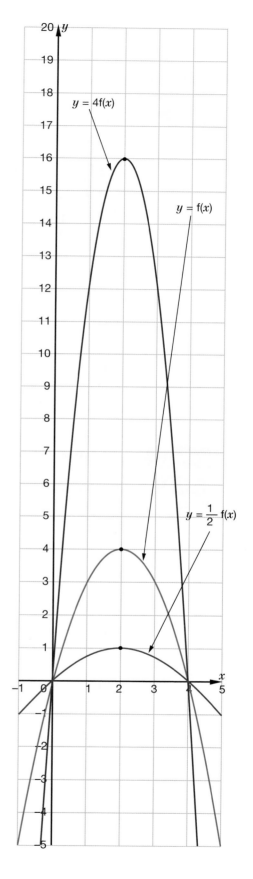

STRETCHES PARALLEL TO THE x-AXIS

The two graphs below show $y = f(x)$ and the $y = f(2x)$ plotted on the same axes.

In each case the transformation from $y = f(x)$ to $y = f(2x)$ is a stretch scale factor $\frac{1}{2}$ parallel to the x-axis.

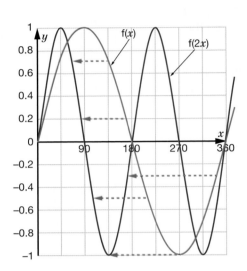
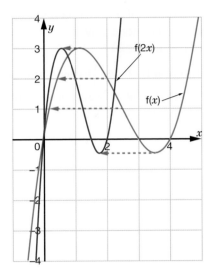

The two graphs below show a function $f(x)$ and the function $f(\frac{1}{2}x)$ plotted on the same axes.

In each case the transformation from $y = f(x)$ to $y = f(\frac{1}{2}x)$ is a stretch scale factor 2 parallel to the x-axis.

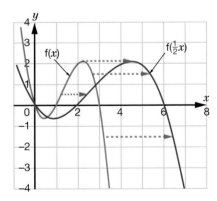
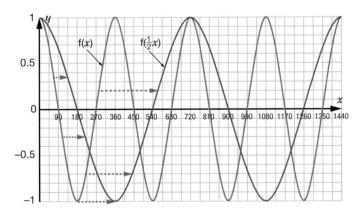

EXAMPLE 14

SKILLS

ANALYSIS

The diagram shows the graph of $y = f(x)$ (red curve) which has a maximum point at (2, 4).

a Find the maximum point of $y = f(4x)$

b Find the maximum point of $y = f(\frac{1}{4}x)$

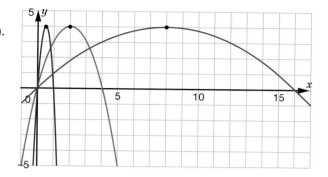

a The graph of $y = f(4x)$ is a stretch of the graph $y = f(x)$ in the x direction by a scale factor of $\frac{1}{4}$, so the maximum point is $(\frac{1}{2}, 4)$.

b The graph of $y = f(\frac{1}{4}x)$ is a stretch of the graph $y = f(x)$ in the x direction by a scale factor of 4, so the maximum point is (8, 4).

KEY POINTS

- The graph of $y = kf(x)$ is a stretch of the graph $y = f(x)$ with a scale factor of k parallel to the y-axis (all y-co-ordinates are multiplied by k).

  - If $k > 1$ the graph is stretched by k.

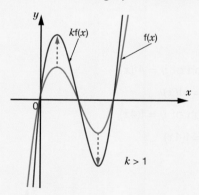

$k > 1$

- If $0 < k < 1$ the graph is compressed by k.

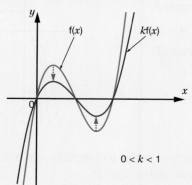

$0 < k < 1$

- The graph of $y = f(kx)$ is a stretch of the graph $y = f(x)$ with a scale factor of $\frac{1}{k}$ parallel to the x-axis (all x-co-ordinates are multiplied by $\frac{1}{k}$).

  - If $k > 1$ the graph is compressed by $\frac{1}{k}$

  - If $0 < k < 1$ the graph is stretched by $\frac{1}{k}$

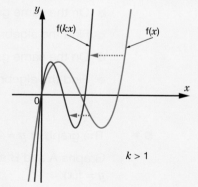

$k > 1$

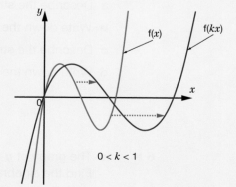

$0 < k < 1$

EXERCISE 5

1 ▶ The graph shows $y = f(x)$ which has a maximum point at $(-1, 2)$.

Find the co-ordinates of the maximum point of

a $y = 2f(x)$ **b** $y = f(2x)$

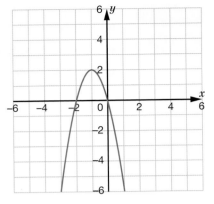

2 ▶ Describe the transformation that maps the graph of $y = f(x)$ to the graph of

a $y = f(3x)$ **b** $y = 3f(x)$

3 ▶ The diagram shows the graph of $y = \sin x$ for $-180° \le x \le 180°$

a Sketch the graph of $y = 2 \sin x$ for $-180° \le x \le 180°$

b Sketch the graph of $y = \sin(2x)$ for $-180° \le x \le 180°$

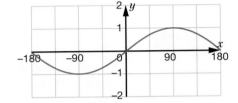

4 ▶ $f(x) = 1 + x$

a Sketch the graph of $y = f(x)$

b On the same grid sketch the graph of $y = 4f(x)$

c Find the algebraic equation of $y = 4f(x)$

d On the same grid sketch the graph of $y = f(4x)$

e Find the algebraic equation of $y = f(4x)$

5 ▶ The graph of $y = f(x)$ is shown.

Graphs A and B are stretches of the graph of $y = f(x)$

a Describe the stretch from $f(x)$ to graph A.

b Write down the equation of graph A.

c Describe the stretch from $f(x)$ to graph B.

d Write down the equation of graph B.

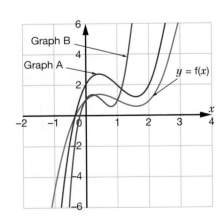

6 ▶ **a** The graph of $y = x^2 + x + 1$ is stretched in the y direction with a scale factor of 5. Find the algebraic equation of the stretched graph.

b The graph of $y = 9x^2 - 6x + 2$ is stretched in the x direction with a scale factor of 3. Find the algebraic equation of the stretched graph.

EXERCISE 5*

1 ▶ The graph shows $y = f(x)$

Sketch these graphs.

a $y = f(\tfrac{1}{2}x)$ **c** $y = \tfrac{1}{2}f(x)$

b $y = 2f(x)$ **d** $y = f(2x)$

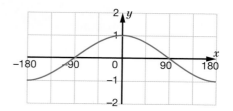

2 ▶ Describe the transformation that maps the graph of $y = f(x)$ to the graph of

a $y = \tfrac{1}{4}f(x)$ **c** $y = 7f(x)$

b $y = f(5x)$ **d** $y = f(\tfrac{1}{2}x)$

3 ▶ The graph shows $f(x) = x^2 + x$. Sketch these graphs.

a $f(x) = 3x^2 + 3x$ **b** $f(x) = 4x^2 + 2x$

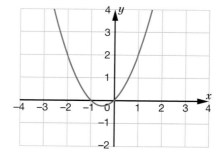

4 ▶ $f(x) = \dfrac{8 - x}{8}$

a Sketch the graph of $y = f(x)$

b On the same grid sketch the graph of $y = 5f(x)$

c Find the algebraic equation of $y = 5f(x)$

d On the same grid sketch the graph of $y = f(8x)$

e Find the algebraic equation of $y = f(8x)$

5 ▶ The graph of $y = f(x)$ is shown. Graphs A and B are stretches of the graph of $y = f(x)$

a Describe the stretch from $f(x)$ to graph A.

b Write down the equation of graph A.

c Describe the stretch from $f(x)$ to graph B.

d Write down the equation of graph B.

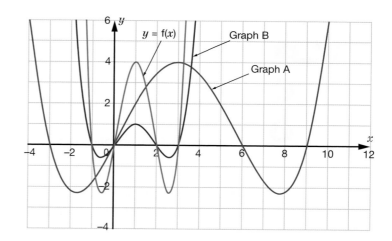

6 ▶ **a** The graph of $y = x^3 + 2x^2 - 3x + 2$ is stretched in the x direction by a scale factor of $\tfrac{1}{4}$ followed by a translation of $\begin{pmatrix} 0 \\ -2 \end{pmatrix}$. Find the algebraic equation of the new graph.

b The graph of $y = 4x^2 - 4x + 5$ is stretched in the x direction by a scale factor of 2 followed by a translation of $\begin{pmatrix} 1 \\ 0 \end{pmatrix}$. Find the algebraic equation of the new graph.

ACTIVITY 2

You can use a graphics calculator or computer to check your answers to this activity.

Find transformations of $y = x^2$ which will produce this pattern.

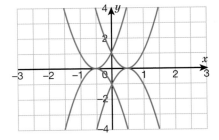

Find transformations of $y = x^3$ and $y = \dfrac{1}{x}$ which will produce this pattern.

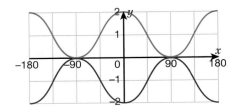

Find transformations of $y = \cos x$ that will produce this pattern.

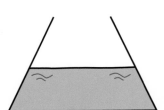

EXERCISE 6

REVISION

1 ▶   **a** Draw the curve $y = x(x - 5)$ for $0 \le x \le 6$

   b By drawing suitable tangents, find the gradient of the curve at the point
      $x = 1$, $x = 2.5$ and $x = 5$

   c Find the equation of the tangent to the curve at the point where $x = 1$

2 ▶   The depth, d mm, of fluid poured into a conical beaker
   at a constant rate, after t seconds is shown in this table.

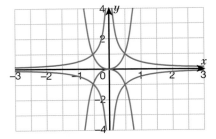

t (s)	0	4	8	12	16	20
d (mm)	1	2.5	6.3	15.8	39.8	100

   a Draw the graph of depth against time from this table.

   b Estimate the rate of change of the depth in mm/s when
      $t = 10$ s and when $t = 15$ s.

3 ▶ Here is the graph of $y = \sin x$ for $-180° \le x \le 180°$

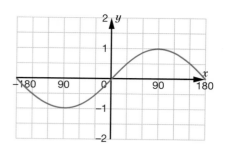

Match the equations to the graphs below.

a $y = \sin x + 0.5$ **b** $y = \sin(x + 30)$

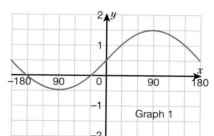

Graph 1

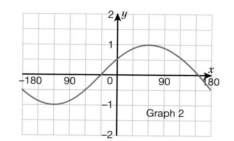

Graph 2

4 ▶ The diagram shows the graph of $y = f(x) = 3x + 2$

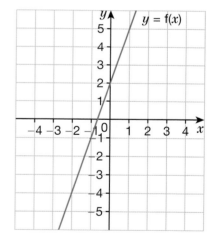

The graph has been transformed in different ways.
Match the function notation to the graphs.

a f(2x)   **c** f(−x)

b 2f(x)   **d** −f(x)

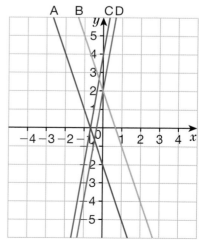

5 ▶   The diagram shows the graph of $y = f(x)$

Sketch the following graphs.

a $y = f(x) + 2$

b $y = f(x + 2)$

c $y = 2f(x)$

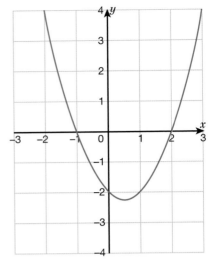

6 ▶   The diagram shows the graph of $y = f(x)$

The turning point of the curve is A(2, 4).

Write down the co-ordinates of the turning points of the curves with these equations.

a $y = -f(x)$

b $y = f(-x)$

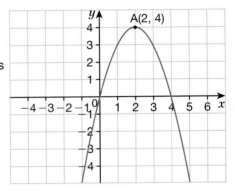

7 ▶   Mr Gauss, a mathematics teacher, is trying to draw a **quadratic curve** with a maximum at the point (0, 4).

a   Suggest an equation that Mr Gauss might use.

b   Where does this curve **intersect** the x-axis?

Mr Gauss wants the curve to intersect the x-axis at (−1, 0) and (1, 0).

c   Find the equation that will satisfy these conditions.

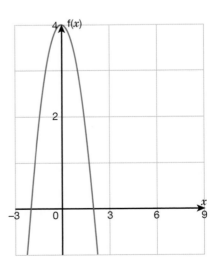

EXERCISE 6*　　REVISION

1 ▶　**a** Plot the graph of $y = 3^x$ for $0 \leq x \leq 4$ by first copying and completing the table.

x	0	1	2	3	4
y	1		9		

　　b Find the gradient of the curve at $x = 1$ and $x = 2$

　　c Find the equation of the tangent to the graph where $x = 1$

2 ▶　A catapult fires a stone vertically upwards. The height, h metres, of the stone, t seconds after firing, is given by the formula $h = 40t - 5t^2$

　　a Draw the graph of h against t for $0 \leq t \leq 8$

　　b Copy and complete this table by drawing tangents to this curve and measuring their gradients.

t (s)	0	1	2	3	4	5	6	7	8
VELOCITY (m/s)	40		20			−10			−40

　　c Draw the velocity–time graph for the stone for $0 \leq t \leq 8$

　　d What can you say about the stone's acceleration?

3 ▶　The diagram shows the graph of $y = \mathrm{f}(x) = 5 - \frac{1}{2}x$ and the graphs of some transformations of $\mathrm{f}(x)$

　　Match the function notation to the graphs.

　　a $\mathrm{f}(2x)$　　　　**c** $\mathrm{f}(-x)$

　　b $2\mathrm{f}(x)$　　　　**d** $-\mathrm{f}(x)$

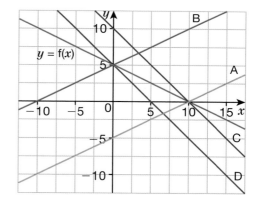

4 ▶　Here is a sketch of $y = \mathrm{f}(x) = (x - 3)^2 + 2$

　　The graph has a minimum point at (3, 2).

　　It intersects the y-axis at (0, 11).

　　a Sketch the graph of $y = \mathrm{f}\left(\frac{1}{2}x\right)$

　　b Write down the minimum value of $\mathrm{f}\left(\frac{1}{2}x\right)$

　　c Write down the co-ordinates of the minimum point of $y = \mathrm{f}(2x)$

　　d Explain why the graphs of $y = \mathrm{f}(x)$, $y = \mathrm{f}\left(\frac{1}{2}x\right)$ and $y = \mathrm{f}(2x)$ all intersect the y-axis at the same point.

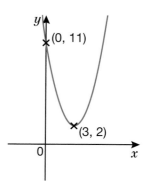

5 ▶ The curve $y = f(x)$ passes through the three points A (0, 2), B (3, 0) and C (2, −4). Find the corresponding points that the following curves pass through.

   a $y = 2f(x)$ **b** $y = f(-x)$ **c** $y = f(x + 1)$

6 ▶ The graph shows $y = f(x)$
Sketch the following graphs.

   a $y = f(x) + \dfrac{1}{2}$ **c** $y = \dfrac{1}{2}f(x)$

   b $y = f(x + \dfrac{1}{2})$ **d** $y = f(\dfrac{1}{2}x)$

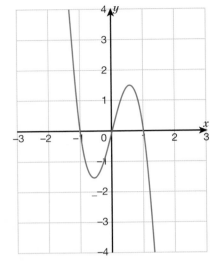

7 ▶ Dave is a keen sea fisherman and wants to predict the depth of water at his favourite fishing site using the sine function. The diagram shows his first attempt, where the x-axis represents time in minutes and y is the depth in metres.

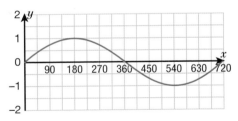

   a Describe the transformation of the sine function that will produce this graph.

   b Write down the equation that will produce this graph.

The water depth actually varies between 1 m and 5 m as shown in the next diagram, so Dave tries to improve his model.

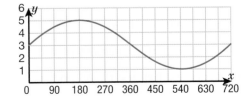

   c Describe the transformations from the first graph to the second graph.

   d Write down the equation that will produce the second graph.

   e What depth of water does this model predict for a time of 2 hours?

EXAM PRACTICE: GRAPHS 8

1 **a** For the graph shown, find, by drawing tangents, the gradients at
   i $x = -1$ **ii** $x = 1$

 b Where is the gradient zero?

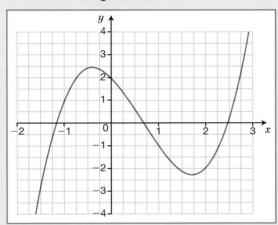

[6]

2 The temperature of Sereena's bath, t °C, m minutes after the bath has been run, is given by the equation

$t = \dfrac{1100}{m + 20}$, valid for $0 \le m \le 20$

 a Copy and complete the following table of t against m, giving t to 2 s.f.

m	1	2	4	6	10	15	20
t	52			42			28

 b Draw the graph of t against m for $0 \le m \le 20$.
 c Find the rate of change of temperature in °C/minute when $m = 7$.   [5]

3 Here is a sketch of $y = f(x)$.

 a Draw sketches of the graphs
   i $y = f(x) + 3$
   ii $y = f(x - 3)$

 b Write down where (0, 0) is mapped to for both graphs.

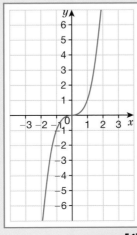

[4]

4 The graph of $y = f(x)$ is shown.
Copy the diagram and sketch the graph of

 a $y = 2f(x)$ **b** $y = f(2x)$

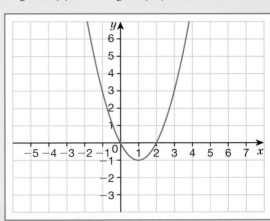

[4]

5 The diagram shows the graph of $y = f(x)$

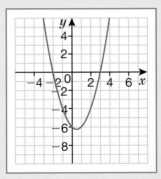

 a Sketch a copy of the graph.
On the same axes sketch the graphs of
   i $y = -f(x)$ **ii** $y = f(-x)$

 b Describe the transformation of the graph of $y = f(x)$ to the graph of $y = -f(x)$

 c Describe the transformation of the graph of $y = f(x)$ to the graph of $y = f(-x)$ [6]

[Total 25 marks]

CHAPTER SUMMARY: GRAPHS 8

GRADIENT OF A CURVE AT A POINT

Use a ruler to draw a tangent at the point.

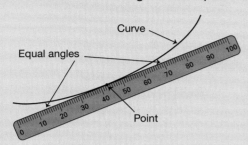

The gradient of the tangent is an estimate of the gradient of the curve at the point.

TRANSLATING GRAPHS

TRANSLATING GRAPHS IN THE y-DIRECTION

The graph of $y = f(x) + a$ is a translation of the graph of $y = f(x)$ by $\begin{pmatrix} 0 \\ a \end{pmatrix}$

The graph of $y = f(x) - a$ is a translation of the graph of $y = f(x)$ by $\begin{pmatrix} 0 \\ -a \end{pmatrix}$

TRANSLATING GRAPHS IN THE x-DIRECTION

The graph of $y = f(x + a)$ is a translation of the graph of $y = f(x)$ by $\begin{pmatrix} -a \\ 0 \end{pmatrix}$

The graph of $y = f(x - a)$ is a translation of the graph of $y = f(x)$ by $\begin{pmatrix} a \\ 0 \end{pmatrix}$

REFLECTING GRAPHS

The graph of $y = -f(x)$ is a reflection of $y = f(x)$ in the x-axis.

The graph of $y = f(-x)$ is a reflection of $y = f(x)$ in the y-axis.

STRETCHING GRAPHS

STRETCHING GRAPHS IN THE y-DIRECTION

The graph of $y = kf(x)$ is a stretch of the graph $y = f(x)$ with a scale factor of k parallel to the y-axis (all y-co-ordinates are multiplied by k).

If $k > 1$ the graph is stretched by k.

If $0 < k < 1$ the graph is compressed by k.

STRETCHING GRAPHS IN THE x-DIRECTION

The graph of $y = f(kx)$ is a stretch of the graph $y = f(x)$ with a scale factor of $\frac{1}{k}$ parallel to the x-axis (all x co-ordinates are multiplied by $\frac{1}{k}$)

If $k > 1$ the graph is compressed by $\frac{1}{k}$

If $0 < k < 1$ the graph is stretched by $\frac{1}{k}$

SHAPE AND SPACE 9

Hipparchus of Nicaea (190–120 BC) was a Greek mathematician and astronomer who is widely understood to have been the first scientist to produce trigonometric tables. Today we use calculators to find these ratios, but for many years these ratios were found from tables that required precise calculation. Astronomers use trigonometry to find the distances to nearby stars and other celestial objects.

LEARNING OBJECTIVES

- Use Pythagoras' Theorem in 3D
- Use trigonometry in 3D to solve problems

BASIC PRINCIPLES

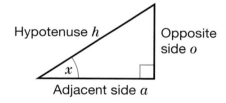

- Trig ratios:

$$\tan x = \frac{\text{opp}}{\text{adj}} \qquad \sin x = \frac{\text{opp}}{\text{hyp}} \qquad \cos x = \frac{\text{adj}}{\text{hyp}}$$

- Identify the **hypotenuse**. This is the longest side: the side opposite the **right angle**. Then the opposite side is opposite the angle. And the **adjacent** side is adjacent to (next to) the angle.

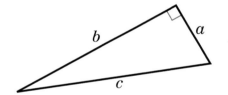

- Pythagoras' Theorem: $a^2 + b^2 = c^2$

3D TRIGONOMETRY

SOLVING PROBLEMS IN 3D

The angle between a line and a **plane** is identified by dropping a **perpendicular** line from a point on the line onto the plane and by joining the point of contact to the point where the line **intersects** the plane.

In the diagram, XYZ is a plane and AB is a line that meets the plane at an angle θ. BP is perpendicular from the top of the line, B, to the plane. P is the point of contact. Join AP. The angle θ, between the line and the plane, is angle BAP in the **right-angled triangle** ABP.

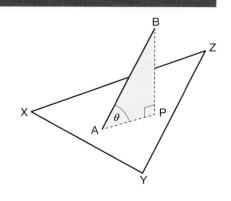

ABCDEFGH is a **cuboid**.

Find to 3 **significant figures**

a length EG

b length CE

c the angle CE makes with plane EFGH (angle CEG).

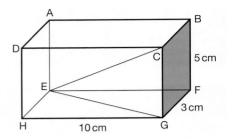

a Draw triangle EGH.

Using Pythagoras' Theorem
$EG^2 = 3^2 + 10^2 = 109$
$EG = \sqrt{109}$
$EG = 10.4\,cm$ (3 s.f.)

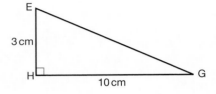

b Draw triangle CEG.

Using Pythagoras' Theorem
$CE^2 = 5^2 + 109 = 134$
$CE = \sqrt{134}$
$CE = 11.6\,cm$ (3 s.f.)

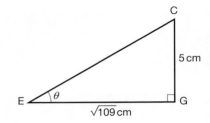

c Let angle CEG $= \theta$

$\tan\theta = \dfrac{5}{\sqrt{109}} \Rightarrow \theta =$ angle CEG $= 25.6°$ (3 s.f.)

When solving problems in 3D:

■ Draw clear, large diagrams including all the facts.

■ Redraw the appropriate triangle (usually right-angled) including all the facts. This simplifies a 3D problem into a 2D problem using Pythagoras' Theorem and trigonometry to solve for angles and lengths.

■ Use all the **decimal places** shown on your calculator at each stage in your working to avoid errors in your final answer caused by **rounding** too soon.

VWXYZ is a solid regular pyramid on a rectangular base WXYZ. WX = 8 cm and XY = 6 cm.
The **vertex** V is 12 cm vertically above the centre of the base.

Find

a VX

b the angle between VX and the base WXYZ (angle VXZ)

c the area of pyramid **face** VWX.

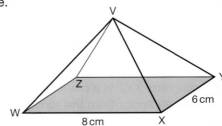

a Let M be the mid-point of ZX.

Draw WXYZ in 2D.

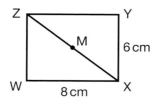

Z ——— Y

M   6 cm

W   8 cm   X

By Pythagoras' Theorem on triangle ZWX

$ZX^2 = 6^2 + 8^2 = 100$

$ZX = 10\,cm \Rightarrow MX = 5\,cm$

Draw triangle VMX.

V is vertically above the mid-point M of the base.

By Pythagoras' Theorem on triangle VMX

$VX^2 = 5^2 + 12^2 = 169$

$VX = 13\,cm$

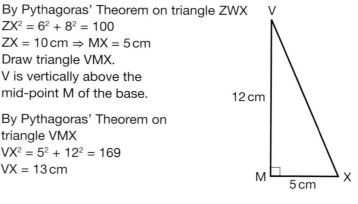

V

12 cm

M   5 cm   X

b Angle VXZ = Angle VXM = θ

$\tan\theta = \dfrac{12}{5} \Rightarrow \theta = 67.4°$ (3 s.f.)

$\Rightarrow$ Angle VXZ = 67.4° (3 s.f.)

c Let N be the **mid-point** of WX.

Area of triangle VWX $= \dfrac{1}{2} \times$ base $\times$ perpendicular height

$= \dfrac{1}{2} \times WX \times VN$

$= \dfrac{1}{2} \times 8 \times VN$

Draw triangle VNX in 2D.

By Pythagoras' Theorem on triangle VNX

$13^2 = 4^2 + VN^2$

$VN^2 = 13^2 - 4^2 = 153$

$VN = \sqrt{153} \Rightarrow$ Area of VWX $= \dfrac{1}{2} \times 8 \times \sqrt{153} = 49.5\,cm^2$ (3 s.f.)

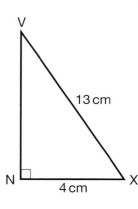

V

13 cm

N   4 cm   X

EXERCISE 1

Give all answers to 3 significant figures.

1 ▶ ABCDEFGH is a cuboid.

Find

a EG

b AG

c the angle between AG and plane EFGH (angle AGE).

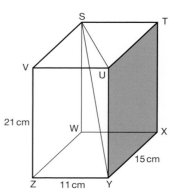

A

B

D

C   8 cm

E

F

H   11 cm   G   4 cm

2 ▶ STUVWXYZ is a rectangular cuboid.

Find

a SU

b SY

c the angle between SY and plane STUV (angle YSU).

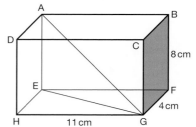

S   T

V   U

21 cm   W   X

Z   11 cm   Y   15 cm

3 ▶ LMNOPQRS is a **cube** of side 10 cm.

Find

a PR

b LR

c the angle between LR and plane PQRS (angle LRP).

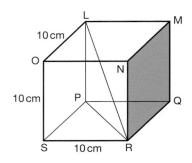

4 ▶ ABCDEFGH is a cube of side 20 cm.

Find

a CF

b DF

c the angle between DF and plane BCGF (angle DFC)

d the angle MHA, if M is the mid-point of AB.

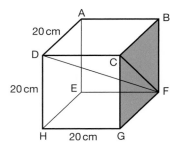

5 ▶ ABCDEF is a **prism**. The **cross-section** is a right-angled triangle. All the other faces are rectangles.

Find

a AC

b AF

c angle FAB

d the angle between AF and plane ABCD (angle FAC).

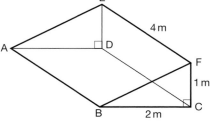

6 ▶ PQRSTU is an artificial ski-slope where PQRS and RSTU are both rectangles and perpendicular to each other.

Find

a UP

b PR

c the angle between UP and plane PQRS

d the angle between MP and plane PQRS, if M is the mid-point of TU.

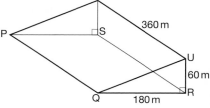

7 ▶ ABCD is a solid on a horizontal triangular base ABC. Edge AD is 25 cm and vertical. AB is perpendicular to AC. Angles ABD and ACD are equal to 30° and 20° respectively.

Find

a AB

b AC

c BC.

8 ▶ PQRS is a solid on a horizontal triangular base PQR. S is vertically above P. Edges PQ and PR are 50 cm and 70 cm respectively. PQ is perpendicular to PR. Angle SQP is 30°.

Find

a SP

b RS

c angle PRS.

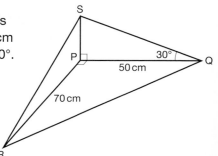

EXERCISE 1*

1 ▶ PABCD is a solid regular pyramid on a rectangular base ABCD. AB = 10 cm and BC = 7 cm. The vertex P is 15 cm vertically above the centre of the base.

Find
a PA
b the angle between PA and plane ABCD
c the area of pyramid face PBC.

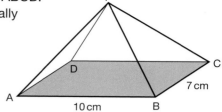

2 ▶ PQRST is a solid regular pyramid on a square base QRST where QR = 20 cm and edge PQ = 30 cm.

Find
a the height of P above the base QRST
b the angle that PS makes with the base QRST
c the total surface area of the pyramid including the base.

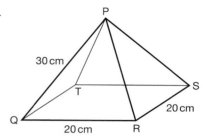

3 ▶ STUVWXYZ is a cuboid. M and N are the mid-points of ST and WZ respectively.

Find angle
a SYW
b TNX
c ZMY.

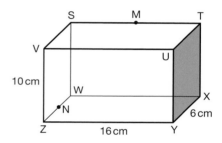

4 ▶ ABCDEFGH is a solid cube of volume 1728 cm³. P and Q are the mid-points of FG and GH respectively.

Find
a angle QCP
b the total surface area of the solid remaining after pyramid PGQC is removed.

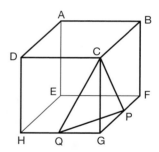

5 ▶ A church is made from two solid rectangular blocks with a regular pyramidal roof above the tower, with V being 40 m above ground level.

a Find VA.
b Find the **angle of elevation** of V from E.

Tiles cost £250 per m².
c How much will it cost to cover the tower roof in tiles?

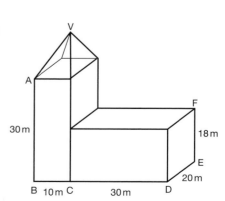

6 ▶ A **hemispherical** lampshade of **diameter** 40 cm is hung from a point by four chains that are each 50 cm in length. If the chains are equally spaced on the circular edge of the hemisphere, find

   a the angle that each chain makes with the horizontal
   b the angle between two adjacent chains.

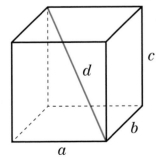

7 ▶ The angle of elevation to the top of a church tower is measured from A and from B.

From A, due South of the church tower VC, the angle of elevation VAC = 15°. From B, due East of the church, the angle of elevation VBC = 25°. AB = 200 m. Find the height of the tower.

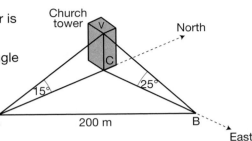

8 ▶ An aircraft is flying at a constant height of 2000 m. It is flying due East at a constant speed. At T, the aircraft's angle of elevation from O is 25°, and on a **bearing** from O of 310°. One minute later, it is at R and due North of O.

RSWT is a rectangle and the points O, W and S are on horizontal ground.

Find
   a the lengths OW and OS
   b the angle of elevation of the aircraft, ∠ROS
   c the speed of the aircraft in km/h.

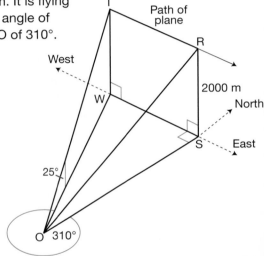

ACTIVITY 1

SKILLS

ANALYSIS

The diagram shows a cuboid.
The diagonal of the cuboid has length d.

a Show that

   i $d^2 = a^2 + b^2 + c^2$

   ii $\sin\theta = \dfrac{c}{\sqrt{a^2 + b^2 + c^2}}$ where θ is the angle between the diagonal and the base.

b For a cube of side a and diagonal d, show that

   i the total surface area $A = 2d^2$

   ii the volume $V = \dfrac{d^3}{3\sqrt{3}}$

   iii $\sin\phi = \sqrt{\dfrac{2}{3}}$ where ϕ is the angle between the diagonal and the front edge.

EXERCISE 2   REVISION

1 ▶   ABCDEFGH is a cuboid.

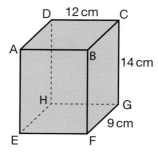

     a Calculate the length of diagonals
       i FH   **ii** BH   **iii** FC   **iv** CE.
     b Find the angle between the diagonal DF and the plane EFGH.
     c Find the angle between the diagonal GA and the plane ABCD.
     d Find the angle between the diagonal CE and the plane AEHD.

2 ▶   The diagram shows a cuboid.
     The base EFGH is in a horizontal plane and triangle
     AEG is in a vertical plane.

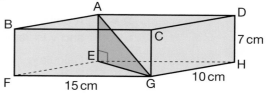

     a Work out the length of AG.
     b Calculate the angle that AG makes with
       the plane EFGH.

     Give all your answers **correct to** 3 s.f.

3 ▶   The diagram shows a pyramid with a square base. The vertex, V,
     is vertically above the centre of the base.

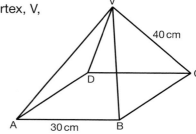

     Calculate
     a the length AC
     b the height of the pyramid
     c the angle that VC makes with BC
     d the angle that VC makes with ABCD.

4 ▶   The diagram shows a ski slope.
     DE = 300 m, AD = 400 m, CE = 100 m.

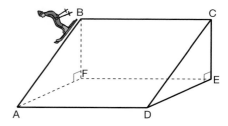

     Find
     a the angle that CD makes with ADEF
     b AE
     c the angle that CA makes with ADEF.

5 ▶   A vertical pole, CP, stands at one corner of a rectangular, horizontal field.
     AB = 40 m, AD = 30 m, angle PDC = 25°

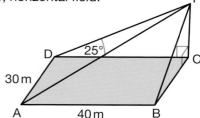

     Calculate
     a the height of the pole, CP
     b the angle of elevation of P from B
     c the angle of elevation of P from A.

     Give all your answers correct to 3 s.f.

REVISION

1 ▶ The diagram shows a prism. The cross-section is a right-angled triangle. All of the other faces are rectangles.
AB = 12 cm, BC = 9 cm, CD = 16 cm

Calculate

a the length of AD
b the angle that AD makes with the plane BCDE
c the angle that AD makes with the plane ABEF.

Give all your answers correct to 3 s.f.

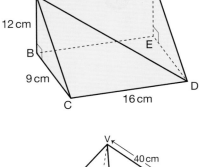

2 ▶ A doll's house has a horizontal square base ABCD and V is vertically above the centre of the base.

Calculate

a the length AC
b the height of V above ABCD
c the angle VE makes with the horizontal
d the total volume.

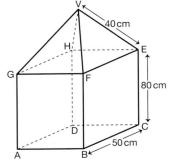

3 ▶ The diagram shows a rectangular-based pyramid. The vertex, V, is vertically above M, the centre of the base, WXYZ. The base lies in a horizontal plane.
WX = 32 cm, XY = 24 cm and VW = VX = VY = VZ = 27 cm.
N is the mid-point of XY.

Calculate

a the length of WY
b the length of VM
c the angle that VY makes with the base WXYZ
d the length of VN
e the angle that VN makes with the base WXYZ.

Give all your answers correct to 3 s.f.

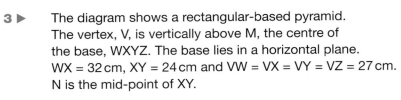

4 ▶ The diagram shows a wedge. AB = 14 cm, and BC = 18 cm.
ABCD is a rectangle in a horizontal plane.
ADEF is a rectangle in a vertical plane.
P is one-third of the way from E to F.
PQ is perpendicular to AD.
The angle between PC and the plane ABCD is 30°.

Calculate the angle between PB and the plane ABCD.
Give your answer correct to 3 s.f.

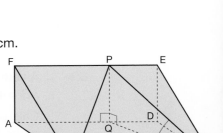

5 ▶ A cube of side 8 cm stands on a horizontal table. A hollow cone of height 20 cm is placed over the cube so that it rests on the table and touches the top four corners of the cube.

a Show that $x = 8\sqrt{2}$ cm.
b Find the vertical angle (the angle of the vertex) of the cone.

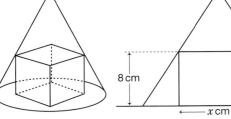

EXAM PRACTICE: SHAPE AND SPACE 9

1 ABCDEFGH is a cube.

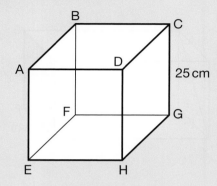

25 cm

Find
a length AG   **[3]**
b the angle between AG and the plane EFGH.   **[3]**

2 ABCDE is a square-based pyramid.
The base BCDE lies in a horizontal plane.
AB = AC = AD = AE = 18 cm
AM is perpendicular to the base.

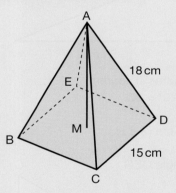

18 cm

15 cm

a Calculate the length of
i BD   **[2]**
ii BM   **[1]**
iii AM.   **[3]**

b Calculate the angle that AD makes with
the base, correct to the nearest degree.   **[3]**

c Calculate the angle between AM and the
face ABC, correct to the nearest degree.   **[4]**

3 The diagram shows part of the roof of a new
out-of-town superstore. The point X is vertically
above A, and ABCD is a horizontal rectangle in
which CD = 5.6 m, BC = 6.4 m. The line XB is
inclined at 70° to the horizontal.

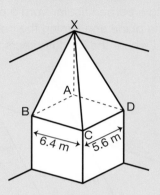

6.4 m   5.6 m

Calculate the angle that the ridge XC makes
with the horizontal.   **[6]**

[Total 25 marks]

CHAPTER SUMMARY: SHAPE AND SPACE 9

3D TRIGONOMETRY

A plane is a flat surface.

The angle between a line and a plane is identified by dropping a perpendicular from a point on the line onto the plane and joining the point of contact to the point where the line intersects the plane.

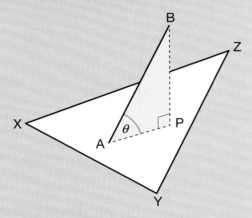

In the diagram, XYZ is a plane and AB is a line that meets the plane at an angle θ. BP is perpendicular from the top of the line, B, to the plane. P is the point of contact. Join AP. The angle θ, between the line and the plane, is angle BAP in the right-angled triangle ABP.

When solving problems in 3D:

■ Draw clear, large diagrams including all the facts.

■ Redraw the appropriate triangle (usually right-angled) including all the facts. This simplifies a 3D problem into a 2D problem using Pythagoras' Theorem and trigonometry to solve for angles and lengths.

■ Use all the decimal places shown on your calculator at each stage in your working to avoid errors in your final answer caused by rounding too soon.

ABCDEFGH is a cuboid.

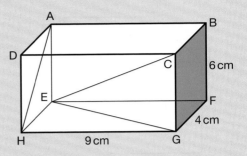

Find

a length EG

b length CE

c the angle CE makes with plane EFGH (angle CEG).

---

a Draw triangle EGH.

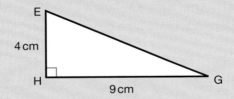

By Pythagoras' Theorem
$EG^2 = 4^2 + 9^2 = 97$
$EG = \sqrt{97}$
$EG = 9.85\,cm$ (3 s.f.)

b Draw triangle CEG.

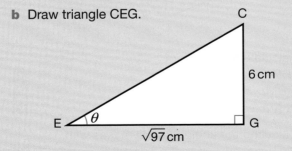

By Pythagoras' Theorem
$CE^2 = 6^2 + 97 = 133$
$CE = \sqrt{133}$
$CE = 11.5\,cm$ (3 s.f.)

c Let angle CEG $= \theta$

$\tan\theta = \dfrac{6}{\sqrt{97}} \Rightarrow \theta =$ angle CEG $= 31.4°$ (3 s.f.)

HANDLING DATA 6

The origins of the word 'histogram' are uncertain. Some believe it is a combination of the Greek words 'histos', which describes something standing upright (like the masts of a ship) and 'gramma' meaning 'drawing or writing'. However, it is also said that the English mathematician Karl Pearson, who introduced the term in 1891, derived the word from 'historical diagram'.

Karl Pearson 1857–1936 ▶

LEARNING OBJECTIVES

■ Draw and interpret histograms

BASIC PRINCIPLES

■ Draw and interpret **bar charts** and frequency diagrams (for equal class intervals).

■ Work out the width of class intervals.

■ Work out the mid-point of class intervals.

■ Write down the **modal class**, and the interval that contains the **median** from a grouped frequency table.

■ Estimate the **range** and work out an estimate for the **mean** from a grouped frequency table.

DRAWING HISTOGRAMS

Histograms appear similar to bar charts, but there are clear differences.

Bar charts have **frequency** on the vertical axis and the frequency equals the height of the bar.

Histograms have **frequency density** on the vertical axis, which makes the frequency proportional to the area of the bar. When data is presented in groups of different class widths (such as $0 \leq x < 2$, $2 \leq x < 10$ etc.) a histogram is drawn to display the information.

EXAMPLE 1

SKILLS

ANALYSIS

Ella records the time of 40 phone calls. The results are shown in the table on the next page.

Show the results on a bar chart.

TIME, t (mins)	FREQUENCY f
$0 \leq t < 1$	2
$1 \leq t < 2$	5
$2 \leq t < 3$	8
$3 \leq t < 4$	9
$4 \leq t < 5$	7
$5 \leq t < 6$	4
$6 \leq t < 7$	3
$7 \leq t < 8$	2

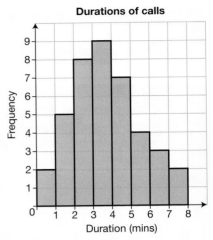

Durations of calls

HINT

The frequency densities are worked out in a calculation table. Four columns are needed.

Ella decides to group the same results as shown in the following table.
Show the results on a bar chart and on a histogram.

TIME, t (mins)	FREQUENCY	CLASS WIDTH	FREQUENCY DENSITY
$0 \leq t < 3$	15	3	$15 \div 3 = 5$
$3 \leq t < 4$	9	1	$9 \div 1 = 9$
$4 \leq t < 8$	16	4	$16 \div 4 = 4$

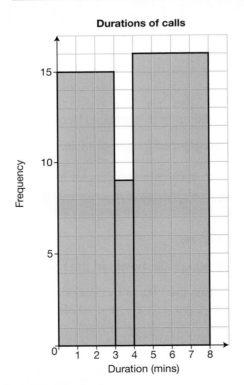

Durations of calls

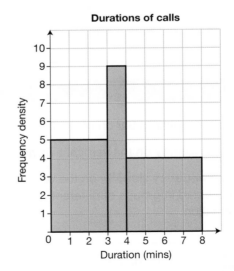

Durations of calls

The bar chart displays the **frequencies**.

The histogram displays the **frequency densities**.

The bar chart with groups of different widths gives a very misleading impression of the original data.

The histogram gives a good impression of the original data although some of the fine detail has been lost in the process of grouping.

- For data grouped in unequal class intervals, you need a histogram.

- In a histogram, the area of the bar represents the frequency.

- The height of each bar is the frequency density.

- Frequency density = $\dfrac{\text{frequency}}{\text{class width}}$

Data with different class intervals is often used to estimate information such as means, proportions and probabilities.

EXAMPLE 2

SKILLS

INTERPRETATION

Use Ella's grouped data from Example 1 to

a calculate an estimate of the mean time of a phone call

b estimate the number of phone calls that are more than $5\frac{1}{2}$ minutes.

HINT

Extend the calculation table to include the mid-point (x) of each group and fx for each group.

a

TIME, t (mins)	FREQUENCY, f	WIDTH	FREQUENCY DENSITY	MID-POINT, x	fx
$0 \le t < 3$	15	3	$15 \div 3 = 5$	1.5	22.5
$3 \le t < 4$	9	1	$9 \div 1 = 9$	3.5	31.5
$4 \le t < 8$	16	4	$16 \div 4 = 4$	6	96
	$\sum f = 40$				$\sum fx = 150$

Estimate of the mean = $\dfrac{\sum fx}{\sum f} = \dfrac{150}{40} = 3.75$ mins

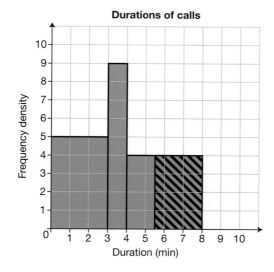

Durations of calls

The shaded area on the histogram represents the calls that are more than $5\frac{1}{2}$ minutes. This area is $4 \times 2.5 = 10$. As the area represents frequency, an estimate of the number of phone calls over $5\frac{1}{2}$ minutes is 10.

Note: The original ungrouped data suggests that only seven calls were more than $5\frac{1}{2}$ minutes, showing that accuracy is lost when data is grouped.

EXERCISE 1

1 ▶ The table shows the ages of 60 patients.

AGE, a (years)	FREQUENCY	CLASS WIDTH	FREQUENCY DENSITY
$0 < a \leq 10$	3	10	$\frac{3}{10} = 0.3$
$10 < a \leq 20$	14		
$20 < a \leq 40$	17		
$40 < a \leq 60$	19		
$60 < a \leq 80$	7		

a Work out each class width.
b Work out the frequency density for each class.

2 ▶ This table shows the time that 100 people spent watching TV one evening.

TIME, t (hours)	FREQUENCY
$0 \leq t \leq 0.5$	5
$0.5 < t \leq 1$	35
$1 < t \leq 2$	56
$2 < t \leq 3$	4

Draw a histogram for this data.

3 ▶ This table contains data for the weights of 65 women.

WEIGHT, w (kg)	FREQUENCY
$40 < w \leq 45$	2
$45 < w \leq 55$	17
$55 < w \leq 65$	31
$65 < w \leq 70$	11
$70 < w \leq 90$	4

Draw a histogram for this data.

4 ▶ Mrs Morris records the results in the Year 11 maths exam.

MARK, m (%)	FREQUENCY
$0 < m \leq 40$	2
$40 < m \leq 50$	4
$50 < m \leq 60$	12
$60 < m \leq 80$	63
$80 < m \leq 100$	9

Q4a HINT
Use the mid-point of each class interval.

a Work out an estimate for the mean mark of Year 11.
b Draw a histogram for the information given in the table.

1 ▶ The table shows the quantities of water, in litres, used by 300 people per day.

VOLUME, s (litres)	FREQUENCY
$75 < s \le 125$	45
$125 < s \le 150$	50
$150 < s \le 175$	70
$175 < s \le 225$	90
$225 < s \le 300$	45

a Work out an estimate for the mean volume of water used by each person per day.
b Draw a histogram to represent the data.

2 ▶ The histogram shows the test scores of 320 children in a school.

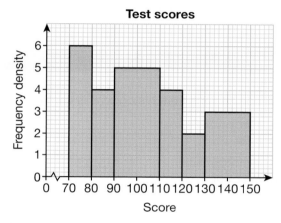

Test scores

Estimate the median test score.

3 ▶ The histogram shows patients' waiting times in a doctors' surgery during one week.

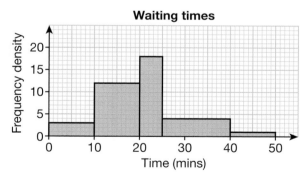

Waiting times

a Rachel says, 'More than half the patients waited less than 20 minutes.'
Is Rachel correct? Explain your answer.

b Estimate the number of patients who waited for more than 33 minutes.

4 ▶ The histogram and the frequency table show the same information about the time that vehicles spent in a car park.

TIME, t (minutes)	FREQUENCY
$0 < t \le 30$	45
$30 < t \le 60$	54
$60 < t \le 100$	
$100 < t \le 120$	50
$120 < t \le 180$	30

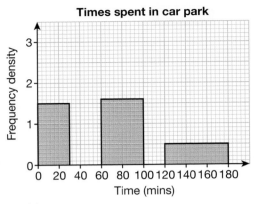

Times spent in car park

a Copy and complete the histogram and frequency table.

b 50 vehicles were in the car park for more than T minutes. Work out an estimate of the value of T.

INTERPRETING HISTOGRAMS

Making accurate and logical conclusions from histograms is commonly required after they have been drawn.

EXAMPLE 3

SKILLS

INTERPRETATION

The histogram shows the **masses** of pumpkins in a farm shop.

Work out an estimate for the median mass.

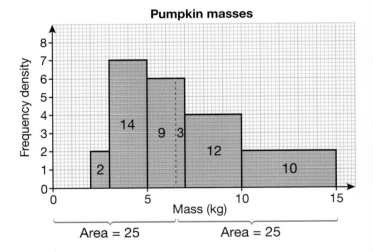

Pumpkin masses

Total frequency = total areas of all the bars

$$= 1 \times 2 + 2 \times 7 + 2 \times 6 + 3 \times 4 + 5 \times 2 = 50$$

The median is the 25.5th value and lies in the class $5 < m \le 7$

Frequency = area = 9, frequency density = $\dfrac{\text{frequency}}{\text{class width}} = 6$

Class width = $\dfrac{\text{frequency}}{\text{frequency density}} = 9 \div 6 = 1.5$

An estimate for the median is found by adding the class width to the lower class boundary:
$5 + 1.5 = 6.5 \, \text{kg}$

EXERCISE 2

1 ▶ The histogram shows the heights of a **sample** of students.

    a How many students were between 150 and 155 cm tall?

    b How many students were between 170 and 180 cm tall?

    c How many students were measured in total?

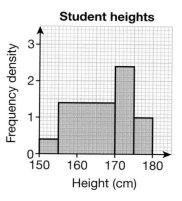

2 ▶ The histogram shows the distances that people in a theatre audience had travelled to the show.

    a How many people in the audience travelled less than 10 miles?

    b Estimate how many people in the audience travelled less than 15 miles.

    c Estimate how many people travelled between 15 and 40 miles.

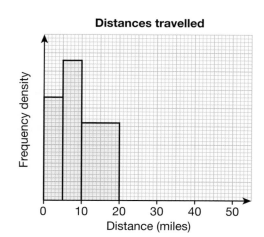

3 ▶ The incomplete table and histogram give some information about the distances people travel to work.

    a Use the information in the histogram to complete the frequency table below.

DISTANCE (d) IN MILES	FREQUENCY
$0 < d \leq 5$	140
$5 < d \leq 10$	
$10 < d \leq 20$	
$20 < d \leq 35$	120
$35 < d \leq 50$	30

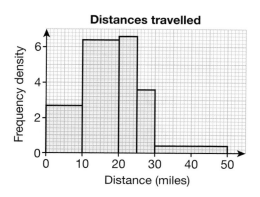

    b Copy and complete the histogram.

4 ▶ The histogram shows the weights of all the babies born in one day in a maternity hospital.

Q4a HINT
Calculate the
frequency density
scale.

a Five babies weighed between 2.8 and
3 kg. Work out the frequency density
for that class.

b How many babies were born in total?

c Work out an estimate for the median
weight.

d Draw a frequency table for the data in
the histogram.

e Work out an estimate for the mean
weight from your frequency table.

f How many of the babies were heavier
than the mean weight?

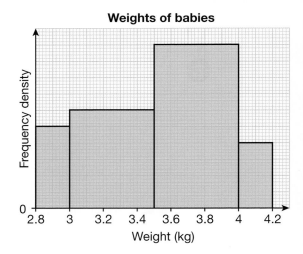

Weights of babies

EXERCISE 2*

1 ▶ The histogram shows the heights of some sunflowers.

a How many sunflowers are there in total?

b Work out an estimate of the median height.

c Estimate how many sunflowers are taller
than 110 cm.

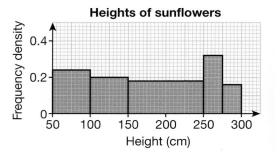

Heights of sunflowers

---

2 ▶ The histogram represents the heights of plants, in centimetres, at a garden centre.

a How many plants are represented
by the histogram?

b Estimate the median height of the
plants.

c Estimate how many plants are
more than 36 cm tall.

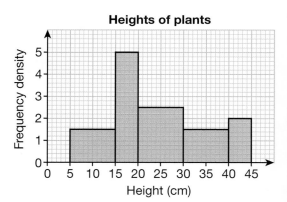

Heights of plants

3 ▶ The histogram shows the distribution of the masses of 80 young finches.

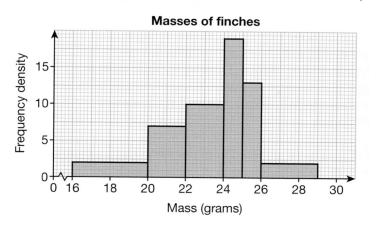

Masses of finches

a Estimate the median mass of a young finch.

b Create a grouped frequency table using the information in the histogram.

c Use your grouped frequency table to calculate an estimate of the mean mass of a young finch.

4 ▶ The histogram shows the finishing times of the runners in a 10 km cross-country race.

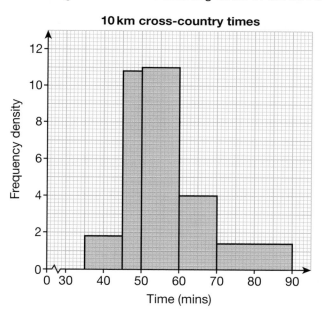

10 km cross-country times

a How many runners were in the race?

b There was a target time of 67 minutes. How many runners failed to meet the target time?

c i Create a grouped frequency table using the information in the histogram.
   ii Use your table to calculate an estimate of the mean time.

ACTIVITY 1

Comparison of data: Brain training

A company wishes to investigate how effective their brain training programme is. They first test how long it takes to solve a logic puzzle, then they allow the participants to use the programme for a week, and then they test the participants again.

Two sets of data were gathered from the same group of adults. The data shows the time taken to solve a puzzle before and after they have taken the course of 'Brain Training'.

TIME TAKEN (t seconds)	FREQUENCY (BEFORE), f_1	FREQUENCY (AFTER), f_2
$12 \le t < 18$	10	13
$18 \le t < 21$	9	14
$21 \le t < 24$	16	16
$24 \le t < 27$	10	14
$27 \le t < 33$	15	3

Study the data carefully and use it to draw two separate histograms.

Use statistical methods to decide if this sample shows whether or not the 'Brain Training' programme makes a difference to a person's ability to solve the puzzle.

REVISION

1 ▶   This table shows the times taken for 55 runners to complete a charity run.

TIME, t (minutes)	$40 < t \le 45$	$45 < t \le 50$	$50 < t \le 60$	$60 < t \le 80$
FREQUENCY	4	17	22	12

Draw a histogram for this data.

2 ▶   This table contains data on the heights of 76 students.

HEIGHT, h (m)	$1.50 < h \le 1.52$	$1.52 < h \le 1.55$	$1.55 < h \le 1.60$	$1.60 < h \le 1.65$	$1.65 < h \le 1.80$
FREQUENCY	4	18	25	15	14

Draw a histogram for this data.

3 ▶ The histogram shows the distance a group of football fans have to travel to a match.

a How many fans travelled less than 5 km?

Q3b HINT
How many fans travelled between 10 and 15 km?

b Estimate how many fans travelled less than 15 km.

c Estimate how many fans travelled between 25 and 32 km.

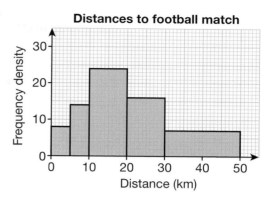

Distances to football match

4 ▶ The histogram shows the masses of some elephants.

a How many elephants are there in total?

Q4b HINT
First find the class containing the median mass.

b Work out an estimate of the median mass.

c Estimate how many elephants weigh more than 5.2 tonnes.

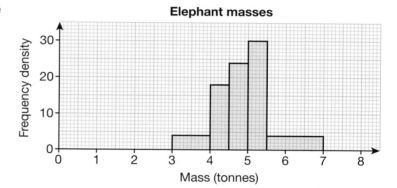

Elephant masses

EXERCISE 3*   **REVISION**

1 ▶ The lengths of some caterpillars are shown in the table.

LENGTH, l (mm)	FREQUENCY
$10 < l \le 15$	2
$15 < l \le 20$	8
$20 < l \le 30$	15
$30 < l \le 40$	12
$40 < l \le 60$	5

Draw a histogram for this data.

2 ▶ The incomplete table shows the times that vehicles spent in an out-of-town shopping centre car park.

TIME, t (minutes)	FREQUENCY
$0 < t \le 30$	18
$30 < t \le 60$	
$60 < t \le 150$	180
$150 < t \le 210$	90
$210 < t \le 240$	57

The incomplete histogram shows the same information.

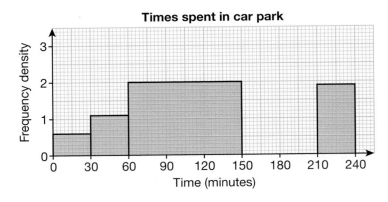

Times spent in car park

a Work out the missing number in the frequency table, and copy and complete the histogram.

b Estimate how many vehicles were in the car park for 120–180 minutes.

3 ▶ The histogram gives information about the areas of 285 farms.

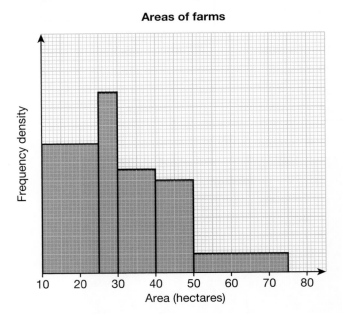

Areas of farms

Work out an estimate for the number of these farms with an area greater than 38 hectares.

4 ▶   The incomplete frequency table and incomplete histogram show the same data for the heights of 104 adults.

HEIGHT, h (cm)	FREQUENCY
$140 < h \leq 160$	22
$160 < h \leq 165$	
$165 < h \leq 170$	
$170 < h \leq 185$	27
$185 < h \leq 200$	12

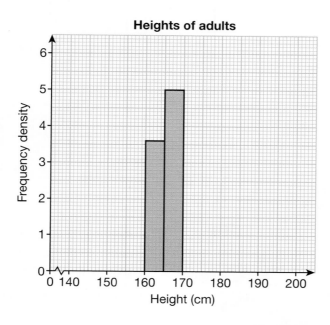

Heights of adults

a   Copy and complete the frequency table and the histogram.

b   Clare says that an estimate of the median height will be between 167 cm and 168 cm. Is she correct? Show **working** to justify your answer.

c   Calculate an estimate of the mean height. Give your answer **correct to** 1 d.p.

EXAM PRACTICE: HANDLING DATA 6

1 The lengths of 150 dolphins are recorded in this table.

LENGTH, l (M)	FREQUENCY
$1.5 < l \leq 2.0$	7
$2.0 < l \leq 2.5$	19
$2.5 < l \leq 2.8$	31
$2.8 < l \leq 3.0$	52
$3.0 < l \leq 3.5$	34
$3.5 < l \leq 5.0$	7

a Draw a histogram for this data.   **[2]**

b Estimate
  i the mean length of the dolphins   **[3]**
  ii the median length of the dolphins   **[2]**
  iii how many dolphins are over 3.25 m long. **[2]**

c State the modal class.   **[1]**

2 The histogram shows the heights of Year 7 students.

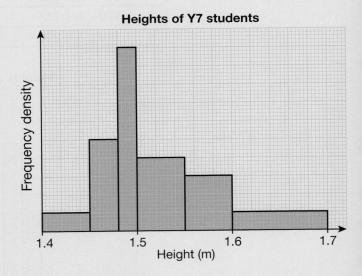

a Five Year 7 students are between 1.4 metres and 1.45 metres tall. Work out the frequency density for that class.   **[2]**

b How many Year 7 students are there in total?   **[2]**

c Work out an estimate for the median height.   **[2]**

d Draw a frequency table for the data in the histogram.   **[3]**

e Work out an estimate for the mean height from your frequency table.   **[3]**

f How many of the students are taller than the mean height?   **[3]**

[Total 25 marks]

CHAPTER SUMMARY: HANDLING DATA 6

DRAWING AND INTERPRETING HISTOGRAMS

Histograms are used to display continuous data.

The area of each bar is proportional to the frequency.

The height of each bar is the frequency density.

$$\text{Frequency density} = \frac{\text{frequency}}{\text{class width}}$$

The lengths of 48 worms are recorded in the table.

Draw a histogram to display this data.

Estimate the mean and median length of worm from this sample.

LENGTH, L (mm)	FREQUENCY, f	CLASS WIDTH, w	FREQUENCY DENSITY $= f \div w$	MID-POINT, x	fx
$15 \le L < 20$	6	5	$6 \div 5 = 1.2$	17.5	105
$20 \le L < 30$	14	10	$14 \div 10 = 1.4$	25	350
$30 \le L < 40$	26	10	$26 \div 10 = 2.6$	35	910
$40 \le L < 60$	2	20	$2 \div 20 = 0.1$	50	100

$$\sum f = 48 \qquad\qquad\qquad \sum fx = 1465$$

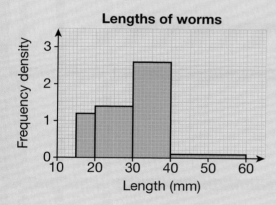

Lengths of worms

Estimated mean worm length $= \dfrac{\sum fx}{\sum f} = \dfrac{1465}{48} = 30.5\,\text{cm}$ (3 s.f.)

Median = 24.5th value, so this lies in the $30 \le L < 40$ class.

Total frequency of the first two bars is 20.

Class width of class from 30 to the median at 24:

Frequency = area = 4, frequency density = 2.6

Class width $= 4 \div 2.6 = 1.54$

Estimate of median worm length $= 30 + 1.54 = 31.5\,\text{cm}$ (3 s.f.)

UNIT 10

A polygon with ten sides is a decagon.

A googol is defined as 10^{10^2} which is 1 followed by 100 zeros.

The first three prime numbers sum to 10. Ten is the base of our decimal numeral system and as a result many financial currencies around the world also have a base of ten (dollars, euros, pounds…).

NUMBER 10

The first humans only needed the numbers 1, 2, 3, … (called the set of natural numbers) to count objects. Later negative numbers and zero (called the set of integers), as well as fractions, were needed to cope with developing civilisations. Early civilisations thought that all numbers could be expressed as fractions, which are the ratio of two integers (this is where the word rational comes from, which describes this set of numbers).

In the sixth century BC, the ancient Greeks proved that $\sqrt{2}$ could not be written as a fraction. They soon found other numbers, such as $\sqrt{3}$ and $\sqrt{5}$, that could not be written as fractions. These numbers are called irrational numbers. In 1761 it was shown that π is also an irrational number.

LEARNING OBJECTIVES

- Understand the difference between rational and irrational numbers
- Simplify surds
- Simplify expressions involving surds
- Expand expressions involving surds
- Rationalise the denominator of a fraction

BASIC PRINCIPLES

- Recognise that $\sqrt{}$ corresponds to the positive square root.
- Recognise that $\sqrt{a} \times \sqrt{a} = a$
- Understand the dot notation used for **recurring decimals**.
- Convert **terminating** and recurring decimals to exact fractions.
- Find **factor** pairs of a number (with one factor also being a square number).
- **Expand** (and **simplify**) the **product** of two brackets.
- Use Pythagoras' Theorem.
- Use the trigonometric ratios.

ACTIVITY 1

The search for the value of π has fascinated people for centuries. By the beginning of the 21st century π was known to 1.33×10^{13} digits, and someone had memorised it correctly to 67 000 digits! The expansion starts

π = 3.141 592 653 589 793 238 462 643 383 279 502 884 197 169 399 375 ...

a Work out the percentage error (to 3 s.f.) in using the values of π in the table compared to using the value of π stored in your calculator.

APPROXIMATION OF π	SOURCE
3	**Bible** 1 Kings 7:23
$3\frac{1}{8}$	**Babylon** 2000 BC; found on a clay tablet in 1936
$\frac{256}{81}$	**Egypt** 2000 BC; found on the 'Rhind Papyrus'
$\frac{22}{7}$	**Syracuse** 250 BC; Archimedes
$\frac{377}{120}$	**Greece** 140 BC; Hipparchus
$\frac{355}{113}$	**China** AD 450; Tsu Chung-Chih
$\sqrt{10}$	**India** AD 625; Brahmagupta
$\frac{864}{275}$	**Italy** AD 1225; Fibonacci
$\sqrt[4]{9^2 + \frac{19^2}{22}}$	**India** AD 1910; Ramanujan

b By measuring the length of the expansion of π given above, estimate in km what length the 1.33×10^{13} digits of the expansion of π would be in full.

RATIONAL AND IRRATIONAL NUMBERS

The decimal expansion of **irrational numbers** is infinite with no pattern. An infinite decimal expansion where the digits recur is rational because it can be written as a fraction (see Number 7).

Together the **rational numbers** and the irrational numbers form the set of **real numbers**. All these sets can be shown on a Venn diagram where

■ $\mathbb{N}$ is the set of **natural numbers** or positive integers {1, 2, 3, 4, ...}

■ $\mathbb{Z}$ is the set of integers {..., −2, −1, 0, 1, 2, ...}

■ $\mathbb{Q}$ is the set of rational numbers

■ $\mathbb{R}$ is the set of real numbers.

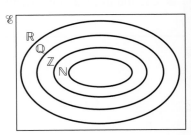

ACTIVITY 2

If $\mathbb{W}$ is the **set** of whole numbers $\{0, 1, 2, 3, \ldots\}$

$\mathbb{Z}^+$ is the set of positive **integers** $\{1, 2, 3, \ldots\}$

$\mathbb{Z}^-$ is the set of negative integers $\{\ldots, -3, -2, -1\}$

draw **Venn diagrams** to show the relation between

a $\mathbb{W}$, $\mathbb{Z}^+$ and $\mathbb{Q}$

b $\mathbb{Z}^+$, $\mathbb{Z}^-$ and $\mathbb{R}$

EXAMPLE 1

Which of these numbers are rational and which are irrational?

Give reasons for your answers.

$2 \qquad 0.789 \qquad 0.\dot{6}\dot{7} \qquad \sqrt{2} \times \sqrt{2} \qquad \sqrt{2} + \sqrt{2}$

$2 = \dfrac{2}{1}$ so it is rational.

$0.789 = \dfrac{789}{1000}$ so it is rational.

$0.\dot{6}\dot{7} = \dfrac{67}{99}$ so it is rational.

$\sqrt{2} \times \sqrt{2} = 2$ so it is rational.

$\sqrt{2} + \sqrt{2} = 2\sqrt{2}$ so it is irrational.

EXAMPLE 2

a Find an irrational number between 4 and 5.

b Find a rational number between $\sqrt{2}$ and $\sqrt{3}$.

---

a $4^2 = 16$ and $5^2 = 25$, so the square root of any integer between 17 and 24 inclusive will be irrational, for example $\sqrt{19}$.

b $\sqrt{2} = 1.414213\ldots$ and $\sqrt{3} = 1.732050\ldots$ so any terminating or recurring decimal between these numbers will be rational, for example 1.5

KEY POINTS

■ Rational numbers can be written as a fraction in the form $\dfrac{a}{b}$

where a and b are integers and $b \neq 0$

  ■ 4 is rational as it can be written as $\dfrac{4}{1}$

  ■ $0.\dot{4}$ is rational as it can be written as $\dfrac{4}{9}$

■ An irrational numbers cannot be written as a fraction.

  ■ $\sqrt{2}$ and are π both irrational.

  ■ The decimal expansion of irrational numbers is infinite and shows no pattern.

EXERCISE 1

For Questions 1–8, state which of the numbers are rational and which are irrational. Express the rational numbers in the form $\frac{a}{b}$ where a and b are integers.

1 ▶   5.7

2 ▶ $0.\dot{4}\dot{7}$

3 ▶ $\sqrt{49}$

4 ▶ 2π

5 ▶ $\sqrt{3} \times \sqrt{3}$

6 ▶ $\sqrt{3} \div \sqrt{3}$

7 ▶ $\sqrt{5} + \sqrt{5}$

8 ▶ $\sqrt{3} - \sqrt{3}$

9 ▶   Find a rational number between $\sqrt{5}$ and $\sqrt{7}$.

10 ▶   Find an irrational number between 7 and 8.

11 ▶   The circumference of a circle is 4 cm. Find the **radius**, giving your answer as an irrational number.

12 ▶   The area of a circle is 9 cm². Find the radius, giving your answer as an irrational number.

EXERCISE 1*

For Questions 1–7, state which of the numbers are rational and which are irrational. Express the rational numbers in the form $\frac{a}{b}$ where a and b are integers.

1 ▶ $\pi + 2$

2 ▶ $\sqrt{\dfrac{4}{25}}$

3 ▶ $\sqrt{0.36}$

4 ▶ $\sqrt{2\dfrac{1}{4}}$

5 ▶ $\sqrt{5} - \sqrt{3}$

6 ▶ $\sqrt{13} \div \sqrt{13}$

7 ▶ $\sqrt{3} \times \sqrt{3} \times \sqrt{3}$

8 ▶   Find a rational number between $\sqrt{11}$ and $\sqrt{13}$.

9 ▶   Find an irrational number between 2 and 3.

10 ▶   Find two different irrational numbers whose product is rational.

11 ▶   The **circumference** of a circle is 6 cm. Find the area, giving your answer as an irrational number.

12 ▶   **Right-angled triangles** can have sides with lengths that are rational or irrational numbers of units. Give an example of a right-angled triangle to fit each description below. Draw a separate triangle for each part.

   a All sides are rational.

   b The **hypotenuse** is rational and the other two sides are irrational.

   c The hypotenuse is irrational and the other two sides are rational.

   d The hypotenuse and one of the other sides is rational and the remaining side is irrational.

SURDS

A **surd** is a number written exactly using roots. Surds are used to express irrational numbers in exact form. For example, $\sqrt{3}$ and $\sqrt[3]{5}$ are surds.

$\sqrt{4}$ and $\sqrt[3]{27}$ are not surds, because $\sqrt{4} = 2$ and $\sqrt[3]{27} = 3$

Note that $\sqrt{2}$ means the positive square root of 2.

When an answer is given as a surd, it is an exact answer. For example, the diagonal of a square of side 1 cm is exactly $\sqrt{2}$ cm long, or $\sin 60°$ is exactly $\dfrac{\sqrt{3}}{2}$

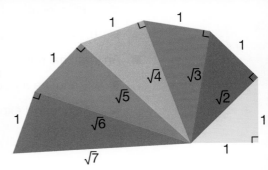

SIMPLIFYING SURDS

EXAMPLE 3

Simplify these, giving the exact answer.

a $2\sqrt{3} + 3\sqrt{3}$

b $(2\sqrt{3})^2$

c $2\sqrt{3} \times 3\sqrt{3}$

d $(\sqrt{3})^3$

a $2\sqrt{3} + 3\sqrt{3} = 5\sqrt{3}$

b $(2\sqrt{3})^2 = 2\sqrt{3} \times 2\sqrt{3} = 2 \times 2 \times \sqrt{3} \times \sqrt{3} = 4 \times 3 = 12$

c $2\sqrt{3} \times 3\sqrt{3} = 2 \times 3 \times \sqrt{3} \times \sqrt{3} = 6 \times 3 = 18$

d $(\sqrt{3})^3 = \sqrt{3} \times \sqrt{3} \times \sqrt{3} = 3\sqrt{3}$

EXERCISE 2

Simplify these, giving the exact answer.

1 ▶ $2\sqrt{5} + 4\sqrt{5}$ **4** ▶ $(2\sqrt{5})^2$ **7** ▶ $(\sqrt{2})^3$

2 ▶ $5\sqrt{3} - \sqrt{3}$ **5** ▶ $4\sqrt{2} \times \sqrt{2}$ **8** ▶ $4\sqrt{2} \div \sqrt{2}$

3 ▶ $(4\sqrt{2})^2$ **6** ▶ $5\sqrt{7} \times 3\sqrt{7}$ **9** ▶ $8\sqrt{5} \div \sqrt{5}$

EXERCISE 2*

Simplify these, giving the exact answer.

1 ▶ $4\sqrt{11} + \sqrt{11}$ **4** ▶ $8\sqrt{3} \times 4\sqrt{3}$ **7** ▶ $(\sqrt{3})^4$

2 ▶ $6\sqrt{7} - 2\sqrt{7}$ **5** ▶ $(2\sqrt{7})^3$ **8** ▶ $4\sqrt{7} \div \sqrt{7}$

3 ▶ $(3\sqrt{11})^2$ **6** ▶ $3\sqrt{2} \times 4\sqrt{2} \times 5\sqrt{2}$ **9** ▶ $6\sqrt{11} \div \sqrt{11}$

Example 3 part **b** showed that $(2\sqrt{3})^2 = 4 \times 3 = 12$

This means that $\sqrt{12} = \sqrt{4 \times 3} = 2\sqrt{3}$

This is because $\sqrt{a \times b} = \sqrt{a} \times \sqrt{b}$

EXAMPLE 4

Simplify   **a** $\sqrt{18}$ **b** $\sqrt{35}$ **c** $\sqrt{72}$

a Find two factors of 18, one of which is a **perfect square**, for example $18 = 9 \times 2$.

Then $\sqrt{18} = \sqrt{9} \times \sqrt{2} = 3\sqrt{2}$

b In this case a perfect square factor cannot be found.

As $35 = 5 \times 7$, then $\sqrt{35} = \sqrt{5} \times \sqrt{7} = \sqrt{5}\sqrt{7}$

c $72 = 36 \times 2$, so $\sqrt{72} = \sqrt{36} \times \sqrt{2} = 6\sqrt{2}$

In Example 4 part **c** you might have written $72 = 9 \times 8$ so $\sqrt{72} = 3\sqrt{8}$.

This is correct, but it is not in its simplest form because $\sqrt{8}$ can be simplified as $\sqrt{8} = \sqrt{4} \times \sqrt{2} = 2\sqrt{2}$.

This gives $\sqrt{72} = 3 \times 2\sqrt{2} = 6\sqrt{2}$ as before.

So when finding factors, try to find the largest possible factor that is a perfect square.

EXAMPLE 5

Simplify $\sqrt{18} + 2\sqrt{2}$

$\sqrt{18} = 3\sqrt{2}$ (see Example 4)
So $\sqrt{18} + 2\sqrt{2} = 3\sqrt{2} + 2\sqrt{2} = 5\sqrt{2}$

EXAMPLE 6

Express $5\sqrt{6}$ as the square root of a single number.

$5\sqrt{6} = \sqrt{25} \times \sqrt{6} = \sqrt{25 \times 6} = \sqrt{150}$

To simplify $\sqrt{\dfrac{16}{25}}$ note that $\left(\dfrac{4}{5}\right)^2 = \dfrac{4^2}{5^2} = \dfrac{16}{25}$ so this means that $\sqrt{\dfrac{16}{25}} = \dfrac{4}{5}$

This is because $\sqrt{\dfrac{a}{b}} = \dfrac{\sqrt{a}}{\sqrt{b}}$

EXAMPLE 7

Simplify $\sqrt{\dfrac{81}{49}}$

$\sqrt{\dfrac{81}{49}} = \dfrac{\sqrt{81}}{\sqrt{49}} = \dfrac{9}{7}$

KEY POINTS

- $\sqrt{a \times b} = \sqrt{a} \times \sqrt{b}$
- $\sqrt{\dfrac{a}{b}} = \dfrac{\sqrt{a}}{\sqrt{b}}$

EXERCISE 3

For Questions 1–7, simplify

1 ▶ $\sqrt{12}$ 　　　　　　5 ▶ $\sqrt{12} + \sqrt{3}$
2 ▶ $\sqrt{18}$ 　　　　　　6 ▶ $\sqrt{32} - 2\sqrt{2}$
3 ▶ $2\sqrt{48}$ 　　　　　　7 ▶ $\sqrt{8} + 3\sqrt{32}$
4 ▶ $3\sqrt{45}$

For Questions 8–10, express as the square root of a single number

8 ▶ $5\sqrt{2}$ 　　　　　9 ▶ $3\sqrt{3}$ 　　　　　10 ▶ $3\sqrt{6}$

For Questions 11–13, simplify

11 ▶ $\sqrt{\dfrac{1}{4}}$ 　　　　12 ▶ $\sqrt{\dfrac{4}{25}}$ 　　　　13 ▶ $\sqrt{\dfrac{36}{81}}$

14 ▶ A rectangle has sides of length $\sqrt{12}$ and $\sqrt{27}$. Find the area, the **perimeter** and the length of a diagonal, expressing each answer as a surd in its simplest form.

EXERCISE 3*

For Questions 1–7, simplify

1 ▶ $\sqrt{28}$ 　　　　　　5 ▶ $\sqrt{48} + \sqrt{3}$
2 ▶ $\sqrt{99}$ 　　　　　　6 ▶ $\sqrt{27} - \sqrt{12}$
3 ▶ $5\sqrt{80}$ 　　　　　　7 ▶ $4\sqrt{63} - \sqrt{28} + 3\sqrt{112}$
4 ▶ $3\sqrt{117}$

For Questions 8–10, express as the square root of a single number

8 ▶ $5\sqrt{3}$

9 ▶ $4\sqrt{5}$

10 ▶ $3\sqrt{7}$

For Questions 11–13, simplify

11 ▶ $\sqrt{\dfrac{1}{36}}$

12 ▶ $\sqrt{\dfrac{81}{100}}$

13 ▶ $\sqrt{\dfrac{49}{169}}$

14 ▶   A piece of wire is bent into the shape shown in the diagram. All dimensions are in cm. Find the length x, the perimeter and the area of the shape, expressing each answer as a surd in its simplest form.

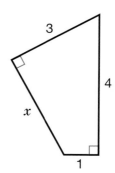

ACTIVITY 3

a The diagram shows an **isosceles** right-angled triangle.

The two shorter sides are 1 cm in length.

i   What is the exact value of x (the hypotenuse)?

ii What is the size of the angle marked a?

iii Use your answers to parts **i** and **ii** to fill in the table below, expressing each answer as a surd in its simplest form.

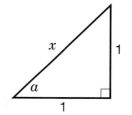

$\sin 45° =$	$\cos 45° =$	$\tan 45° =$

b The diagram shows half an **equilateral triangle** ABC with sides of length 2 cm. D is the **mid-point** of AC.

i   Find the exact value of the length BD.

ii What are the sizes of the angles BAD and ABD?

iii Use your answers to parts **i** and **ii** to fill in the table below, expressing each answer as a surd in its simplest form.

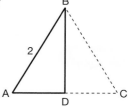

$\sin 30° =$	$\cos 30° =$	$\tan 30° =$
$\sin 60° =$	$\cos 60° =$	$\tan 60° =$

To simplify expressions like $(2 + 3\sqrt{2})^2$, expand the brackets using FOIL and then simplify.

EXAMPLE 8

Expand and simplify $(2 + 3\sqrt{2})^2$.

$(2 + 3\sqrt{2})^2 = (2 + 3\sqrt{2})(2 + 3\sqrt{2}) = 4 + 6\sqrt{2} + 6\sqrt{2} + 3\sqrt{2} \times 3\sqrt{2}$

$= 4 + 12\sqrt{2} + 18$

$= 22 + 12\sqrt{2}$

EXERCISE 4

Expand and simplify

1 ▶ $(1 + \sqrt{2})^2$

2 ▶ $(1 - \sqrt{3})^2$

3 ▶ $(3 + 2\sqrt{3})^2$

4 ▶ $(3 - 3\sqrt{2})^2$

5 ▶ $(1 + \sqrt{5})(1 - \sqrt{5})$

6 ▶ $(\sqrt{2} + \sqrt{3})^2$

7 ▶ $(\sqrt{5} - \sqrt{2})^2$

8 ▶ $(1 + \sqrt{2})(1 - \sqrt{5})$

9 ▶ $(1 - \sqrt{3})(1 + \sqrt{2})$

10 ▶ A rectangle has sides of length $\sqrt{3} + 1$ and $\sqrt{3} - 1$. Find the perimeter, the area and the length of a diagonal, expressing each answer as a surd in its simplest form.

EXERCISE 4*

Expand and simplify

1 ▶ $(2 + \sqrt{5})^2$

2 ▶ $(4 - \sqrt{2})^2$

3 ▶ $(4 - \sqrt{3})(4 + \sqrt{3})$

4 ▶ $(2 + 4\sqrt{2})^2$

5 ▶ $(4 - 5\sqrt{3})^2$

6 ▶ $(\sqrt{7} - \sqrt{3})^2$

7 ▶ $(\sqrt{7} - \sqrt{5})(\sqrt{7} + \sqrt{5})$

8 ▶ $(3 + 2\sqrt{2})(5 - 2\sqrt{7})$

9 ▶ $(2 + 3\sqrt{3})(4 - 3\sqrt{5})$

10 ▶ A right-angled triangle has a hypotenuse of length $2 + \sqrt{2}$ and one other side of length $1 + \sqrt{2}$. Find the length of the third side and the area, expressing each answer as a surd in its simplest form.

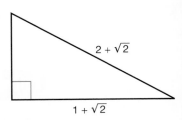

RATIONALISING THE DENOMINATOR

When writing fractions it is not usual to write surds in the **denominator**. The surds can be cleared by multiplying the top and bottom of the fraction by the same number. This is equivalent to multiplying the fraction by 1 and so does not change its value. The process is called '**rationalising the denominator**'.

EXAMPLE 9

Rationalise the denominator of these fractions.

a $\dfrac{14}{\sqrt{7}}$ **b** $\dfrac{\sqrt{5} + 5}{\sqrt{5}}$

---

a Multiply the fraction by $\dfrac{\sqrt{7}}{\sqrt{7}}$

$$\frac{14}{\sqrt{7}} = \frac{14}{\sqrt{7}} \times \frac{\sqrt{7}}{\sqrt{7}}$$

$$= \frac{14\sqrt{7}}{7}$$

$$= 2\sqrt{7}$$

b Multiply the fraction by $\dfrac{\sqrt{5}}{\sqrt{5}}$

$$\frac{\sqrt{5} + 5}{\sqrt{5}} = \frac{\sqrt{5} + 5}{\sqrt{5}} \times \frac{\sqrt{5}}{\sqrt{5}}$$

$$= \frac{(\sqrt{5} + 5) \times \sqrt{5}}{5}$$

$$= \frac{5 + 5\sqrt{5}}{5}$$

$$= 1 + \sqrt{5}$$

KEY POINT

■ To rationalise the denominator of $\dfrac{a}{\sqrt{b}}$ multiply by $\dfrac{\sqrt{b}}{\sqrt{b}}$

ACTIVITY 4

SKILLS

ANALYSIS

a Expand and simplify

   i $(1 - \sqrt{2})(1 + \sqrt{2})$ **ii** $(4 + \sqrt{3})(4 - \sqrt{3})$ **iii** $(3 - 2\sqrt{2})(3 + 2\sqrt{2})$

b Are your answers to part **a** rational or irrational?

c What must you multiply   **i** $(2 + \sqrt{2})$ **ii** $(3 - 2\sqrt{5})$ by to get a rational answer?

d What must you multiply   **i** $(a + \sqrt{b})$ **ii** $(a - c\sqrt{b})$ by to get a rational answer?

KEY POINTS

■ To rationalise the denominator of $\dfrac{1}{a - \sqrt{b}}$ multiply by $\dfrac{a + \sqrt{b}}{a + \sqrt{b}}$

■ To rationalise the denominator of $\dfrac{1}{a + \sqrt{b}}$ multiply by $\dfrac{a - \sqrt{b}}{a - \sqrt{b}}$

EXAMPLE 10

Rationalise the denominator of these fractions.

a $\dfrac{1}{1 + \sqrt{2}}$ **b** $\dfrac{7}{3 - \sqrt{2}}$

---

a Multiply by $\dfrac{1 - \sqrt{2}}{1 - \sqrt{2}}$

$$\frac{1}{1 + \sqrt{2}} = \frac{1}{1 + \sqrt{2}} \times \frac{1 - \sqrt{2}}{1 - \sqrt{2}} = \frac{1 - \sqrt{2}}{1 - 2} = \frac{1 - \sqrt{2}}{-1} = -1 + \sqrt{2}$$

b Multiply by $\dfrac{3 + \sqrt{2}}{3 + \sqrt{2}}$

$$\frac{7}{3 - \sqrt{2}} = \frac{7}{3 - \sqrt{2}} \times \frac{3 + \sqrt{2}}{3 + \sqrt{2}} = \frac{7(3 + \sqrt{2})}{9 - 2} = \frac{7(3 + \sqrt{2})}{7} = 3 + \sqrt{2}$$

ACTIVITY 5

Use your calculator to check that

a $\dfrac{1}{1 + \sqrt{2}} = -1 + \sqrt{2}$ **b** $\dfrac{7}{3 - \sqrt{2}} = 3 + \sqrt{2}$

EXERCISE 5

Rationalise the denominator of these fractions and simplify if possible.

1 ▶ $\dfrac{1}{\sqrt{5}}$

2 ▶ $\dfrac{3}{\sqrt{3}}$

3 ▶ $\dfrac{4}{\sqrt{2}}$

4 ▶ $\dfrac{2}{\sqrt{8}}$

5 ▶ $\dfrac{10}{\sqrt{20}}$

6 ▶ $\dfrac{4}{\sqrt{12}} + \dfrac{3}{\sqrt{27}}$

7 ▶ $\dfrac{3}{2\sqrt{2}}$

8 ▶ $\dfrac{1 + \sqrt{2}}{\sqrt{2}}$

9 ▶ $\dfrac{5 + 2\sqrt{5}}{\sqrt{5}}$

10 ▶ $\dfrac{1}{1 - \sqrt{5}}$

11 ▶ $\dfrac{1}{2 - \sqrt{3}}$

12 ▶ $\dfrac{1 - \sqrt{2}}{3 + \sqrt{2}}$

EXERCISE 5*

Rationalise the denominator of these fractions and simplify if possible.

1 ▶ $\dfrac{1}{\sqrt{13}}$

2 ▶ $\dfrac{a}{\sqrt{a}}$

3 ▶ $\dfrac{6 - \sqrt{3}}{\sqrt{3}}$

4 ▶ $\dfrac{\sqrt{8}}{\sqrt{12}}$

5 ▶ $\dfrac{14 + 3\sqrt{7}}{\sqrt{7}}$

6 ▶ $\dfrac{15}{\sqrt{18}} - \dfrac{12}{\sqrt{32}}$

7 ▶ $\dfrac{1}{2 + \sqrt{7}}$

8 ▶ $\dfrac{22}{6 - \sqrt{3}}$

9 ▶ $\dfrac{2 + \sqrt{3}}{2 - \sqrt{3}}$

10 ▶ $\dfrac{-7\sqrt{8}}{\sqrt{2} - 3}$

11 ▶ $\dfrac{10}{\sqrt{5} + \sqrt{3}}$

12 ▶ Given $\dfrac{8 - \sqrt{18}}{\sqrt{2}} = a + b\sqrt{2}$ where a and b are integers, find a and b.

ACTIVITY 6

a Show that $\dfrac{1}{\sqrt{1} + \sqrt{2}} + \dfrac{1}{\sqrt{2} + \sqrt{3}} + \dfrac{1}{\sqrt{3} + \sqrt{4}} = 1$

b Find $\dfrac{1}{\sqrt{1} + \sqrt{2}} + \dfrac{1}{\sqrt{2} + \sqrt{3}} + \dfrac{1}{\sqrt{3} + \sqrt{4}} + \ldots + \dfrac{1}{\sqrt{8} + \sqrt{9}}$

c If $\dfrac{1}{\sqrt{1} + \sqrt{2}} + \dfrac{1}{\sqrt{2} + \sqrt{3}} + \dfrac{1}{\sqrt{3} + \sqrt{4}} + \ldots + \dfrac{1}{\sqrt{n - 1} + \sqrt{n}} = 9$, find n.

EXERCISE 6

REVISION

1 ▶ Which of $0.\dot{3}$, π, $\sqrt{25}$ and $\sqrt{5}$ are rational?

2 ▶ Find a rational number between $\sqrt{3}$ and $\sqrt{5}$.

3 ▶ Find an irrational number between 3 and 4.

4 ▶ Write $3\sqrt{5}$ as the square root of a single number.

For Questions 5–12, simplify

5 ▶ $3\sqrt{3} + 2\sqrt{3}$

6 ▶ $3\sqrt{3} - 2\sqrt{3}$

7 ▶ $3\sqrt{3} \times 2\sqrt{3}$

8 ▶ $3\sqrt{3} \div 2\sqrt{3}$

9 ▶ $\sqrt{63}$

10 ▶ $3\sqrt{32}$

11 ▶ $3\sqrt{8} - \sqrt{18}$

12 ▶ $\dfrac{6}{\sqrt{8}} + \dfrac{3}{\sqrt{18}}$

13 ▶ Expand $(3 + 5\sqrt{2})^2$. Express your answer in the form $a + b\sqrt{2}$

14 ▶ Expand and simplify $(1 + \sqrt{2})(1 - \sqrt{2})$.

For Questions 15–19, rationalise the denominator.

15 ▶ $\dfrac{6}{\sqrt{3}}$ **17 ▶** $\dfrac{3+\sqrt{3}}{\sqrt{3}}$ **19 ▶** $\dfrac{4}{\sqrt{5}-2}$

16 ▶ $\dfrac{\sqrt{27}}{\sqrt{12}}$ **18 ▶** $\dfrac{6}{1+\sqrt{7}}$

20 ▶ A rectangle has sides of length $3\sqrt{2}$ and $5\sqrt{2}$. Find the perimeter, the area and the length of a diagonal, expressing each answer as a surd, if appropriate, in its simplest form.

EXERCISE 6*   **REVISION**

1 ▶ Which of $(\sqrt{3})^2$, $\sqrt{13}$, $\sqrt{5}+\sqrt{5}$ and $0.2\dot{3}$ are rational?

2 ▶ Find a rational number between $\sqrt{7}$ and $\sqrt{11}$.

3 ▶ Find an irrational number between 6 and 7.

4 ▶ Write $4\sqrt{11}$ as the square root of a single number.

For Questions 5–11, simplify

5 ▶ $5\sqrt{5}-3\sqrt{5}$

6 ▶ $5\sqrt{5}\times 3\sqrt{5}$ **9 ▶** $\dfrac{3}{\sqrt{27}}-\dfrac{2}{\sqrt{48}}$

7 ▶ $\sqrt{242}$

8 ▶ $7\sqrt{54}-\sqrt{24}$ **10 ▶** $(\sqrt{18}-\sqrt{2})^2$

                     11 ▶ $(\sqrt{7}-2\sqrt{2})(\sqrt{7}+3\sqrt{2})$

12 ▶ Expand $(5-2\sqrt{3})^2$. Express your answer in the form $a+b\sqrt{3}$

13 ▶ Expand and simplify $(4+2\sqrt{3})(1+\sqrt{2})(4-2\sqrt{3})$. Express your answer as a surd in its simplest form.

For Questions 14–20, rationalise the denominator.

14 ▶ $\dfrac{1}{2\sqrt{5}}$

15 ▶ $\dfrac{12}{\sqrt{6}}$ **21 ▶** The two shorter sides of a right-angled triangle are $3\sqrt{3}$ and $4\sqrt{3}$. Find the values of the length of the hypotenuse, the perimeter and the area. Express each answer as a surd, if appropriate, in its simplest form.

16 ▶ $\dfrac{6+2\sqrt{6}}{\sqrt{6}}$

17 ▶ $\dfrac{\sqrt{112}}{\sqrt{28}}$

18 ▶ $\dfrac{9}{5+\sqrt{7}}$ **22 ▶** Find the exact height of an equilateral triangle with side length 2 units. From this, find the values of $\cos 60°$ and $\sin 60°$, expressing each answer as a surd in its simplest form.

19 ▶ $\dfrac{(4-\sqrt{2})(4+\sqrt{2})}{\sqrt{11}-\sqrt{7}}$

20 ▶ $\sqrt{\dfrac{1}{2}}+\sqrt{\dfrac{1}{4}}+\sqrt{\dfrac{1}{8}}$

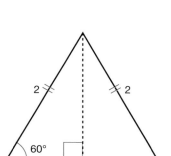

EXAM PRACTICE: NUMBER 10

1 Simplify fully   **a** $\sqrt{45} - \sqrt{5}$ **b** $\sqrt{\dfrac{49}{64}}$ **[4]**

2 Simplify fully   **a** $\sqrt{12} + 2\sqrt{27}$ **b** $3\sqrt{50} - 2\sqrt{32}$ **[4]**

3 Simplify fully   **a** $(2 - 2\sqrt{3})^2$ **b** $(1 + \sqrt{3})(1 - \sqrt{3})$ **[4]**

4 Rationalise the denominator and simplify fully   **a** $\dfrac{1 + \sqrt{5}}{\sqrt{5}}$ **b** $\dfrac{-7}{2 - \sqrt{11}}$ **[4]**

5 Work out, writing each answer in the form $\dfrac{a\sqrt{2}}{b}$ **a** $\dfrac{5}{\sqrt{2}} + \dfrac{8}{\sqrt{32}}$ **b** $\dfrac{7}{\sqrt{72}} - \dfrac{3}{\sqrt{8}}$ **[4]**

6 The diagram shows an isosceles right-angled triangle.

   a Use this diagram to work out values of $\sin 45°$ and $\cos 45°$.
     Express your answers as surds in their simplest form.

   b Carla sails 36 km on a **bearing** of 045°. How far east has
     she travelled? Express your answer as a surd in its simplest form.   **[2]**

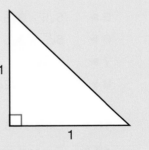

7 The diagram shows a right-angled triangle. Find the value of x.

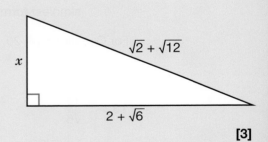

[3]

[Total 25 marks]

CHAPTER SUMMARY: NUMBER 10

RATIONAL AND IRRATIONAL NUMBERS

- Rational numbers can be written as a fraction in the form $\frac{a}{b}$ where a and b are integers and $b \neq 0$.

  - 4 is rational as it can be written as $\frac{4}{1}$

  - $0.\dot{4}$ is rational as it can be written as $\frac{4}{9}$

- Irrational numbers cannot be written as a fraction.

  - $\sqrt{2}$ and π are both irrational.

  - The decimal expansion of irrational numbers is infinite and shows no pattern.

SURDS

- A surd is a root that is irrational. The roots of all prime numbers are surds.

- $\sqrt{2}$ means the positive square root of 2.

- When an answer is given using a surd, it is an exact answer.

- These rules can be used to simplify surds:

  - $\sqrt{a \times b} = \sqrt{a} \times \sqrt{b}$

  - $\sqrt{\frac{a}{b}} = \frac{\sqrt{a}}{\sqrt{b}}$

- Surds can be cleared from the bottom of a fraction by multiplying the top and bottom of the fraction by a suitable surd. To rationalise the denominator of

  - $\frac{a}{\sqrt{b}}$ multiply by $\frac{\sqrt{b}}{\sqrt{b}}$

  - $\frac{1}{a - \sqrt{b}}$ multiply by $\frac{a + \sqrt{b}}{a + \sqrt{b}}$

  - $\frac{1}{a + \sqrt{b}}$ multiply by $\frac{a - \sqrt{b}}{a - \sqrt{b}}$

EXAMPLES OF MANIPULATING SURDS

$\sqrt{2} \times \sqrt{2} = 2$

$3\sqrt{2} + 5\sqrt{2} = 8\sqrt{2}$

$(3 + \sqrt{2})(3 + \sqrt{2}) = 9 + 6\sqrt{2} + 2 = 11 + 6\sqrt{2}$

$\sqrt{18} = \sqrt{9 \times 2} = \sqrt{9} \times \sqrt{2} = 3\sqrt{2}$

$\frac{6}{\sqrt{3}} = \frac{6}{\sqrt{3}} \times \frac{\sqrt{3}}{\sqrt{3}} = \frac{6\sqrt{3}}{3} = 2\sqrt{3}$

ALGEBRA 10

Formulae involving algebraic fractions occur frequently in science, for example $\frac{1}{f} = \frac{1}{u} + \frac{1}{v}$ is a formula in optics while $\frac{1}{R} = \frac{1}{R_1} + \frac{1}{R_2}$ is a formula in electronics. Being able to simplify these formulae is very important to a scientist.

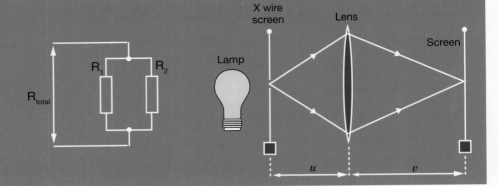

LEARNING OBJECTIVES

■ Simplify more complex algebraic fractions

■ Add and subtract more complex algebraic fractions

■ Multiply and divide more complex algebraic fractions

■ Solve equations that involve more complex algebraic fractions

BASIC PRINCIPLES

■ Find the lowest common **multiple** of two numbers.

■ Add and subtract fractions.

■ **Simplify** fractions by 'cancelling down'.

■ Multiply and divide fractions.

■ **Expand** brackets and simplify algebraic expressions.

■ Solve linear equations.

■ Solve equations involving fractions including those with x in the **denominator**.

■ Solve **quadratic equations**.

■ Simplify algebraic fractions.

■ To simplify a number fraction it is easiest to express the numbers in **prime factor** form, and then cancel.

■ Simplifying algebraic fractions is done in the same way by **factorising** as much as possible and then cancelling common terms.

SIMPLIFYING ALGEBRAIC FRACTIONS

EXAMPLE 1

Simplify $\dfrac{2x^2 - 6x}{x^2 - 2x - 3}$

SKILLS

ANALYSIS

Factorise the **numerator** and denominator

$$\frac{2x^2 - 6x}{x^2 - 2x - 3} = \frac{2x(x - 3)}{(x + 1)(x - 3)}$$

Divide the numerator and denominator by the **common factor** $(x - 3)$

$$= \frac{2x}{x + 1}$$

EXAMPLE 2

SKILLS

ANALYSIS

Simplify $\dfrac{x^2 + x - 2}{x^2 - 4}$

Factorise the numerator and denominator ($x^2 - 4$ is the difference of two squares)

$$\frac{x^2 + x - 2}{x^2 - 4} = \frac{(x + 2)(x - 1)}{(x + 2)(x - 2)}$$

Divide the numerator and denominator by the common factor ($x + 2$)

$$= \frac{x - 1}{x - 2}$$

KEY POINTS

- To simplify, factorise as much as possible before 'cancelling'.
- Whole brackets can be cancelled, not the individual terms in the brackets.

EXERCISE 1

Simplify fully

1 ▶ $\dfrac{3x + 12}{2x + 8}$

2 ▶ $\dfrac{10x - 25}{4x^2 - 10x}$

3 ▶ $\dfrac{xy + y^2}{x^2 + xy}$

4 ▶ $\dfrac{x^2 - x - 6}{x - 3}$

5 ▶ $\dfrac{x^2 - x - 6}{5(x + 2)}$

6 ▶ $\dfrac{x + 1}{x^2 + 3x + 2}$

7 ▶ $\dfrac{x^2 + 7x + 10}{x + 5}$

8 ▶ $\dfrac{2x^2 + 4x}{x^2 + 3x + 2}$

9 ▶ $\dfrac{x^2 - y^2}{(x - y)^2}$

10 ▶ $\dfrac{4x^2 + 24x}{x^2 - 36}$

11 ▶ $\dfrac{x^2 - x - 2}{x^2 - 5x + 6}$

12 ▶ $\dfrac{x^2 - x - 12}{x^2 - 2x - 8}$

EXERCISE 1*

Simplify fully

1 ▶ $\dfrac{6a + 9b}{10a + 15b}$

2 ▶ $\dfrac{4 - x^2}{(x + 2)^2}$

3 ▶ $\dfrac{x^3y + xy}{xy^2 + x^3y^2}$

4 ▶ $\dfrac{x^2 - 9x + 20}{x^2 - 2x - 15}$

5 ▶ $\dfrac{x^2 + 3x - 18}{x^2 + 11x + 30}$

6 ▶ $\dfrac{x^2 - x - 12}{x^2 - 16}$

7 ▶ $\dfrac{3x^2 - 2x}{9x^2 - 4}$

8 ▶ $\dfrac{3r^2 + 6r - 45}{3r^2 + 18r + 15}$

9 ▶ $\dfrac{2t^2 + 10t - 28}{2t^2 + 18t + 28}$

10 ▶ $\dfrac{a^3 - ab^2}{a^3 + 2a^2b + ab^2}$

11 ▶ $\dfrac{2x^2 + 12x - 32}{xy - 2y}$

12 ▶ $\dfrac{3x^3 - 3x^2 - 18x}{x^2y - 3xy}$

ADDING AND SUBTRACTING ALGEBRAIC FRACTIONS

Algebraic fractions are added and subtracted in the same way as number fractions. In the same way as with number fractions, find the lowest common denominator, otherwise the working can become very complicated.

EXAMPLE 3

SKILLS

ANALYSIS

Express $\dfrac{3(4x-1)}{2} - \dfrac{2(5x+3)}{3}$ as a single fraction.

The lowest common multiple of 2 and 3 is 6, so

$$\frac{3(4x-1)}{2} - \frac{2(5x+3)}{3} = \frac{9(4x-1) - 4(5x+3)}{6}$$
$$= \frac{36x - 9 - 20x - 12}{6}$$
$$= \frac{16x - 21}{6}$$

Note: The brackets in the numerator are not multiplied out until the second step. This will help avoid mistakes with signs, especially when the fractions are subtracted.

KEY POINTS

■ Find the lowest common denominator when adding or subtracting.
■ Multiply out the numerator after adding or subtracting.

EXERCISE 2

Express as a single fraction

1 ▶ $\dfrac{x}{3} + \dfrac{x+1}{2}$

2 ▶ $\dfrac{x}{2} - \dfrac{x+2}{4}$

3 ▶ $\dfrac{x+4}{5} + \dfrac{x}{3}$

4 ▶ $x + \dfrac{x+3}{4}$

5 ▶ $\dfrac{x+1}{5} - x$

6 ▶ $\dfrac{x-1}{3} + \dfrac{x+2}{4}$

7 ▶ $\dfrac{x-3}{2} - \dfrac{x+1}{5}$

8 ▶ $\dfrac{x+1}{3} + \dfrac{x+2}{2}$

9 ▶ $\dfrac{3x-2}{6} - \dfrac{3x+2}{9}$

10 ▶ $\dfrac{2}{3} - \dfrac{2-3x}{6}$

11 ▶ $\dfrac{2x+5}{4} - \dfrac{2(x-3)}{3}$

12 ▶ $\dfrac{2(2x+1)}{5} + \dfrac{3(x-1)}{2}$

EXERCISE 2*

Express as a single fraction

1 ▶ $\dfrac{2x-1}{5} + \dfrac{x+3}{2}$

2 ▶ $\dfrac{x+1}{3} - \dfrac{2x+1}{4}$

3 ▶ $\dfrac{2x+1}{7} - \dfrac{x-2}{2}$

4 ▶ $\dfrac{3x-1}{5} + \dfrac{1-3x}{7}$

5 ▶ $\dfrac{4x+2}{2} - \dfrac{2(5x-3)}{4}$

6 ▶ $\dfrac{2(x-5)}{3} - \dfrac{3(x+20)}{5}$

7 ▶ $\dfrac{3(4x-1)}{2} - \dfrac{2(5x+3)}{3}$

8 ▶ $\dfrac{x-3}{18} - \dfrac{x-2}{24}$

9 ▶ $\dfrac{3x}{4} + \dfrac{x-1}{5} - \dfrac{x+1}{6}$

10 ▶ $\dfrac{3x-2}{4} - \dfrac{x}{12} + \dfrac{x-5}{6}$

11 ▶ $\dfrac{x-2}{3} - \dfrac{x-7}{6} + \dfrac{10x-1}{9}$

12 ▶ $\dfrac{2(3x-4)}{5} - \dfrac{5(4x-3)}{2} + x$

FRACTIONS WITH x IN THE DENOMINATOR

Use the same method as for fractions with numbers in the denominator. It is even more important to use the lowest common denominator, otherwise the working can become very complicated.

EXAMPLE 4

SKILLS

ANALYSIS

Express $\dfrac{3}{x-1} - \dfrac{2}{x+1}$ as a single fraction.

The lowest **common denominator** is $(x-1)(x+1)$.

Write the two fractions as two separate equivalent fractions with a common denominator.

$$\frac{3}{x-1} - \frac{2}{x+1} = \frac{3(x+1)}{(x-1)(x+1)} - \frac{2(x-1)}{(x+1)(x-1)} \qquad \text{(Add the fractions)}$$

$$= \frac{3(x+1) - 2(x-1)}{(x-1)(x+1)} \qquad \text{(\textbf{Multiply out} the numerator)}$$

$$= \frac{3x+3 - 2x+2}{(x-1)(x+1)} \qquad \text{(Simplify the numerator)}$$

$$= \frac{x+5}{(x-1)(x+1)}$$

HINT
it is better to leave the denominator in a factorised form.

EXAMPLE 5

SKILLS

ANALYSIS

Express $\dfrac{x+1}{x+2} - \dfrac{x-2}{x-1}$ as a single fraction.

The lowest common denominator is $(x+2)(x-1)$.

Write the two fractions as two separate equivalent fractions with a common denominator.

$$\frac{x+1}{x+2} - \frac{x-2}{x-1} = \frac{(x+1)(x-1)}{(x+2)(x-1)} - \frac{(x-2)(x+2)}{(x-1)(x+2)} \qquad \text{(Add the fractions)}$$

$$= \frac{(x-1)(x+1) - (x-2)(x+2)}{(x+2)(x-1)} \qquad \text{(Multiply out the numerator)}$$

$$= \frac{(x^2-1) - (x^2-4)}{(x+2)(x-1)} \qquad \text{(Simplify the numerator)}$$

$$= \frac{x^2-1-x^2+4}{(x+2)(x-1)} \qquad \text{(Simplify the numerator)}$$

$$= \frac{3}{(x+2)(x-1)}$$

EXAMPLE 6

SKILLS

ANALYSIS

Express $\dfrac{2}{x-2} - \dfrac{6}{x^2 - x - 2}$ as a single fraction.

To find the lowest common denominator, factorise $x^2 - x - 2$ as $(x-2)(x+1)$, giving the lowest common denominator as $(x-2)(x+1)$. Leave the denominator in a factorised form to make it easier to simplify the resulting expression.

Write the two fractions as two separate equivalent fractions with a common denominator.

$$\dfrac{2}{x-2} - \dfrac{6}{x^2 - x - 2} = \dfrac{2(x+1)}{(x-2)(x+1)} - \dfrac{6}{(x-2)(x+1)} \qquad \text{(Add the fractions)}$$

$$= \dfrac{2(x+1) - 6}{(x-2)(x+1)} \qquad \text{(Multiply out and simplify the numerator)}$$

$$= \dfrac{2x - 4}{(x-2)(x+1)} \qquad \text{(Factorise the numerator)}$$

$$= \dfrac{2(x-2)}{(x-2)(x+1)} \qquad \text{('Cancel' } (x-2))$$

$$= \dfrac{2}{x+1}$$

EXERCISE 3

Express as single fractions

1 ▶ $\dfrac{1}{2x} + \dfrac{1}{3x}$

2 ▶ $\dfrac{3}{4x} - \dfrac{1}{2x}$

3 ▶ $\dfrac{1}{2} - \dfrac{1}{x-2}$

4 ▶ $\dfrac{4}{x^2} - \dfrac{2}{xy}$

5 ▶ $\dfrac{1}{x+1} + \dfrac{1}{x-1}$

6 ▶ $\dfrac{2}{x-4} - \dfrac{1}{x+3}$

7 ▶ $\dfrac{3}{x-1} - \dfrac{2}{x+2}$

8 ▶ $\dfrac{1}{x} + \dfrac{x}{x-3}$

9 ▶ $\dfrac{2}{x+4} + \dfrac{10}{x^2 + 3x - 4}$

10 ▶ $\dfrac{3}{x+2} - \dfrac{3}{x^2 + 5x + 6}$

EXERCISE 3*

Express as single fractions

1 ▶ $\dfrac{1}{2x} - \dfrac{1}{4x} + \dfrac{1}{6x}$

2 ▶ $\dfrac{5}{x+1} - \dfrac{2}{x+4}$

3 ▶ $1 - \dfrac{1}{x+1}$

4 ▶ $\dfrac{3x}{4y} - \dfrac{y}{2x}$

5 ▶ $\dfrac{1}{x} - \dfrac{1}{x(x+1)}$

6 ▶ $\dfrac{x+3}{x+2} - \dfrac{x-3}{x-2}$

7 ▶ $\dfrac{1}{x^2 + 3x + 2} + \dfrac{1}{x+2}$

8 ▶ $\dfrac{3x-7}{x^2 + 2x - 3} + \dfrac{1}{2x-2}$

9 ▶ $\dfrac{x+1}{x^2 - 4x + 3} - \dfrac{x-3}{x^2 - 1}$

10 ▶ $\dfrac{x-2}{x^2 - 3x - 4} - \dfrac{x-4}{x^2 - 6x + 8}$

MULTIPLYING AND DIVIDING ALGEBRAIC FRACTIONS

Factorise as much as possible and then simplify.

EXAMPLE 7

Simplify $\dfrac{x^2 - 2x}{x + 3} \times \dfrac{3x + 9}{x - 2}$

First factorise as much as possible.

$$\dfrac{x^2 - 2x}{x + 3} \times \dfrac{3x + 9}{x - 2} = \dfrac{x(x - 2)}{(x + 3)} \times \dfrac{3(x + 3)}{(x - 2)} \qquad \text{'Cancel' } (x + 3) \text{ and } (x - 2)$$

$$= 3x$$

If dividing, 'turn the second fraction upside down and multiply' in exactly the same way as number fractions are dealt with.

EXAMPLE 8

Simplify $\dfrac{x^2 - 3x}{2x^2 + 7x + 3} \div \dfrac{x^2 - 5x + 6}{2x^2 - 3x - 2}$

$$\dfrac{x^2 - 3x}{2x^2 + 7x + 3} \div \dfrac{x^2 - 5x + 6}{2x^2 - 3x - 2} = \dfrac{x(x - 3)}{(2x + 1)(x + 3)} \times \dfrac{(2x + 1)(x - 2)}{(x - 3)(x - 2)} = \dfrac{x}{x + 3}$$

KEY POINTS

- Factorise and then multiply or divide.
- To divide, turn the second fraction upside down and multiply.

EXERCISE 4

Simplify

1 ▶ $\dfrac{3x + 3}{2} \times \dfrac{x + 1}{3}$

5 ▶ $\dfrac{x^2 + 2x}{x^2 - x} \times \dfrac{6x - 6}{x + 2}$

9 ▶ $\dfrac{r^2 - r - 2}{r^2 - 3r + 2} \times \dfrac{r + 2}{r + 1}$

2 ▶ $\dfrac{x + 2}{x - 1} \times (x - 1)^2$

6 ▶ $\dfrac{x - y}{x^2 - xy} \times \dfrac{xy + y^2}{x + y}$

10 ▶ $\dfrac{x + 2}{x + 4} \div \dfrac{x^2 - 2x - 8}{x^2 + 2x - 8}$

3 ▶ $\dfrac{3x - 9}{4} \div \dfrac{x - 3}{x + 3}$

7 ▶ $\dfrac{2a + 4b}{ab + 2b^2} \times \dfrac{2a^2 - ab}{2a - b}$

11 ▶ $\dfrac{x + 3}{x + 5} \div \dfrac{x^2 - 2x - 15}{x^2 + 2x - 15}$

4 ▶ $\dfrac{4x + 4}{x - 3} \times \dfrac{x^2 - 3x}{2x + 2}$

8 ▶ $\dfrac{p^2 + p - 2}{p^2 - p - 6} \times \dfrac{p - 3}{p - 2}$

12 ▶ $\dfrac{x^2 - 9}{x^2 - 6x + 9} \div \dfrac{x + 4}{x - 3}$

EXERCISE 4*

Simplify

1 ▶ $\dfrac{4x + 28}{x^2 + 2x} \times \dfrac{x^2 - 3x}{2x + 14}$

7 ▶ $\dfrac{x^2 + x - 2}{x^2 + 3x - 4} \times \dfrac{x^2 + x - 12}{x^2 - 5x + 6}$

2 ▶ $\dfrac{3x^2 - 12x}{3x + 18} \times \dfrac{2x + 12}{x^2 - 3x}$

8 ▶ $\dfrac{x^2 - x - 2}{x^2 + x - 6} \times \dfrac{x^2 - x - 12}{x^2 + 3x + 2}$

3 ▶ $\dfrac{x^2 - 9}{4} \times \dfrac{8}{x^2 + 5x + 6}$

9 ▶ $\dfrac{p^2 + 7p + 12}{p^2 - 7p + 10} \div \dfrac{p + 3}{p - 2}$

4 ▶ $\dfrac{x^2 + 9x + 20}{x^2 - 16} \times \dfrac{x - 4}{x^2 + 6x + 5}$

10 ▶ $\dfrac{q^2 - 5q - 14}{q^2 + 3q - 18} \div \dfrac{q - 7}{q - 3}$

5 ▶ $\dfrac{x + 1}{x^2 + 6x + 8} \div \dfrac{x + 3}{x^2 + 5x + 4}$

11 ▶ $\dfrac{x^2 + xy - 2y^2}{x^2 - 4y^2} \div \dfrac{x^2 + 2xy - 3y^2}{xy - 2y^2}$

6 ▶ $\dfrac{x^2 - x - 6}{x^2 - 9} \times \dfrac{x^2 + 4x + 3}{x + 1}$

12 ▶ $\dfrac{x^2 - 2xy + y^2}{x^2 - y^2} \div \dfrac{x^2 + xy}{x^2 + 2xy + y^2}$

SOLVING EQUATIONS WITH ALGEBRAIC FRACTIONS

EQUATIONS WITH NUMERICAL DENOMINATORS

If an equation involves fractions, **clear** the fractions by multiplying both sides of the equation by the LCD. Then simplify and solve in the usual way.

EXAMPLE 9

Solve $\dfrac{2x - 1}{9} = \dfrac{x + 1}{6}$

SKILLS

ANALYSIS

The LCM of 9 and 6 is 18.

$\dfrac{2x - 1}{9} = \dfrac{x + 1}{6}$ (Multiply both sides by 18)

$18 \times \dfrac{2x - 1}{9} = 18 \times \dfrac{x + 1}{6}$ (Simplify)

$2(2x - 1) = 3(x + 1)$ (Multiply out brackets)

$4x - 2 = 3x + 3$ (Collect terms)

$4x - 3x = 3 + 2$ (Simplify)

$x = 5$

Check: $\dfrac{2 \times 5 - 1}{9} = \dfrac{5 + 1}{6}$

EXAMPLE 10

SKILLS

ANALYSIS

Solve $\dfrac{x+1}{3} - \dfrac{x-3}{4} = 1$

The LCM of 3 and 4 is 12.

$$\dfrac{x+1}{3} - \dfrac{x-3}{4} = 1 \qquad \text{(Multiply both sides by 12)}$$

$$12 \times \dfrac{x+1}{3} - 12 \times \dfrac{x-3}{4} = 12 \times 1 \qquad \text{(Notice everything is multiplied by 12)}$$

$$4(x+1) - 3(x-3) = 12 \qquad \text{(Multiply out, note \textbf{sign} change in 2nd bracket)}$$

$$4x + 4 - 3x + 9 = 12 \qquad \text{(Collect terms)}$$

$$4x - 3x = 12 - 4 - 9 \qquad \text{(Simplify)}$$

$$x = -1$$

Check: $\dfrac{-1+1}{3} - \dfrac{-1-3}{4} = 1$

KEY POINTS

- Clear fractions by multiplying both sides by the lowest common denominator.
- Always check your answer by substituting it into the original equation.

EXERCISE 5

Solve these equations.

1 ▶ $\dfrac{x}{3} = 7$

2 ▶ $\dfrac{x}{2} = \dfrac{x+2}{4}$

3 ▶ $\dfrac{2-x}{3} = x$

4 ▶ $x = \dfrac{x+3}{2}$

5 ▶ $\dfrac{x-4}{6} = \dfrac{x+2}{3}$

6 ▶ $\dfrac{2-3x}{6} = \dfrac{2}{3}$

7 ▶ $\dfrac{x+1}{3} = \dfrac{2x+1}{4}$

8 ▶ $\dfrac{2(x-5)}{3} = \dfrac{2x+1}{4}$

9 ▶ $\dfrac{2x-3}{6} + \dfrac{x+2}{3} = \dfrac{5}{2}$

10 ▶ $\dfrac{x+1}{7} - \dfrac{3(x-2)}{14} = 1$

11 ▶ $\dfrac{3x-6}{5} + 2x = 4$

12 ▶ $\dfrac{6-3x}{3} - \dfrac{5x+12}{4} = -1$

EXERCISE 5*

For Questions 1–12, solve the equation.

1 ▶ $\dfrac{3x}{7} = \dfrac{6}{35}$

2 ▶ $\dfrac{x-3}{3} = 4$

3 ▶ $\dfrac{x+1}{2} = 2x$

4 ▶ $\dfrac{3-x}{3} = \dfrac{2+x}{2}$

5 ▶ $\dfrac{x-3}{12} + \dfrac{x}{5} = 4$

6 ▶ $\dfrac{2x+1}{3} = x - 2$

7 ▶ $\dfrac{7x-1}{6} + 5x = 6$

8 ▶ $\dfrac{1+x}{2} = \dfrac{2-x}{3} + 1$

9 ▶ $\dfrac{x+1}{5} + \dfrac{2x-7}{2} = \dfrac{3}{2}$

10 ▶ $\dfrac{2x-3}{4} - \dfrac{3x-8}{3} = \dfrac{5}{12}$

11 ▶ $4 - \dfrac{x-2}{2} = 3 + \dfrac{2-3x}{3}$

12 ▶ $\dfrac{1-x}{2} - \dfrac{2+x}{3} + \dfrac{3-x}{4} = 1$

13 ▶ Pedro does one-sixth of his journey to school by car and two-thirds by bus. He then walks the final kilometre. How long is his journey to school?

14 ▶ Meera is competing in a triathlon. She cycles half the course then swims one-thirtieth of the course. She then runs 14 km to the finish. How long is the course?

EQUATIONS WITH x IN THE DENOMINATOR

These are solved in the same way as equations with numbers in the denominator. Clear fractions by multiplying both sides by the LCD, then simplify and solve in the usual way.

EXAMPLE 11

Solve $x + 2 = \dfrac{8}{x}$

SKILLS

ANALYSIS

The LCD is x

$x + 2 = \dfrac{8}{x} \Rightarrow x \times x + x \times 2 = x \times \dfrac{8}{x}$ (Multiply both sides by x and simplify)

$\Rightarrow \quad x^2 + 2x = 8$ (Rearrange into standard form)

$\Rightarrow \quad x^2 + 2x - 8 = 0$ (Solve by factorising)

$\Rightarrow (x - 2)(x + 4) = 0$

$\Rightarrow \qquad x = 2 \text{ or } x = -4$

EXAMPLE 12

Solve $\dfrac{x}{x - 2} - \dfrac{2}{x + 1} = 3$

SKILLS

ANALYSIS

The LCD is $(x - 2)(x + 1)$

$\dfrac{x}{x - 2} - \dfrac{2}{x + 1} = 3 \Rightarrow (x - 2)(x + 1) \times \dfrac{x}{x - 2} - (x - 2)(x + 1) \times \dfrac{2}{x + 1} = (x - 2)(x + 1) \times 3$ (Simplify)

$\Rightarrow x(x + 1) - 2(x - 2) = 3(x - 2)(x + 1)$ (Expand brackets)

$\Rightarrow \quad x^2 + x - 2x + 4 = 3x^2 - 3x - 6$ (Rearrange into standard form)

$\Rightarrow \quad 2x^2 - 2x - 10 = 0$ (Divide both sides by 2)

$\Rightarrow \quad x^2 - x - 5 = 0$ (Solve using the quadratic formula)

$\Rightarrow \qquad x = \dfrac{1 \pm \sqrt{1 + 20}}{2}$

$\Rightarrow \qquad x = 2.79 \text{ or } x = -1.79 \text{ to 3 s.f.}$

EXERCISE 6

Solve these equations.

1 ▶ $x + 5 = \dfrac{14}{x}$

2 ▶ $\dfrac{1}{2} - \dfrac{1}{x - 2} = \dfrac{1}{4}$

3 ▶ $\dfrac{1}{x + 4} - \dfrac{1}{3} = -\dfrac{1}{6}$

4 ▶ $\dfrac{4}{x} = \dfrac{3x - 7}{5}$

5 ▶ $\dfrac{x}{x + 2} - \dfrac{1}{x} = 1$

6 ▶ $\dfrac{1}{x} + \dfrac{x}{x - 3} = 1$

7 ▶ $\dfrac{4}{x - 1} + \dfrac{7}{x - 1} = 11$

8 ▶ $\dfrac{6}{x - 2} - \dfrac{6}{x + 1} = 1$

9 ▶ $\dfrac{2x}{x - 3} - \dfrac{4}{x + 1} = 3$

10 ▶ $\dfrac{x - 2}{x - 1} = \dfrac{x + 4}{2x + 4}$

11 ▶ $\dfrac{2x - 3}{x + 1} = \dfrac{x + 3}{x + 5}$

12 ▶ $\dfrac{2x - 1}{x + 2} = \dfrac{4x + 1}{5x + 2}$

EXERCISE 6*

For Questions 1–10, solve the equation.

1 ▶ $2x + 5 = \dfrac{7}{x}$

5 ▶ $\dfrac{7}{x-1} - \dfrac{4}{x+4} = 1$

9 ▶ $\dfrac{3x+2}{x+1} + \dfrac{x+2}{2x-5} = 4$

2 ▶ $1 - \dfrac{3}{x} = \dfrac{10}{x^2}$

6 ▶ $\dfrac{6x}{x+1} - \dfrac{5}{x+3} = 3$

10 ▶ $\dfrac{4x-5}{x+2} + \dfrac{2x-4}{3x-8} = 3$

3 ▶ $\dfrac{7}{9} - \dfrac{x}{x+5} = \dfrac{1}{3}$

7 ▶ $\dfrac{2x+1}{x+1} = \dfrac{x+2}{2x+1}$

4 ▶ $\dfrac{1}{2} + \dfrac{x}{x+7} = \dfrac{4}{5}$

8 ▶ $\dfrac{x-3}{3x+1} = \dfrac{x+2}{x-5}$

11 ▶ Mala drives the first 30 km of her journey at x km/h. She then increases her speed by 20 km/h for the final 40 km of her journey. Her journey takes 1 hour.

  a Show that $\dfrac{30}{x} + \dfrac{40}{x+20} = 1$
  b Find x.

12 ▶ Lucas is running a race. He runs the first 800 m at x m/s. He then slows down by 3 m/s for the final 400 m. His total time is 130 seconds. Find x.

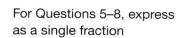

For Questions 13–15, solve the equation.

13 ▶ $\dfrac{3}{x+2} + \dfrac{4}{x+3} = \dfrac{7}{x+6}$

15 ▶ $y - 1 = \dfrac{y^2+3}{y-1} + \dfrac{y-2}{y-6}$

14 ▶ $\dfrac{4}{x-3} + \dfrac{3x-3}{x^2-x-6} = \dfrac{2-20x}{2x+4}$

EXERCISE 7

REVISION

For Questions 1–4, simplify

1 ▶ $\dfrac{3x+6}{x+2}$

2 ▶ $\dfrac{x^2+7x+10}{x+5}$

3 ▶ $\dfrac{x^2+6x+9}{x^2-9}$

4 ▶ $\dfrac{x^2-2x+1}{2x^2+x-3}$

For Questions 5–8, express as a single fraction

5 ▶ $\dfrac{x-2}{3} + \dfrac{x}{4}$

6 ▶ $\dfrac{3x-2}{6} - \dfrac{3x+2}{9}$

7 ▶ $\dfrac{3}{x-1} + \dfrac{2}{x+1}$

8 ▶ $\dfrac{x+3}{x+2} - \dfrac{x+1}{x+4}$

For Questions 9–12, solve the equation.

9 ▶ $\dfrac{x+2}{3} - \dfrac{x-2}{4} = 1$

10 ▶ $\dfrac{x}{x-1} + \dfrac{1}{x} = 1$

11 ▶ $\dfrac{3}{x-1} - \dfrac{8}{x+2} = 1$

12 ▶ $\dfrac{x+3}{x+5} = \dfrac{x+1}{x+2}$

EXERCISE 7*

REVISION

For Questions 1–4, simplify

1 ▶ $\dfrac{2x+14}{3x+21}$

2 ▶ $\dfrac{x^2-12x+11}{x^2+4x-5}$

3 ▶ $\dfrac{x^2+11x+28}{x^2-49}$

4 ▶ $\dfrac{x^2-11x+10}{3x^2-29x-10}$

For Questions 5–8, express as a single fraction

5 ▶ $\dfrac{3(x-1)}{6} + \dfrac{2(x+1)}{9}$

6 ▶ $\dfrac{x+1}{2} - \dfrac{x+2}{3} + \dfrac{x+3}{4}$

7 ▶ $\dfrac{x+2}{x+1} - \dfrac{x+1}{x+2}$

8 ▶ $\dfrac{x}{x^2-3x-4} - \dfrac{1}{x-4}$

For Questions 9–12, solve the equation.

9 ▶ $\dfrac{3x-1}{7} = \dfrac{2x+1}{11} + 1$

10 ▶ $\dfrac{2x-1}{x+3} = \dfrac{x}{2x+2}$

11 ▶ $\dfrac{x+1}{x-3} - \dfrac{1}{x} = 2$

12 ▶ $\dfrac{2x+3}{x-5} = \dfrac{x-4}{x-3}$

EXAM PRACTICE: ALGEBRA 10

1 Simplify

a $\dfrac{x^2 + x}{x + 1}$

b $\dfrac{x^2 - x - 6}{2x - 6}$

c $\dfrac{x^2 + 5x - 6}{x^2 + 8x + 12}$ **[4]**

2 Express as a single fraction

a $\dfrac{2(x + 1)}{6} - \dfrac{3(x - 2)}{4}$

b $\dfrac{x}{x^2 - 3x - 4} - \dfrac{1}{x - 4}$ **[6]**

3 Simplify

a $\dfrac{6x}{3x + 9} \times \dfrac{x^2 + 4x + 3}{x^2}$

b $\dfrac{x^2 + x - 6}{x + 1} \div \dfrac{x + 3}{2x^2 + x - 1}$ **[6]**

4 Solve these equations.

a $\dfrac{3x - 6}{5} + 2x = 4$

b $\dfrac{2}{x} + \dfrac{2x}{x - 1} = 5$

c $\dfrac{x + 1}{x - 2} - \dfrac{x + 1}{x + 4} = 1$ **[9]**

[Total 25 marks]

CHAPTER SUMMARY: ALGEBRA 10

SIMPLIFYING ALGEBRAIC FRACTIONS

To simplify, factorise as much as possible and then cancel.

$$\frac{x^2 + 3x + 2}{x^2 - x - 5} = \frac{(x + 2)(x + 1)}{(x + 2)(x - 3)} = \frac{x + 1}{x - 3}$$

ADDING AND SUBTRACTING ALGEBRAIC FRACTIONS

Add or subtract in the same way as number fractions.

To find the LCD you may need to factorise the denominators first.

$$\frac{x}{3} + \frac{x - 1}{6} = \frac{2x + x - 1}{6} = \frac{2x - 1}{6}$$

$$\frac{1}{x} - \frac{1}{x - 1} = \frac{x - 1 - x}{x(x - 1)} = \frac{-1}{x(x - 1)}$$

MULTIPLYING AND DIVIDING ALGEBRAIC FRACTIONS

First factorise, then multiply or divide.

To divide, turn the second fraction upside down and multiply.

$$\frac{x^2 + 5x + 4}{x^2 - x - 12} \times \frac{x + 3}{x + 4} = \frac{(x + 4)(x + 1)}{(x - 4)(x + 3)} \times \frac{(x + 3)}{(x + 4)} = \frac{x + 1}{x - 4}$$

$$\frac{x - 6}{x + 3} \div \frac{x^2 - 36}{x^2 + 10x + 21} = \frac{(x - 6)}{(x + 3)} \times \frac{(x + 3)(x + 7)}{(x - 6)(x + 6)} = \frac{x + 7}{x + 6}$$

SOLVING EQUATIONS WITH ALGEBRAIC FRACTIONS

Multiply both sides by the LCD to clear the fractions.

$$\frac{x + 1}{3} = \frac{x - 1}{2} \Rightarrow 2(x + 1) = 3(x - 1) \Rightarrow 2x + 2 = 3x - 3 \Rightarrow x = 5$$

$$\frac{1}{x} - \frac{x}{x - 2} = 2 \Rightarrow x - 2 - x^2 = 2x^2 - 4x \Rightarrow 3x^2 - 5x + 2 = 0$$

$$\Rightarrow (3x - 2)(x - 1) = 0 \Rightarrow x = 1 \text{ or } \frac{2}{3}$$

Always check your answer by substituting it into the original equation.

GRAPHS 9

Calculus is the study of rates of change and is especially useful when the rate of change is not constant. Isaac Newton (1642–1727) and Gottfried Leibniz (1646–1716) both studied rates of change and it is commonly agreed that their ideas formed the bedrock for further scientific studies to develop into the modern era.

Isaac Newton ▲

Gottfried Leibniz ▲

LEARNING OBJECTIVES

- Understand the relationship between the gradient of a function and its rate of change
- Differentiate integer powers of x
- Find the gradient of a tangent at a point by differentiation
- Find the co-ordinates of the maximum and minimum points on a curve
- Find expressions for the velocity and acceleration of a particle
- Apply calculus to real-life problems

BASIC PRINCIPLES

- Find (an estimate for) the **gradient** of a **tangent** at a point (by drawing the tangent by hand).

- Identify the **maximum** and **minimum points** on a graph (visually).

- Recognise the general form of the graphs of **quadratic** and **cubic functions**.

- The gradient of a graph can often produce useful quantities such as **velocity**, **acceleration** and many others. If a graph is available it is possible to estimate the gradient by drawing the best-fitting tangent by eye. Calculus is a process that enables the exact value of this gradient to be calculated.

- The graph of $y = \sin x$ for $0° \le x \le 180°$ shows tangents drawn every 2°. It is clear that the gradient of this curve is changing as x changes.

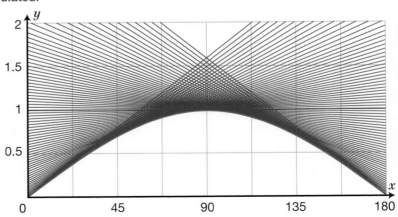

THE GRADIENT OF A FUNCTION

The gradient of a curve at any point is equal to the gradient of the tangent to the curve at that point. An estimate for the gradient of this tangent can be found by drawing a tangent by hand. The exact gradient can be calculated by using calculus.

Calculus involves considering very small values (or increments) in x and y: δx and δy. δx is pronounced 'delta x' and means a very small distance (increment) along the x-axis. It does not mean δ multiplied by x. δx must be considered as one symbol.

The process of calculating a gradient (or a **rate** of change) is shown below.

Consider the graph $y = x^2$

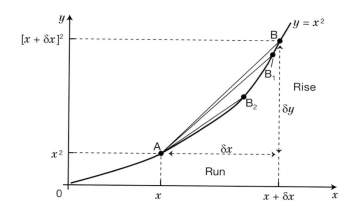

The gradient of the curve at point A is equal to the tangent to the curve at this point.

A good 'first attempt' in finding this gradient is to consider a point B up the curve from A. The gradient of **chord** AB will be close to the required gradient if the steps along the x and y axes (δx and δy respectively) are small.

Let the exact gradient of $y = x^2$ at A be m. An estimate of m is found here.

$$\frac{\delta y}{\delta x} = \frac{\text{rise}}{\text{run}} = \frac{(x + \delta x)^2 - x^2}{\delta x} \qquad \textbf{(Expand } (x + \delta x)^2)$$

$$= \frac{(x^2 + 2x\delta x + (\delta x)^2) - x^2}{\delta x}$$

$$= \frac{2x\delta x + (\delta x)^2}{\delta x}$$

$$= \frac{\delta x(2x + \delta x)}{\delta x} = 2x + \delta x$$

So the estimate of m is $\dfrac{\delta y}{\delta x} = 2x + \delta x$

This estimate of m improves as point B slides down the curve to B1, B2, etc., closer and closer to A, resulting in δx and δy becoming smaller and smaller until, at point A, $\delta x = 0$.

What happens to $\dfrac{\delta y}{\delta x}$ as δx approaches zero?

Clearly, as δx gets smaller, $2x + \delta x$ approaches $2x$, and eventually, at A, $m = 2x$.
This is a beautifully simple result proving that the gradient of the curve $y = x^2$ at any point x is given by $2x$. So, at $x = 10$, the gradient of $y = x^2$ is 20, and so on.

DIFFERENTIATION

DIFFERENTIATING $y = kx^n$

The method of calculating the gradient is called **differentiation**. It can be applied to $y = kx^n$ when x is any real value and k is a constant. The result of this process is called the gradient function or **derivative** of the **function**.

FUNCTION	GRADIENT FUNCTION $\dfrac{dy}{dx}$	GRADIENT OF FUNCTION AT POINT WHERE $x = 2$
$y = x^1$	$\dfrac{dy}{dx} = 1 \times x^0 = 1$	$\dfrac{dy}{dx} = 1$
$y = x^2$	$\dfrac{dy}{dx} = 2 \times x^1 = 2x$	$\dfrac{dy}{dx} = 2 \times 2 = 4$
$y = x^3$	$\dfrac{dy}{dx} = 3 \times x^2 = 3x^2$	$\dfrac{dy}{dx} = 3 \times 2^2 = 12$
$y = kx^n$	$\dfrac{dy}{dx} = nkx^{n-1}$	$\dfrac{dy}{dx} = nk \times 2^{n-1}$

EXAMPLE 1

SKILLS

ANALYSIS

Calculate the value of the gradient of the function $y = 10$ at $x = 2$

$y = 10$ can be written as $y = 10x^0$

$\dfrac{dy}{dx} = 0 \times 10x^{-1} = 0$

The gradient of the function at $x = 2$ will always be 0.

This should not be a surprise as $y = 10$ is a horizontal line!

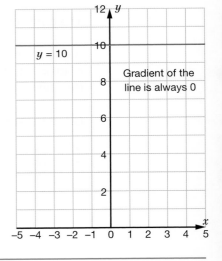

EXAMPLE 2

SKILLS

ANALYSIS

Calculate the value of the gradient of the function $y = 10x$ at $x = 2$.

$y = 10x$ can be written as $y = 10x^1$

$\dfrac{dy}{dx} = 1 \times 10x^0 = 10$

The gradient of the function at $x = 2$ will always be 10.

This should not be a surprise as $y = 10x$ is a line with a gradient of 10!

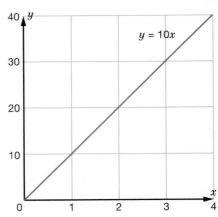

EXAMPLE 3

Calculate the value of the gradient of the function $y = 10x^2$ at $x = 2$

SKILLS

ANALYSIS

$\dfrac{dy}{dx} = 2 \times 10x^1 = 20x$

The gradient of the function at $x = 2$, $\dfrac{dy}{dx} = 20x = 40$

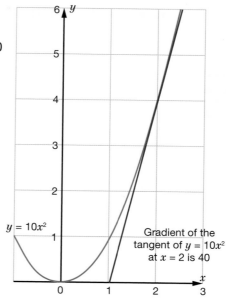

$y = 10x^2$

Gradient of the tangent of $y = 10x^2$ at $x = 2$ is 40

KEY POINT

■ First write the function in index notation.

■ If $y = kx^n$, $\dfrac{dy}{dx} = nk x^{n-1}$

■ If $y = kx$, $\dfrac{dy}{dx} = k$

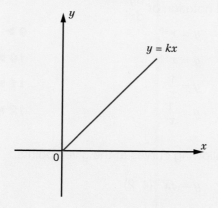

$y = kx$

■ If $y = k$, $\dfrac{dy}{dx} = 0$

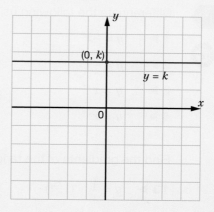

$(0, k)$

$y = k$

EXERCISE 1

Differentiate the following using the correct notation.

1 ▶ $y = 2$

2 ▶ $y = 2x$

3 ▶ $y = x^3$

4 ▶ $y = x^4$

5 ▶ $y = x^5$

6 ▶ $y = x^{10}$

7 ▶ $y = 2x^3$

8 ▶ $y = 2x^4$

9 ▶ $y = 2x^5$

10 ▶ $y = 2x^{10}$

Find the gradients of the tangents to the following curves at the given points.

11 ▶ $y = 2$ (1, 2)

12 ▶ $y = k$ (1, k)

13 ▶ $y = 2x$ (1, 2)

14 ▶ $y = x^3$ (2, 8)

15 ▶ $y = 2x^3$ (2, 16)

16 ▶ $y = 2x^{10}$ (1, 2)

EXAMPLE 4

SKILLS

ANALYSIS

Calculate the value of the gradient of the function $y = \dfrac{10}{x}$ at $x = 2$

$y = \dfrac{10}{x}$ can be written as $y = 10x^{-1}$

$\dfrac{dy}{dx} = -1 \times 10x^{-2} = -10x^{-2} = -\dfrac{10}{x^2}$

The gradient of the function at $x = 2$, $\dfrac{dy}{dx} = -\dfrac{10}{2^2} = -\dfrac{5}{2}$

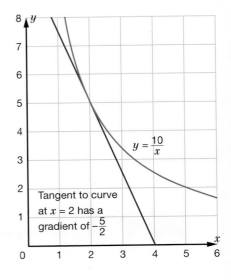

$y = \dfrac{10}{x}$

Tangent to curve at $x = 2$ has a gradient of $-\dfrac{5}{2}$

EXERCISE 1*

Differentiate the following using the correct notation of $\dfrac{dy}{dx}$.

1 ▶ $y = -x^2$

2 ▶ $y = -x^3$

3 ▶ $y = x^{-1}$

4 ▶ $y = x^{-2}$

5 ▶ $y = x^{-3}$

6 ▶ $y = x^{-4}$

7 ▶ $y = \dfrac{1}{x}$

8 ▶ $y = \dfrac{1}{x^2}$

9 ▶ $y = x^{\frac{1}{2}}$

10 ▶ $y = \sqrt{x}$

11 ▶ $y = x^{\frac{1}{3}}$

12 ▶ $y = \sqrt[3]{x}$

Find the gradients of the tangents to the following curves at the given points.

13 ▶ $y = \dfrac{1}{x}$ (1, 1)

14 ▶ $y = \dfrac{1}{x^2}$ $(2, \dfrac{1}{4})$

15 ▶ $y = \sqrt{x}$ (4, 2)

16 ▶ $y = \sqrt[3]{x}$ (8, 2)

DIFFERENTIATING OTHER FUNCTIONS

The process of differentiation also applies when terms are added or multiplied by a constant. Work through the expression differentiating each term in turn.

If the expression involves brackets, multiply out the brackets and simplify first.

EXAMPLE 5

SKILLS

ANALYSIS

Calculate the value of the gradient of the function
$y = (3x + 1)(2x - 3)$ at $x = 2$

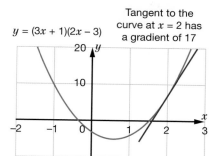

Tangent to the curve at $x = 2$ has a gradient of 17

$y = (3x + 1)(2x - 3)$

$y = 6x^2 - 7x - 3$

$\dfrac{dy}{dx} = 12x - 7$

The gradient of the function at $x = 2$, $\dfrac{dy}{dx} = 12 \times 2 - 7 = 17$

EXAMPLE 6

SKILLS

ANALYSIS

Calculate the value of the gradient of the function $y = (3x + 1)^2$ at $x = 2$

$y = 9x^2 + 6x + 1$

$\dfrac{dy}{dx} = 18x + 6$

The gradient of the function at $x = 2$, $\dfrac{dy}{dx} = 18 \times 2 + 6 = 42$

EXERCISE 2

Differentiate using the correct notation.

1 ▶　$y = x^2 + x$

2 ▶　$y = x^2 + 2x + 1$

3 ▶　$y = x^3 + x^2$

4 ▶　$y = x^4 + x^3 + x^2 + x + 1$

5 ▶　$y = 2x^3 - 3x^2 + 4$

6 ▶　$y = 10x^5 + 5x - 3$

7 ▶　$y = x^{-2} + x^{-1}$

8 ▶　$y = 2x^{-2} + 3x^{-1}$

9 ▶　$y = 3x^{-2} - 2x^{-1} + 1$

10 ▶　$y = 1 + 2x^{-3} - 3x^{-4}$

11 ▶　$y = 10x^{10} + 5x^5$

12 ▶　$y = 10x^{-10} - 5x^{-5}$

Find the gradients of the tangents to the following curves at the given points.

13 ▶　$y = x^2 + x$　　　(1, 2)

14 ▶　$y = x^2 + 2x + 1$　(2, 9)

15 ▶　$y = x^3 + x^2$　　(2, 12)

16 ▶　$y = 10x^{10} + 5x^5$ (1, 15)

17 ▶ Find the gradient of the tangent to the curve shown at $x = 2$.

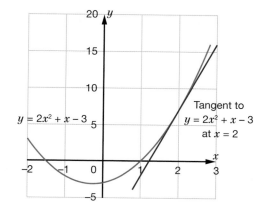

$y = 2x^2 + x - 3$

Tangent to $y = 2x^2 + x - 3$ at $x = 2$

18 ▶ Find the gradient of the tangent to the curve shown at $x = -1$.

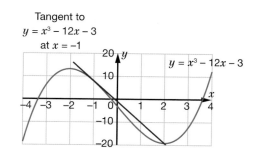

Tangent to $y = x^3 - 12x - 3$ at $x = -1$

$y = x^3 - 12x - 3$

EXAMPLE 7

SKILLS

ANALYSIS

Calculate the value of the gradient of the function $y = x + \sqrt{x}$ at $x = 4$

$$y = x + x^{\frac{1}{2}}$$

$$\frac{dy}{dx} = 1 + \frac{1}{2}x^{-\frac{1}{2}} = 1 + \frac{1}{2\sqrt{x}}$$

The gradient of the function at $x = 4$, $\frac{dy}{dx} = 1 + \frac{1}{2 \times 2} = 1\frac{1}{4}$

EXAMPLE 8

SKILLS

ANALYSIS

Calculate the value of the gradient of the function

$$y = \frac{(x+1)(2x+1)}{x} \text{ at } x = 1.$$

$$y = \frac{2x^2 + 3x + 1}{x} = 2x + 3 + \frac{1}{x} = 2x + 3 + x^{-1}$$

$$\frac{dy}{dx} = 2 + 0 - x^{-2} = 2 - \frac{1}{x^2}$$

The gradient of the function at $x = 1$, $\frac{dy}{dx} = 2 - \frac{1}{1^2} = 1$

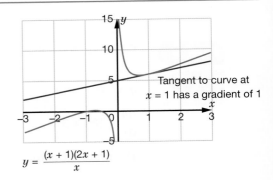

Tangent to curve at $x = 1$ has a gradient of 1

$$y = \frac{(x+1)(2x+1)}{x}$$

EXERCISE 2*

Differentiate the following using the correct notation.

1 ▶ $y = x(x + 3)$

2 ▶ $y = x^2(x + 3)$

3 ▶ $y = (x + 3)(x + 5)$

4 ▶ $y = (2x + 1)(x + 5)$

5 ▶ $y = (x + 4)^2$

6 ▶ $y = (2x - 3)^2$

7 ▶ $y = (1 - 3x)^2$

8 ▶ $y = x(1 - 3x)^2$

9 ▶ $y = \left(3x + \dfrac{1}{x}\right)^2$

10 ▶ $y = \left(3x - \dfrac{1}{x}\right)^2$

11 ▶ $y = \dfrac{1}{x} + \dfrac{1}{x^2}$

12 ▶ $y = \dfrac{1}{x} - \dfrac{1}{x^2} + \pi$

13 ▶ $y = \dfrac{2}{x} + \dfrac{3}{x^2}$

14 ▶ $y = \dfrac{2x^3 + 4x^2 - x}{x}$

15 ▶ $y = \dfrac{(x+3)(3x-1)}{x}$

16 ▶ $y = x + \sqrt{x}$

17 ▶ $y = x + \dfrac{1}{\sqrt{x}}$

18 ▶ $y = x - \dfrac{1}{\sqrt{x}}$

Find the gradients of the tangents to the following curves at the given points.

19 ▶ $y = x(x + 3)$ (1, 4)

20 ▶ $y = (2x + 1)(x + 5)$ (1, 18)

21 ▶ $y = (2x - 3)^2$ (2, 1)

22 ▶ $y = \dfrac{2}{x} + \dfrac{3}{x^2}$ (1, 5)

23 ▶   Find the gradient of the tangent to the curve shown at $x = 0$.

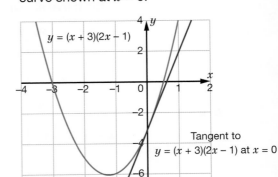

$y = (x + 3)(2x - 1)$

Tangent to $y = (x + 3)(2x - 1)$ at $x = 0$

24 ▶ Find the gradient of the tangent to the curve shown at $x = -2$

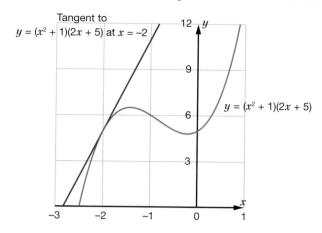

EQUATION OF THE TANGENT TO A CURVE

Using the general equation of a straight line ($y = mx + c$ where m is the gradient and c is the **y-intercept**) it is possible to find the equation of the tangent to a curve at a given point.

EXAMPLE 9

SKILLS

ANALYSIS

Find the equation of the tangent to the curve $y = x^3 - 3x^2 + 2$ at point P where $x = 3$.

The y-value on the curve at $x = 3$ is given by

$y = 27 - 27 + 2 = 2$, so P is (3, 2)

The gradient function $\dfrac{dy}{dx} = 3x^2 - 6x$

The gradient at point P $= 3 \times 3^2 - 6 \times 3 = 9$

If the gradient of the curve at P is the same as the gradient of the tangent at P, the equation of this line must be $y = 9x + c$

As P (3, 2) is on the line it must **satisfy** its equation so $2 = 9 \times 3 + c \Rightarrow c = -25$

The equation is $y = 9x - 25$

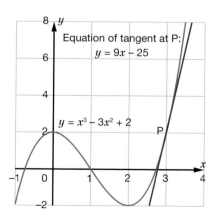

APPLIED RATES OF CHANGE

If a mathematical model exists for a real-life situation, it may be useful to calculate the rate of change at any given moment using the gradient function.

The table below gives some examples.

VERTICAL AXIS	HORIZONTAL AXIS	RATE OF CHANGE (GRADIENT)
Distance (m)	Time (s)	Speed (m/s)
Speed (m/s)	Time (s)	Acceleration (m/s²)
Temperature (°C)	Time (mins)	Temperature change (°C/min)
Profit ($)	Time (hour)	Profit change ($/hour)

EXAMPLE 10

SKILLS

MODELLING

The depth of water, y m, in a tidal harbour entrance t hours after midday is given by the formula

$y = 4 + 3t - t^2$ where $0 \leq t \leq 4$

a Find the rate of change of the depth of sea water in m/hr at
   i 13:00   **ii** 15:00   **iii** 14:30

b Find the rate of change of the depth of sea water in m/hr when the harbour entrance is dry.

a If $y = 4 + 3t - t^2$ the gradient function $\dfrac{dy}{dt} = 3 - 2t$

  i   At 13:00 $t = 1, \dfrac{dy}{dt} = 3 - 2 \times 1 = 1$ m/hr (increasing depth)

  ii   At 15:00 $t = 3, \dfrac{dy}{dt} = 3 - 2 \times 3 = -3$ m/hr (decreasing depth)

  iii At 14:30 $t = 2.5, \dfrac{dy}{dt} = 3 - 2 \times 2.5 = -2$ m/hr (decreasing depth)

b The harbour entrance is dry when

$y = 0 = 4 + 3t - t^2$

$\Rightarrow t^2 - 3t - 4 = 0$

$\Rightarrow (t - 4)(t + 1) = 0$

$t = -1$ or 4

Ignore $t = -1$ as it is outside the **domain** of the model.

At 16:00 $t = 4, \dfrac{dy}{dt} = 3 - 2 \times 4 = -5$ m/hr (decreasing depth)

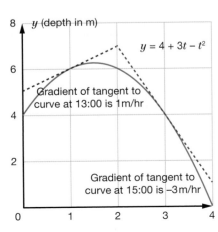

EXERCISE 3

1 ▶ Find the equation of the tangent to the curve $y = x^2 + 3x + 1$ at the point where $x = 2$.

2 ▶ Find the equation of the tangent to the curve $y = 3 + 4x - x^2$ at the point where $x = 1$.

3 ▶ The height above the ground, y m, of a Big-Dipper carriage at a funfair, t seconds after the start is given by
$y = 0.5t^2 - 3t + 5$ for $0 \le t \le 6$

a Find $\dfrac{dy}{dt}$

b Calculate the rate at which the height is changing in m/s after
    i 1 s   **ii** 3 s   **iii** 6 s

4 ▶ The population of P (millions) of bacteria on a piece of cheese t days after it is purchased is given by the equation
$P = \dfrac{1}{2}t^2 + t + 1$ for $0 \le t \le 4$

a Find $\dfrac{dP}{dt}$

b Calculate the rate at which the number of bacteria is changing in millions/day after
    i 1 day   **ii** 3 days   **iii** 4 days

5 ▶ The temperature of a chemical compound, T °C, t minutes after the start of a heating process is given by
$T = 3t^2 + 5t + 15$ for $0 \le t \le 10$

a Find $\dfrac{dT}{dt}$

b Calculate the rate at which the temperature is changing in °C/min after
    i 1 min   **ii** 5 mins   **iii** 10 mins

6 ▶ The depth of sea water at a small port, h m, t hours after midnight is given by
$h = 6 + 11t - 2t^2$ for $0 \le t \le 6$

a Find $\dfrac{dh}{dt}$

b Calculate the rate at which the depth is changing in m/hr at
    i 01:00   **ii** 03:00   **iii** 04:30

1 ▶　Find the equation of the tangent to the curve $y = x^3 - 5x + 1$ at the point where $x = 2$.

2 ▶　Find the equation of the tangent to the curve $y = (2x - 1)^2 + 1$ at the point where $x = -2$.

3 ▶　The number of people, N, at a shopping centre t hours after 08:00 is modelled by

$N = 20t^2 + 80t + 1000$ for $0 \le t \le 4$

　　a Find $\dfrac{dN}{dt}$

　　b Calculate the rate at which people are arriving at the shopping centre in people/hr at
　　　i 08:00　　**ii** 10:00　　**iii** 11:45

4 ▶　The flow $Q\,\text{m}^3$ of a small river t hours after midnight is monitored after a storm and is given by the equation

$Q = t^3 - 8t^2 + 24t + 10$ for $0 \le t \le 5$

　　a Find $\dfrac{dQ}{dt}$

　　b Calculate the rate of flow of the river in m^3/s at
　　　i 01:00　　**ii** 02:30　　**iii** 03:15

5 ▶　The temperature of a cup of coffee, $T\,°\text{C}$, m minutes after it has been made is given by the equation

$T = \dfrac{400}{m}$ for $5 \le t \le 10$

　　a Find $\dfrac{dT}{dm}$

　　b Calculate the rate at which the temperature of the cup of coffee is changing in $°\text{C}/\text{min}$ when
　　　i $m = 5$　　**ii** $m = 10$

6 ▶ The population, P, of a certain type of spider over t months is modelled by the equation

$$P = 2t^2 + \frac{180}{t} \text{ for } 1 \le t \le 12$$

a Find $\dfrac{dP}{dt}$

b Calculate the rate at which the spider population is changing in spiders/month when
　　i $t = 1$　　　**ii** $t = 12$

STATIONARY POINTS

The point where a curve has zero gradient is where $\dfrac{dy}{dx} = 0$

These are called **stationary points** or **turning points** and can be classified as either maximum points or minimum points.

Maximum point

Minimum point

Gradient $= \dfrac{dy}{dx} = 0$

Gradient is **decreasing** from +ve, through zero, to −ve

Gradient is positive just before and negative just after.

Gradient $= \dfrac{dy}{dx} = 0$

Gradient is **increasing** from −ve, through zero, to +ve

Gradient is negative just before and positive just after.

Knowing how to classify each turning point is important.

When the curve is drawn in the question it can be used to describe the nature of the points where $\dfrac{dy}{dx} = 0$

The following four curves should be recognised and used to classify turning points.

Quadratic curves of type $y = ax^2 + bx + c$ have the following shapes depending on the value of the **coefficient** a.

Cubic curves of type $y = ax^3 + bx^2 + cx + d$ have the following shapes depending on the value of the coefficient a.

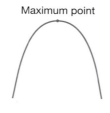

Maximum point

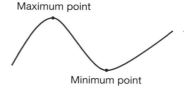

Maximum point

Maximum point

Minimum point

Minimum point

Minimum point

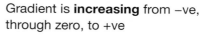

A quadratic function with $a > 0$

A quadratic function with $a < 0$

A cubic function with $a > 0$

A cubic function with $a < 0$

EXAMPLE 11

SKILLS

ANALYSIS

Find the turning points of the curve $y = 2x^3 - 3x^2 - 12x + 1$ and classify their nature.

This cubic curve has the shape

as the coefficient, $a > 0$

The turning point with the lower x-value will be the maximum point.

$\dfrac{dy}{dx} = 6x^2 - 6x - 12 = 0$ at turning points.

$0 = 6(x^2 - x - 2)$

$0 = 6(x - 2)(x + 1)$

So the gradient is zero when $x = 2$ and $x = -1$.

The co-ordinates of these points are $(2, -19)$ and $(-1, 8)$.

Method 1 – Shape of curve

By careful inspection of the curve it is clear that the maximum point has the smaller x-value.

$(-1, 8)$ is the maximum point.

$(2, -19)$ is the minimum point.

Method 2 – Local gradient

$(-1, 8)$ Maximum point

x	−1.1	−1	−0.9
$\dfrac{dy}{dx}$	+ve	0	−ve
CURVE SKETCH	/	−	\

$(2, -19)$ Minimum point

x	1.9	2	2.1
$\dfrac{dy}{dx}$	−ve	0	+ve
CURVE SKETCH	\	−	/

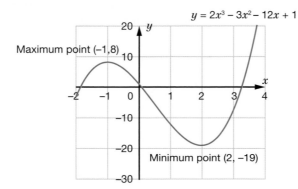

$y = 2x^3 - 3x^2 - 12x + 1$

Maximum point (−1,8)

Minimum point (2, −19)

KEY POINTS

- At a stationary point, the gradient = 0 and is found at the point where $\dfrac{dy}{dx} = 0$
- It can be classified as a maximum or a minimum point by
  - Knowing the shape of the curve
  - Finding the gradient close to the stationary point on either side of it using the gradient function $\dfrac{dy}{dx}$

EXERCISE 4

For Questions 1–8 solve $\dfrac{dy}{dx} = 0$ and find the turning points.

Classify them by considering the shape of the curve.

1 ▶ $y = x^2 - 2x + 3$ 5 ▶ $y = 2x^2 - 4x + 7$

2 ▶ $y = x^2 + 4x - 1$ 6 ▶ $y = 8 - 12x - 2x^2$

3 ▶ $y = 5 + 6x - x^2$ 7 ▶ $y = (x - 3)(x + 1)$

4 ▶ $y = 12 - 8x - x^2$ 8 ▶ $y = (1 + 2x)(1 - 2x)$

9 ▶ The number of people, N, entering a concert hall, t minutes after the doors are opened at 8 pm is modelled by the equation

$N = 50t^2 - 300t + 500$ for $0 \le t \le 5$

a Find $\dfrac{dN}{dt}$

b Calculate when the minimum flow of people entering the concert hall occurs.

10 ▶ The number of leaves, N, on a small oak tree, for the first t days in September is modelled by the equation

$N = 1000 + 200t - 10t^2$ for $0 \le t \le 20$

a Find $\dfrac{dN}{dt}$

b Calculate the maximum number of leaves in this period and the date when this occurs.

EXERCISE 4*

For Questions 1–8 solve $\dfrac{dy}{dx} = 0$ and find the turning points.

Classify them by considering the shape of the curve

1 ▶ $y = x^3 - 6x^2$ 5 ▶ $y = x^3 + 3x^2 - 9x + 5$

2 ▶ $y = x^3 + 3x^2$ 6 ▶ $y = 11 - 18x - 12x^2 - 2x^3$

3 ▶ $y = x^3 - 9x^2 + 3$ 7 ▶ $y = x(2x^2 + 9x - 24)$

4 ▶ $y = 4 - 3x^2 - x^3$ 8 ▶ $y = x(2x - 1)^2$

9 ▶ An open box is made from a thin square metal sheet measuring 10 cm by 10 cm.

Four squares of side x cm are cut away and the remaining four sides are folded upwards to make an open box of depth x cm.

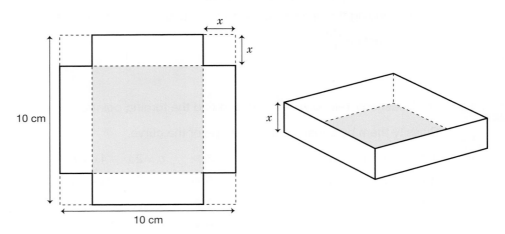

a Show that the volume V cm³ of the box is given by

$$V = 100x - 40x^2 + 4x^3 \text{ for } 0 \le x \le 5$$

b Find $\dfrac{\mathrm{d}V}{\mathrm{d}x}$

c Calculate the maximum volume of the box and state its dimensions.

10 ▶ The temperature, $T°C$, of the sea off Nice in France, t months after New Year's Day is given by

$$T = 5t + \frac{20}{t} - 5 \text{ for } 1 \le t \le 6$$

a Find $\dfrac{\mathrm{d}T}{\mathrm{d}t}$

b Calculate the coolest sea temperature in this period and state when this occurs.

MOTION OF A PARTICLE IN A STRAIGHT LINE

Consider a particle moving in a straight line which produces the following graphs against time.

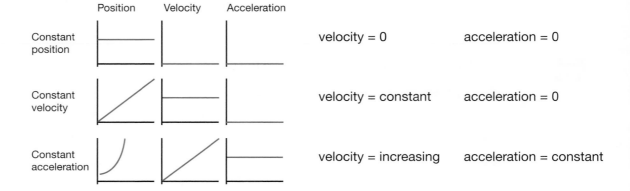

If the particle starts at O and moves away from it in a straight line, this direction is taken to be positive. If the particle moves towards O, the direction is taken to be negative.

0

$\vdash$···$\rightarrow$ +ve direction

$s = 0$

Displacement, s: distance from a fixed point 0

Velocity and acceleration are examples of rates of change:

velocity, v: rate of change of **displacement**; $\pm$ indicates particle's direction.

acceleration, a: rate of change of velocity.

EXAMPLE 12

SKILLS

MODELLING

A bicycle travels along a straight line. Its displacement, s m, from its starting position, O, after time t seconds is given by the equation

$s = 4t + 3t^2 - t^3$ m

After a time of 2 seconds, find the bicycle's

a velocity in m/s

b acceleration in m/s²

---

a $s = 4t + 3t^2 - t^3$

$v = \dfrac{ds}{dt} = 4 + 6t - 3t^2$ m/s

At $t = 2$, $v = 4 + 6(2) - 3(2)^2$ $v = 4$ m/s   (moving away from O)

b $v = 4 + 6t - 3t^2$

$a = \dfrac{dv}{dt} = 6 - 6t$ m/s²

At $t = 2$, $a = 6 - 6(2)$ $a = -6$ m/s²   (slowing down)

KEY POINTS

- Velocity is the rate at which displacement changes with time.

- $v = \dfrac{ds}{dt}$ (Gradient of distance–time graph is velocity)

- Acceleration is the rate at which velocity changes with time.

- $a = \dfrac{dv}{dt}$ (Gradient of velocity–time graph is acceleration)

- displacement   *differentiate*   velocity   *differentiate*   acceleration

  (s) $\rightarrow$ $(v = \dfrac{ds}{dt})$ $\rightarrow$ $(a = \dfrac{dv}{dt})$

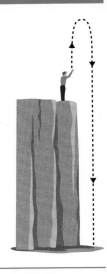

EXAMPLE 13

SKILLS

INTERPRETATION

A pebble is thrown vertically upwards such that its height, s m, above the top of the cliff after t s is given by

$$s = 10t - 5t^2$$

Find the maximum height reached by the pebble and when this occurs.

$s = 10t - 5t^2$

$$v = \frac{ds}{dt} = 10 - 10t$$

At the maximum height the velocity is zero $\Rightarrow$ $10 - 10t = 0$ $\Rightarrow$ $t = 1$

At $t = 1\,\text{s}$ $\qquad s_{max} = 10(1) - 5(1)^2$ $\qquad s_{max} = 5\,\text{m}$

EXERCISE 5

12th

1 ▶ The displacement, s m, of a particle after t s from a fixed point O is given by

$$s = 10t^2 - 30t + 1$$

After a time of 2 s find the particle's

a velocity in m/s   **b** acceleration in m/s²

2 ▶ The displacement, s m, of a particle after t s from a fixed point O is given by

$$s = 10 + 7t - t^2$$

After a time of 3 s find the particle's

a velocity in m/s   **b** acceleration in m/s²

3 ▶ The displacement, s m, of a particle after t s from a fixed point O is given by

$$s = t^3 + 2t^2 - 3t + 1$$

After a time of 2 s find the particle's

a velocity in m/s   **b** acceleration in m/s²

4 ▶ The displacement, s m, of a particle after t s from a fixed point O is given by

$$s = 20 + 12t + 3t^2 - t^3$$

After a time of 2 s find the particle's

a velocity in m/s   **b** acceleration in m/s²

5 ▶ A tennis ball is projected vertically upwards such that its height, s m, after t s is given by

$$s = 16t - 4t^2$$

Find the maximum height reached by the ball and when this occurs.

6 ▶ The velocity of a stone, v m/s, t s after it is thrown upwards is given by

$$v = 4 + 12t - t^2$$

Calculate the stone's

Q6c HINT
Maximum velocity
occurs when
acceleration is
zero.

a velocity after 2 s

b acceleration after 2 s

c maximum velocity.

EXERCISE 5*

1 ▶ The displacement, s m, of a particle after t s from a fixed point O is given by

$$s = 5t^2 - \frac{4}{t} + 1$$

After a time of 2 s find the particle's

a velocity in m/s   **b** acceleration in m/s²

2 ▶ The displacement, s m, of a particle after t s from a fixed point O is given by

$$s = 10\sqrt{t} + 1$$

After a time of 4 s find the particle's

a velocity in m/s   **b** acceleration in m/s²

3 ▶ A small rock falls off the top of a cliff by the sea. Its height, s m, above the sea t s later is modelled by

$$s = 50 - 5t^2$$

a Find the cliff's height.

b When does the rock hit the sea?

c Find the impact velocity of the rock as it enters the sea.

4 ▶ Alec is batting in a cricket match. He hits a ball such that its distance from him, s m, after t s is given by

$$s = 40t - 5t^2 \text{ for } 0 \le t \le 8$$

a Find the speed of the ball as it leaves Alec's bat.

b How far does the ball travel before it stops?

5 ▶ The velocity of a particle, v km/s , after t s is given by

$$v = 20 - t - \frac{25}{t} \text{ for } 1 \le t \le 10$$

a Find an expression for the acceleration of the particle in km/s², in terms of t.

b Calculate the particle's maximum speed in km/s.

6 ▶ A tennis ball is projected vertically upwards from the top of a 55 m cliff by the sea. Its height from the point of projection, s m, t s later is given by

$$s = 50t - 5t^2$$

a When does the tennis ball hit the sea?

b Find the impact velocity of the ball as it enters the sea.

c Find the **mean** speed of the ball for its entire time of flight.

ACTIVITY 1

The Dimox paint factory wants to store $50\,\text{m}^3$ of paint in a closed **cylindrical** tank. To reduce costs, it wants to use the minimum possible surface area (including the top and bottom).

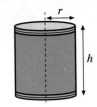

a If the total surface area of the tank is $A\,\text{m}^2$, and the **radius** is r m, show that the height h m of the tank is given by $h = \dfrac{50}{\pi r^2}$

and use this to show that $A = 2\pi r^2 + \dfrac{100}{r}$

b Show that $\dfrac{\mathrm{d}A}{\mathrm{d}r} = 4\pi r - \dfrac{100}{r^2}$

c Solve $\dfrac{\mathrm{d}A}{\mathrm{d}r} = 0$ to find the dimensions of the tank that will produce the minimum surface area of the tank and state this area.

EXERCISE 6

12ᵗʰ

REVISION

1 ▶ Find the gradient of the tangent for $y = 2x^3 + x^2 - 4x + 1$ at $x = 1$

2 ▶ The profit made by a company, p ($1000's), after t years is given by

$p = 60t - 5t^2$ for $0 \le t \le 12$

Find the rate of profit made by the company in $1000s per year after

a 2 years   b 10 years.

3 ▶ Find and classify the turning point on the curve $y = 2(x + 1)(x - 1)$

4 ▶ The displacement, s m, of a particle after t s from a fixed point O is given by $s = (t + 5)(3t - 2)$

After a time of 2 s find the particle's

a velocity in m/s   b acceleration in m/s².

EXERCISE 6*

12ᵗʰ

REVISION

1 ▶ Find the equation of the tangent to the curve $y = (x + 1)(x^2 - 5)$ at $x = 2$

2 ▶ The number of hairs on a man's head, n, depends on his age, t years, and is given by the formula

$n = 125t^2 - \dfrac{5t^3}{3}$ for $0 \le t \le 75$

Find the greatest number of hairs on the man's head and at what age this occurs.

3 ▶ Find and classify the turning points on the curve $y = 2x^3 - 15x^2 + 24x + 3$

4 ▶ The velocity of a particle, v m/s, after t s is given by

$v = 30 - t - \dfrac{144}{t}$ for $1 \le t \le 12$

a Find an expression for the acceleration of the particle in m/s² in terms of t.
b Calculate the particle's maximum speed in m/s.

EXAM PRACTICE: GRAPHS 9

1 Find the gradient of the tangents to the following curves at $x = 1$.

a $y = 4x^3 + 7x - 3$ **b** $y = (x + 1)(5x - 1)$ **c** $y = 1 - \dfrac{2}{x^2}$ **[7]**

2 Find and classify the stationary points of the curve $y = 2x^3 - 9x^2 + 12x - 1$ **[6]**

3 The petrol consumption (p km/litre) of a car is related to its speed (x km/hr) by the equation

$$p = 70 - \frac{x}{3} - \frac{2400}{x} \text{ for } 60 \le x \le 130$$

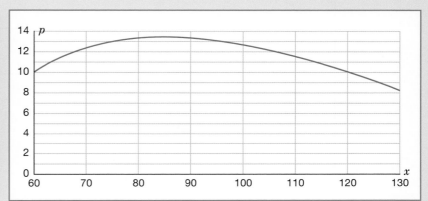

a Calculate an expression for $\dfrac{\mathrm{d}p}{\mathrm{d}x}$

b Find the most economical speed for the car and the petrol consumption at this speed.   **[6]**

4 The displacement, s m, of a particle after t s from a fixed point O is given by

$$s = 10\sqrt{t} + t + \pi \text{ for } 0 \le t \le 4$$

After a time of 1 s find the particle's

a velocity in m/s   **b** acceleration in m/s²   **[6]**

[Total 25 marks]

CHAPTER SUMMARY: GRAPHS 9

GRADIENTS AND DIFFERENTIATION

If $y = x^n$, the gradient function $\dfrac{dy}{dx} = n\,x^{n-1}$, where n can be any real value.

If $y = kx^n$ $\qquad \dfrac{dy}{dx} = nk\,x^{n-1}$

If $y = kx$ $\qquad \dfrac{dy}{dx} = k$

If $y = k$ $\qquad \dfrac{dy}{dx} = 0$

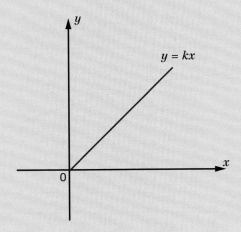

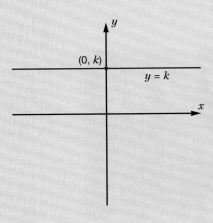

STATIONARY POINTS

At a stationary point, the gradient = 0 and is found at the point where $\dfrac{dy}{dx} = 0$

It can be classified as a maximum or a minimum point by

■ Knowing the shape of the curve
■ Finding the gradient close to the stationary point on either side of it using the gradient function $\dfrac{dy}{dx}$

Quadratic curves of type $y = ax^2 + bx + c$ have the following shapes depending on the value of the coefficient a.

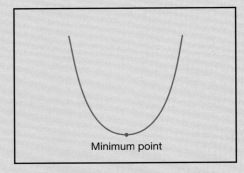

A quadratic function with $a > 0$

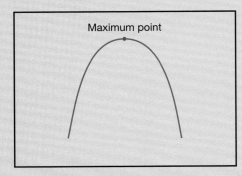

A quadratic function with $a < 0$

Cubic curves of type $y = ax^3 + bx^2 + cx + d$ have the following shapes depending on the value of the coefficient a.

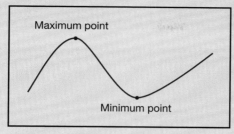

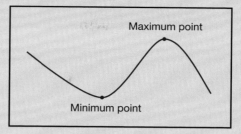

A cubic function with $a > 0$ A cubic function with $a < 0$

MOTION OF A PARTICLE IN A STRAIGHT LINE

s	displacement	Distance from a fixed point O.
v	velocity	Rate of change of displacement; ± indicates particle's direction.
a	acceleration	Rate of change of velocity.

Velocity is the rate at which displacement changes with time.

$v = \dfrac{ds}{dt}$ (Gradient of distance–time graph is velocity)

Acceleration is the rate at which velocity changes with time.

$a = \dfrac{dv}{dt}$ (Gradient of velocity–time graph is acceleration)

Displacement	*differentiate*	Velocity	*differentiate*	Acceleration
(s)	$\rightarrow$	$(v = \dfrac{ds}{dt})$	$\rightarrow$	$(a = \dfrac{dv}{dt})$

Maximum velocity occurs when acceleration = 0   i.e., v_{max} occurs when $\dfrac{dv}{dt} = 0$

SHAPE AND SPACE 10

Graphs of trigonometric functions have many real-life applications. Their natural curves often apply to situations when a repeated process occurs.

A tidal chart of Sydney Harbour in Australia for April 2016 is shown below.

Sine and cosine functions can be used to model many real-life situations including electric currents, musical tones, radio waves, tides and weather patterns.

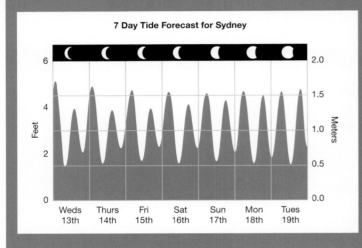

LEARNING OBJECTIVES

■ Recognise, plot and draw the graphs of the trigonometric functions: $y = \sin\theta$, $y = \cos\theta$ and $y = \tan\theta$

■ Use the graphs of the trigonometric functions to solve equations

■ Use the sine and cosine rules

■ Work out the area of a triangle using, Area $= \frac{1}{2}ab\sin C$

BASIC PRINCIPLES

■ Trig. ratios:

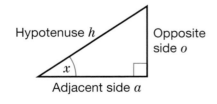

■ $\sin x = \dfrac{o}{h}$

■ $\cos x = \dfrac{a}{h}$

■ $\tan x = \dfrac{o}{a}$

Identify the **hypotenuse**. This is the longest side: the side opposite the **right angle**. Then the opposite side is opposite the angle. And the **adjacent** side is adjacent to (next to) the angle.

■ **Bearings** are measured clockwise from North.

   ■ When drawing bearings, start by drawing an arrow to indicate North.

   ■ When calculating angles on a bearings diagram, look for '**alternate angles**'.

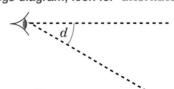

■ e is the **angle of elevation**.

■ d is the **angle of depression**.

GRAPHS OF SINE, COSINE AND TANGENT

TRIGONOMETRIC GRAPHS ($0° \leq \theta \leq 360°$)

Drawing the graphs of $y = \sin\theta$, $y = \cos\theta$ and $y = \tan\theta$ is done by first generating the values from a calculator. The shapes of the graphs are beautiful and simple to recognise. This section extends these ratios beyond angles of 90° and first looks at the graphs where $0° \leq \theta \leq 360°$.

To understand how these graphs are produced it is important to understand some principles. Look at the circle of **radius** 1 unit with its centre at the origin of a set of co-ordinate axes.

$x = 1 \times \cos\theta$ and $y = 1 \times \sin\theta$

P has co-ordinates $(\cos\theta, \sin\theta)$

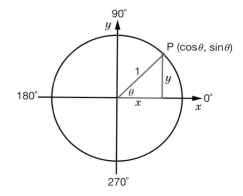

P can move around the circle's **circumference** in an anticlockwise direction from the x-axis where $\theta = 0°$ to $\theta = 360°$ to produce a single revolution. As P moves, the value of the angle from the positive x-axis, θ, changes.

Clearly, the co-ordinates of P will also change. The height of P above the x-axis $= \sin\theta$.

The London Eye is an attraction in London which is a 135 m high wheel, with 32 'pods' that people can ride in. A single pod on the London Eye makes a similar motion to the above point P as it rotates around the axis of the wheel.

Angles are measured from the positive x-axis.

The graphs of $y = \sin\theta$, $y = \cos\theta$ and $\tan\theta$ can be plotted for $0° \leq \theta \leq 360°$.

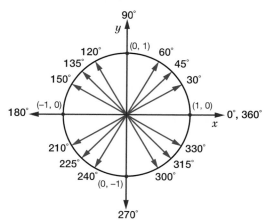

ACTIVITY 1, THE SINE GRAPH

SKILLS

ANALYSIS

Use a calculator to copy and complete the table below for $y = \sin\theta$ for $0° \leq \theta \leq 360°$ to exact or 2 s.f. values.

$\theta°$	0	30	60	90	120	150	180	210	240	270	300	330	360
$\sin\theta°$	0						0						0

Use the table to draw the graph of $y = \sin\theta$.

The sine graph has a maximum value of 1 at 90° and a minimum value of −1 at 270°.

It repeats itself every 360°.

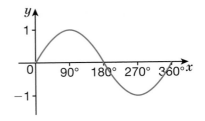

ACTIVITY 2, THE COSINE GRAPH

SKILLS

ANALYSIS

Use a calculator to copy and complete the table below for $y = \cos\theta$ for $0° \leq \theta \leq 360°$ to exact or 2 sig.fig. values.

$\theta°$	0	30	60	90	120	150	180	210	240	270	300	330	360
$\cos\theta°$	1			0						0			

Use the table to draw the graph of $y = \cos\theta$.

The cosine graph has a maximum value of 1 at 0° and 360° and a minimum value of −1 at 180°.

It repeats itself every 360°.

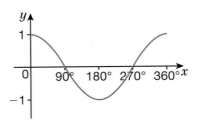

ACTIVITY 3, THE TANGENT GRAPH

SKILLS

ANALYSIS

The unit circle diagram shows that $\tan\theta = \dfrac{\text{opposite}}{\text{adjacent}} = \dfrac{y}{x} = \dfrac{\sin\theta}{\cos\theta}$ and as P moves around the circle θ changes and this will change $\tan\theta$.

Use a calculator to copy and complete the table below for $y = \tan\theta$ for $0° \leq \theta \leq 360°$ to exact or 2 s.f. values.

You might wish to add more values of $\tan\theta$ close to 90° and 270° to improve the graph.

$\theta°$	0	30	60	90	120	150	180	210	240	270	300	330	360
$\tan\theta°$	0						0						0

Use the table to draw the graph of $y = \tan\theta$.

The tangent graph approaches, but never touches, $\theta = 90°$ and $\theta = 270°$.

It repeats itself every 180°.

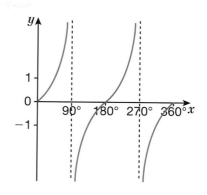

TRIGONOMETRIC GRAPHS (ANY ANGLE)

The unit circle with point P can be used again to generate the graphs of $y = \sin\theta$, $y = \cos\theta$ and $y = \tan\theta$ for any angle P that moves around the circle's circumference

■ in an anticlockwise direction from $\theta = 0°$ any number of times generating angles beyond 360°

■ in a clockwise direction from $\theta = 0°$ any number of times, however these angles are negative.

As the point on the end of the blue arrow rotates, the value of the angle from the positive x-axis, θ, changes. The co-ordinates of P will also change, therefore generating graphs that repeat their pattern.

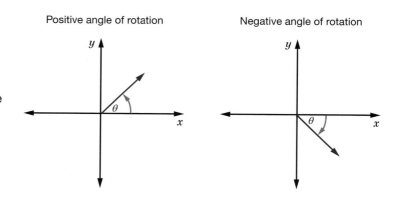

Positive angle of rotation   Negative angle of rotation

EXAMPLE 1

SKILLS

ANALYSIS

Draw the graph of $y = \sin\theta$ for values of θ from −360° to 720°.

The maximum and minimum values are 1 and −1.

The graph repeats itself every 360° in both directions.

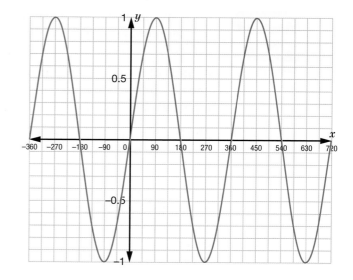

EXAMPLE 2

SKILLS

ANALYSIS

Draw the graph of $y = \cos\theta$ for values of θ from $-360°$ to $720°$.

The maximum and minimum values are 1 and -1.

The graph repeats itself every $360°$ in both directions.

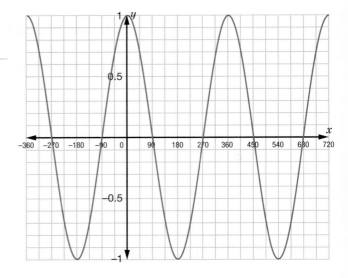

EXAMPLE 3

SKILLS

ANALYSIS

Draw the graph of $y = \tan\theta$ for values of θ from $-360°$ to $720°$.

The graph has values of θ where it cannot be defined; these are called **asymptotes**.

The graph repeats itself every $180°$ in both directions.

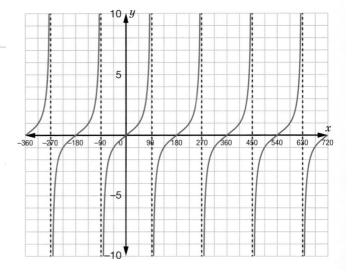

EXAMPLE 4

SKILLS

ANALYSIS

Solve the equation $\sin\theta = 0.4$ for all values of θ in the **domain** $-360° \le \theta \le 720°$ to 1 d.p.

A graph of $y = \sin\theta$ in the required domain can be used with the line $y = 0.4$ to find the repeated pattern of solutions.

The first positive angle at A is $23.6°$ (1 d.p.) and the second positive angle at B is $156.4°$ (1 d.p.).

By careful use of the symmetry of the curve it is easy to see that all the values in this domain to 1 d.p. are $-336.4°$, $-203.6°$, $23.6°$, $156.4°$, $383.6°$ and $516.4°$.

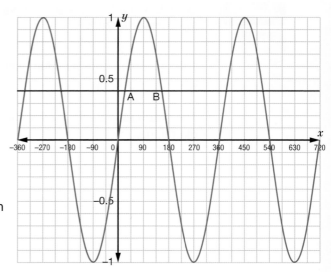

EXERCISE 1

For all questions in this exercise use a calculator and the symmetry of the trigonometry graphs of sine, cosine and tangent on pages 398–399.

1 ▶ Solve the following equations for all values of θ in the domains stated for $0° \le \theta \le 360°$.

　　a $\sin\theta = 0$　　**b** $\cos\theta = 0$　　**c** $\tan\theta = 0$　　**d** $\sin\theta = 1$　　**e** $\cos\theta = 1$　　**f** $\tan\theta = 1$

2 ▶ Solve the following equations for all values of θ in the domains stated for $0° \le \theta \le 360°$.

　　a $\sin\theta = 0.5$　　**b** $\cos\theta = 0.5$　　**c** $\tan\theta = \dfrac{1}{\sqrt{3}}$　　**d** $\sin\theta = \dfrac{1}{\sqrt{2}}$　　**e** $\cos\theta = \dfrac{1}{\sqrt{2}}$　　**f** $\tan\theta = \sqrt{3}$

EXERCISE 1*

For all questions in this exercise use a calculator and the symmetry of the trigonometry graphs of sine, cosine and tangent on pages 399–400.

1 ▶ Solve the following equations for all values of θ in the domains stated for $-360° \le \theta \le 720°$

　　a $\sin\theta = 0$　　**b** $\cos\theta = 0$　　**c** $\tan\theta = 0$　　**d** $\sin\theta = 1$　　**e** $\cos\theta = 1$　　**f** $\tan\theta = 1$

2 ▶ Solve the following equations for all values of θ in the domains stated for $-360° \le \theta \le 720°$.

　　a $\sin\theta = -0.5$　　**b** $\cos\theta = -0.5$　　**c** $\tan\theta = -\dfrac{1}{\sqrt{3}}$　　**d** $\sin\theta = -\dfrac{1}{\sqrt{2}}$　　**e** $\cos\theta = -\dfrac{1}{\sqrt{2}}$　　**f** $\tan\theta = -\sqrt{3}$

THE SINE RULE

The sine rule is a method of calculating sides and angles for any triangle. It is useful for finding the length of one side when all angles and one other side are known, or finding an angle when two sides and the **included angle** are known.

The sine rule can be used in any triangle to find a length or an angle.

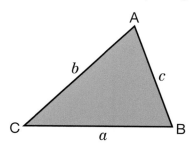

■ a is the side opposite angle A

■ b is the side opposite angle B

■ c is the side opposite angle C

To use the sine rule to find a length, you need to know any two angles and a side.

$$\frac{a}{\sin A} = \frac{b}{\sin B} = \frac{c}{\sin C}$$

To use the sine rule to find an angle, you need to know two sides and the non-included angle.

$$\frac{\sin A}{a} = \frac{\sin B}{b} = \frac{\sin C}{c}$$

EXAMPLE 5

In triangle ABC, find the length of side a, **correct to** 3 **significant figures**.

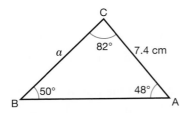

$$\frac{a}{\sin A} = \frac{b}{\sin B}$$

$$\frac{a}{\sin 48°} = \frac{7.4}{\sin 50°} \qquad \text{(Multiply both sides by } \sin 48°\text{)}$$

$$a = \frac{7.4}{\sin 50°} \times \sin 48°$$

$$a = 7.1787... \text{ cm}$$

$$a = 7.18 \text{ cm (to 3 s.f.)}$$

EXAMPLE 6

In triangle ABC, find angle B correct to 3 significant figures.

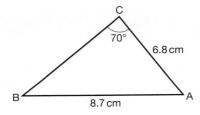

$$\frac{\sin B}{b} = \frac{\sin C}{c}$$

$$\frac{\sin B}{6.8} = \frac{\sin 70°}{8.7} \qquad \text{(Multiply both sides by 6.8)}$$

$$\sin B = \frac{\sin 70°}{8.7} \times 6.8$$

$$B = 47.262...$$

$$B = 47.3° \text{ (to 3 s.f.)}$$

KEY POINT

■ The sine rule:

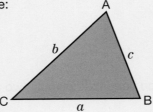

■ $\dfrac{a}{\sin A} = \dfrac{b}{\sin B} = \dfrac{c}{\sin C}$ Use this to calculate an unknown side.

■ $\dfrac{\sin A}{a} = \dfrac{\sin B}{b} = \dfrac{\sin C}{c}$ Use this to calculate an unknown angle.

Write your answers correct to 3 significant figures.

1 ▶ Find x.

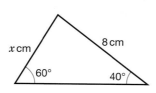

2 ▶ Find y.

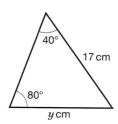

3 ▶ Find MN.

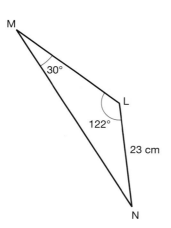

4 ▶ Find RT.

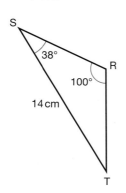

5 ▶ Find AC.

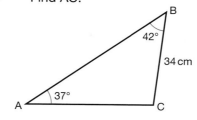

6 ▶ Find YZ.

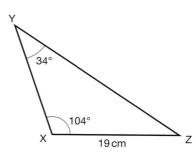

7 ▶ Find x.

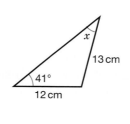

8 ▶ Find y.

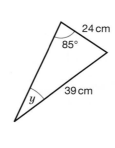

9 ▶ Find ∠ABC.

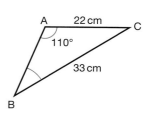

10 ▶ Find ∠XYZ.

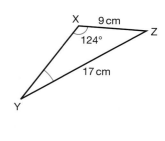

11 ▶ Find ∠ACB.

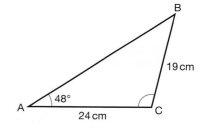

12 ▶ Find ∠DCE.

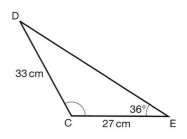

ACTIVITY 4

THE 'AMBIGUOUS CASE'

For a triangle, when the lengths of two sides and the size of a non-included angle are known it is sometimes possible to find two values for the length of the third side. This is called the 'ambiguous case'.

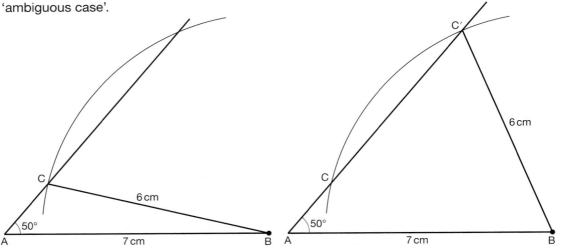

Note that for the triangle ABC, with AB = 7 cm, BC = 6 cm and ∠BAC = 50°, two triangles can be constructed: ABC and ABC′. The **arc** centred at B cuts the line from A at C and C′. This is an example of the 'ambiguous case' where two triangles can be constructed from the same facts.

Show by calculation that AC ≃ 1.8 cm and AC′ ≃ 7.2 cm.

EXAMPLE 7

A yacht crosses the start line of a race at C, on a bearing of 026°. After 2.6 km, it goes around a buoy (a floating navigation device), marked B, and sails on a bearing of 335°. When is it due North of the starting point at A, how far has it sailed in total?

Draw a diagram and include all the facts.

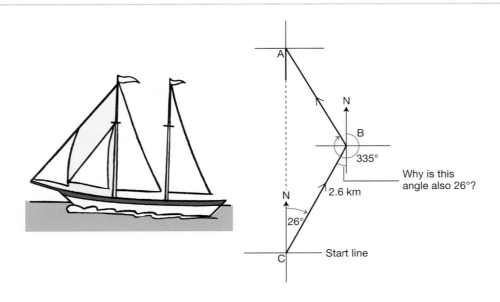

Work out any necessary angles and redraw the triangle and include only the relevant facts.

∠CBA = 335° − 26° − 180° = 129°

So ∠BAC = 25°

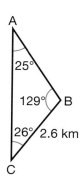

Use the sine rule: in triangle ABC, the length AB has to be calculated.

$$\frac{AB}{\sin 26°} = \frac{2.6}{\sin 25°}$$ (Multiply both sides by $\sin 26°$)

$$AB = \frac{2.6}{\sin 25°} \times \sin 26°$$

$$AB = 2.6969 \ldots \text{km}$$

Total distance travelled = CB + BA
= 2.6 km + 2.697 km = 5.30 km (to 3 s.f.)

EXERCISE 2*

Write your answers correct to 3 significant figures.

1 ▶   Find x.

3 ▶   Find ∠LMN.

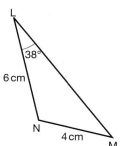

5 ▶   Find EF, ∠DEF and ∠FDE.

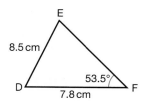

2 ▶   Find y.

4 ▶   Find ∠RST.

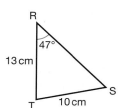

6 ▶   Find MN, ∠MLN and ∠LNM.

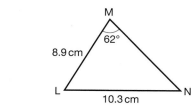

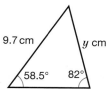

7 ▶ A yacht crosses the start line of a race on a bearing of 031°. After 4.3 km, it goes around a buoy and sails on a bearing of 346°. When it is due North of the starting point, how far has it sailed in total?

8 ▶ A boat crosses the start line of a race on a bearing of 340°. After 700 m, it goes around a buoy and sails on a bearing of 038°. When it is due North of the starting point, how far has it sailed in total?

9 ▶ The bearing of B from A is 065°. The bearing of C from B is 150°, and the bearing of A from C is 305°. If AC = 300 m, find BC.

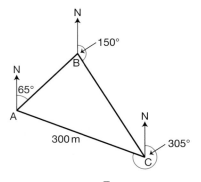

10 ▶ From two points X and Y, the angles of elevation of the top T of a tower Z are 14° and 19°, respectively. If XYZ is a horizontal straight line and XY = 120 m, find YT and the height of the tower.

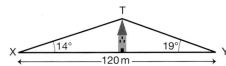

11 ▶

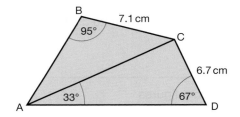

   a Work out the length of AC.

     Give your answer correct to 3 significant figures.

   b Work out the size of angle BAC.

     Give your answer correct to 1 **decimal place**.

12 ▶ In triangle ABC, AB = 8 cm, BC = 6 cm and angle BAC = 40°.

   Work out the size of angle ACB.

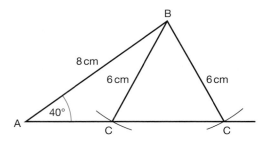

   The diagram shows that there are two possible triangles. Therefore there are two possible answers for angle ACB. Calculate both possible answers, correct to 1 decimal place

THE COSINE RULE

The cosine rule is another method of calculating sides and angles of any triangle. It is used either to find the third side when two sides and the included angle are given, or to find an angle when all three sides are given.

The cosine rule can be used in any triangle to find a length or an angle.

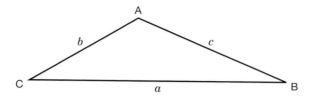

To use the cosine rule to find a length, you need to know two sides and the included angle.

$$a^2 = b^2 + c^2 - 2bc \cos A$$

To use the cosine rule to find an angle, you need to know all three sides.

$$\cos A = \frac{b^2 + c^2 - a^2}{2bc}$$

EXAMPLE 8

In triangle ABC find a correct to 3 significant figures.

$$a^2 = b^2 + c^2 - 2bc \cos A$$
$$= 6^2 + 9^2 - 2 \times 6 \times 9 \times \cos 115°$$
$$= 36 + 81 - 108 \times (-0.4226)$$
$$a^2 = 162.642...$$
$$a = \sqrt{162.642...}$$
$$= 12.8 \text{ (to 3 s.f.)}$$

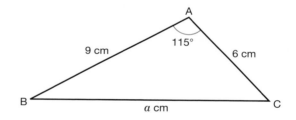

A common error in questions like Example 8 is to **round** too early, resulting in an inaccurate answer. You should only round your answer at the end of a calculation.

EXAMPLE 9

In triangle ABC, find angle A correct to 3 significant figures.

$$\cos A = \frac{b^2 + c^2 - a^2}{2bc}$$
$$= \frac{8^2 + 11^2 - 9^2}{2 \times 8 \times 11}$$
$$= \frac{104}{176}$$

$$A = 53.8° \text{ (to 3 s.f.)}$$

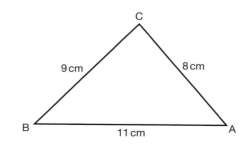

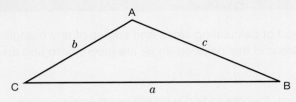

■ The cosine rule:

$$a^2 = b^2 + c^2 - 2bc \cos A$$ Use this to calculate an unknown side.

$$\cos A = \frac{b^2 + c^2 - a^2}{2bc}$$ Use this to calculate an unknown angle.

EXERCISE 3   Write your answers correct to 3 significant figures.

1 ▶ Find a.

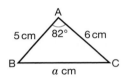

2 ▶ Find b.

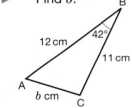

3 ▶ Find AB.

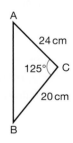

4 ▶ Find AB.

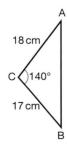

5 ▶ Find RT.

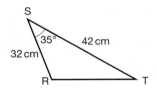

6 ▶ Find MN.

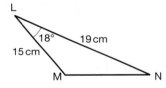

7 ▶ Find x.

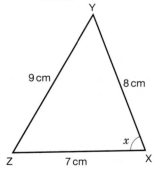

8 ▶ Find y.

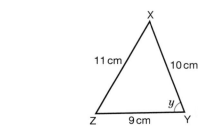

9 ▶ Find ∠ABC.

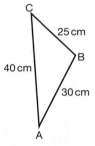

10 ▶ Find ∠XYZ.

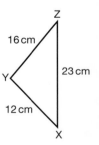

EXERCISE 3*

Write your answers correct to 3 significant figures.

1 ▶ Find x.

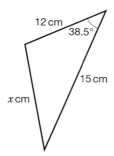

12 cm
38.5°
15 cm
x cm

2 ▶ Find y.

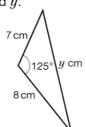

7 cm
125° y cm
8 cm

3 ▶ Find ∠XYZ.

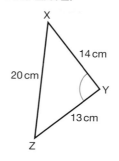

X
14 cm
20 cm
Y
13 cm
Z

4 ▶ Find ∠ABC.

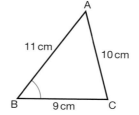

A
11 cm
10 cm
B   9 cm   C

5 ▶ Find ∠BAC.

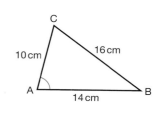

C
16 cm
10 cm
A   14 cm   B

6 ▶ Find ∠RST.

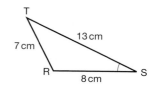

T
13 cm
7 cm
R   8 cm   S

7 ▶ A ship leaves a port P and sails for 12 km on a bearing of 068°. It then sails a further 20 km on a bearing of 106°, to reach port Q.

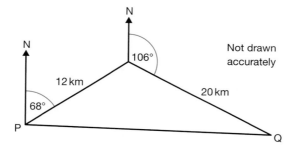

N
N
106°
12 km
20 km
68°
P
Q

Not drawn accurately

a What is the direct distance from P to Q?

Give your answer correct to 3 s.f.

b What is the bearing of Q from P?

Give your answer correct to the nearest degree.

8 ▶ A car drives 15 km on a bearing of 114° from P to Q. It then changes direction and drives 18 km on a bearing of 329° from Q to R.

a Calculate the distance and bearing of P from R.

b As the car drives from Q to R, what is the shortest distance between the car and point P?

Give all your answers correct to 3 s.f.

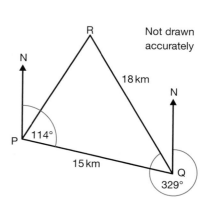

R
Not drawn accurately
18 km
N
N
114°
P
15 km
Q
329°

9 ▶ From S, a yacht sails on a bearing of 040°. After 3 km, at point A, it sails on a bearing of 140°. After another 4 km, at point B, it goes back to the start S. Calculate the total length of the journey.

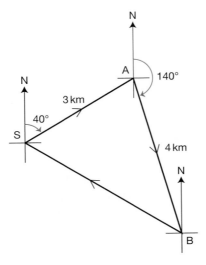

10 ▶ Copy and label this diagram using the facts:

VW = 50 km, WU = 45 km, VU = 30 km.

a Find ∠VWU.

b If the bearing of V from W is 300°, find the bearing of U from W.

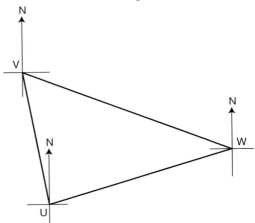

USING TRIGONOMETRY TO SOLVE PROBLEMS

Sometimes both the sine and cosine rules have to be used to solve a problem. Or it is necessary to choose the method which will produce the most efficient solution.

If the triangle is **right-angled**, it is quicker to solve using ordinary trigonometry rather than using the sine rule or cosine rule.

EXAMPLE 10

A motorboat, M, is 8 km from a harbour, H, on a bearing of 125°, while at the same time a rowing boat, R, is 16 km from the harbour on a bearing of 074°.

Find the distance and bearing of the rowing boat from the motorboat.

Two sides and an included angle are given.

Cosine rule:

∠RHM = 125° − 74° = 51°

$h^2 = 8^2 + 16^2 - 2 \times 8 \times 16 \times \cos 51°$

$\quad = 158.89...$

$\Rightarrow h = 12.6$ km (3 s.f.)

Two sides and the non-included angle are given.

Sine rule:

$$\frac{\sin 51°}{12.6} = \frac{\sin M}{16}$$

$\sin M = 0.986...$

$M = 80.7°$ (3 s.f.)

Bearing of R from M = 80.7° − 55° (180 − 125 = 55)

$\quad\quad\quad = 025.7°$ (3 s.f.)

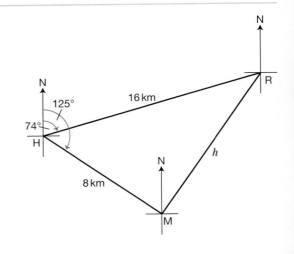

KEY POINT

■ When the problem involves
- ■ two sides and two angles (SASA) use the sine rule
- ■ three sides and one angle (SSSA) use the cosine rule.

EXERCISE 4

Write your answers correct to 3 significant figures.

1 ▶ Find
- **a** length a
- **b** angle C.

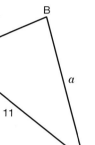

2 ▶ Find
- **a** length r
- **b** angle Q.

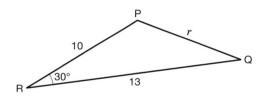

3 ▶ Find
- **a** angle M
- **b** angle N
- **c** angle O.

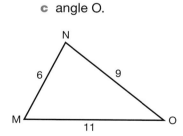

4 ▶ Find
- **a** angle X
- **b** angle Y
- **c** angle Z.

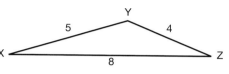

5 ▶ Find
- **a** length a
- **b** length b.

6 ▶ Find
- **a** length p
- **b** length r.

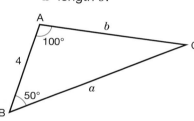

7 ▶ Find
- **a** angle O
- **b** length m.

8 ▶ Find
- **a** angle Z
- **b** length x.

9 ▶ Point P is on a bearing of 060° from port O.

Point Q is on a bearing of 130° from port O.

OQ = 17 km, OP = 11 km

Find the
- **a** distance PQ
- **b** bearing of Q from P.

10 ▶ Towns B and C are on bearings of 140° and 200° respectively from town A.

AB = 7 km and AC = 10 km.

Find the
- **a** distance BC
- **b** bearing of B from C.

Write answers to 3 significant figures.

1 ▶ The diagram shows the positions of three railway stations, A, B and C.

Calculate the bearing of C from A.

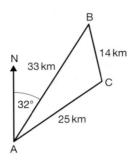

5 ▶ Calculate

a angle BAE

b length CD

c angle ACD.

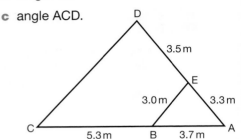

Q2b HINT

The North lines are parallel. Use this to find an angle inside the triangle.

2 ▶ A ship leaves port P and sails for 40 km on a bearing of 041°. It then sails a further 31 km on a bearing of 126° to reach port R.

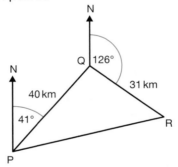

a What is the direct distance between P and R?

b What is the bearing of R from P?

3 ▶ Napoli is 170 km from Rome on a bearing of 130°.

Foggia is 130 km from Napoli on a bearing of 060°.

Find the distance and bearing of Rome from Foggia.

4 ▶ At 12:00, a ship is at X where its bearing from Trondheim, T, is 310°.

At 14:00, the ship is at Y where its bearing from T is 063°.

If XY is a straight line, TX = 14 km and TY = 21 km, find the ship's speed in km/h and the bearing of the ship's journey from X to Y.

6 ▶ The diagonals of a **parallelogram** have lengths 12 cm and 8 cm, and the angle between them is 120°. Find the side lengths of the parallelogram.

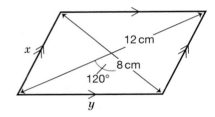

7 ▶ Two circles, centres X and Y, have radii 10 cm and 8 cm and **intersect** at A and B. XY = 13 cm. Find ∠BXA.

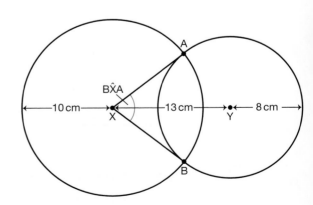

8 ▶ SBC is a triangular running route, with details in the table below.

	BEARING	DISTANCE
START S TO B	195°	2.8 km
B TO C	305°	1.2 km

Find the distance CS, and the bearing of S from C.

9 ▶ Find

a ∠XZY

b WX.

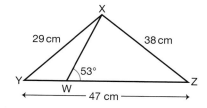

10 ▶ ABCD is a **tetrahedron**.

a Work out the length of AC.

b Work out the length of CD.

c Given that BD = 17 cm, calculate angle BCD.

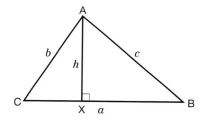

AREA OF A TRIANGLE

The diagram shows triangle ABC.

The area of the triangle $= \frac{1}{2} \times$ base $\times$ **perpendicular** height

$$= \frac{1}{2} \times a \times h$$

(In the right-angled triangle ACX, $h = b \sin C$)

$$= \frac{1}{2}\, a(b \sin C)$$

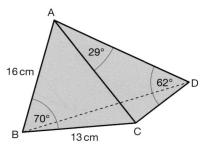

EXAMPLE 11   Find the area of the triangle ABC.

Area $= \frac{1}{2}\, ab \sin C$

$= \frac{1}{2} \times 11.5 \times 7.3 \times \sin 110°$

$= 39.4\,\text{cm}^2$ (3 s.f.)

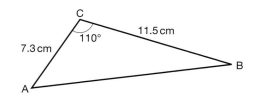

KEY POINT

■ The area of this triangle $= \frac{1}{2}\, ab \sin C$

A simple formula to remember this is

'Area of a triangle = half the **product** of two sides × sine of the included angle'

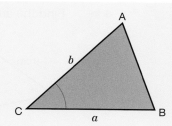

EXERCISE 5

1 ▶   Find the area of triangle ABC.

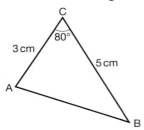

2 ▶   Find the area of triangle XYZ.

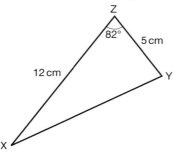

5 ▶   Find the area of the triangle ABC if AB = 15 cm, BC = 21 cm and B = 130°

6 ▶   Find the area of an **equilateral triangle** of side 20 cm.

3 ▶   Find the area of triangle PQR.

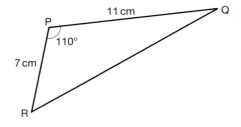

4 ▶   Find the area of triangle LMN.

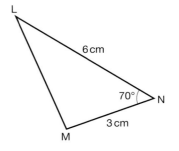

EXERCISE 5*

1 ▶   Find the side length of an equilateral triangle of area 1000 cm².

2 ▶   Find side length a if the area of triangle ABC is 75 cm².

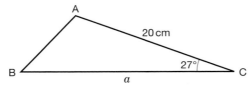

Q6 HINT
Use the upper bounds for all parts of the calculation.

3 ▶   Find the angle ABC if the area of the isosceles triangle is 100 cm².

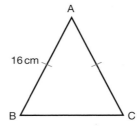

4 ▶   Find the area of triangle ABC.

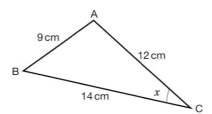

5 ▶   Find the **perimeter** of triangle ABC if its area is 200 cm².

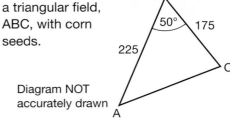

6 ▶   Jerry wants to cover a triangular field, ABC, with corn seeds.

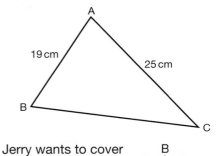

Diagram NOT accurately drawn

Here are the measurements that Jerry makes:

angle ABC = 50° correct to the nearest degree

BA = 225 m correct to the nearest 5 m

BC = 175 m correct to the nearest 5 m.

Work out the **upper bound** for the area of the field.

ACTIVITY 5

SKILLS
ANALYSIS

Proof of Sine Rule

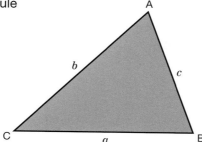

a Write an expression for the area of the triangle using angle C.

b Write two more expressions for the area of the triangle using angles B and A.

c Using your answers to parts **a** and **b**, show that

$$\frac{a}{\sin A} = \frac{b}{\sin B} = \frac{c}{\sin C}$$

Note: Proof of the sine rule is not required by the specification.

ACTIVITY 6

SKILLS
ANALYSIS

Proof of Cosine Rule

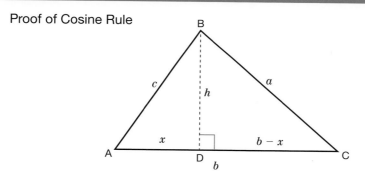

a In triangle BCD, write a^2 in terms of h, b and x and **expand** the brackets.

b In triangle ABD, write c^2 in terms of h and x.

c Use your answers to parts **a** and **b** to write a^2 in terms of b, c and x.

d Show that $a^2 = b^2 + c^2 - 2bc\cos A$

Note: Proof of the cosine rule is not required by the specification.

REVISION

1 ▶ Here is a **sketch** of $y = \sin x$

Write down the co-ordinates of each of the labelled points.

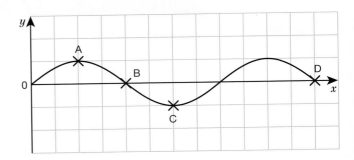

Q1 HINT
Check your answers by seeing if $\sin x = y$.

2 ▶ The diagram shows a sketch of the graph $y = \cos x°$

Write down the co-ordinates of points A, B and C.

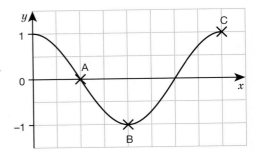

3 ▶ Here is the graph of $y = \tan x$ for $0° \le x \le 360°$

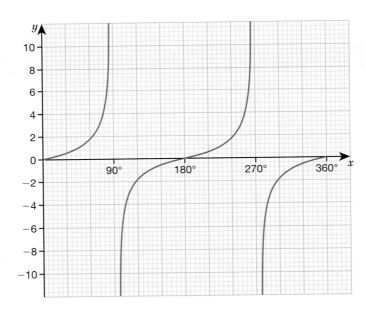

a How does the graph repeat?

b Use the graph to estimate the value of

   i $\tan 60°$ **ii** $\tan 300°$.

c Describe the symmetry of the curve.

d Copy and complete, inserting numbers greater than 180.

   i $\tan 60° = \tan$____°

   ii $\tan 100° = \tan$____°

   iii $\tan 120° = \tan$____°

4 ▶ Find

   a angle C

   b *b*.

Give your answers to 3 significant figures.

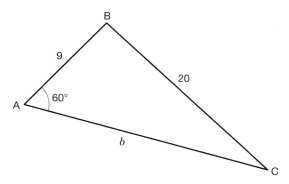

5 ▶ Find the area of △XYZ.

Give your answer to 3 significant figures.

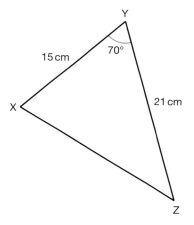

6 ▶ The bearing of B from A is 150°, the bearing of C from B is 280°, the bearing of A from C is 030°.

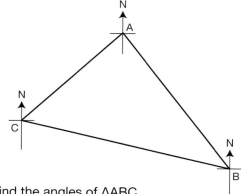

   a Find the angles of △ABC.

   b The distance AC is 4 km. Use the sine rule to find the distance AB.

Give your answers to 3 significant figures.

7 ▶

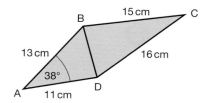

   a Work out the length of BD. Give your answer correct to 3 significant figures.

   b Work out the size of angle CBD.

      Give your answer correct to 1 decimal place.

   c Work out the area of **quadrilateral** ABCD. Give your answer correct to 3 significant figures.

8 ▶ The diagram shows the positions of three towns, A, B and C.

Calculate

   a the bearing of C from A

   b the area of triangle ABC.

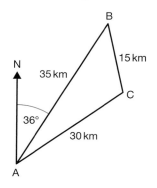

EXERCISE 6*

REVISION

1 ▶   Here is a sketch of $y = \sin x$

Write down the co-ordinates of each of the labelled points.

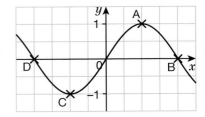

2 ▶   The diagram shows a sketch of the graph $y = \cos x°$

Write down the co-ordinates of points A, B, C and D.

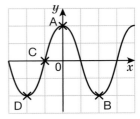

3 ▶   **a**   Sketch the graph of $y = \tan x$ in the interval 0° to 720°.

      b   Given that $\tan 60° = \sqrt{3}$ solve the equation $\sqrt{3}\,\tan x = 3$ in the interval 0° to 720°.

4 ▶   A yacht at point A is due west of a cliff C. It sails on a bearing of 125°, for 800 m, to a point B. If the bearing of C from B is 335°, find the distance BC.

5 ▶   The sides of a parallelogram are 4.6 cm and 6.8 cm, with an included angle of 116°. Find the length of each diagonal.

6 ▶   Find the area of ΔXYZ.

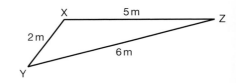

7 ▶   **a**   Work out the length of BD. Give your answer correct to 3 significant figures.

      b   Work out the size of angle BCD. Give your answer correct to 1 decimal place.

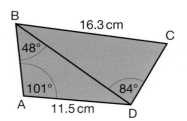

8 ▶   In triangle ABC, AB = 12 cm, BC = 7 cm and angle BAC = 35°

Work out the size of angle ACB.

The diagram shows that there are two possible triangles. Hence there are two possible answers. Give both, correct to 1 decimal place.

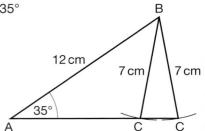

EXAM PRACTICE: SHAPE AND SPACE 10

1   **a** Sketch the graph of $y = \sin\theta$ for $0° \le \theta \le 360°$.   **[3]**

     b Use your graph to solve the equation $\sin\theta = \dfrac{\sqrt{3}}{2}$ to find all the values of θ for $0° \le \theta \le 360°$.   **[4]**

2   Find

     a length a **[3]**

     b angle B.   **[3]**

[Triangle ABC: B at top, 12 cm between A and B, a between B and C, 60° angle at A, 7 cm between A and C.]

3   Find

     a angle C   **[3]**

     b the area of triangle ABC.   **[3]**

[Triangle ABC: A at top, 21 cm between A and C, 80° angle at B, 17 cm between C and B.]

4   The diagram shows the positions of towns A, B and C.

     a Calculate the direct distance from A to C.   **[3]**

     b Calculate the bearing of C from A.   **[3]**

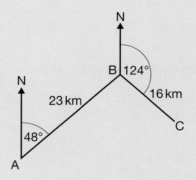

[Total 25 marks]

CHAPTER SUMMARY: SHAPE AND SPACE 10

GRAPHS OF SINE, COSINE AND TANGENT

$y = \sin\theta$

The sine graph repeats every 360° in both directions.

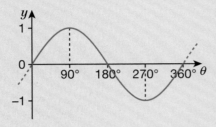

$y = \cos\theta$

The cosine graph repeats every 360° in both directions.

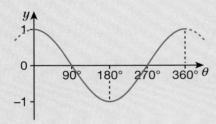

$y = \tan\theta$

The tangent graph repeats every 180° in both directions.

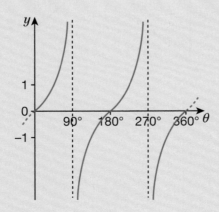

SINE RULE AND COSINE RULE

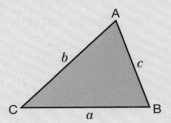

- a is the side opposite angle A
- b is the side opposite angle B
- c is the side opposite angle C

When the problem involves two sides and two angles (SASA) use the sine rule

- $\dfrac{a}{\sin A} = \dfrac{b}{\sin B} = \dfrac{c}{\sin C}$ to calculate an unknown side

- $\dfrac{\sin A}{a} = \dfrac{\sin B}{b} = \dfrac{\sin C}{c}$ to calculate an unknown angle.

When the problem involves three sides and one angle (SSSA) use the cosine rule

- $a^2 = b^2 + c^2 - 2bc \cos A$ to calculate an unknown side

- $\cos A = \dfrac{b^2 + c^2 - a^2}{2bc}$ to calculate an unknown angle.

AREA OF A TRIANGLE

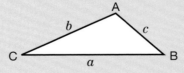

Area of this triangle $= \dfrac{1}{2}ab \sin C$

A simple formula to remember this is

'Area = half the product of two sides × sine of the included angle'

HANDLING DATA 7

'Actuaries' are professionals who use probability theory to measure how likely it is that life events such as motor vehicle crashes, house fires, losing your computer at school and numerous others will occur.

These figures are used by insurance companies to decide how much to charge their customers for insurance cover.

Edmund Halley constructed his own life table in 1693 showing how it

Edmund Halley 1656–1742 ▲

could be used to calculate how much someone of a particular age should pay to buy a pension (monetary income) after they stop working.

LEARNING OBJECTIVE

■ Draw and use more complex tree diagrams

BASIC PRINCIPLES

■ P(E) means the probability of event E occurring.

■ P(E′) means the probability of event E not occurring.

■ All probabilities have values between 0 and 1 inclusive, therefore $0 \leq P(E) \leq 1$

■ $P(E) + P(E′) = 1$, so $P(E′) = 1 - P(E)$

■ Multiplication 'and' rule:
  ■ If two events A and B can occur without being affected by each another (for example, a die is thrown and it starts to rain), they are independent.
  ■ For two **independent events** A and B, $P(A \text{ and } B) = P(A) \times P(B)$

■ Addition 'or' rule:
  ■ If two events A and B cannot occur at the same time (for example, a card drawn from a pack cannot be an Ace and a Queen) they are called mutually exclusive.
  ■ For **mutually exclusive events** A and B, $P(A \text{ or } B) = P(A) + P(B)$

MORE COMPOUND PROBABILITY

EXAMPLE 1

SKILLS

ANALYSIS

A coin is tossed and a die is thrown. Find the probability that the result is a tail and a **multiple** of three.

Event T is tail appearing: $P(T) = \frac{1}{2}$

Event M is a multiple of three appearing: $P(M) = \frac{2}{6}$

Probability of T and M: $P(T \text{ and } M) = P(T) \times P(M) = \frac{1}{2} \times \frac{2}{6} = \frac{1}{6}$

EXAMPLE 2

SKILLS

ANALYSIS

A card is selected from a pack of 52 playing cards at **random**.

Find the probability that either an Ace or a Queen is selected.

Event A is an Ace is picked out: $P(A) = \frac{4}{52}$

Event Q is a Queen is picked out: $P(Q) = \frac{4}{52}$

Probability of A or Q: $P(A \text{ or } Q) = P(A) + P(Q)$

$$= \frac{4}{52} + \frac{4}{52}$$

$$= \frac{8}{52}$$

$$= \frac{2}{13}$$

4 Aces

4 Queens

Note: The rules of probability are quite simple, but more challenging questions require a deeper understanding and an efficient use of these laws.

MORE TREE DIAGRAMS

It is not always necessary to draw all the branches of a tree diagram as this can make things look more complicated than they are.

EXAMPLE 3

SKILLS

REAL

Female elephants are called 'cows'; male elephants are called 'bulls'.

The probability that a particular elephant gives birth to a cow is $\frac{3}{5}$

Use a tree diagram to find the probability that from three births she has produced two bulls.

Let event C be the birth of a cow, and event B be the birth of a bull. Consider a series of three births, being careful only to draw the branches that are required. There are three combinations that result in exactly two bulls from three births: BBC, BCB and CBB.

Draw a **tree diagram** to show all the possible outcomes. The events are independent.

P(2 bulls from 3 births) = P(BBC) + P(BCB) + P(CBB)

$$= \left(\frac{2}{5} \times \frac{2}{5} \times \frac{3}{5}\right) + \left(\frac{2}{5} \times \frac{3}{5} \times \frac{2}{5}\right) + \left(\frac{3}{5} \times \frac{2}{5} \times \frac{2}{5}\right)$$

$$= \frac{12}{125} + \frac{12}{125} + \frac{12}{125}$$

$$= \frac{36}{125}$$

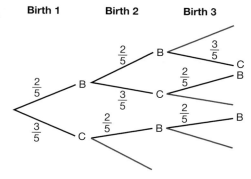

Alternatively, because multiplication is commutative (independent of the order), one combination could have been considered (for example, BBC) and multiplied by 3, as the fractions do not change.

MORE CONDITIONAL PROBABILITY

The probability of an event happening can depend on the occurrence and outcome of a previous event.

EXAMPLE 4

SKILLS

ANALYSIS

Onya-Birri is an albino koala in San Diego Zoo. Albinos are born without melanin in their skin, which results in them having a very pale complexion and white fur in the case of koalas.

The probability of a koala having the albino gene is $\frac{1}{75}$. The albino gene must come from *both* parents, but then only one out of four cubs (baby koalas) produced by a pair of albino-gene carriers is an albino.

Find the probability of two randomly paired koalas producing an albino cub like Onya-Birri.

Let C represent an albino-gene carrier and C′ represent a non-albino-gene carrier.

Let event A be that an albino koala cub is born. Then

$$P(A) = P(C) \times P(C) \times \frac{1}{4}$$

$$= \frac{1}{75} \times \frac{1}{75} \times \frac{1}{4}$$

$$= \frac{1}{22\,500}$$

Clearly, Onya-Birri is a rare creature!

Note: Only part of the tree diagram needed to be drawn to work out the probabilities.

EXERCISE 1

1 ▶ Two six-sided dice have the letters A, B, C, D, E and F on the faces, with one letter on each **face**.

    a If the two dice are thrown, calculate the probability that two vowels appear.
      (Vowels are a, e, i, o, u.)

    b What is the probability of throwing a vowel and a consonant?
      (Consonants are non-vowels.)

2 ▶ A pack of 20 cards is formed using the ace, ten, jack, queen and king of each of the four suits from an ordinary full pack of playing cards. This reduced pack is shuffled and then dealt one at a time *without replacement*. Calculate

    a P(the first card dealt is a king)
    b P(the second card dealt is a king)
    c P(at least one king is dealt in the first three cards).

3 ▶ A box contains two white balls and five red balls.
A ball is randomly selected and its colour is noted.
It is then put back in the box *together with two more balls of the same colour*.

    a If a second ball is now randomly taken from the box, calculate the probability that it is the same colour as the first ball.
    b What is the probability that the second ball is a different colour from the first ball?
    c Find the probability that the second ball is white.
    d Nick says, 'After the first ball is taken out, if we then add to the box any number of balls of the same colour as the first ball, the probability that the second ball is white will not change!'
    Is this statement true or false? Justify your answer.

4 ▶ The probability that a washing machine will break down in the first 5 years of use is 0.27.
The probability that a television will break down in the first 5 years of use is 0.17.
Mr Khan buys a washing machine and a television on the same day.
By using a tree diagram or otherwise, calculate the probability that, in the five years after that day

    a both the washing machine and the television will break down
    b at least one of them will break down.

Use tree diagrams in these questions where appropriate.

5 ▶ Teresa loves playing games. If the day is sunny, the probability that she plays tennis is 0.7. If it is not a sunny day, the probability that she plays tennis is 0.4. Also, the probability that Saturday will not be sunny is 0.15. Use a tree diagram to find the probabilities of these outcomes.

    a Teresa plays tennis on Saturday.
    b Teresa does not play tennis on Saturday.

6 ▶ The probability of the Bullet Train to Tokyo Station arriving late is 0.3.
The probability that it arrives on time is 0.6. What is the probability that the train

    a is early
    b is not late
    c is on time for two days in a row
    d is late on one Friday, but not late on the next two days
    e is late once over a period of three days in a row?

7 ▶ Two six-sided dice have the letters U, V, W, X, Y and Z on the faces, with one letter on each face.
If the two dice are thrown, calculate the probability that they show

a two vowels

b a vowel and a consonant.

(Remember: Vowels are a, e, i, o, u and a consonant is a non-vowel.)

8 ▶ Six small plastic tiles are placed in a bag and jumbled up. Each tile has a single letter on it, the letters being B, O and Y. There are two 'B's, two 'O's and two 'Y's.

a If one tile is taken out one by one *without replacing* any letters and the letters are placed in order, show that the probability of obtaining each of the following words is $\frac{1}{15}$

i BOY   **ii** YOB

b If two letter 'S's are put in the bag from the start, what is the probability that after four tiles are taken out, the word BOYS is *not* made?

9 ▶ A box contains five clearly different pairs of gloves. Kiril is in a hurry and randomly takes out two gloves *without replacing* any gloves. What is the probability that

a the gloves are both right-handed

b the gloves will be a right- and a left-handed glove

c they will be a matching pair?

10 ▶ A drawer contains four pairs of black socks, three pairs of blue, two pairs of green, one pair of yellow and one red sock. Two socks are randomly selected *without replacing* any socks. What is the probability that

a they are both black

b they are both the same colour

c one of the socks is a red one?

EXAMPLE 5

SKILLS

ANALYSIS

A vet has a 90% probability of finding a particular virus in a horse.

If the virus is found, the operation has an 80% success rate the first time that it is attempted.

If this operation is unsuccessful, it can be repeated, but with a success rate of only 40%.

Any operations after that have such a low chance of success that a vet will not attempt further operations.

Let event D be that the virus is found.

Let event S be that the operation succeeds.

Let event F be that the operation fails.

a What is the probability that a horse with the virus will be operated on successfully once?

b What is the probability that a horse with the virus will be operated successfully on by the second operation?

c What is the probability that an infected horse is cured?

a P(the first operation is successful)
= P(D and S)
= P(D) × P(S)
= 0.90 × 0.80 = 0.72

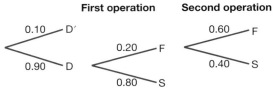

First operation **Second operation**

0.10 — D′

0.90 — D

0.20 — F

0.80 — S

0.60 — F

0.40 — S

b P(the second operation is successful)
= P(D and F and S)
= P(D) × P(F) × P(S)
= 0.90 × 0.20 × 0.40 = 0.072

c P(infected horse is cured)
= P(the first operation is successful or the second operation is successful)
= P(the first operation is successful) + P(the second operation is successful)
= 0.72 + 0.072 = 0.792

EXERCISE 1*

1 ▶ A virus is present in 1 in 250 of a group of sheep. To make testing for the virus possible, a quick test is used on each individual sheep. However, the test is not completely reliable. A sheep with the virus tests positive in 85% of cases and a healthy sheep tests positive in 5% of cases.

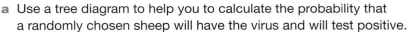

a Use a tree diagram to help you to calculate the probability that a randomly chosen sheep will have the virus and will test positive.
b What is the probability that a randomly chosen sheep will have the virus and will test negative?
c What is the probability that a sheep will test positive?

2 ▶ A new technique is 80% successful in finding cancer in patients with cancer. If cancer is found, a suitable operation has a 75% success rate the first time it is attempted. If this operation is unsuccessful, it can be repeated only twice more, with success rates of 50% and 25% respectively.

What is the probability of a patient with cancer
a being cured from the first operation
b being cured from the third operation
c being cured?

3 ▶ Two opera singers, Mario and Clarissa, both sing on the same night, in separate performances. The independent probabilities that two newspapers X and Y publish reviews of their performances are given in this table.

	MARIO'S PERFORMANCE	CLARISSA'S PERFORMANCE
PROBABILITY OF REVIEW IN NEWSPAPER X	$\frac{1}{2}$	$\frac{2}{3}$
PROBABILITY OF REVIEW IN NEWSPAPER Y	$\frac{1}{4}$	$\frac{2}{5}$

a If Mario buys both newspapers, find the probability that both papers review his performance.

b If Clarissa buys both newspapers, find the probability that only one paper reviews her performance.

c Mario buys one of the newspapers at random. What is the probability that it has reviewed *both* performances?

4 ▶ If a clock is 'slow', it shows a time that is earlier than the correct time. If a clock is 'fast', it shows a time that is later than the correct time. A school has an unreliable clock. The probabilities that the clock becomes 1 minute fast or slow in any 24-hour period are shown in this table.

	BECOMES 1 MINUTE FAST	NO CHANGE	BECOMES 1 MINUTE SLOW
PROBABILITY	$\frac{1}{2}$	$\frac{1}{3}$	$\frac{1}{6}$

If the clock is set to the correct time at noon on Sunday, find the probability of these events.

a The clock is correct at noon on Tuesday.

b The clock is *not* slow at noon on Wednesday.

5 ▶ A bag contains three bananas, four pears and five kiwi-fruits. One piece of fruit is randomly taken out from the bag and *eaten before the next one is taken*. Use a tree diagram to find the probability that

a the first two fruits taken out are pears

b the second fruit taken out is a pear

c the first three fruits taken out are all different

d neither of the first two fruits taken out is a banana, but the third fruit is a banana.

6 ▶ An office block has five floors (ground, 1st, 2nd, 3rd and 4th), that are all connected by a lift. When the lift goes up and reaches any floor (except the 4th), the probability that it continues to go up after it has stopped is $\frac{3}{4}$

When it goes down and reaches any floor (except the ground floor), the probability that it continues to go down after it has stopped is $\frac{1}{4}$ The lift stops at any floor that it passes.

The lift has gone down and is now at the first floor. Calculate the probability of these events.

a Its second stop is the third floor.

b Its third stop is the fourth floor.

c Its fourth stop is the first floor.

7 ▶ A box contains three $1 coins, five 50-cent coins and four 20-cent coins. Two are selected randomly from a bag *without replacing* any coins. What is the probability that

a the two coins are of equal value
b the two coins will total less than $1
c at least one of the coins will be a 50-cent coin?

8 ▶ Mr and Mrs Hilliam plan to have a family of four children. If babies of either sex are equally likely to be born and assuming that only single babies are born, what is the probability of the Hilliam children being

a four girls
b three girls
c at least one girl?

9 ▶ A card is randomly taken from an ordinary pack of cards and not replaced. This process is repeated again and again. Explain, with calculations, how these probabilities are found.

a P(first card is a heart) = $\frac{1}{4}$

b P(second card is a heart) = $\frac{1}{4}$

c P(third card is a heart) = $\frac{1}{4}$

10 ▶ A bag X contains ten coloured discs of which four are white and six are red. A bag Y contains eight coloured discs of which five are white and three are red. A disc is taken out at random from bag X and placed in bag Y. A second disc is now taken out at random from bag X and placed in bag Y.

a Using a tree diagram, show that the probability that the two discs taken out are both red is $\frac{1}{3}$

b Copy and complete this table.

OUTCOME		PROBABILITY
BAG X	**BAG Y**	
4R + 4W	5R + 5W	$\frac{1}{3}$
5R + 3W		
	3R + 7W	

c A disc is now taken out at random from the ten discs in bag Y and placed in bag X, so that there are now nine discs in each bag. Find the probability that there are
i four red discs in bag X
ii seven red discs in bag X
iii six red discs in bag X
iv five red discs in bag X.

EXAMPLE 6

SKILLS

ANALYSIS

Show that in a room of only 23 people, the probability of two people having the same birthday is just over $\frac{1}{2}$

Let E be the event that there are two people in the room with the same birthday.

$P(E) + P(E') = 1$

$P(E) = 1 - P(E')$

If a person is chosen, the probability that the next person chosen does not have the same birthday as the first person is $\frac{364}{365}$

If a third person is chosen, the probability of this person not having the same birthday as the previous two people is $\frac{363}{365}$ and so on.

$P(E) = 1 - 1 \times \frac{364}{365} \times \frac{363}{365} \times \frac{362}{365} \times \ldots \times \frac{343}{365} \approx 0.51$ (just over $\frac{1}{2}$)

ACTIVITY 1

SKILLS

INTERPRETATION

The randomised response questionnaire technique was started in America to find out the proportion of people who have participated in antisocial, embarrassing or illegal activities. Most people would not answer such questions truthfully, but this method involves responding truthfully depending on the scores on two dice. There is no reason for anyone not to tell the truth, because each person can keep their dice score hidden from the other people.

Two dice are used for each person: one black, one white. The black die is rolled first. If it is even, the survey question must be answered truthfully. If it is odd, the white die is rolled and the respondent must answer the question, 'Is the white die number even?' truthfully.

Let p be the probability that the 'Awkward Question' is answered as Yes.

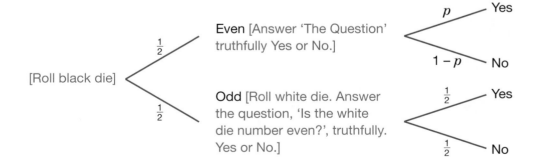

Let the event Y be that the response to both questions is Yes.

$P(Y) = P(\text{even and Yes}) + P(\text{odd and Yes}) = \frac{1}{2} \times p + \frac{1}{2} \times \frac{1}{2} = \frac{1}{4}(2p + 1)$

Sixty students were questioned using this technique for the question:
'Have you ever copied homework assignments from the Internet?'

Of this group of 60 students, 36 responded with a yes. Show that the probability of this **sample** of 60 students copying their homework from the Internet is $\frac{7}{10}$

Select a suitable question (not an embarrassing one!) which people might not want to openly answer truthfully and then carry out your own randomised response survey with as large a sample as possible.

ACTIVITY 2

INTERPRETATION

Omar and Yosef decide to play a simple game with a single fair die. The winner is the first person to roll a six. They take turns to roll the die.

Show that is it an advantage to Omar if he goes first in this game.

Let O be the event that Omar wins the game.

Let O_1 be event that Omar wins on his first throw, O_2 be event that Omar wins on his second throw etc.

There are an infinite (unlimited) number of ways that Omar can win the game.

$$P(O) = P(O_1) + P(O_2) + P(O_3) + \dots \text{ This forms an infinite sequence.}$$

$$= \frac{1}{6} + \frac{5}{6} \times \frac{5}{6} \times \frac{1}{6} + \frac{5}{6} \times \frac{5}{6} \times \frac{5}{6} \times \frac{5}{6} \times \frac{1}{6} + \dots$$

$$= \frac{1}{6}\left[1 + \left(\frac{5}{6}\right)^2 + \left(\frac{5}{6}\right)^4 + \dots\right] = \dots$$

Show that this sequence is greater than a half. (It is in fact $\frac{6}{11}$)

It is clearly an advantage to go first as this probability is larger than $\frac{1}{2}$

EXERCISE 2

REVISION

1 ▶ Fran has a box containing a very large number of postage stamps. Two-thirds are Cuban; the rest are from Brazil. She randomly picks out two without replacing any stamps.

    a Calculate an estimate of the probability that she has picked out
      i two Cuban stamps
      ii a Cuban and a Brazilian stamp.
    b A third stamp is randomly picked out from the box. Calculate the probability that Fran has at least two Brazilian stamps from her three selections.

2 ▶ Gina and Iona are taking part in a Ski Instructor Test. They are only allowed one re-test if they fail the first. Their probabilities of passing are shown below.

	GINA	IONA
P(PASS ON 1ST TEST)	0.7	0.4
P(PASS ON 2ND TEST)	0.9	0.6

These probabilities are independent of each other. Calculate the probability that
a Iona becomes a ski instructor at the second attempt
b Gina passes at her first attempt whilst Iona fails her first test
c only one of the girls becomes a ski instructor, assuming that a re-test is taken if the first test is failed.

3 ▶ Melissa is growing apple and pear trees for her orchard. She plants seeds in her greenhouse, but forgets to label the seedling pots. She knows the quantities for each variety of apple and pear seeds that she planted. The quantities are in the table below.

	EATING VARIETY	COOKING VARIETY
APPLE	24	76
PEAR	62	38

She picks out an apple seed pot at random.
a Calculate the probability that it is an eating apple variety.
b Calculate the probability that it is a cooking apple variety.
c She picks out an eating variety seed pot at random. Calculate the probability that it is
 i an apple
 ii a pear.
d She picks out three seed pots at random. Find the probability that she has at least one eating apple variety.

4 ▶ The probability that a seagull lays a certain number of eggs is shown.

NUMBER OF EGGS	PROBABILITY
0	0.1
1	0.2
2	0.3
3	0.2
4	0.1
5 or more	0.1

What is the probability that
a a seagull lays more than four eggs
b a seagull lays no eggs
c two seagulls lay a total of at least four eggs?

5 ▶ A football player calculates from his previous season's results the probabilities that he will score goals in a game.

NUMBER OF GOALS	PROBABILITY
0	0.4
1	0.3
2	0.2
3 or more	0.1

Assume that these probabilities apply to the current season.
a Find the probability that in a game he scores at most two goals.
b Find the probability that in a game he scores at least two goals.
c What is the probability that in two games he scores at least three goals?

EXERCISE 2*

REVISION

1 ▶ Christmas lights are produced in a large quantity and as a result one-fifth are faulty. If individual lights are taken out one by one from this large quantity, find the probability that when the first three lights are taken out

a there is one faulty light
b there are two faulty lights
c there is at least one faulty light.

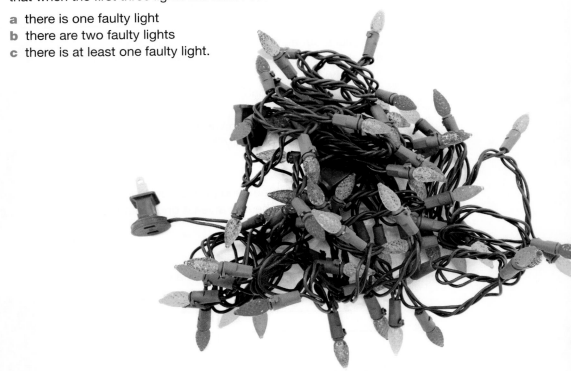

2 ▶ Batteries are made on a factory production line such that 15% are faulty. If batteries are removed for quality control, find the probability from the first three batteries of

a no defective
b two defectives
c at least two defectives.

3 ▶ A vet has three independent tests A, B and C to find a virus in a cow. Tests are carried out in the order A, B, C.

| PROBABILITIES OF A POSITIVE TEST DEPENDING ON THE PRESENCE OF THE VIRUS | | |
TEST A	TEST B	TEST C
VIRUS PRESENT $\frac{2}{3}$	$\frac{4}{5}$	$\frac{5}{6}$
VIRUS *NOT* PRESENT $\frac{1}{5}$	$\frac{1}{6}$	$\frac{1}{7}$

a Find the probability that two tests will be positive and one test will be negative on a cow with the virus.

b To conclude that a cow has the virus, at least two tests must be positive. Find the probability that

i after three tests on a cow with the virus, the virus is not declared to be present

ii after three tests on a cow without the virus, the virus is declared to be present.

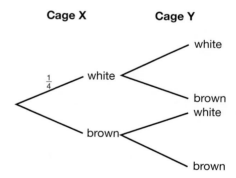

4 ▶ Mimi has not revised before a multiple choice test. She has no idea of the correct answers. She guesses the answers and has a probability of p of obtaining the right option. From the first three questions the probability of her getting only one correct is $\frac{3}{8}$
Find p.

5 ▶ Cage X contains four hamsters, one white and three brown. Meanwhile cage Y has three hamsters, two white and one brown. One hamster is taken from cage X and placed in cage Y. Hamsters are then taken one by one from cage Y.

a Copy and complete this tree diagram.

b Use your tree diagram to find the probability that the first hamster taken from cage Y is
i white
ii brown.

c What is the probability that, of the first two hamsters taken from cage Y, both are white?

EXAM PRACTICE: HANDLING DATA 7

1 A pack of yoghurts contains the following flavours: 3 strawberry, 5 black cherry and 4 pineapple.

Penny eats two yoghurts from this pack, picking them at random.

Work out the probability that
a both yoghurts are the same flavour [3]
b at least one is strawberry [3]
c only one of them is black cherry. [3]

2 Bill and Jo play some games of table tennis.

The probability that Bill wins the first game is 0.7

When Bill wins a game, the probability that he wins the next game is 0.8

When Jo wins a game, the probability that she wins the next game is 0.5

The first person to win two games wins the match.

a Complete the tree diagram. [3]

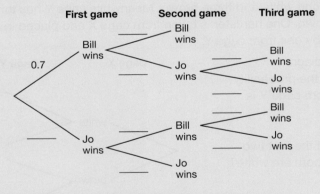

b Calculate the probability that Bill wins the match. [4]

3 Peter wants to pass his driving test.

The probability that he passes at his first attempt is 0.7

When Peter passes his driving test, he does not take it again.

If he fails, the probability that he passes at the next attempt is 0.8

a Complete the tree diagram for Peter's first two attempts. [2]

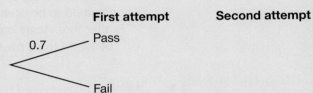

b Calculate the probability that Peter needs exactly two attempts to pass his driving test. [3]

c Calculate the probability that Peter passes his driving test at his third or fourth attempt. [4]

[Total 25 marks]

CHAPTER SUMMARY: HANDLING DATA 7

PROBABILITY

The probability of an event not happening is P(E′).

$$P(E) + P(E') = 1$$

INDEPENDENT EVENTS

If A and B are independent events

$$P(A \text{ and } B) = P(A) \times P(B)$$

MUTUALLY EXCLUSIVE EVENTS

If A and B are mutually exclusive events

$$P(A \text{ or } B) = P(A) + P(B)$$

COMBINED EVENTS

To calculate the probability of an event occurring, it is necessary to consider all the ways in which that event can happen.

A tree diagram shows two or more events and their probabilities. It is not always necessary to draw all the branches of a tree diagram as this can make things look more complicated than they are.

A bag contains two red balls, three green balls and four white balls. Two balls are taken out of the bag without replacing any balls.

1st draw **2nd draw**

	R	RR $\frac{2}{72}$
R $\frac{2}{9}$	$\frac{1}{8}$ R, $\frac{3}{8}$ G, $\frac{4}{8}$ W	RG $\frac{6}{72}$
		RW $\frac{8}{72}$
G $\frac{3}{9}$	$\frac{2}{8}$ R, $\frac{2}{8}$ G, $\frac{4}{8}$ W	GR $\frac{6}{72}$
		GG $\frac{6}{72}$
		GW $\frac{12}{72}$
W $\frac{4}{9}$	$\frac{2}{8}$ R, $\frac{3}{8}$ G, $\frac{3}{8}$ W	WR $\frac{8}{72}$
		WG $\frac{12}{72}$
		WW $\frac{12}{72}$

$$P(\text{both the same colour}) = P(RR) + P(GG) + P(WW) = \frac{2}{72} + \frac{6}{72} + \frac{12}{72} = \frac{20}{72} = \frac{5}{18}$$

CONDITIONAL PROBABILITY

The probability of an event based on the occurrence of a previous event is called a conditional probability. The sample space has been changed by a previous event.

If a red disc is taken out from a bag containing 3 red and 8 blue discs, the probability that the next disc taken out is red

$$= \frac{\text{number of red discs after 1 is removed}}{\text{total number of discs left in the bag after 1 is removed}} = \frac{2}{10} = \frac{1}{5}$$

FACT FINDER: GOTTHARD BASE TUNNEL

The Gotthard Base Tunnel is an amazing success in engineering. Opened on 1st June **2016**, it is the longest railway tunnel in the world. Nearly **100 miles** of tunnels were drilled into the Swiss Alpine mountains to create a route between Erstfeld and Bodio. The length of each of the two parallel tunnels is **35 miles**, with the remaining length used as safety and service tunnels. Each tunnel covers a length of **21 Golden Gate Bridges**. The greatest diameter of each of the two single-track tubes is **9.58 m**.

It is the product of **17 years** of work, including cutting through solid rock at depths of up to **2250 m**, making it the world's deepest tunnel. By comparison, Japan's Seikan Tunnel – the second deepest and longest – is only **237 m** deep. Special drills were used to bore through more than **31 million tonnes** of rock. **5 Great Giza pyramids'** weight of rock had to be excavated. **4 million m³** of concrete was used in the construction, which is about the amount needed for **84 Empire State Buildings** in New York. And almost **2000 miles** – the distance from Madrid to Moscow – of copper cable was used.

Its main purpose is to transport people and freight directly across the Alps, to reduce traffic accidents and environmental damage from large trucks on Alpine road systems. The new rail track is virtually flat and straight, so it will enable passenger trains to travel at up to **155 miles per hour**, reducing the journey time between Zurich, Switzerland, and Milan, Italy, from **four hours to two-and-a-half hours**.

2600 people were employed on the project which ended when a golden sleeper was placed at the end of the rail track. The total cost of the project was **$12.5 billion**.

Source: Based on data from http://www.gottardo2016.ch/en

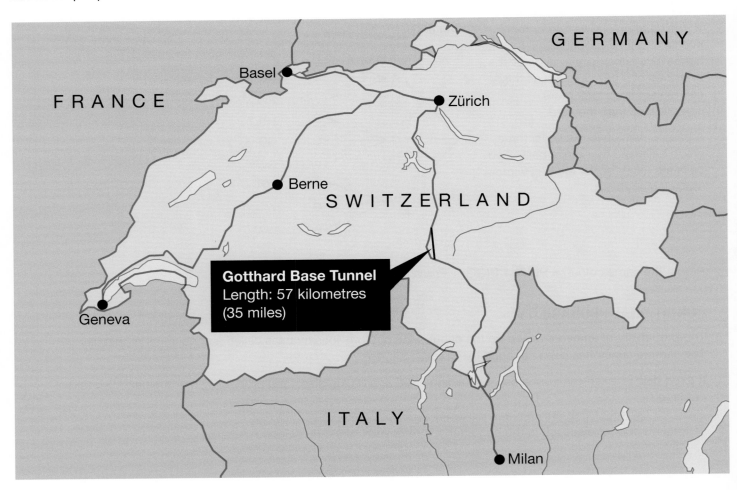

Gotthard Base Tunnel
Length: 57 kilometres
(35 miles)

Unless otherwise stated, give all answers correct to 3 significant figures.

EXERCISE 1

1 ▶ What year did construction work on the Gotthard Base Tunnel start?

2 ▶ Estimate the cost per day of the project in dollars in standard form.

3 ▶ Estimate the length of the Golden Gate Bridge in metres in standard form.
(1 mile ≈ 1609.34 m)

4 ▶ Estimate the mass of rock of the Great Giza pyramid in kg in standard form.
(1 tonne = 1000 kg)

5 ▶ The greatest depth of the Gotthard Base Tunnel is k × the greatest depth of Japan's Seikan Tunnel.
Find k as an exact fraction in its lowest possible terms.

EXERCISE 1*

1 ▶ **a** Calculate the total volume of the two tunnels in m³ in standard form.

     b Compare this to the volume of a 10 m × 10 m × 2.5 m classroom.
(Volume of a cylinder, $v = \pi r^2 h$ and 1 mile ≈ 1609.34 m)

2 ▶ If the speed of the air-flow ventilation system for each tunnel is 7 m/s, calculate the rate of air flow in m³/s in standard form.

3 ▶ A train leaves Erstfeld at 08:00. For the first half of the journey, the train travels at its maximum speed. The final half is completed at half its maximum speed. Find the train's arrival time at Bodio.

4 ▶ The volume of the Great Giza pyramid is 2 583 200 m³. Calculate the density of the pyramid in kg/m³ in standard form.

5 ▶ If p m³ is the volume of concrete used per metre of construction of ALL the tunnels, and q m the length of copper wire used per metre, find p^q in standard form.

FACT FINDER: MOUNT VESUVIUS

In the world's history, there have been many volcanic eruptions and earthquakes that have been far more powerful than Italy's Mount Vesuvius. Their impact has also been much greater.

In **1815**, **92 000** people died because of the Tambora volcano's eruption in Indonesia, and in **1976**, **242 000** people lost their lives when an earthquake shook Tangshan in eastern China.

However, the eruption of Vesuvius in **AD 79** has become famous worldwide because of the extraordinary documentation of the event and its impact on the towns of Pompeii and Herculaneum. The instability of the area is caused by the African tectonic plate moving North at **3 cm per year**. The eruption lasted three days, during which time the sky was dark with dust and ash was thrown **17 km** into the air, covering nearby villages and

towns in just a few hours. Pompeii (population **20 000**) was covered with **7 m** depth of hot ash, while Herculaneum (population **5000**) was hit by a rapid mud-flow **25 m** deep, which captured people and animals as they tried to escape. During the eruption, **4 km³** of magma (hot fluid contained in a volcano) and **10⁶ m³** of lapilli (fine ash) were thrown out of the volcano in **24 hours**, causing **3600** lives to be lost.

Many hand-crafted objects were perfectly preserved in Herculaneum, because of the protection of the deep mud-flow. These are kept in the Naples Museum, where they have inspired poets, philosophers and scientists over the years.

The present-day peak of **1281 m** is much lower than the volcano in the year **AD 79** when it reached **3000 m**. Since **1631** there have been major eruptions from Vesuvius in **1760**, **1794**, **1858**, **1861**, **1872**, **1906**, **1929**, **1933** and **1944**.

The top of the crater (the mouth of the volcano) today is an enormous circular opening **600 m** in diameter and **200 m** deep. Walking around the peak takes about **one hour** and you can still see vapour escaping from fumaroles (cracks), some of which have temperatures of **500 °C**, showing that Vesuvius is still active. If the volcano erupted today at the same strength as in **AD 79**, about **600 000** people would have to be evacuated (taken to safety).

Vesuvius is modelled as a right-circular pyramid whose top has been blown off. Its cross-section is as shown.

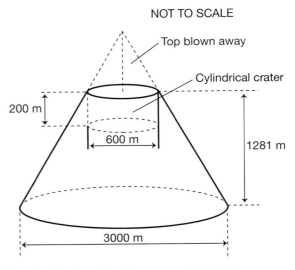

NOT TO SCALE

Source: Based on data from http://www.aroundrometours. com/30interestingfactsaboutpompeiiandmountvesuviusart39uid1.htm and http://geology.com/volcanoes/vesuvius/

Unless otherwise stated, give all answers correct to 3 significant figures.

EXERCISE 1

1 ▶ How many years were there between the volcanic destruction of Pompeii and the Tambora eruption?

2 ▶ What percentage of the population of Pompeii and Herculaneum died because of the volcanic activity?

3 ▶ How many years will the African tectonic plate take to move one mile North?
(1 mile ≈ 1600 m)

4 ▶ What is the percentage decrease in the height of Vesuvius from its height in the year AD 79?

5 ▶ Given that C degrees Celsius is related to F degrees Fahrenheit by the formula $C = \dfrac{5(F - 32)}{9}$, find the temperature around some of the fumaroles (cracks) in the crater of Vesuvius today in degrees Fahrenheit.

6 ▶ Calculate the speed that a tourist travels, in m/s, walking around the top of the crater of Vesuvius.

EXERCISE 1*

1 ▶ Given that the approximate areas of Herculaneum and Pompeii are half a square mile and four square miles respectively, calculate an estimate in m³ in standard form of

    a the volume of hot mud-flow that filled Herculaneum

    b the volume of hot ash that fell on Pompeii.
    (1 mile ≈ 1600 m)

2 ▶ **a** Since 1631, what is the average number of years between major volcanic activities in Vesuvius?

    b Using these data, when would you reasonably expect the next eruption of Vesuvius to be?

3 ▶ Calculate the rate at which magma was thrown out from Vesuvius in AD 79 in m³/s in standard form.

4 ▶ **a** Given that the average density of magma from Vesuvius in AD 79 was approximately 4500 kg/m³, calculate the mass of magma that came out in tonnes per second.

    b Compare your answer in **a** to an average family car of mass 1 tonne.
    (1 tonne = 1000 kg)

5 ▶ Calculate an estimate of the volume of Mount Vesuvius in m³ in standard form.
(Volume of cone = $\dfrac{1}{3}\pi r^2 h$)

6 ▶ Calculate an estimate of the area of the slopes of Mount Vesuvius in km².
(Curved surface area of cone = $\pi r l$)

FACT FINDER: THE SOLAR SYSTEM

Our Solar System consists mainly of eight planets, some with moons, orbiting the Sun, together with rocky objects called asteroids. Here are some facts about the Solar System.

BODY	DIAMETER (KM)	DISTANCE FROM THE SUN (KM)
Sun	1 390 000	
Earth	12 800	1.5×10^8
Mars	6790	2.28×10^8
Jupiter	143 000	7.78×10^8

Our Solar System, with the Sun at its centre, is only one small part of our galaxy, which is called the Milky Way. A galaxy is a system of stars, and the Milky Way, which is shaped like a thin disc, has over **100 000 million** of them. The largest known star has a diameter of $\mathbf{2.5 \times 10^8}$ **km**.

Because the distances between stars are so large, they are measured in light years, which is the distance that light travels in one year. Light travels at **300 000 km/s**; in contrast, a fighter jet travels at around **2000 km/h**. The Milky Way is **120 000 light years in diameter**, and our Sun is **30 000 light years** from the centre. The nearest star to our Sun is **4.2 light years** away, while the Andromeda Galaxy, the most distant object in our skies that is (just) visible without a telescope is $\mathbf{2.2 \times 10^6}$ **light years** away.

Many countries have worked together so as to explore the Solar System for the benefit of people on Earth. The International Space Station is a large spacecraft on which astronauts live and conduct experiments to find out what happens to people when they live in space. The space station's orbit is about **250 miles** above the

Major Tim Peake (UK) on the International Space Station

surface of the Earth. In **2015–2016**, Major Tim Peake (the first British European Space Agency astronaut) spent **185 days** aboard the space station, which travels at an average speed of **17 100 mph**.

Source: Based on data from http://spacefacts.com/solarsystem/

Unless otherwise stated, give all answers correct to 3 significant figures.

EXERCISE 1

1 ▶ Imagine making a scale model of the Solar System, with the Sun represented by an orange 8 cm in diameter. Show that the Earth could be represented by a ball of clay 0.7 mm in diameter placed 8.6 m away from the orange.

2 ▶ For the model, work out the diameter and distance from the orange of the other planets given in the table.

3 ▶ What is the diameter of the largest star in our model?

4 ▶ Work out how long it would take a fighter jet to fly from the Earth to the Sun. (Assume that this is possible!)

5 ▶ How long does it take light to travel from the Sun to the Earth?

EXERCISE 1*

1 ▶ How many kilometres are there in a light year?

2 ▶ The Sun is represented by an orange 8 cm in diameter in a scale model of the Solar System. How far away from the Earth is the nearest star to the Sun in this model in km?

3 ▶ How long would it take a fighter jet to fly to the nearest star?

4 ▶ How far away is the Andromeda Galaxy in km?

5 ▶ How many revolutions around the Earth did Major Tim Peake travel over his time in the International Space Station?
Express your answer in standard form. (1 mile ≈ 1600 m)

FACT FINDER: THE WORLD'S POPULATION

Our planet is a rocky sphere, with a diameter of **12 800 km** and a mass of **6.6 × 10²¹** tonnes, spinning through space on its path round the Sun. Viewed from space the Earth is a mainly white and blue planet, covered with clouds, water and some land. It is a watery planet, as **70%** of the surface is covered with water.

The dominant species is humans, primates with an average mass of **50 kg**. Humans are only able or willing to live on **12%** of the available land area, as the rest is considered either too cold, too dry or too steep.

Until the modern era, the population of the world grew slowly. It is estimated that the population grew at about **0.05%** per year, reaching about **300 million** in AD 1. During the following **16 centuries** the growth rate varied greatly, partly because of diseases such as the Black Death. The world's population reached over one billion in **1804**. The **second billion** took another **123 years**, but the **third billion** only took another **33 years**.

The world population growth rate peaked at about **2% per year** in **1960**. The population of humans grew to about **6 billion** by the year **2000**. The population at the start of 2016 reached a record high of **7.3 billion**, with a growth rate of **1.13% per year**.

The recent global population explosion is not only the consequence of increased birth rates but also the result of a decrease in death rate (now down to two people every second) because of significant advances in public health and medicine.

However, the United Nations estimates that the world population will stabilise at **11.2 billion** by the year **2100**, assuming that effective family planning will result in a universally low birth rate. Global average birth rates are now **2.3 per woman**, down from **5** in **1950**. Education plays a key role, as globally **26.1%** of people were **under 14 years old** in **2016**.

Source: Based on data from https://www.compassion.com/poverty/population.htm and http://www.huffingtonpost.com/normschriever/25factsfortheunitedn_b_10694424.html

Unless otherwise stated, give all answers correct to 3 significant figures.

EXERCISE 1

1 ▶ What year did the world population first reach

    a two billion

    b three billion?

2 ▶ **a** What was the increase in world population from the year 2016 to 2017?

    b What was the average increase in the population per second during 2016?

    c How many people were under 14 in 2016?

3 ▶ What percentage of the total surface area of the Earth is suitable for humans to live on?

4 ▶ The surface area, S, of a sphere is given by $S = 4\pi r^2$
Work out the area available for humans to live on, giving your answer in km^2.

5 ▶ Work out the average number of humans per km^2 of habitable land (population density) in the year 2016.

EXERCISE 1*

1 ▶ **a** What factor do you multiply by to increase a quantity by 1.13%?

    b Work out the world population at the start of
      i 2017   **ii** 2018   **iii** 2020

2 ▶ Assuming that the growth rate of 1.13% per year at the start of 2016 remains constant, show that the world population, P, n years after 2016, is given by
$P = 7.3 \times 10^9 \times 1.0113^n$

3 ▶ **a** Given that $P = 7.3 \times 10^9 \times 1.0113^n$, copy and complete the table below for n years after 2016. Then draw a graph of P (billions) against n for the years 2016 to 2100.

YEAR	2016	2020	2040		2080	2100
n	0	4		44		84
P ($\times 10^9$)	7.3	7.6				

    b Use your graph to estimate
      i when the population will be double that in 2016
      ii the difference in population between this model and the United Nations estimate for the year 2100.

4 ▶ Work out the population density (people/km^2) in areas of habitable land at the start of the year 2100 using the figures from the model in question 3.

5 ▶ Using the model in question 3, work out the mass of humans on the planet in the year 4733. Compare this to the mass of the Earth.

FACT FINDER: THE TOUR DE FRANCE 2015

The Tour de France is the world's most famous annual multiple-stage bicycle race, taking place mostly in France. The race has been held since **1903** except when it was stopped for both World Wars.

In **2015** there were **22 teams** each with **9 rider**s. **1400** bikes, **3000** wheels and **4850** tyres were used by the teams in total, with the average price of each bike being **$15000**. The total tour distance was **3664 km** with each rider having an average of **486000 pedal strokes**. The average speed of each cyclist over the whole race was **40.69 km/hr**. It is an extraordinary feat of human endurance and each rider consumed a total of **123 900 calories** over for the whole race.

The Col du Giaber was the highest stage of the Tour at **2645 m** above sea level. The climb up to Mont Ventoux is one of the toughest in professional cycling, but this was not used in **2015**. The wind, unrelenting gradient and exposure to intense sunlight make Mont Ventoux a brutal finish.

The horizontal distance is **21.765 km** and the vertical ascent is **1617 m**, whilst the actual distance ridden along the ascent road is **21.825 km**. The fastest ever ascent was in **2004** by Iban Mayo (Spain) who clocked a time of **55 minutes 51 seconds**.

The main prize is the yellow jersey, awarded to the rider who has completed all the stages in the lowest overall time. In **2015**, the winner, Chris Froome (GB), had an aggregate time of **84 hours 46 minutes 14 seconds**. Over the history of the race the winners of the yellow jersey have come from a variety of countries: France (**36**), Belgium (**18**), Spain (**12**), Italy (**10**), Luxembourg (**5**), USA (**3**) and GB (**2**).

Chris Froome also won **$610000**, with the total prize money for the **2015** tour being **$4.3 million**. The Tour's first race in **1903** had a prize of **$34000** for the overall winner.

The Tour is the third most watched sports event in broadcasting in the world after the Football World Cup and the Olympics. **3.9 billion people** watched this event in **2015** on **121 TV stations**, whilst there were **12 million** road-side spectators.

Source: Based on data from http://www.letour.com/us/ and http://www.active.com/cycling/articles/23funfactsyoudidntknowaboutthetourdefrance

Unless otherwise stated, give all answers correct to 3 significant figures.

EXERCISE 1

1 ▶ How many riders started the Tour de France in 2015?

2 ▶ What was Chris Froome's average speed in km/hr over the whole race?

3 ▶ What was the average time taken by the cyclists to complete the Tour in hours minutes and seconds?

4 ▶ Given that the population of the world in 2015 was 7.3 billion, calculate the percentage of the population who viewed the Tour

    a on TV

    b as road-side spectators.

5 ▶ Given that a beef burger contains 230 calories, find the number of beef burgers equivalent to the total calories consumed by all the riders on the Tour.

EXERCISE 1*

1 ▶ What is the average gradient of the Mont Ventoux ascent?

2 ▶ Find the average speed of the record ascent of Mont Ventoux by Iban Mayo in

    a m/s along the road

    b m/s vertically.

3 ▶ Find the percentage increase in winner's prize money from the Tour's first race to 2015.

4 ▶ Calculate the average money earned by each rider in pedal strokes per dollar.

5 ▶ Given that the diameter of a racing bike's wheels is 622 mm, calculate the mean revolutions per minute of the rider's wheels on the Tour.

Froome and Mont Ventoux
(Rob Ijbema)

CHALLENGES

1 ▶ How many squares are there on a chessboard?

2 ▶ Three cards are chosen randomly from a pack of 52 playing cards. What is the probability that the third card is a heart?

3 ▶ What is the angle between the hands of a clock at 14:18?

4 ▶ If a hot-air balloon flies around the Equator at a constant height of x m, how much further than the circumference of the Earth will it travel?

5 ▶ A, B, C and D are four different digits. If ABCD × 9 = DCBA, find ABCD.

Q5 HINT
0 is a digit.

6 ▶ Find the exact value of $666\,666\,666^2 - 333\,333\,333^2$

7 ▶ If $x - y = 3$ and $x^2 - y^2 = 63$, what is $x^2 + y^2$?

8 ▶ A teenager wrote the digits of his fathers' age and then wrote the digits of his own age after that of his father. From this four-digit number he subtracted the difference of their ages and the answer was 4289.

Find the sum of the father's and the son's ages.

9 ▶ A rock climber climbed up a rock face at a uniform rate. At 9 o'clock he was one-sixth of the way up, and at 11 o'clock he was three-quarters of the way up.

What fraction of the total rock face had he completed at 10 o'clock?

10 ▶ The two circles are concentric. The line shown is 20 cm long and is a tangent to the inner circle. Find the shaded area.

11 ▶ The diagram shows a sailing boat with two parallel masts of height 12 m and 8 m.
Use similar triangles to find the height h m above the deck where the two supporting wires cross.

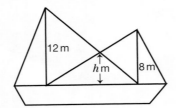

12 ▶ The diagram shows a cube of side 4 cm. M and N are the mid-points of the sides AB and AC respectively. Using surds, find the exact area of the triangle DMN.

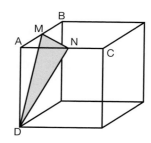

13 ▶ What is wrong with the following proof?
Let $a = b + c$ where a, b, c are all greater than zero. Thus $a > b$
$$a = b + c$$
$$\Rightarrow a(a - b) = (b + c)(a - b)$$
$$\Rightarrow a^2 - ab = ab + ac - b^2 - bc$$
$$\Rightarrow a^2 - ab - ac = ab - b^2 - bc$$
$$\Rightarrow a(a - b - c) = b(a - b - c)$$
$$\Rightarrow a = b$$
Substituting $a = 3, b = 2, c = 1$ gives $3 = 2$

14 ▶ A two-digit number is increased by 36 when the digits are reversed. The sum of the digits is 10. Find the original number.

15 ▶ A sheet of paper 8 cm by 6 cm is folded so that one corner is over a diagonally opposite corner. What is the length of the fold?

16 ▶ The diagram shows 17 matches forming squares in a 3 by 2 rectangle.

Remove five matches to leave three of the original squares.

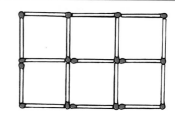

17 ▶ The diagram shows a square PQRS inside a right-angled isosceles triangle ABC. PQ = QR = 2 cm.

Find the area of the triangle ABC.

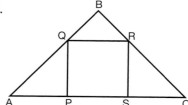

18 ▶ Gertie the goat is tied outside a 6 m by 10 m rectangular fenced area by a 6 m long rope which is tied to the middle of the shorter side as shown in the diagram.

a What area can Gertie move around in?

Q19a HINT
The rope cannot go over the fenced area, it must go round the corners.

b Gertie thinks the grass is greener inside the fenced area so she jumps over the fence. What area inside the fenced area can she move around in now?

19 ▶ I eat 'Choc-o-bars' at the rate of 40 per month. The 'Amazing New Choc-o-bar' is introduced and contains 20% less chocolate than the old one. How many extra bars must I purchase each month to keep my chocolate consumption the same?

20 ▶ After throwing a dart n times at a dartboard, my percentage of hits is p%. With my next throw I make a hit and my percentage of hits is now $(p + 1)$%. Prove that $n + p = 99$.

21 ▶ Two concentric circles have radii of 10 cm and 26 cm. What is the length of the longest straight line that can be drawn between them?

22 ▶ a Assuming a water molecule is a cube of side 3×10^{-10} m, find the volume of a water molecule in m³.

     b The volume of a cup of tea is 200 cm³. How many molecules of water are there in the cup of tea?

     c If all these molecules are placed end to end in a straight line, how many times would the line go around the Earth? Assume that the circumference of the Earth is 40 000 km.

23 ▶ A tetromino is a two-dimensional shape consisting of four squares joined together along common edges. One tetromino is shown in the diagram.

Two tetrominoes are the same if one is simply a rotation of the other in its plane.

     a Draw diagrams of the six other possible tetrominoes.

     b Is it possible to fit these seven tetrominoes into a 7 × 4 rectangle so that none of them overlap? Give reasons. (You might find it helpful to shade the tetrominoes to look like the squares on a chessboard.)

24 ▶ Recently the Sun, the Earth and the planet Mars were all in a straight line with the Earth between the Sun and Mars as shown in the diagram.

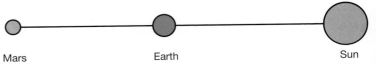

Mars   Earth   Sun

How many days will it take before the Sun, Earth and Mars are in a straight line again, with the Earth between the Sun and Mars? Assume that the orbits are coplanar and the Earth travels at a constant speed in a circular orbit around the Sun in 365 days and that Mars also travels at constant speed in a circular orbit, taking 687 days to complete one orbit.

25 ▶ A pond contains 300 tadpoles. f are frog tadpoles and the rest are toad tadpoles. If 100 more frog tadpoles are added to the pond, the probability of catching a frog tadpole is doubled. Find f.

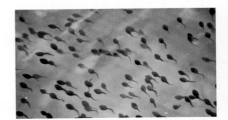

26 ▶ ABCD is a square of side 6 cm.
AFB and DEC are equilateral triangles.

Calculate the distance EF exactly.

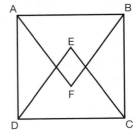

27 ▶ Show that the exact length of the diagonal of a regular pentagon ABCDE of unit

side length is given by $\dfrac{1 + \sqrt{5}}{2}$

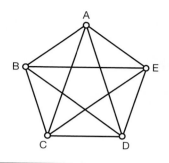

---

28 ▶ AB is a diameter of a circle of radius 1 cm.

Two circular arcs of equal radius are drawn with centres A and B meeting on the circumference of the circle as shown.

Calculate the shaded area exactly.

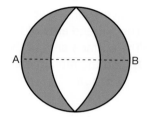

---

29 ▶ A regular hexagon has a perimeter of 60 m measured to the nearest 10 m.

Find to 3 significant figures

a the upper-bound area

b the lower-bound area.

---

30 ▶ Show that the point of intersection of the line $y = 1 - x$ and the curve $y^4 = x^4 - 5$ is $\sqrt{\dfrac{5}{2}}$ from the origin O.

Q30 HINT

$x^2 - y^2 = (x + y)(x - y)$

EXAMINATION PRACTICE PAPERS

PAPER 1

1 The surface area of the Earth is $5.1 \times 10^8 \, km^2$.
The surface area of the Pacific Ocean is $1.8 \times 10^8 \, km^2$.

a Express the area of the Pacific Ocean as a percentage of the area of the surface area of the Earth
Give your answer correct to 3 significant figures. **[2]**
The surface area of the Arctic Ocean is $1.4 \times 10^7 \, km^2$.
The surface area of the Southern Ocean is $3.5 \times 10^7 \, km^2$.

b Find the ratio of the surface area of the Arctic Ocean to the surface area of the Southern Ocean in the
form $1 : n$. **[2]**

2 a Expand $5(2y - 3)$ **[1]**   **d** Factorise $2x^2 - 22x$ **[2]**
  b Expand and simplify $(2x - 1)(x + 5)$ **[2]**   **e** Solve $3(3x - 2) = 21$ **[2]**
  c Factorise $4y + 24yz$ **[2]**

3 Two points, A and B, are plotted on a centimetre graph.
A and B have co-ordinates (2, 1) and (8, 5) respectively. Calculate the:

a co-ordinates of the mid-point, M, of the line joining A and B **[1]**
b length of AB to 3 significant figures **[2]**
c equation of the line passing through M that is perpendicular to AB, expressing your answer in the form
$ax + by + c = 0$ where a, b and c are integers. **[2]**

4 a Solve the inequality $7x - 11 > 3$ **[1]**   **b** Represent the solution to part **a** on a number line. **[2]**

5 Robin fires 15 arrows at a target.
The table shows information about his scores.

a Find his median score. **[1]**   **b** Work out his mean score correct to 3 significant figures. **[2]**

SCORE	FREQUENCY
1	5
2	4
3	1
4	4
5	1

6 The probability that a person chosen at random has brown eyes is 0.45
The probability that a person chosen at random has green eyes is 0.12

a Work out the probability that two people chosen at random both have either brown eyes **or** green eyes. **[2]**

500 people are to be chosen at random.

b Calculate an estimate for the number of people who will have green eyes. **[2]**

7 The diagram shows a prism.
The cross-section of the prism is a trapezium.
The lengths of the parallel sides of the trapezium are 9 cm and 5 cm.
The distance between the parallel sides of the trapezium is 6 cm.
The length of the trapezium is 15 cm.

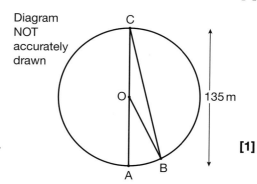

Work out
a the area of the trapezium in mm² **[2]** **b** the volume of the prism in mm³. **[2]**

Give your answers in standard form.

8 A = {1, 2, 3, 4}
B = {1, 3, 5}

a List the members of the set **b** Explain clearly the meaning of 5 ∈ A′ **[1]**
   i A ∩ B   **ii** A ∪ B **[1,1]**

9 The table gives information about the ages, in years, of the 80 members of a swimming club.

AGE t (years)	FREQUENCY
$10 < t \le 20$	8
$20 < t \le 30$	28
$30 < t \le 40$	30
$40 < t \le 50$	10
$50 < t \le 60$	4

a Work out an estimate for the mean age of the 80 members. **[2]**
b Complete a cumulative frequency table for this data. **[2]**
c Draw a cumulative frequency graph for your table. **[2]**
d Use your graph to find an estimate for the median age and inter-quartile range of the members of the club. **[2]**

10 The diagram represents part of the London Eye.
A, B and C are points on a circle, centre O.
A, B and C represent three capsules.
The capsules A and B are next to each other.
A is at the bottom of the circle and C is at the top.
The London Eye has 32 equally spaced capsules on the circle.

Diagram NOT accurately drawn

135 m

a Find the size of the smaller angle between BC and the horizontal. **[1]**

The capsule moves in a circle of diameter 135 m.

b Calculate the distance moved by a capsule in making a complete revolution. Give your answer correct to 3 significant figures. **[2]**

The capsule moves at an average speed of 0.26 m/s.

c Calculate the time taken for a capsule to make a complete revolution. Give your answer in minutes, correct to the nearest second. **[2]**

11 a Find the gradient of the line with equation $3x - 4y = 15$ [1]

 b Work out the co-ordinates of the point of intersection of the line with equation $3x - 4y = 15$
 and the line with equation $5x + 6y = 6$ [2]

12 $f(x) = 3x - 2$, $g(x) = \dfrac{1}{x}$

 a Find the value of
  i fg(3)   **ii** gf(3) [1,1]

 b i Show that the composite function $fg(x) = \dfrac{3 - 2x}{x}$ [2]

  ii Which value of x must be excluded from the domain of fg(x)? [1]

 c Solve the equation $f(x) = fg(x)$ [2]

13 The unfinished table and histogram show information from a survey of women about the number of calories in the food they eat in one day.

NUMBER OF CALORIES (n)	FREQUENCY
$0 < n \le 1000$	90
$1000 < n \le 2000$	
$2000 < n \le 2500$	140
$2500 < n \le 4000$	

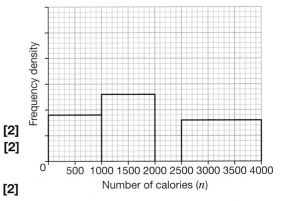

 a i Use the information in the table to complete the histogram. [2]
  ii Use the information in the histogram to complete the table. [2]

 b Find an estimate for the upper quartile of the number of calories. You must make your method clear. [2]

14 An electrician has wires of the same length made from the same material.
 The electrical resistance, R ohms, of a wire is **inversely** proportional to the square of its radius, r mm.
 When $r = 2$, $R = 0.9$

 a i Express R in terms of r. [1]
  ii Sketch the graph of R against r. [1]

 One of the electrician's wires has a resistance of 0.1 ohms.

 b Calculate the radius of this wire in mm. [2]

15 A rectangular piece of card has length $(x + 4)$ cm and
 width $(x + 1)$ cm.
 A rectangle 5 cm by 3 cm is cut from the corner of the
 piece of card.
 The remaining piece of card, shown shaded in the
 diagram, has an area of 35 cm².
 Find x correct to 3 significant figures.

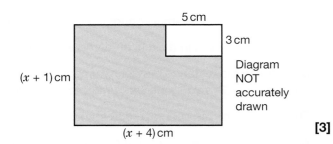

[3]

16 Part of the graph of $y = x^2 - 2x - 4$ is shown on the grid.

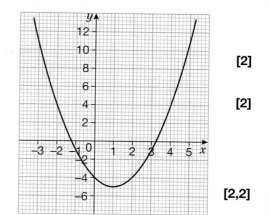

    a Use the graph to find estimates of the solutions to the equation $x^2 - 2x - 4 = 0$. Give your answers correct to 1 decimal place.　**[2]**

    b Draw a suitable straight line on the grid to find estimates of the solutions of the equation $x^2 - 3x - 6 = 0$
    Give your answer correct to 1 decimal place.　**[2]**

    c For $y = x^2 - 2x - 4$ find

      i $\dfrac{dy}{dx}$ **ii** the solution to $\dfrac{dy}{dx} = 0$ and state the co-ordinates of the minimum point of the curve.　**[2,2]**

17 In the triangle shown, BC = 7.8 cm
Angle ABC = 42°
Angle BAC = 110°

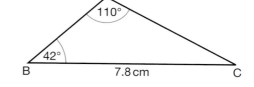

Calculate the
a length of AB　**[2]**
b area of triangle ABC.　**[2]**

Give your answers correct to 3 significant figures.

18 The diagram shows six counters.
Each counter has a letter on it.
Bishen puts the six counters in a bag.

He takes a counter at random from the bag. He records the letter which is on the counter and does **not** replace the counter in the bag. He then takes a second counter at random and records the letter which is on the counter.

    a Calculate the probability that the first letter will be A and the second letter will be N.　**[2]**
    b Calculate the probability that both letters will **not** be the same.　**[2]**

19 A cylindrical tank has a radius of 30 cm and a height of 45 cm.
The tank contains water to a depth of 36 cm.
A metal sphere is dropped into the water and is completely covered.
The water level rises by 5 cm.
Calculate the surface area of the sphere in cm² correct to 3 significant figures.　**[3]**

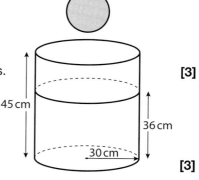

20 Solve for x, $\dfrac{3}{x+2} + \dfrac{x+17}{x^2-x-6} = 1$　**[3]**

21 Show that the sum of the first n odd numbers is n^2.　**[3]**

22 If $f(x) = \sin x$ for $0° \le x \le 360°$, sketch the graph of $-2f(x)$ and state the co-ordinates of its minimum and maximum points.　**[4]**

Total Marks = 100

PAPER 2

1 Krishnan used 620 units of electricity.
The first 180 units cost £0.0825 per unit.
The remaining units cost £0.0705 per unit.
Tax is added at 15% of the total amount.
Complete Krishnan's bill.

180 units at £0.0825 per unit	£........
.... units at £0.0705 per unit	£........
Total amount	£........
Tax at 15% of the total amount	£........
Amount to pay	£........

[4]

2 In the diagram, ABC and ADE are straight lines.
CE and BD are parallel.
AB = AD
Angle BAD = 38°

Work out the value of p and angle BDE.
Give a reason for each step in your working.

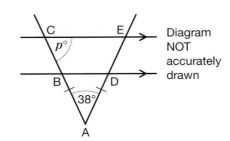

Diagram NOT accurately drawn

[2]

3 Arul has x sweets.
Nikos had four times as many sweets as Arul.
Nikos gave 6 of his sweets to Arul.
Now they both have the same number of sweets.
Form an equation in x and solve it to find the number of sweets that Arul had at the start. **[3]**

4 The mean height of a group of four girls is 158 cm.
Sienna joins the group and the mean height of the five girls is 156 cm.
Work out Sienna's height. **[2]**

5 Plumbers' solder mix is made from tin and lead.
The ratio of the weight of tin to the weight of lead is 1 : 2

a Work out the weight of tin and the weight of lead in 240 grams of plumbers' solder. **[2]**
b What weight of plumbers' solder contains 75 grams of tin? **[2]**

6 a Find the highest common factor of 48 and 180. **[2]**
b Find the lowest common multiple of 48 and 180. **[2]**

7 This formula is used in science, $v = \sqrt{2gh}$

a Hanif uses the formula to work out an estimate for the value of v without using a calculator when $g = 9.81$ and $h = 0.82$

Write down approximate values for g and h that Hanif could use. **[2]**

b Make g the subject of the formula $v = \sqrt{2gh}$ **[2]**

8 a The universal set, $\mathscr{E}$ = {Angela's furniture}
A = {chairs}
B = {kitchen furniture}
Describe fully the set A ∩ B **[1]**

b P = {2, 4, 6, 8}
Q = {odd numbers less than 10}
i List the members of the set P ∪ Q. **[2]**
ii Is it true that P ∩ Q = ∅? Explain your answer. **[2]**

9 Triangle PQR is right-angled at R.
PR = 4.7 cm and PQ = 7.6 cm.

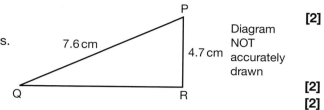

Diagram NOT accurately drawn

  a Calculate the size of angle PQR to 1 decimal place.   **[2]**

The length, 7.6 cm, of PQ is correct to 2 significant figures.

  b Calculate to 3 significant figures
    i the upper bound of the length of QR   **[2]**
    ii the lower bound of the length of QR.   **[2]**

10 A mobile phone company makes a special offer.
Usually one minute of call time costs 12 cents.
For the special offer, this call time is increased by 20%.

  a Calculate the call time which costs 12 cents during this special offer.
    Give your answer in seconds.   **[2]**
  b Calculate the cost per minute for the special offer.   **[2]**
  c Calculate the percentage decrease in the cost per minute for the special offer.   **[2]**

11 Show by completing the square that the equation $2x^2 = 8x - 7$ has solutions $\dfrac{2\sqrt{2} \pm 1}{\sqrt{2}}$ **[2]**

12 Quadrilateral P is mathematically similar to quadrilateral Q.

  a Calculate the value of x.   **[2]**
  b Calculate the value of y.   **[2]**

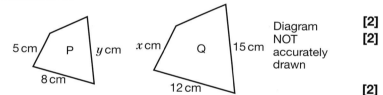

Diagram NOT accurately drawn

The area of quadrilateral P is 80 cm².

  c Calculate the area of quadrilateral Q.   **[2]**

13 Two chords, AB and CD, of a circle intersect at right angles at X.
AX = 2.8 cm
BX = 1.6 cm
CX = 1.2 cm

Calculate the length AD and angle DAX correct to 3 significant figures.   **[3]**

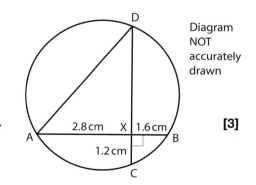

Diagram NOT accurately drawn

14 OABC is a parallelogram.

$\overrightarrow{OA} = \begin{pmatrix} 1 \\ 2 \end{pmatrix}$, $\overrightarrow{OC} = \begin{pmatrix} 4 \\ 0 \end{pmatrix}$

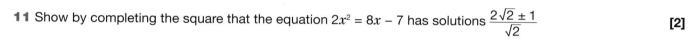

Diagram NOT accurately drawn

X is the point on OB such that OX = kOB, where $0 < k < 1$

  a Find, in terms of k, the vectors
    i $\overrightarrow{OX}$ **ii** $\overrightarrow{AX}$ **iii** $\overrightarrow{XC}$ **[2,2,2]**

  b Find the value of k for which $\overrightarrow{AX} = \overrightarrow{XC}$ **[2]**
  c Prove that the diagonals of the parallelogram OABC bisect one another.   **[2]**

15 a Megan invests $1200 in SafeBank earning her compound interest at 3% per year for 5 years.
Calculate the value of Megan's investment after this period. **[2]**

b Liam buys a new pair of running shoes for $98 **after** they have been reduced by 30%.
Calculate the original price of the shoes. **[2]**

16 A farmer wants to make a rectangular pen for keeping sheep.
He uses a wall, AB, for one side.
For the other three sides, he uses 28 m of fencing.
He wants to make the area of the pen as large as possible.
The width of the pen is x metres.
The length parallel to the wall is $(28 - 2x)$ metres.

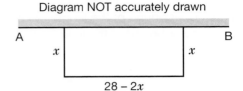

Diagram NOT accurately drawn

a The area of the pen is $y\,\text{m}^2$. Show that $y = 28x - 2x^2$ **[2]**
b Use calculus to find the largest possible area of the pen. **[3]**

17 A fan is shaped as a sector of a circle, radius 12 cm, with angle
110° at the centre.

a Calculate the area and perimeter of the fan to
3 significant figures. **[2]**

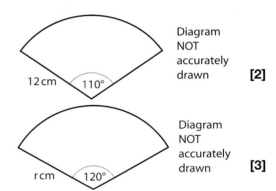

Diagram NOT accurately drawn

Another fan is shaped as a sector of a circle, radius r cm,
with angle 120° at the centre.

b Prove that if the **total** perimeter of this fan is 200 cm,
its radius, $r = \dfrac{300}{3 + \pi}$ cm **[3]**

Diagram NOT accurately drawn

18 In order to start a course, Bae has to pass a test.
He is allowed only two attempts to pass the test.
The probability that Bae will pass the test at his first attempt is $\dfrac{2}{5}$
If he fails at his first attempt, the probability that he will pass
at his second attempt is $\dfrac{3}{4}$

First attempt   Second attempt

Pass

Fail

a Copy and complete the probability tree diagram. **[2]**
b Calculate the probability that Bae will be allowed to start the course. **[3]**

19 Solve the simultaneous equations
$$y = x - 1$$
$$2x^2 + y^2 = 2$$ **[4]**

20 A particle moves along a line.
For $t \geq 1$, the distance of the particle from O at time t seconds is x metres, where $x = 3t - t^2 + \dfrac{8}{t}$

a Find an expression for the velocity of the particle. **[3]**
c Find at what time the acceleration is $0\,\text{m/s}^2$. **[3]**
b Find an expression for the acceleration of the particle. **[3]**

21 Solve the equation $(2x - 1)(x + 1)(x + 2) - (x + 1)(x + 2)(x + 3) = 0$ **[3]**

22 Show that the area of an equilateral triangle of side $4n$ is given by $4\sqrt{3}\,n^2$ **[4]**

Total marks = 100

PAPER 3

1 Work out the value of $\dfrac{3.23 \times 10^4}{\sqrt{1.8 \times 10^6}}$

Give your answer in standard form correct to 3 significant figures.   **[2]**

2 Rectangular tiles have width x cm and height $(x + 7)$ cm.

Some of these tiles are used to form a shape.
The shape is 6 tiles wide and 4 tiles high.

The width and height of this shape are equal.
Form and solve an equation to find the value of x.   **[3]**

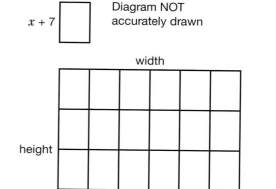

x

$x + 7$

Diagram NOT accurately drawn

width

height

Diagram NOT accurately drawn

3 Illustrate the following inequalities on a graph and show the region that satisfies them all. Shade the regions that are **not** required.
$x \geq 2$
$y > 1$
$2x + 3y \leq 13$ **[3]**

4 The length of flower stems is given in the table below.

LENGTH, x (cm)	FREQUENCY
$0 \leq x < 2$	4
$2 \leq x < 6$	15
$6 \leq x < 8$	18
$8 \leq x < 12$	p

An estimate of the mean length of flower stems = 5.5 cm.
Calculate the value of p.   **[2]**

5 A, B, C and D are points on the circumference of a circle with centre O.
Work out the sizes of the angles marked x and y.
Give reasons for all statements you make.   **[3]**

Not drawn accurately

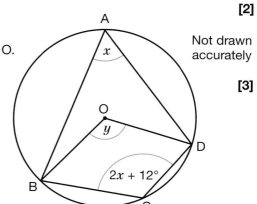

6 a Simplify the following expressions fully.

    i $(x - 3)(x + 7)$ **[2]**   **iii** $(3x - 1)(2x - 1)(x - 1)$.   **[2]**

    ii $(5x - 3)^2$ **[2]**   **iv** $(\sqrt{7} - 1)(2\sqrt{7} + 1)$ **[2]**

b Rationalise the denominators of

    i $\dfrac{\sqrt{7} + 1}{\sqrt{7}}$ **[2]**   **ii** $\dfrac{\sqrt{7}}{\sqrt{7} + 1}$ **[2]**

7 ABC is a triangle.
AB = AC = 13 cm
BC = 10 cm
M is the mid-point of BC.
Angle AMC = 90°

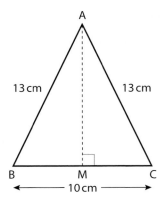

Diagram NOT accurately drawn

A solid has five faces.
Four of the faces are triangles identical to triangle ABC.
The base of the solid is a square of side 10 cm.

Calculate the total surface area of this solid.   **[3]**

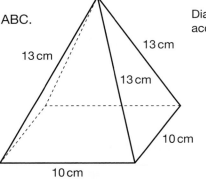

Diagram NOT accurately drawn

8 The size of each interior angle of a regular polygon with n sides is 120°.
Find the interior angle of a regular polygon with $4n$ sides.   **[2]**

9 Draw a co-ordinate grid from −6 to 6 on both axes.
Draw triangle A at (−2, −1), (−1, −1) and (−1, 1).   **[1]**

    a Rotate A 90° clockwise about (1, 2). Label the image B.   **[2]**

    b Enlarge B by a scale factor of 2, centre of enlargement (−4, 6). Label your image C.   **[2]**

    c Translate C by the vector $\begin{pmatrix} 2 \\ -6 \end{pmatrix}$. Label the image D.   **[2]**

    d Rotate D 90° anticlockwise about (−3, −5). Label the image E.   **[2]**

    e Describe the single transformation that maps E onto A.   **[1]**

10 The table gives information about the ages, in years, of 100 aeroplanes.

AGE (t years)	FREQUENCY
$0 < t \le 5$	41
$5 < t \le 10$	26
$10 < t \le 15$	20
$15 < t \le 20$	10
$20 < t \le 25$	3

a Copy and complete the cumulative frequency table.   **[2]**

AGE (t years)	CUMULATIVE FREQUENCY
$0 < t \le 5$	
$0 < t \le 10$	
$0 < t \le 15$	
$0 < t \le 20$	
$0 < t \le 25$	

b On a copy of the grid, draw a cumulative frequency graph for your table.   **[2]**

c Use your graph to find an estimate for the inter-quartile range of the ages.   **[2]**

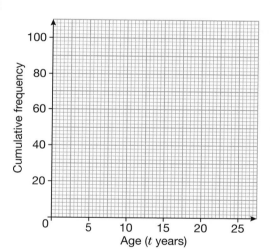

11 The straight line, L, passes through the points (0, −1) and (2, 3).

a Write down the equation of L.   **[2]**

b Calculate the equation of the line that is perpendicular to L that passes through point (2, 3).   **[2]**

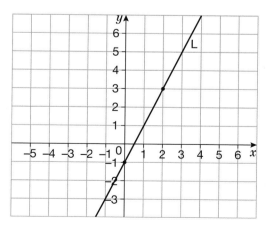

12 $p = 3^8$

a Express $p^{\frac{1}{2}}$ in the form 3^k, where k is an integer.   **[2]**

$q = 2^9 \times 5^{-6}$

b Express $q^{-\frac{1}{3}}$ in the form $2^m \times 5^n$, where m and n are integers.   **[2]**

13 a For the equation $y = 5000x - 625x^2$, find the co-ordinates of the turning point on the graph of
$y = 5000x - 625x^2$ [2]

b State with a reason whether this turning point is a maximum or a minimum. [1]

c A publisher has set the price for a new book.
The profit, £y, depends on the price of the book, £x, where $y = 5000x - 625x^2$
State with a reason, the price you would advise the publisher to set for the book. [2]

14 A solid hemisphere has a **total** surface area of 3600 cm².

Find the hemisphere's
a diameter [2]
b volume correct to 3 significant figures. [2]

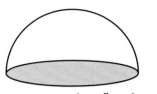

15 Oil is stored in either small drums or large drums.
The shapes of the drums are mathematically similar.
A small drum has a volume of 6000 cm³ and a surface area of 2000 cm².
The **height** of a large drum is 3 times the height of a small drum.

a Calculate the volume of a large drum in cm³ in standard form. [2]
b The cost of making a drum is $1.20 for each m² of surface area.
A company wants to store 3240 m³ of oil in large drums.
Calculate the cost of making enough drums to store this oil. [2]

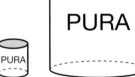

Diagram NOT
accurately drawn

16 The function f(x) is defined as f(x) = $\dfrac{x}{x-1}$

a Find the value of f(3). [1]
b State which value of x must be excluded from the domain of f(x). [1]
c i Find ff(x). Give your answer in its simplest form. [2]
   ii What does your answer to part **c i** show about the function f? [1]

17 A boat at point A is due West of Granite Island (G).
It sails on a bearing of 120°, for 1500 m, to point B.
If the bearing of G from B is 320°, calculate the distance BG.
Give your answer correct to 3 significant figures. [3]

18 Each student in a group plays at least one of hockey, tennis and football.
10 students play hockey only.
9 play football only.
13 play tennis only.
6 play hockey and football but not tennis.
7 play hockey and tennis.
8 play football and tennis.
x play all three sports.

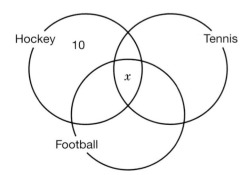

a Write down the expression, in terms of x, for the number
of students who play hockey and tennis, but not football. [2]

There are 50 students in the group.

b Find the value of x. [2]

19 The number of men in a club is n and $\frac{1}{3}$ of the people in this club are men.

   a Write down an expression, in terms of n, for the number of people in the club.   **[2]**

Two of the people in the club are chosen at random.

The probability that both these people are men is $\frac{1}{10}$

   b Calculate the number of people in the club.   **[2]**

20 The expression $x^2 - 8x + 21$ can be written in the form $(x - a)^2 + b$ for all values of x.

   a Find the values of a and b.   **[3]**

The equation of a curve is $y = f(x)$ where $f(x) = x^2 - 8x + 21$

The minimum point of the curve is M.

   b Write down the co-ordinates of M.   **[2]**

21 Prove algebraically that the difference between the squares of two consecutive integers is equal to the sum of the integers.   **[3]**

22 The diagram shows a sketch graph of $y = f(x) = 7 + 6x - x^2$

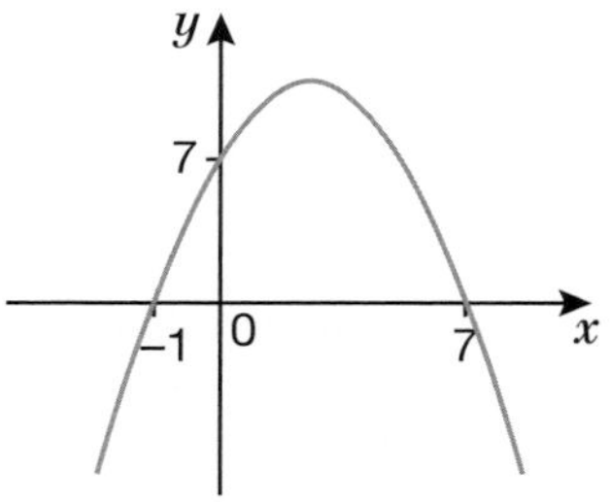

   a Copy the diagram and, on the same axes, draw sketch graphs of

     i $y = -f(x)$ **[1]**

     ii $y = \frac{1}{2}f(x)$ **[1]**

     iii $y = f(2x)$ **[2]**

   b Describe fully each transformation.   **[1,1,1]**

   c Work out the algebraic equation of each transformed graph. Simplify your answer fully.   **[1,1,1]**

Total marks = 100

PAPER 4

1 The bearing of town P from town Q is 070°.
Find the bearing of town Q from town P.   **[2]**

2 a Expand these and simplify where appropriate.

   i $(3x - 4)(3x + 4)$ **ii** $(3x - 4)^2$ **[2,2]**

   b Factorise these.

   i $2xy - 12x^2y^3$ **ii** $3x^2y - 9x^3y^2 + 15x^4y^3$ **[2,2]**

   c Simplify these.

   i $xy \times x^3y^5$ **ii** $\dfrac{x^2\,y^7\,z^5}{x\,y^2\,z^3}$ **[2,2]**

3 a Write the number 37 000 000 000 in standard form.   **[2]**

   b Write 7.5×10^{-5} as an ordinary number.   **[1]**

   c Calculate the value of $\dfrac{2.5 \times 10^{-3}}{1.25 \times 10^7}$ **[2]**

   Give your answer in standard form.

4 In the diagram, PAQ is the tangent at A to a circle with centre O.
Angle BOC = 112°
Angle CAQ = 73°

Work out the size of angle OBA.
Show your working, giving reasons for any statements you make.   **[2]**

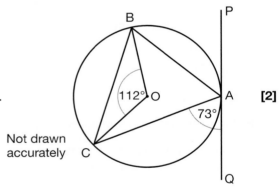

Not drawn accurately

5 Marco invests €1200 in Big Bank for three years at a compound
interest rate of 2% per year.
He then moves **all** of this money into Small Bank for 2 years at a
compound interest rate of 3% per year.
Calculate Marco's percentage profit over the 5-year period correct to 3 significant figures.   **[2]**

6 Solve the simultaneous equations
$3x + 4y = 5$
$2x - 5y = 11$ **[3]**

7 In the diagram, PQR is a straight line and PQ:QR = 2:3
$\overrightarrow{OP} = 2\mathbf{a} + 5\mathbf{b}$

$\overrightarrow{OQ} = 4\mathbf{a} + \mathbf{b}$

Write $\overrightarrow{OR}$ in terms of **a** and **b**. Give your answer in its simplest form.   **[3]**

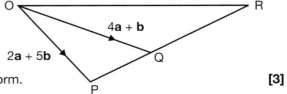

8 The Venn diagram shows the sports played by a group of 80 students.
The three most popular sports were tennis (T), golf (G) and cricket (C).
A student is picked at random.
Work out

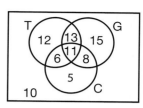

a P(T ∩ C ∩ G) **[2]** **c** P(T|(G ∪ C)) **[2]**
b P((T ∩ G ∩ C)|T) **[2]** **d** P(G|C′) **[2]**

9 In an arithmetic progression the 10th term is 39 and the 5th term is 19.
Find the

a first term **b** common difference **c** sum to ten terms. **[2,2,2]**

10 $y = \dfrac{a}{b - c}$ and $a = 7.8 \pm 0.5$, $b = 7.1 \pm 0.5$ and $c = 2.1 \pm 0.5$

Find

a the maximum value of y **b** the minimum value of y. **[2,2]**

Give your answers correct to 2 significant figures.

11 $g(x) = \dfrac{2}{3x - 1}$

Find

a g(2) **[1]** **b** $g^2(2)$ **[2]** **c** the solution(s) of $x = g(x)$ **[2]** **d** $g^{-1}(x)$ **[2]**

12 The sides of an equilateral triangle ABC and two **regular** polygons
meet at the point A.
AB and AD are adjacent sides of a regular 10-sided polygon.
AC and AD are adjacent sides of a regular n-sided polygon.
Work out the value of n.

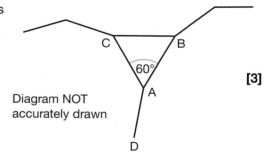

Diagram NOT
accurately drawn

[3]

13 The lengths of the sides of a triangle are in the ratio of 4 : 5 : 6
The perimeter of the triangle is 150 cm.
Find the area of the triangle correct to 3 significant figures. **[3]**

14 ABC is an equilateral triangle of side 8 cm.
With the vertices A, B and C as centres, arcs of radius 4 cm are drawn
to cut the sides of the triangle at P, Q and R.
The shape formed by the arcs is shaded.
Calculate the

a perimeter of the shaded shape in cm to 1 decimal place **[2]**
b area of the shaded shape in cm² to 1 decimal place. **[3]**

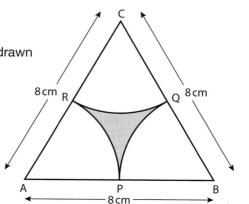

Diagram NOT
accurately drawn

15 a Copy and complete the table of values for $y = x^2 - \dfrac{3}{x}$ **[2]**

x	0.5	1	1.5	2	3	4	5
y	−5.75	−2					24.4

b On a copy of the grid, draw the graph of

$y = x^2 - \dfrac{3}{x}$ for $0.5 \le x \le 5$ **[2]**

c Use your graph to find an estimate for a solution of the equation

$x^2 - \dfrac{3}{x} = 0$ **[2]**

d Draw a suitable straight line on your graph to find an estimate for a solution of the equation

$x^2 - 2x - \dfrac{3}{x} = 0$ **[2]**

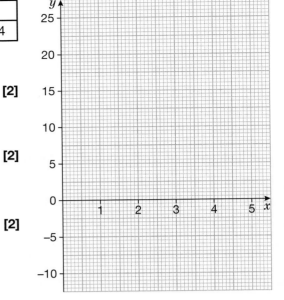

16 The unfinished table and histogram show information about the weights, in kg, of some babies.

WEIGHT (w kg)	FREQUENCY
$0 < w \le 2$	
$2 < w \le 3.5$	150
$3.5 < w \le 4.5$	136
$4.5 < w \le 6$	

a Use the histogram to complete the table.   **[2]**
b Use the table to complete the histogram.   **[2]**

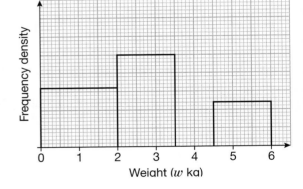

17 The diagram shows one disc with centre A and radius 4 cm and another disc with centre B and radius x cm.
The two discs fit exactly into a rectangular box 10 cm long and 9 cm wide.
The two discs touch at P.
APB is a straight line.

a Use Pythagoras' Theorem to show that $x^2 - 30x + 45 = 0$ **[2]**
b Find the value of x.   **[2]**

Give your value correct to 3 significant figures.

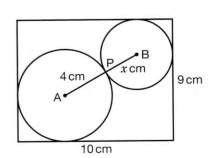

Diagram NOT
accurately drawn

18 A bicycle moves in a straight line.
From a fixed point O, its distance, x m, t seconds later is given by $x = \frac{2}{3}t^3 - \frac{9}{2}t^2 + 4t$.
Find

    a when the bicycle is stationary   **[2]**
    b the acceleration of the bicycle at $t = 3$.   **[2]**

19 There are 48 beads in a bag.
Some of the beads are red and the rest of the beads are blue.
Shan is going to take a bead at random from the bag.
The probability that she will take a red bead is $\frac{3}{8}$

    a Work out the number of red beads in the bag.   **[2]**

    Shan adds some **red** beads to the 48 beads in the bag.
    The probability that she will take a red bead is now $\frac{1}{2}$

    b Work out the number of red beads she adds.   **[2]**

20 A right regular square pyramid of base length x is shown.
All the four edges are also of length x.
Prove that

    a the perpendicular height of the pyramid, $h = \frac{x}{\sqrt{2}}$ **[2]**
    b the angle between an edge and the square base = 45°.   **[2]**

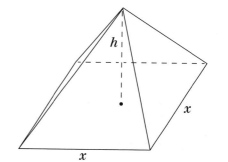

21 A farmer wishes to put a fence around a rectangular field with
one side formed by a stone wall.
If the total length of the fence is 150, calculate the maximum area
and the value of x at which this occurs by using differentiation.   **[4]**

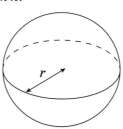

22 A solid metal sphere of radius r is melted down and recast
into a solid circular right cone of base radius r and height h.
Prove that the **total** surface area, A, of the solid cone
is given by $A = \pi r^2 (1 + \sqrt{17})$ **[3]**

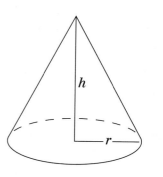

Total marks = 100

GLOSSARY

acceleration (*noun*) the rate at which **velocity** changes with time

adjacent (*adjective*) next to

alternate angles (*noun*) (*plural*) two equal angles formed on opposite sides and at opposite ends of a line that crosses two parallel lines

angle of depression (*noun*) the angle measured downwards from the horizontal

angle of elevation (*noun*) the angle measured upwards from the horizontal

arc (*noun*) part of a curve or circle

arithmetic sequence (*noun*) a sequence of numbers in which the terms increase (or decrease) by a fixed number (= **common difference**); for example, 2, 5, 8, 11, 14 (common difference = + 3)

arithmetic series (*noun*) an arithmetic sequence with the signs + or – placed between each number; for example, 2 + 5 + 8 + 11 + 14

asymptote (*noun*) a line that a curve gets very close to but never touches

bar chart (*noun*) a diagram using rectangles of equal width (bars) whose heights represent an amount or quantity

bearing (*noun*) an angle measured clockwise from north, used to describe a direction

bisect (*verb*) to divide something into two equal parts

bisector (*noun*) a straight line that divides an angle or another line into two equal parts

cancel down (*verb*) to simplify a fraction by dividing the **numerator** (= top number) and **denominator** (= bottom number) by the same number (a **common factor**)

chord (*noun*) a straight line joining two points on a curve

circumference (*noun*) the total distance around a circle

class (*noun*) a group of data in a collection of grouped data

clear (*verb*) to remove fractions in an equation by multiplying both sides of the equation by the lowest **common denominator**

coefficient (*noun*) a number in front of a **variable** (= a letter); for example, the 3 in $3x$

collinear (of points) (*adjective*) three or more points that all lie on the same straight line

column vector (*noun*) a vector such as $\begin{pmatrix} 3 \\ 8 \end{pmatrix}$ used to describe a translation; the top number gives the movement parallel to the horizontal axis (to the right or left); the bottom number gives the movement parallel to the vertical axis (up or down)

common denominator (*noun*) a number that can be divided exactly by all the **denominators** (= bottom numbers) in a set of fractions; for example 12 is the common denominator of 3, 4 and 6

common difference (*noun*) the difference between each term in an **arithmetic sequence**

common factor (*noun*) a number (a **factor**) that two or more other numbers can be exactly divided by; for example, 2 is a common factor of 4, 6 and 8

compasses (*noun*) a V-shaped instrument with one sharp point and a pencil at the other end used for drawing circles

complement (*noun*) the complement of a **set** is all those members (objects) which are not in that set but which are in the universal set (all members being considered)

compound interest (*noun*) interest is the extra money that you must pay back when you borrow money or money paid to you by a bank or financial institution when you keep money in an account there; compound interest is calculated on both the sum of money lent or borrowed or saved and on the unpaid interest already earned or charged

compound measure (*noun*) a measure made up of two or more different measurements; for example, speed, density and pressure

conditional probability (*noun*) the probability of a **dependent event**; that is, the probability of event A given that event B has already occurred

congruent (*adjective*) exactly the same size and shape

consecutive (*adjective*) numbers follow each other in order, with no gaps, from smallest to largest

correct to (*adjective*) accurate to; a number given correct to '…' has been rounded, e.g. 3.592 is 3.6 correct to 1 decimal place and 3.59 correct to 2 decimal places

corresponding angles (*noun*) (*plural*) two equal angles that are formed when a line crosses two parallel lines; each angle is on the same side of the two parallel lines and on the same side of the line which they cross

counter-example (*noun*) an example that does not fit a statement

cross-section (*noun*) something that has been cut in half so that you can look at the inside, or a drawing of this

cube (*noun*) a solid object that has 6 identical square **faces** (= flat surfaces)

cubic function (*noun*) a function where the highest power of x is 3; for example $y = 2x^3 + 3x^2 + 5x - 4$

cubic graph (*noun*) the graph of a cubic function of the form $y = ax^3 + bx^2 + cx + d$ where a, b, c and d are numbers and the highest power of x is 3

cuboid (*noun*) a solid object that has 6 rectangular **faces** (= flat surfaces)

cyclic quadrilateral (*noun*) a flat shape with 4 sides (= a **quadrilateral**) with all 4 vertices on the **circumference** of a circle

cylinder (*noun*) a solid with straight parallel sides and a circular cross-section; an example of a **prism**

decimal place (or **d.p.**) (*noun*) the position of a digit to the right of a decimal point

degree mode (*noun*) the angle mode setting on a scientific calculator

degree of accuracy (*noun*) tells you how accurate a number is; a number could be given to '...' decimal places or '...' significant figures

denominator (*noun*) the number below the line in a fraction

dependent event (*noun*) an event that depends on the outcome of another event

derivative (*noun*) the rate of change or gradient of a **function**

diameter (*noun*) a straight line passing through the centre of a circle and joining two points that lie on the circle; a chord that passes through the centre of a circle

differentiate (*verb*) to find the gradient of a **function**

direct proportion (*noun*) a relationship between two quantities such that as one quantity increases the other increases at the same rate

displacement (*noun*) the distance and direction of an object from a fixed point

domain (*noun*) the set of all input values of a **function**

equilateral triangle (*noun*) a triangle whose 3 sides are all the same length and all the angles are 60°

expand (an expression) (*verb*) to **multiply out** brackets (= remove brackets)

exterior angle (*noun*) the angle formed outside a **polygon** between any one edge and the extended adjacent edge

face (*noun*) the flat surface of a solid object

factor (*noun*) a number that divides exactly into another number

factorial (*noun*) (symbol !) the result of multiplying a sequence (set of numbers) of descending **integers** (= whole numbers); for example, 5 factorial = 5! = 5 × 4 × 3 × 2 × 1

factorise (*verb*) to put an expression into brackets; the reverse of expand

frequency (of a piece of data) (*noun*) the number of times each piece of data is found

frequency density (*noun*) a measure of the frequency divided by the class width: frequency density $= \dfrac{\text{frequency}}{\text{class width}}$; the height of each bar in a **histogram**

function (*noun*) a set of rules for turning one number into another

gradient (*noun*) the measure of the slope of a straight line relative to the horizontal; for a straight line gradient $= \dfrac{\text{change in the } y \text{ coordinates}}{\text{change in the } x \text{ coordinates}} = \dfrac{\text{'rise'}}{\text{'run'}}$

happy number (*noun*) starting with any positive integer, find the sum of the squares of its digits; repeat the process until a 1 is obtained, when the original number is described as happy; for example, 23 is a happy number ($23 \rightarrow 2^2 + 3^2 = 13 \rightarrow 1^2 + 3^2 = 10 \rightarrow 1^2 + 0^2 = 1$); if a 1 is never obtained, the original number is an unhappy number

hemisphere (*noun*) a half of a sphere

histogram (*noun*) a diagram similar to a bar chart where the width of each bar is equal to the class width, the area of the bar represents the frequency, and the height of each bar is the **frequency density**

hyperbola (*noun*) a special curve; the graph of a reciprocal function of the form $y = \dfrac{a}{x}$ where a is a constant

hypotenuse (*noun*) the longest side of a **right-angled triangle**; it is the side opposite the right angle

improper fraction (*noun*) a fraction such as $\dfrac{107}{8}$ in which the **numerator** (= top number) is larger than the **denominator** (= bottom number)

included angle (*noun*) the angle between two sides

independent event (*noun*) an event that does not affect the probability of another event

indices (*noun*) (*plural*) (*singular* **index**) (= **powers**) a number which tells you how many times to multiply the given number or term by itself; for example, the ^{2} in 4^2 is the index

inequality (*noun*) (*plural* **inequalities**) an expression containing one or more of the symbols <, >, ≤ or ≥

integer (*noun*) a whole number

intercept (*noun*) the point on a graph where the graph crosses an axis

interior angle (*noun*) an angle formed inside a **polygon** between two adjacent edges

intersect (*verb*) if two lines intersect, they meet or cross over each other

intersection (*noun*) the intersection of two **sets** A and B is the set of elements which are in both A and B

inverse (*noun*) opposite in effect or the reverse of; for example, addition (+) is the inverse of subtraction (−)

inverse proportion (*noun*) a relationship between two quantities such that one quantity increases at the same rate as the other quantity decreases

irrational number (*noun*) a number that cannot be written as a fraction; for example, $\sqrt{2}$ and π

isosceles triangle (*noun*) a triangle with two equal sides and two equal angles

like terms (*noun*) terms whose **variables** (such as x or y) and **powers** (such as ^{2} or ^{3}) are exactly the same; for example, $2x$ and $5x$ are like terms, and $4x^2$ and $2x^2$ are like terms

linear simultaneous equations (*noun*) two equations, whose graphs are straight lines, that are solved together to find two unknowns

lower bound (*noun*) the value half a unit less than the rounded measurement

magnitude (of a vector) (*noun*) the length of a **vector**

map (*noun*) the connection from one set to another

map (*verb*) to translate, reflect or rotate a shape so that it fits (maps) onto another shape exactly

mapping diagram (*noun*) a diagram that lists two **sets**; arrows are used to show how the members are to be matched

mass (*noun*) a measure of how much matter is in an object

maximum point (*noun*) the highest **turning point** on a graph

mean (*noun*) the numerical value found by adding together all the separate values of a data set and dividing by how many pieces of data there are

median (*noun*) the middle value of a set of values that have been arranged in size order

mid-point (*noun*) a point that is halfway along the line segment joining two points

minimum point (*noun*) the lowest **turning point** on a graph

mixed number (*noun*) a number that consists of a whole number and a fraction; for example $7\frac{1}{4}$

modal class (*noun*) the **class** of grouped data which has the highest frequency

multiple (of a number) (*noun*) is the product of that number and an integer (whole number)

multiply out (*verb*) (= **expand**) to remove brackets; to multiply out a bracket, multiply each term inside the bracket by the term outside the bracket; to multiply out double brackets, multiply each term in one bracket by each term in the other bracket

mutually exclusive events (*noun*) events that cannot happen at the same time

natural numbers (*noun*) the set of positive whole numbers (= **integers**) 1, 2, 3, 4, 5, 6, …

numerator (*noun*) the number above the line in a fraction; for example 5 is the numerator in $\frac{5}{6}$

parabola (*noun*) a special curve, shaped like an arch (or upside-down arch); it is another name for a quadratic graph

parallelogram (*noun*) a flat shape with 4 sides (= a **quadrilateral**) in which opposite sides are parallel and the same length

perfect square (*noun*) a number whose square root is a whole number; for example, 1, 4, 9, 16 and 25

perimeter (*noun*) the total distance around the edge of a shape

perpendicular (*adjective*) if one line is perpendicular to another line, they form an angle of 90° (= a **right angle**)

plane (*noun*) a flat surface

polygon (*noun*) a flat shape with 3 or more straight sides

power (*noun*) (= **index**) the small number written to the right and above another number

prime factor (*noun*) a factor that is a **prime number**

prime number (*noun*) a natural number greater than 1 that can only be divided by 1 and itself; for example 3 and 7. Note that 1 is not a prime number

prism (*noun*) any solid with parallel sides that has a constant cross-section

product (*noun*) the number you get by multiplying two or more numbers

quadrant (of a circle) (*noun*) a quarter of a circle

quadratic curve (*noun*) another name for a **parabola**

quadratic equation (or **quadratic**) (*noun*) an equation where the highest power of x is 2; for example $3x^2 + 5x - 4 = 0$

quadratic graph (*noun*) the graph of a quadratic equation of the form $y = ax^2 + bx + c$ where a, b and c are numbers and the highest power of x is 2; its shape is that of a **parabola**

quadrilateral (*noun*) a flat shape with 4 straight sides

radius (*noun*) (*plural* **radii**) the distance from the centre to the curve which makes a circle; any straight line from the centre to the curve which makes a circle

random (*adjective*) happening or chosen by chance; a result is random if each possible result has the same chance of happening, and a selection is random if each object (or number) has the same chance of being chosen

range (*noun*) the difference between the lowest and highest values of a set of data

range (of a function) (*noun*) the set of all output values of a **function**

rate (of change) (*noun*) how fast something changes over a period of time

ratio (*noun*) a ratio shows the relative sizes of two or more quantities

rationalise (the denominator) (*verb*) to remove the surd from the **denominator** (= bottom number) of a fraction

rational number (*noun*) a number that can be written as a fraction in the form $\frac{a}{b}$ where a and b are **integers** (= whole numbers) and $b \neq 0$; for example, 4 $\left(\frac{4}{1}\right)$ and $0.\dot{4}\left(\frac{4}{9}\right)$

real numbers (*noun*) the set of all **rational** and **irrational** **numbers**

reciprocal graph (*noun*) the graph of a reciprocal function of the form $y = \frac{a}{x}$ where a is a number

recurring decimal (*noun*) a decimal in which one or more digits repeats; for example, 0.111 111…

resultant vector (*noun*) a vector that is equivalent to two or more other vectors

rhombus (*noun*) a flat shape with 4 equal sides (= a **quadrilateral**) in which opposite sides are parallel and opposite angles are equal

right angle (*noun*) an angle that is 90° exactly

right-angled triangle (*noun*) a triangle that has a **right angle** (= 90°)

round (*verb*) to make a number simpler, but keeping its value close to what it was; the result is less accurate, but easier to use

sample (*noun*) a small set of objects chosen from a larger set of objects which is examined to find out something about the larger set

satisfy (*verb*) to make an equation or an inequality true

scalar (*noun*) any **real number** (having only **magnitude**)

scale (*noun*) a set of marks with regular spaces between them on the axes of a graph

scale (*noun*) a **ratio** such as 1 : 100 which shows the relationship between a length on a drawing or map and the actual length in real life

scale factor (*noun*) the scale factor of an enlargement is the number of times by which each original length has been multiplied

sector (*noun*) the part of a circle whose perimeter is an arc and two radii

segment (*noun*) the part of a circle that is separated from the rest of the circle when you draw a straight line across it

semicircle (*noun*) a half of a circle

set (*noun*) a collection of objects (numbers, letters, symbols, etc.)

set-builder notation (*noun*) mathematical notation used to describe what is in a **set**; for example, $\{x : x > 0\}$ is the set of all x such that x is greater than 0

sign (*noun*) a mathematical symbol such as $+$, $-$, $\times$, $\div$

significant figure (or **s.f.**) (*noun*) each of the digits of a number that are used to express it to the required accuracy, starting from the first non-zero digit from the left-hand end of the number

similar (*adjective*) shapes are similar when one shape is an enlargement of the other

similar triangles (*noun*) two triangles are similar if they are the same shape

simplify (an expression) (*verb*) to collect all **like terms** so the expression is a collection of terms connected by $+$ and $-$ signs

simplify (a fraction) (*verb*) to divide the **numerator** (= top number) and **denominator** (= bottom number) by the same number (a **common factor**)

simultaneous equations (*noun*) two (or more) equations that are solved together to find two (or more) unknowns

sketch (*verb*) to make a rough drawing of a diagram or graph or shape

sketch (*noun*) a rough drawing of a diagram or graph which gives a general idea of the shape or relationship, or a rough drawing of a shape showing angles and lengths

standard form (*noun*) a number is written in standard form when it is expressed as $A \times 10^n$ where A is always between 1 and 10 and n is a positive or negative integer

stationary point (= **turning point**) (*noun*) a maximum or minimum point on a graph where the graph turns; the point where a curve has zero gradient

subject (*noun*) the **variable** (= letter) on its own, on one side of the equals sign in a formula

subset (*noun*) a collection of objects (= **set**) which contains part of another set

subtended angle (*noun*) (in a circle) an angle in which the arms start and finish at the ends of an arc

sum (*noun*) the total resulting from the addition of two or more numbers, amounts or items

sum (*verb*) to add two or more numbers or amounts together

surd (*noun*) a number written exactly using roots; for example, $\sqrt{3}$

tangent (*noun*) a straight line that touches a curve at one point only

terminating decimal (*noun*) a decimal which ends; for example, 0.35

tetrahedron (*noun*) a solid object that has 4 triangular **faces** (= flat surfaces)

translation (*noun*) a 'sliding' movement in which all the points on the shape move the same distance in the same direction

trapezium (*noun*) (*British English*) a flat shape with 4 sides (= a **quadrilateral**) where one pair of opposite sides are parallel

tree diagram (*noun*) a diagram that shows two or more events with possible outcomes and their probabilities listed on branches

turning point (= **stationary point**) (*noun*) a maximum or minimum point on a graph where the graph turns; the point where a curve has zero gradient

union (*noun*) the union of two **sets** A and B is the set of elements which are in A or B, or both

universal set (*noun*) the set that contains all objects being considered

upper bound (*noun*) the value half a unit greater than the rounded measurement

variable (*noun*) a letter such as x or y in a term, expression, formula or equation whose value can vary; a letter used to represent an unknown quantity to solve a problem using an equation(s)

vector (*noun*) a quantity that has **magnitude** (= size) and direction

velocity (*noun*) the rate at which **displacement** changes with time

Venn diagram (*noun*) a drawing in which ovals are drawn inside a rectangle to show the relationships between **sets** (= a collection of objects)

vertices (*noun*) (*plural*) (*singular* **vertex**) points where two or more edges meet (a corner)

working (*noun*) a record of the calculations made when solving a problem

UNIT 6 ANSWERS

UNIT 6: NUMBER 6

EXERCISE 1

1 ▶ a Yes, each hotdog costs £1.80

   b No, 5 apples are 32p each, 9 apples are 33p each

   c No, Tom's speed is 13.33 km/hr, Ric's speed is 13.55 km/hr (4 s.f.)

2 ▶ a Yes, distance is $\frac{3}{4}$ of the time

   b $s = \dfrac{3t}{4}$ **c** 18 km

3 ▶ 4 min 30 s (NOT in direct proportion)

4 ▶ Yes, speed ÷ time is a constant

5 ▶ a

t seconds	60	100	140	200	260
V cm³	300	500	700	1000	1300

   b $V = 5t$ **c** 5 minutes

6 ▶ a 55 600 (3 s.f.) **b** 38.6 (3 s.f.)

7 ▶ a $468   **b** 7

8 ▶ a $97.50   **b** 2.3 m

EXERCISE 1*

1 ▶ 3 min 30 s (NOT in direct proportion)

2 ▶ a Yes, miles are $\frac{5}{8}$ of km or km is 1.6 of mile

   b 1 : 1.6   **c** 62.5 miles

3 ▶ Extension at 4.5 N should be 20.25 mm

4 ▶ £300

5 ▶ a

Area, A cm²	25	30	40	70	100
Cost, $C	1125	1350	1800	3150	4500

   b $C = 45A$ **c** 85 cm²

6 ▶ a 2.4 m   **b** 13.3 m

7 ▶ a 360 g   **b** 10

8 ▶ a 5.25 ohms   **b** $R = 0.0035l$ **c** 2.2 m

ACTIVITY 1

Time (t hours)	2	3	4	5	6	8
Speed (v km/hr)	60	40	30	24	20	15
$t \times v$	120	120	120	120	120	120

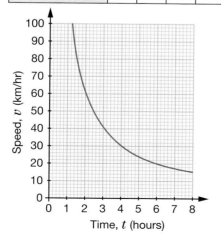

Speed required is 48 km/hr
Yes, calculate 120 ÷ 2.5

EXERCISE 2

1 ▶ a Yes, product of time and speed is 600.

   b 50 m/s

2 ▶ a

Temperature °C	10	16	20	25	30
Number sold	600	375	300	240	200

   b $T \times N = 6000$

   c

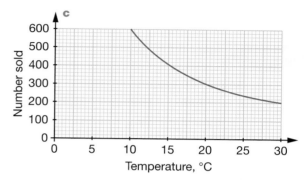

   d The model predicts sales will increase without limit, which is unrealistic

3 ▶ a 20 years   **b** 12 000 men

4 ▶ a 1.25 s   **b** 40 Mb/s

   c $s \times t = 20$

5 ▶ a 16 years   **b** 500

6 ▶ a 4 days

   b Yes, because 3 × 200 is the same as 300 × 2

EXERCISE 2*

1 ▶ a

A	3	4.8	5	6	32
B	16	10	9.6	8	1.5

   b $A \times B = 48$

2 ▶ a 600 parts per million   **b** 1 trillion

3 ▶ a 3 minutes

   b No, if inversely proportional it should take 6 minutes

4 ▶ a 12 amps   **b** $A \times R = 144$

   c 24 ohms

5 ▶ a 24 days   **b** 8

6 ▶ a 20 km/hr   **b** 1 hour 58 mins

ACTIVITY 2

Will depend on the measured reaction time. An average reaction time is around 0.3 seconds ⇒ a speed of 208 km/hr

ACTIVITY 3

a i 2   **ii** 3   **iii** 4   **iv** 5   **v** 6
   $a^{\frac{1}{2}}$ means the square root of a

b i 2   **ii** 3   **iii** 4 $a^{\frac{1}{3}}$ means the cube root of a

EXERCISE 3

1 ▶ 5   2 ▶ 3   3 ▶ 2   4 ▶ $\frac{1}{2}$

5 ▶ 27   6 ▶ 32   7 ▶ $\frac{4}{5}$ 8 ▶ $\frac{8}{27}$

9 ▶ $\frac{9}{4}$ 10 ▶ 128   11 ▶ 3   12 ▶ 2

13 ▶ $x = \frac{1}{3}$ 14 ▶ $x = \frac{5}{2}$

EXERCISE 3*

1 ▶ 12   2 ▶ 30   3 ▶ $\frac{1}{2}$ 4 ▶ $\frac{1}{2}$

5 ▶ 27   6 ▶ 25   7 ▶ $\frac{8}{7}$ 8 ▶ $\frac{27}{64}$

9 ▶ $\frac{81}{16}$ 10 ▶ 4   11 ▶ 5   12 ▶ $\frac{125}{2}$

13 ▶ $x = \frac{3}{4}$ 14 ▶ $x = \frac{1}{10}$

EXERCISE 4

1 ▶ $\frac{1}{9}$ 2 ▶ $\frac{1}{8}$ 3 ▶ $\frac{1}{4}$ 4 ▶ 9

5 ▶ $\frac{1}{3}$ 6 ▶ 1   7 ▶ 2   8 ▶ $\frac{4}{3}$

9 ▶ 81   10 ▶ $\frac{27}{8}$ 11 ▶ $\frac{9}{25}$ 12 ▶ $\frac{1}{2}$

13 ▶ $\frac{1}{5}$ 14 ▶ $\frac{1}{8}$ 15 ▶ 1   16 ▶ 2

17 ▶ 4   18 ▶ $\frac{3}{2}$ 19 ▶ 1   20 ▶ 9

21 ▶ $x = -1$ 22 ▶ $x = -1$ 23 ▶ $x = -1$

24 ▶ $x = -2$ 25 ▶ $-\frac{1}{2}$

EXERCISE 4*

1 ▶ $\frac{1}{25}$ 2 ▶ $\frac{1}{64}$ 3 ▶ $\frac{1}{12}$ 4 ▶ 4

5 ▶ $\frac{49}{4}$ 6 ▶ $\frac{125}{64}$ 7 ▶ $\frac{9}{25}$ 8 ▶ $\frac{1}{8}$

9 ▶ $\frac{1}{2}$ 10 ▶ $\frac{1}{32}$ 11 ▶ 1   12 ▶ 9

13 ▶ 25   14 ▶ $\frac{16}{81}$ 15 ▶ $\frac{8}{27}$ 16 ▶ 4

17 ▶ $\frac{25}{49}$ 18 ▶ 1   19 ▶ $\frac{5}{3}$ 20 ▶ $\frac{25}{8}$

21 ▶ $x = -2$ 22 ▶ $x = -\frac{1}{2}$ 23 ▶ $x = \frac{2}{3}$

24 ▶ $x = -9$ 25 ▶ $x = 4$

EXERCISE 5   REVISION

1 ▶ No, $B = 4A$ except in the last column

2 ▶ a

Depth, d metres	5	8	12	25	40
Pressure, P bars	0.5	0.8	1.2	2.5	4

   b $P = 0.1d$

   c   Yes, pressure is 7.5 bars < 8.5 bars

3 ▶ a   152.5 kg

   b   8 m

4 ▶ a

Number of workers, w	4	8	6	2
Number of days, d	12	6	8	24

   b $dw = 48$ c   48

5 ▶ a   32 hours   b   60 °C

6 ▶ a   6 days   b   5 harvesters

7 ▶ 6   8 ▶ 2   9 ▶ 3   10 ▶ $\frac{1}{3}$

11 ▶ 100   12 ▶ 64   13 ▶ 1   14 ▶ $\frac{4}{5}$

15 ▶ $\frac{1}{64}$ 16 ▶ 8   17 ▶ $\frac{9}{4}$ 18 ▶ $\frac{1}{5}$

19 ▶ $\frac{1}{9}$ 20 ▶ 8   21 ▶ 8   22 ▶ $\frac{1}{4}$

23 ▶ 2   24 ▶ 3   25 ▶ 1   26 ▶ $x = 1$

27 ▶ $x = -1$ 28 ▶ $x = -\frac{1}{2}$

EXERCISE 5*   REVISION

1 ▶ a

Force, F N	1.8	2.52	4.32	10.8	18
Acceleration, a m/s²	0.5	0.7	1.2	3	5

   b $F = 3.6a$ c   25 m/s²

2 ▶ a   6 s   b   162 MB

3 ▶ Yes, $XY = 144$ in all cases.

4 ▶ a

Number of pipes, n	1	4	6	8	10
Time, t hrs	18	4.5	3	2.25	1.8

   b $nt = 18$ c   12

5 ▶ a   126 160   b   50 cm²

6 ▶ a   40 sides   b $\frac{1}{2}$

7 ▶ 11   8 ▶ −5   9 ▶ 4   10 ▶ 1000

11 ▶ 16   12 ▶ $\frac{3}{4}$ 13 ▶ 1   14 ▶ 1

15 ▶ $\frac{25}{16}$ 16 ▶ $\frac{1}{32}$ 17 ▶ $\frac{1}{125}$ 18 ▶ 256

19 ▶ $\frac{1}{3}$ 20 ▶ $-\frac{1}{2}$ 21 ▶ $\frac{1}{81}$ 22 ▶ −3

23 ▶ 27   24 ▶ $\frac{125}{8}$ 25 ▶ $\frac{3}{16}$ 26 ▶ $x = -2$

27 ▶ $x = -\frac{1}{3}$ 28 ▶ $x = -6$

EXAM PRACTICE: NUMBER 6

1 ▶ a

Volume, v litres	7.5	15	30	45
Cost, £C	8.5	17	34	51

   b $C = \frac{17}{15}v$

2 ▶ a   3 pages   b   8

3 ▶ a   8   b   4   c   8   d $\frac{2}{3}$

   e   27   f   8   g $\frac{3}{2}$ h $\frac{1}{81}$

   i   3   j $\frac{8}{27}$ k $\frac{1}{3}$ l $\frac{1}{4}$

   m   6   n $\frac{8}{27}$ o   2   p   4

   q $x = -\frac{1}{3}$

UNIT 6: ALGEBRA 6

ACTIVITY 1   Y Y Y N Y Y N N

EXERCISE 1   1 ▶ a $y = 5x$ b   30   c   5

d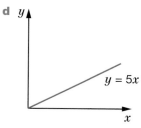

$y = 5x$

2 ▶ a $d = 4t$ **b** 60   **c** 45

3 ▶ a $y = 5x$ **b** $y = 50$

    c $x = 13$

4 ▶ a $y = 6.5x$

    b $y = 91$ **c** $x = 22$

d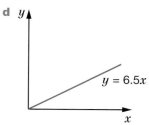

$y = 6.5x$

5 ▶ a $x = 5$ **b** $x = 10.1$ (1 d.p.)

    c $x = 8.125$

6 ▶ a $y = \dfrac{x}{60}$ **b** $y = 9$

7 ▶ a $y = 2x$ **b** 10 cm   **c** 7.5 kg

8 ▶ a $e = \dfrac{M}{20}$ **b** 5 m   **c** 120 kg

9 ▶ a $l = 75t$ **b** 1950 sales

10 ▶ a $N = 7t$

    b Yes, as 210 people would turn up to swim

EXERCISE 1*

1 ▶ a $v = 9.8t$ **b** 49 m/s   **c** 2.5 s

2 ▶ a $c = \dfrac{m}{3}$ **b** $2.50   **c** 600 g

3 ▶ a $d = 150$ m   **b** 1500 m   **c** 266.7 g

4 ▶ a $m = 6.5n$ **b** 975 g

    c 1540 approx

5 ▶ a $h = \dfrac{3y}{2}$ **b** 0.75 m   **c** 4 months

6 ▶ a $d = 500t$ **b** $d = 2500$ km

    c $t = 4.5$ hours

    d **i** The distance doubles.

      ii The distance halves.

7 ▶ a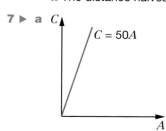

$C = 50A$

b $C = 50A$ **c** £4250

8 ▶ a $y = \dfrac{23x}{3}$ **b** 184   **c** 21

9 ▶ a $d = 10.5t$ **b** 84 km   **c** 0.73 hours

    d **i** It is trebled. **ii** It is divided by 3.

10 ▶ a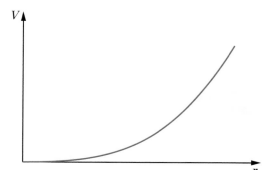

$C = 45t$

    b $C = 45T$ **c** 23 hours

EXERCISE 2

1 ▶ a $y = 4x^2$ **b** 144   **c** 4

2 ▶ a $p = 2q^2$ **b** 18   **c** 7

3 ▶ a $v = 2w^3$ **b** 54   **c** 4

4 ▶ a $m = 10\sqrt{n}$ **b** 20   **c** 25

5 ▶ a $y = 5t^2$

    b 45 m

    c $\sqrt{20}$ s $\simeq 4.47$ s

6 ▶ a $P = \dfrac{h^3}{20}$ **b** $86.40   **c** 8 cm

7 ▶ a $E = 5s^2$ **b** $E = 20$ J   **c** $s = 6.2$ m/s

    d The kinetic energy, E, is multiplied by 4.

8 ▶ a $C = 0.05s^3$ **b** $C = £6.25$

EXERCISE 2*

1 ▶

g	2	4	6
f	12	48	108

2 ▶

n	1	2	5
m	4	32	500

3 ▶ a $T = \dfrac{R^2}{450}$ **b** $T = 50$ minutes

4 ▶ a $d = 4.9t^2$ **b** 490 m   **c** 15 s

    d The distance moved is multiplied by 4.

5 ▶ a $V = 4.188r^3$ **b** 33 504 cm³

    c

V

r

6 ▶ a $R = \left(\dfrac{5}{256}\right)s^2$

b 113 km/h

7 ▶ a $H = 1.5\sqrt[3]{y}$

b 512 years old

8 ▶ $x = 10\sqrt{2}$

ACTIVITY 2

$t^2 \approx 3.95 \times 10^{-20}d^3$

Planet	d (million km)	t (Earth days) (2 s.f.)	t^2/d^3
Mercury	57.9	88	0.04
Jupiter	778	4300	0.04
Venus	108	220	0.04
Mars	228	680	0.04
Saturn	1430	11 000	0.04
Uranus	2870	31 000	0.04
Neptune	4500	60 000	0.04

EXERCISE 3

1 ▶ a $y = \dfrac{12}{x}$ **b** $y = 6$ **c** $x = 4$

d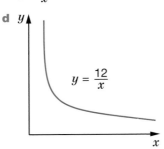

$y = \dfrac{12}{x}$

2 ▶ a $d = \dfrac{250}{t}$ **b** $d = 125$ **c** $t = 5$

3 ▶ a $P = \dfrac{3000}{V}$ **b** $P = 2000$ N/m²

c $V = 2.5$ m³ **d** The volume is halved.

4 ▶ a $t = \dfrac{600\,000}{p}$

b No, it takes 4 minutes. When $p = 2500$ W, $t = 240$ seconds.

5 ▶ a $m = \dfrac{36}{n^2}$ **b** $m = 9$ **c** $n = 6$

6 ▶ a $V = \dfrac{100}{w^3}$ **b** $V = 100$ **c** $w = 5$

7 ▶ a $I = 4 \times \dfrac{10^5}{d^2}$ **b** 0.1 candle power

8 ▶ a $L = \dfrac{1}{4d^2}$ **b** 25 days

EXERCISE 3*

1 ▶ a $C = \dfrac{5000}{t}$ **b** \$277.78 **c** 12.5 °C

2 ▶ a $V = \dfrac{750}{t}$

b No; the balloon pops when $V = 25\,000$ cm³

3 ▶ a Graph showing inverse proportion.

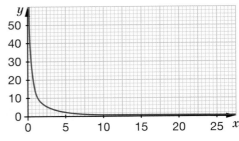

b 12 **c** It is always 12.

4 ▶

b	2	5	10
a	50	8	2

5 ▶ a $r \propto t$, in fact $r = \dfrac{20}{t}$

b

r	1	4	5	10
t	20	5	4	2

6 ▶ a $R = \dfrac{2}{r^2}$ **b** $\dfrac{2}{9}$ ohm

7 ▶ a

Day	N	T
Mon	400	25
Tues	447	20
Wed	500	16

b 407 approx.

8 ▶ 4.5 s

ACTIVITY 3

	Rabbit	Dog	Man	Horse
Pulse (beats/min)	165	135	83	65
Mass (kg)	5	12	70	200

Total heartbeats for a human life-span of 75 years $\approx 3.27 \times 10^9$

According to theory:

	Rabbit	Dog	Man	Horse
Life-span (years)	37.7	46.1	75	95.8

Theory clearly not correct.

EXERCISE 4

1 ▶ a **2 ▶** b^2 **3 ▶** $\dfrac{1}{c^3}$

4 ▶ d^5 **5 ▶** e **6 ▶** f

7 ▶ $\dfrac{1}{a^2}$ **8 ▶** $\dfrac{1}{g^2}$ **9 ▶** a^2

10 ▶ $\dfrac{1}{b}$ **11 ▶** c^3 **12 ▶** d

13 ▶ e^2 **14 ▶** $\dfrac{1}{f}$

15 ▶ a $2^3 \div 2^3 = 2^0$ **b** 8

c $2^3 \div 2^3 = 8 \div 8 = 1$

d $2^3 \div 2^3 = 2^0 = 1$

e $7^5 \div 7^5 = 7^0 = 16\,807 \div 16\,807 = 1$

f $a^0 = 1$

16 ▶ a $\frac{1}{3}$ b $\frac{1}{16}$ c $\frac{1}{100\,000}$

d $\frac{4}{3}$ e $\frac{125}{64}$ f $\frac{4}{5}$

g $\frac{16}{121}$ h $\frac{10}{7}$ i $100\,000$

j $\frac{125}{8}$ k $5^0 = 1$ l $7^1 = 7$

EXERCISE 4*

1 ▶ a^2 2 ▶ b^3 3 ▶ $\frac{2}{c^4}$

4 ▶ $\frac{4}{c^2}$ 5 ▶ $\frac{12}{a}$ 6 ▶ $\frac{8}{b}$

7 ▶ a 8 ▶ 1   9 ▶ $\frac{1}{c}$

10 ▶ $\frac{1}{d}$ 11 ▶ $-9a^6$ 12 ▶ $\frac{1}{a^2}$

13 ▶ $\frac{1}{c^2}$ 14 ▶ e 15 ▶ $k = 2\frac{1}{3}$

16 ▶ $k = -6$ 17 ▶ $x = 3, y = -2$

18 ▶ $x = \frac{4}{5}, y = -\frac{3}{5}$

19 ▶ a 9   b $\frac{125}{12}$ c $\frac{125}{8}$

20 ▶ a 4   b $\frac{1}{3}$ c -2

d $-\frac{1}{2}$ e $\frac{7}{2}$ f $\frac{7}{4}$

EXERCISE 5   REVISION

1 ▶ a $y = 6x$ b $y = 42$ c $x = 11$

2 ▶ a $p = 5q^2$ b $p = 500$ c $q = 11$

3 ▶ a $c = \frac{3}{4}a^2$ b \$675   c $28.3\,\text{m}^2$

4 ▶

n	1	2	4	8
t	80	40	20	10

5 ▶ $g \propto h^3$

6 ▶ $a = 4, b = 2$

7 ▶ a

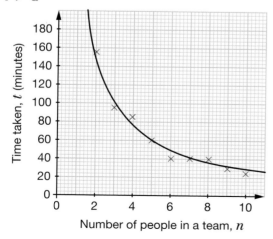

b Answers close to $t = \frac{300}{n}$

c Answer using students' formulae from part b

$t = \frac{300}{n}$ gives $t = 20$ minutes.

8 ▶ a Graph B   b Graph D

c Graph A   d Graph C

9 ▶ a^2 10 ▶ a^4 11 ▶ $\frac{1}{d^2}$

12 ▶ $\frac{1}{b^{\frac{3}{2}}}$

13 ▶ c

14 ▶ a 10   b 2   c $\frac{2}{3}$ d $\frac{5}{7}$

e -3 f $-\frac{1}{3}$ g $-\frac{2}{3}$ h $\frac{1}{10}$

15 ▶ 1

EXERCISE 5*   REVISION

1 ▶ a $y^2 = 50z^3$ b 56.6   c 5.85

2 ▶ 1500 m

3 ▶ a $m = \frac{8839}{\sqrt{n}}$ b 3.23×10^5

c 7.81×10^{-5}

4 ▶

x	0.25	1	4	25
y	20	10	5	2

$y = \frac{10}{\sqrt{x}}$

5 ▶ $w \propto \frac{1}{t^2}$ (rule B)

6 ▶ a $D = \frac{6390}{r^2}$

b $D = 10.2\,\text{cm}$ (1 d.p.)

c $r = 10.0\,\text{cm}$ (1 d.p.)   d $\frac{1}{4}d\,\text{cm}$

7 ▶ a $F = \frac{1.0584 \times 10^{17}}{d^2}$ b 4 N

8 ▶

p	2	4	6
w	7	$1\frac{3}{4}$	$\frac{7}{9}$

9 ▶ a^2 10 ▶ $\frac{3}{c^2}$ 11 ▶ $\frac{2}{a}$

12 ▶ d 13 ▶ $-9a^6$ 14 ▶ 2

15 ▶ 1   16 ▶ $5x^2y^3$

EXAM PRACTICE: ALGEBRA 6

1 ▶ a $y = \frac{x}{2}$ b 2.5   c 1

2 ▶ a $p = \frac{1}{5}q^2$ b 80   c 30

3 ▶ a 9000   b $N = \frac{9000}{d^2}$

c 2250   d 3 mm

4 ▶ a $D = \frac{9540}{r^2}$ b 42.4 cm

c 12.6 cm   d $\frac{x}{\sqrt{3}}$

5 ▶ a $2a^2$ b b^{-2} c c^4 d $x = -\frac{1}{2}$

UNIT 6: SEQUENCES

ACTIVITY 1   Seema; 13 rows with 9 balloons left over

EXERCISE 1

1 ▶ 2, 4, 6, 8
2 ▶ −9, −6, −3, 0
3 ▶ 15, 10, 5, 0
4 ▶ 2, 4, 8, 16
5 ▶ 12, 6, 3, 1.5
6 ▶ Add 4; 19, 23, 27
7 ▶ Subtract 5; −7, −12, −17
8 ▶ Double; 48, 96, 192
9 ▶ Halve; 4, 2, 1
10 ▶ Add 0.3; 1.4, 1.7, 2

EXERCISE 1*

1 ▶ −1, 0.5, 2, 3.5
2 ▶ 3, 1.75, 0.5, −0.75
3 ▶ 1, 2.5, 6.25, 15.625
4 ▶ $3, -1, \frac{1}{3}, -\frac{1}{9}$
5 ▶ 2, 3, 5, 8
6 ▶ Add $2\frac{1}{2}$; 13, $15\frac{1}{2}$, 18
7 ▶ Divide by 3; 3, 1, $\frac{1}{3}$
8 ▶ Square; 65 536, 4.3×10^9, 1.8×10^{19}
9 ▶ Multiply by $-\frac{1}{2}$; $\frac{1}{16}$, $-\frac{1}{32}$, $\frac{1}{64}$
10 ▶ Double then add 1; 63, 127, 255

EXERCISE 2

1 ▶ 3, 5, 7, 9
2 ▶ 4, 9, 14, 19
3 ▶ 30, 27, 24, 21
4 ▶ 2, 5, 10, 17
5 ▶ 3, 6, 9, 12
6 ▶ $2, \frac{3}{2}, \frac{4}{3}, \frac{5}{4}$
7 ▶ 8
8 ▶ 7
9 ▶ 7
10 ▶ 22

EXERCISE 2*

1 ▶ −1, 4, 9, 14
2 ▶ 97, 94, 91, 88
3 ▶ $1, \frac{3}{2}, 2, \frac{5}{2}$
4 ▶ 3, 7, 13, 21
5 ▶ 3, 6, 11, 18
6 ▶ $3, \frac{5}{3}, \frac{7}{5}, \frac{9}{7}$
7 ▶ 8
8 ▶ 12
9 ▶ 10
10 ▶ 51

ACTIVITY 2

a gives the difference between the terms of the sequence

n	1	2	3	4	10
$3n + 2$	5	8	11	14	32

3 = difference between the terms of the sequence

3 + 2 gives the first term of the sequence

a gives the difference between the terms of the sequence. $a + b$ gives the first term of the sequence

EXERCISE 3

1 ▶ 17, 20, 23
2 ▶ −7, −10, −13
3 ▶ 11.5, 13, 14.5
4 ▶ 56, 76, 99
5 ▶ 0, 8, 19
6 ▶ −4, −11, −20

EXERCISE 3*

1 ▶ 58, 78, 101
2 ▶ 1, −6, −15
3 ▶ 1, 8, 17
4 ▶ 35.5, 46.5, 59
5 ▶ 71, 101, 139
6 ▶ −6, −19, −38

ACTIVITY 3

74, 100, 130; 150, 215, 297; −54, −90, −139

ACTIVITY 4

3, 5, 7, 9, 11, 13
$b = 2t + 1$; 201 balloons
$b = 3s + 1$ (s = no. of squares); $b = 5h + 1$ (h = no. of hexagons); $b = 7h + 5$ (h = no. of hexagons)

ACTIVITY 5

4, 8, 12, 16, 20, 24; $4n$

EXERCISE 4

1 ▶ $\frac{1}{n}$
2 ▶ $\frac{1}{2n - 2}$
3 ▶ $3n + 1$
4 ▶ $34 - 4n$
5 ▶ a 1, 3, 5, 7, 9, 11   b $c = 2l - 1$
    c c is always odd, 50 layers.
6 ▶ a 8, 10, 12, 14, 16, 18
    b $s = 2n + 6$ c 47th

EXERCISE 4*

1 ▶ $\frac{n + 1}{n}$
2 ▶ $\frac{2n - 1}{2n + 1}$
3 ▶ $4n - 1$
4 ▶ $9 - 3n$
5 ▶ a 6, 10, 14, 18, 22, 26
    b $s = 4n + 2$ c 202
6 ▶ a 10, 16, 22, 28, 34, 40
    b $s = 6n + 4$ c 32nd

EXERCISE 5

1 ▶ a 3, 6, 9, 12; 300
    b 8, 14, 20, 26; 602
    c 3, 10, 17, 24; 696
    d 18, 15, 12, 9; −279
    e 4, 0, −4, −8; −392
2 ▶ a $5n + 2$ b $4n - 2$ c $21 - 2n$
3 ▶ a $4n - 3 = 101 \Rightarrow n = 26$, yes
    b $5n - 1 = 168 \Rightarrow n = 33.8$, no
    c $45 - 5n = -20 \Rightarrow n = 11$, yes
4 ▶ a 108   b 103   c 103   d 106
5 ▶ 119   6 ▶ 6   7 ▶ 12   8 ▶ 24
9 ▶ $a = 3, d = 5$ 10 ▶ $a = 20, d = -2$
11 ▶ 11 weeks   12 ▶ $36.10

EXERCISE 5*

1 ▶ a 2, 11, 20; 893
    b −2, −11, −20; −893
    c −27, −24, −21; 270
    d 2.5, 3, 3.5; 52
    e 0.25, 0, −0.25; −24.5
2 ▶ a $13 - 3n = -230 \Rightarrow n = 81$, yes
    b $2.5 + 1.5n = 99.5 \Rightarrow n = 64\frac{2}{3}$, no
    c $5n - 36 = 285 \Rightarrow n = 64.2$, no
3 ▶ a 1004   b 1009   c $1000\frac{1}{3}$
4 ▶ $a = 5, d = 3$

5 ▶ −92

6 ▶ $8 + 7n$

7 ▶ 123

8 ▶ $−10 − 4n$

9 ▶ 4

10 ▶ 6700 days or 18 years

11 ▶ 1001 days

12 ▶ 18 weeks

ACTIVITY 6

$$\frac{n(n + 1)}{2}$$

EXERCISE 6

1 ▶ 500 500   **2** ▶ 3320   **3** ▶ 10 100

4 ▶ 40 000   **5** ▶ 920   **6** ▶ 9960

7 ▶ 2385   **8** ▶ 750   **9** ▶ 770

10 ▶ 78   **11** ▶ $2550

12 ▶ a $6n$ b   1261

EXERCISE 6*

1 ▶ 4425   **2** ▶ −16 600   **3** ▶ 2420

4 ▶ 6375   **5** ▶ 140 500   **6** ▶ 0

7 ▶ $n^2 + 4n$ **8** ▶ 6640   **9** ▶ 1504

10 ▶ 25   **11** ▶ $\frac{7}{16}$, $1\frac{9}{16}$ hekats

12 ▶ a   1150   b $33 120

ACTIVITY 7

Number of square	1	2	3	4	5
Number of grains	$1 = 2^0$	$2 = 2^1$	$4 = 2^2$	$8 = 2^3$	$16 = 2^4$
Total number	1	3	7	15	$31 = 2^5 − 1$

Number of square	6	7	64
Number of grains	$32 = 2^5$	$64 = 2^6$	2^{63}
Total number	$63 = 2^6 − 1$	$127 = 2^7 − 1$	$2^{64} − 1$

$2^{64} − 1 \approx 1.84 \times 10^{19}$

4.43×10^{20} mm^3 or 443 km^3

2.77×10^{17} mm or 2.77×10^{11} km

1800:1

EXERCISE 7   REVISION

1 ▶ a   32, 44, 58   b   40, 40, 38

2 ▶ a   1800 m   b $800 + 200n$

    c   36 days

3 ▶ a   Sequence 1; 3072   b $4n − 1$

4 ▶ a   Odd numbers   b   4, 9, 16, 25

    c   121   d n^2 e   29

5 ▶ $a = −2, d = 3$ **6** ▶ $a = 27, d = −4$

7 ▶ 203   **8** ▶ 101

9 ▶ 45 days   **10** ▶ 1072

11 ▶ 1704   **12** ▶ a $4n − 3$ b   9730

EXERCISE 7*   REVISION

1 ▶ a   20, 28, 30   b   10, 3, −7

2 ▶ a   £2.90   b $50 + 20n$ c   12 years

3 ▶ a   17, 21, 25, 29

    b   7, 13   c   1, 3, 7, 13, …; 21st term

4 ▶ a   1, 8, 21, 40, 65

    b $s = 3n^2 − 2n$ c   15th

5 ▶ $a = −8, d = 4$ **6** ▶ $a = 10, d = −2.5$

7 ▶ $35 + 17n$ **8** ▶ 1232

9 ▶ $72.50, $87.50, … $177.50

10 ▶ $n(2n + 7)$ **11** ▶ 8 days   **12** ▶ 270 m

EXAM PRACTICE: SEQUENCES

1 ▶ a   80, 76, 72

    b   3, 10, 21   c $0, \frac{1}{3}, \frac{1}{2}$

2 ▶ a   −9, −13, −17

    b   52, 68, 86   c   45, 60, 77

3 ▶ $a = 2, d = 6$ **4** ▶ 650

5 ▶ 2   **6** ▶ 741

UNIT 6: SHAPE AND SPACE 6

EXERCISE 1

1 ▶ 100° (Angle at centre of circle is twice angle at circumference.)

2 ▶ 30° (Angle at centre of circle is twice angle at circumference.)

3 ▶ ∠CAB = 105° (Angle at centre of circle is twice angle at circumference.)

∠ABO = 45° (Angles in a quadrilateral sum to 360°.)

4 ▶ ∠BAO + ∠CAO = 40° (Angle at centre of circle is twice angle at circumference.)

∠OAC = 30° (Base angles in an isosceles triangle are equal.)

∠BAO = 40 − 30 = 10°

∠ABO = 10° (Base angles in an isosceles triangle are equal.)

5 ▶ ∠ACB = 40° (Angles on a straight line sum to 180°.)

∠AOB = 80° (Angle at centre of circle is twice angle at circumference.)

Reflex ∠AOB = 280° (Angles at a point sum to 360°.)

6 ▶ ∠ABC = 110° (Angles on a straight line sum to 180°.)

Reflex ∠AOC = 220° (Angle at centre of circle is twice angle at circumference.)

∠AOC = 140° (Angles at a point sum to 360°.)

7 ▶ ∠CBD = 30° (Angles in the same segment are equal.)

∠ACB = 60° (Angles in a triangle sum to 180°.)

8 ▶ ∠ACB = 70° (Angles on a straight line sum to 180°.)

∠AEB = 70° (Angles in the same segment are equal.)

∠BED = 110° (Angles on a straight line sum to 180°.)

9 ▶ ∠WYX = 40° (Angles in the same segment are equal.)

∠XYZ = 100° (Opposite angles of a cyclic quadrilateral sum to 180°.)

∠WYZ = 100 − 40 = 60°

10 ▶ ∠ZWY = 24° (Angles in the same segment are equal.)

∠ZWX = 44 + 24 = 68°

∠ZYX = 112° (Opposite angles of a cyclic quadrilateral sum to 180°.)

11 ▶ ∠OAB = ∠OBA = 35° (Base angles of an isosceles triangle are equal.)

∠AOB = 110° (Angles in a triangle sum to 180°.)

∠COB = 180 − 110 = 70° (Angles on a straight line sum to 180°.)

∠OBC = 90° (Angle between the tangent and a radius is 90°.)

x = 180 − 90 − 70 = 20° (Angles in a triangle sum to 180°.)

12 ▶ ∠CBO = ∠CDO = 90° (Angle between the tangent and a radius is 90°.)

∠BOD = 112° (Angle at the centre is twice the angle at the circumference.)

g = 360 − (112 + 90 + 90) = 68° (Angles in a quadrilateral sum to 360°.)

13 ▶ **a** Students' explanations may vary, e.g.

∠GHO = x, ∠HGO = x (Base angles in an isosceles triangle are equal)

∠HOG = 180 − 2x (Angles in a triangle total 180°)

∠FGH = 90° (Angle in a semicircle is 90°)

∠FGO = ∠GFO = 90 − x (Base angles in an isosceles triangle are equal)

∠GOF = 180 − (180 − 2x) = 2x (Angle at the centre is twice the angle at the circumference)

b ∠GFO = 49°, ∠HGO = 41°

14 ▶ a = 180 − 108 = 72° (Angles on a straight line sum to 180°)

b = 180 − 72 = 108° (Opposite angles in a cyclic quadrilateral sum to 180°)

c = 180 − 87 = 93° (Opposite angles in a cyclic quadrilateral sum to 180°)

15 ▶ ∠ADB and ∠BCA are angles in the same segment

16 ▶ ∠ADC + ∠ABC sum to 180°

EXERCISE 1*

1 ▶ ∠AOC = 360 − 220 = 140° (Angle at centre of circle is twice angle at circumference.)

2 ▶ Let ∠BAO = x, extend AB to P and OC to Q

∠PBC = x (Corresponding angles are equal.)

∠BCO = x (Alternate angles are equal.)

∠ABC = 180 − x (Angles on a straight line sum to 180°.)

Reflex ∠AOC = 360 − 2x (Opposite angles of a cyclic quadrilateral sum to 180°.)

∠AOC = 2x (Angles at a point sum to 360°.)

x + x + 180 − x + 2x = 360 (Angles in a quadrilateral sum to 360°.)

x = 60°

3 ▶ Reflex ∠AOC = 230° (Angles at a point sum to 360°.)

∠ABC = 115° (Angle at centre of circle is twice angle at circumference.)

4 ▶ ∠AOC = 360 − reflex ∠AOC (Angles at a point sum to 360°.)

Reflex ∠AOC = 2 × ∠ABC (Angle at centre of circle is twice angle at circumference.)

70 + 40 + ∠ABC + ∠AOC = 360 (Angles in a quadrilateral sum to 360°.)

110 + ∠ABC + (360 − 2 × ∠ABC) = 360

∠ABC = 110°

5 ▶ ∠VRS = ∠RVS = 54° (Base angles in an isosceles triangle are equal.)

∠TUV = 54° (Angles in the same segment are equal.)

6 ▶ ∠TRS = 52° (Alternate angles are equal.)

∠TUS = 52° (Angles in the same segment are equal.)

∠UVT = 76° (Angles in a triangle sum to 180°.)

7 ▶ ∠PZW = 119° (Angles in a triangle sum to 180°.)

∠WZY = 61° (Angles on a straight line sum to 180°.)

∠WXY = 61° (Angles in the same segment are equal.)

∠PXY = 119° (Angles on a straight line sum to 180°.)

8 ▶ ∠XWY = 90° (Angle in a semicircle is 90°.)

∠XWZ = 90 − 68 = 22°

∠XYZ = 22° (Angles in the same segment are equal.)

9 ▶ ∠OCB = x (Base angles in an isosceles triangle are equal.)

∠OBC = 180 − 2x (Angles in a triangle sum to 180°.)

$\angle OBA = 2x$ (Angles on a straight line sum to 180°.)

$\angle OAB = 2x$ (Base angles in an isosceles triangle are equal.)

$\angle BOA = 180 - 4x$ (Angles in a triangle sum to 180°.)

$\angle XOA = 180 - x - (180 - 4x) = 3x$ (Angles on a straight line sum to 180°.)

10 ▶ $\angle ABC = x$ (Angle at centre of circle is twice angle at circumference.)

$\angle BCD = 180 - 90 - x = 90° - x$ (Angles in a triangle sum to 180°.)

11 ▶ $\angle ADB = x°$ (Base angles in an isosceles triangle are equal.)

$\angle BDC = (180 - 4x)°$ (Base angles in an isosceles triangle are equal. Angles in a triangle total 180°).

Therefore $\angle ADC = (180 - 3x)°$

Therefore $\angle ADC + \angle ABC = 180°$ so quadrilateral is cyclic.

12 ▶ $\angle AFE + \angle EDA = 180°$ (opposite angles of a cyclic quadrilateral total 180°)

Similarly, $\angle ABC + \angle CDA = 180°$

But $\angle CDA + \angle EDA = \angle CDE$.

Therefore $\angle ABC + \angle CDE + \angle EFA = 360°$

13 ▶ $\angle BEC = \angle CDB$ (angles in the same segment are equal)

Therefore $\angle CEA = \angle BDA$

14 ▶ Join AO. $\angle OAX = 90°$ (Angle in a semicircle is 90°)

Therefore $\angle OAY = 90°$ (Angles on a straight line sum to 180°)

OX = OY (radii) and OA is common

Therefore triangles OYA and OXA are congruent (RHS) and AX = AY

15 ▶ Let $\angle QYZ = y°$ and $\angle YZQ = x°$.

So $\angle ZWX = y°$ and $\angle PWZ = 180 - y°$

In $\triangle YQZ$: $x + y + 20 = 180$

Therefore $x + y = 160$ (1)

In $\triangle PWZ$: $180 - y + x + 30 = 180$

Therefore $y - x = 30$ (2)

From (1) and (2) $x = 65$ and $y = 95$

Therefore angles of the quadrilateral are 65°, 85°, 95°, 115°

16 ▶ Draw OP and OQ. Let $\angle POQ = 2x°$.

Therefore $\angle PRQ = x°$ and $\angle XOQ = (180 - x)°$

Therefore $\angle XRQ + \angle XOQ = 180°$ and RXOQ is a cyclic quadrilteral

ACTIVITY 1

Circle	$\angle ECB$	$\angle OCB$	$\angle OBC$	$\angle BOC$	$\angle BAC$
C_1	60°	30°	30°	120°	60°
C_2	$x°$	$(90 - x)°$	$(90 - x)°$	$2x°$	$x°$

$\angle ECB = \angle BAC$

EXERCISE 2

1 ▶ $\angle TRS = 40°$ (Alternate segment theorem)

$\angle RTS = 70°$ (Angles in a triangle sum to 180°.)

2 ▶ $\angle TPQ = 30°$ (Angles in a triangle sum to 180°.)

$\angle QTB = 30°$ (Alternate segment theorem)

3 ▶ $\angle CBT = 40°$ (Base angles in an isosceles triangle are equal.)

$\angle BCT = 100°$ (Angles in a triangle sum to 180°.)

$\angle DCT = 80°$ (Angles on a straight line sum to 180°.)

$\angle DTA = 80°$ (Alternate segment theorem)

4 ▶ $\angle YZT = 50°$ (Base angles in an isosceles triangle are equal.)

$\angle YTX = 50°$ (Alternate segment theorem)

5 ▶ $\angle T_1T_2M = 75°$ (Alternate segment theorem)

$\angle T_1T_2M = 75°$ (Base angles in an isosceles triangle are equal.)

$\angle T_2MT_1 = 30°$ (Angles in a triangle sum to 180°.)

6 ▶ $\angle T_1T_2A = 105°$ (Alternate segment theorem)

$\angle T_1T_2B = 75°$ (Angles on a straight line sum to 180°.)

$\angle T_2T_1C = 105°$ (Alternate segment theorem)

$\angle T_2T_1B = 75°$ (Angles on a straight line sum to 180°.)

$\angle T_2BT_1 = 30°$ (Angles in a triangle sum to 180°.)

7 ▶ $\angle TBC = 25°$ (Base angles in an isosceles triangle are equal.)

$\angle ABT = 75°$ (Alternate segment theorem)

$\angle ABC = 75 + 25 = 100°$

8 ▶ $\angle TDC = 60°$ (Alternate segment theorem)

$\angle EDT = 50°$ (Alternate segment theorem)

$\angle EDC = 60 + 50 = 110°$

9 ▶ a 90° (Angle between tangent and radius is 90°)

b (Angles in a triangle sum to 180°.)

c 60° (Base angles in an isosceles triangle are equal.)

d 60° (Alternate segment theorem)

10 ▶ a 90° (Angle between tangent and radius is 90°)

b (Angles in a triangle sum to 180°.)

c 20° (Base angles in an isosceles triangle are equal.)

d 20° (Alternate segment theorem)

11 ▶ a $\angle NTM = \angle NPT$ (Alternate segment theroem)

b $\angle PLT = \angle NTM$ (Corresponding angles are equal)

12 ▶ ∠ATF = ∠FDT (Alternate segment theorem)

∠FDT = ∠BAF (Alternate angles, AC parallel to DT)

13 ▶ **a** ∠ATC = ∠ABT (Alternate segment theorem)

b ∠ABT = ∠BTD (Alternate angles, AB parallel to CD)

14 ▶ ∠CTB = ∠CDT (Alternate segment theorem)

∠CBT is common

Therefore all three angles are equal and the triangles are similar

EXERCISE 2*

1 ▶ ∠BT_1T_2 = 65° (Base angles in an isosceles triangle are equal.)

∠T_1DT_2 = 65° (Alternate segment theorem)

2 ▶ ∠TED =110° (Opposite angles of a cyclic quadrilateral sum to 180°.)

∠EDT = 35° (Base angles in an isosceles triangle are equal.)

∠ATE = 35° (Alternate segment theorem)

3 ▶ ∠T_1T_2C = 60° (Alternate segment theorem)

∠ET_2T_1 = 80° (Alternate segment theorem)

∠ET_2C = 60 + 80 = 140°

4 ▶ ∠TDC = 180 − 2(40 + x) = 100 − 2x (Base angles in an isosceles triangle are equal.)

∠TOC = 200 − 4x (Angle at centre of circle is twice angle at circumference.)

200 − 4x + 2x = 180 (Angles in a triangle sum to 180°.)

x = 10°

5 ▶ ∠ATE = 55° (alternate segment theorem)

∠TBC = 125° (angles on straight line sum to 180°)

∠BTC = 35° (angle sum of triangle is 180°)

∠ATB = 90° (angles on straight line sum to 180°)

Therefore AB is a diameter

6 ▶ ∠ATD = x° (alternate segment theorem), ∠DTC = 90° (CD is a diameter), therefore ∠BTC = (90 − x)° (angles on straight line sum to 180°)

Therefore ∠TBC is a right angle (angle sum of triangle is 180°)

7 ▶ ∠ACT = x + 9° (Alternate segment theorem)

∠CAT = x + 9° (Base angles in an isosceles triangle are equal.)

3x + 18 = 180 (Angles in a triangle sum to 180°.)

x = 54°

8 ▶ **a** ∠BAE = 90° (Angle in a semicircle is 90°.)

∠DAE = 90 − 35 = 55°

b ∠BAE = ∠BDE = 90° (Angle in a semicircle is 90°.)

∠DAE = 90 − 35 = 55°

∠DBE = 55° (Angles in the same segment are equal.)

∠ABE = 180 − 90 − 20 = 70° (Angles in a triangle sum to 180°.)

∠ABD = 70 + 55 = 125°

∠AED = 180 − 125 = 55° (Opposite angles of a cyclic quadrilateral sum to 180°.)

∠BED = 55 − 20 = 35°

c ∠ADE = 70° (angle sum of triangle is 180°)

∠ACD = 35° (angle sum of triangle is 180°)

Therefore ∠ACD is isosceles

9 ▶ **a** ∠BTD = 110° (Angles in a triangle sum to 180°.)

∠DTA = 70° (Angles on a straight line sum to 180°.)

b ∠TCD = 70° (Alternate segment theorem)

∠BCT = 110° (Angles on a straight line sum to 180°.)

c ∠CBT is common and ∠CTB = 40° (alternate segment theorem) therefore all three angles are equal and triangles BCT and BTD are similar

10 ▶ ∠TEB = 80° (Alternate segment theorem)

∠ATD = 80° (Vertically opposite angles are equal.)

∠TFD = 80° (Alternate segment theorem)

∠EFT = 20° (Angles in a triangle sum to 180°.)

11 ▶ ∠TEB = 115° (Alternate segment theorem)

∠ATD = 115° (Vertically opposite angles are equal.)

∠TED = 115° (Alternate segment theorem)

∠DEB = 360 − 115 − 115 = 130° (Angles at a point sum to 180°.)

12 ▶ **a** Triangles ACG and ABF are right-angled

b Angles ACG and ABF are equal and in the same segment of the chord FG

13 ▶ **a** **i** ∠T_2CT_1, T_2T_1B

ii ∠AT_1C, ∠BT_1D, ∠T_1DB, ∠CT_2T_1

b Triangles BT_1T_2 and T_1BD are isosceles, therefore BT_2 = BD

14 ▶ **a** ∠EOC = 2x° (angle at centre is twice angle at circumference)

∠CAE = (180 − 2x)° (opposite angles of a cyclic quadrilateral sum to 180°)

b ∠AEB = x° (angle sum of triangle is 180°)

Therefore triangle ABE is isosceles

c ∠ECB = $x°$ (base angles of isosceles triangle are equal)

Therefore ∠BEC = $(180 - 2x)° = ∠CAE$

Since ∠BAE and ∠BEC are equal and angles in alternate segments, BE must be the tangent to the larger circle at E

ACTIVITY 2

OM is common; OA = OB (radii of same circle);

∠OMA = 90° (angles on a straight line sum to 180°)

Therefore the triangles are congruent (RHS).

OAM and OBM are congruent, so AM = AB.

Therefore M is the mid-point of AB.

EXERCISE 3

1 ▶ OM = 15 cm

2 ▶ a 90°　　　**b** 65°　　　**c** 130°

3 ▶ 12　　　**4 ▶** 3　　　**5 ▶** 8

6 ▶ 8　　　**7 ▶** 4　　　**8 ▶** 3

ACTIVITY 3

∠APD = ∠BPC　　(Vertically opposite angles are equal)

∠CDA = ∠CBA　　(Angles in same segment off chord AC are equal)

∠BAD = ∠BCD　　(Angles in same segment off chord BC are equal)

⇒ Triangles APD and CPB are similar

⇒ $\dfrac{CP}{AP} = \dfrac{BP}{DP}$

⇒ AP × BP = CP × DP

ACTIVITY 4

∠BAD = ∠BCD　　(Angles in same segment off chord BD are equal)

∠P is common to triangle APD and triangle BPC

∠ADP = ∠CBP

⇒ Triangles APD and CPB are similar

⇒ $\dfrac{CP}{AP} = \dfrac{BP}{DP}$

⇒ AP × BP = CP × DP

EXERCISE 3*

1 ▶ a AM = 6 cm (The perpendicular from the centre of a circle to a chord bisects the chord.)

b AO = 10 cm

2 ▶ AB = 20 cm　　**3 ▶** 6　　**4 ▶** 3

5 ▶ 8　　**6 ▶** 5　　**7 ▶** 4　　**8 ▶** 4

EXERCISE 4

1 ▶ 16 cm　　　　**2 ▶** x = 2 or 14

3 ▶ a a = 38° (The angle between the tangent and the chord is equal to the angle in the alternate segment.)

b b = 35° (The angle between the tangent and the chord is equal to the angle in the alternate segment.)

c = 93° (Angles in a triangle sum to 180° or the angle between the tangent and the chord is equal to the angle in the alternate segment.)

d = 93° (Angles on a straight line sum to 180°.)

c e = 62° (The angle between the tangent and the chord is equal to the angle in the alternate segment.)

g = 79° (The angle between the tangent and the chord is equal to the angle in the alternate segment.)

f = 39° (Angles on a straight line sum to 180°.)

4 ▶ a ∠TAB = 58° (Angles in a triangle sum to 180° and tangents drawn to a circle from a point outside the circle are equal in length.)

a = 58° (The angle between the tangent and the chord is equal to the angle in the alternate segment.)

b b = 55° (The angle in a semicircle is a right angle. Angles in a triangle sum to 180°. The angle between the tangent and the chord is equal to the angle in the alternate segment.)

c c = 66° (Alternate angles are equal.)

d = 66° (The angle between the tangent and the chord is equal to the angle in the alternate segment.)

e = 48° (Angles in a triangle sum to 180°.)

5 ▶ a 28° (Angle between the tangent and a radius is 90°.)

b 160° (Angle between the tangent and a radius is 90° and angles in a quadrilateral sum to 360°.)

c 124° (Angles in a triangle sum to 180° and the base angles in an isosceles triangle are equal.)

d 76° (Angles round a point sum to 360°.)

e 52° (Angles in a triangle sum to 180° and the base angles in an isosceles triangle are equal.)

6 ▶ ∠DOB = 114° (Angle between the tangent and a radius is 90° and angles in a quadrilateral sum to 360°.)

Reflex ∠DOB = 246° (Angles round a point sum to 360°)

∠OBC = 20° (Angle between the tangent and a radius is 90°.)

∠DCB = 57° (Angle at the circumference is twice the angle at the centre.)

∠ODC = 37° (Angles in a quadrilateral sum to 360°.)

EXERCISE 4*

1 ▶ a 90°   **b** 45°
     c 90°   **d** All of them

2 ▶ $x = 3$

3 ▶ Angle ABC = 180° − x (angles on a straight line)

Also angle ABC = 180° − angle ADC (opposite angles of a cyclic quadrilateral sum to 180°)

So angle ADC = x

Angle ADC + angle CDT = 180° (angles on a straight line)

So $x + y = 180°$

4 ▶ a Angle on straight line adjacent to 79° and 53° is 48°.

$a = 48°$ (alternate segment theorem)

b $b = 42°$ (angle sum of triangle)

$c = b = 42°$ (alternate segment theorem)

$d = 80°$ (alternate segment theorem, or angles on a straight line)

c $e = 65°$ (alternate segment theorem)

Angle between radius and chord = 90 − 65 = 25° (angle between the tangent and a radius is 90°)

$f = 25°$ (angles subtended by same arc)

d Angles between tangents and chord opposite 62° angle are both 62° (alternate segment theorem)

$g = 180 − 62 − 62 = 56°$ (angle sum of triangle)

e $h = 83 − 42 = 41°$ (exterior angle property)

Angle adjacent to $i = 42°$ (alternate segment theorem)

$i = 55°$ (angles on a straight line)

f $j = 70°$ (alternate segment theorem)

$k = 58°$ (alternate segment theorem)

$l = 110°$ (opposite angles of a cyclic quadrilateral sum to 180°)

$m = 180 − 58 − 58 = 64°$ (isosceles triangle formed by tangents of equal length)

5 ▶ Let angle BAC = x

Angle BAD = 2 × angle BAC = $2x$ (CA bisects angle BAD)

Angle BOD = 2 × angle BAD = $4x$ (angle at centre = 2 × angle at circumference)

Angle DCT = angle DAC = x (alternate segment theorem)

Therefore angle DCT = $\frac{1}{4}$ angle BOD

6 ▶ $x = 360 − (90 + 90 + 132) = 48°$ (Angle between the tangent and a radius is 90°. Angles in a quadrilateral sum to 360°.)

$y = 90 − 58 = 32°$ (Angle between the tangent and a radius is 90°.)

$z = 360 − (228 + 32 + 66) = 34°$ (Angles at a point sum to 360°. Angles in a quadrilateral sum to 360°. Angle at the centre is twice the angle at the circumference.)

EXAM PRACTICE: SHAPE AND SPACE 6

1 ▶ a OA = 6.5 cm

     b AM = 6 cm

     c OM = 2.5 cm

2 ▶ a $x = 16$

     b $x = 10$

3 ▶ a $x = 60°$ (Angles in a triangle sum to 180°), $y = 60°$ (Alternate segment theorem), $z = 55°$ (Alternate segment theorem)

     b $x = 40°$ (Isosceles triangles), $y = 70°$ (Radius perpendicular to tangent), $z = 40°$ (Alternate segment theorem)

4 ▶ Angle OBA = 90° − $3x$ (angle between tangent and radius = 90°)

Angle OAB = 90° − $3x$ (base angle of isosceles triangle OAB, equal radii)

Angle OAC = x (base angle of isosceles triangle OAC, equal radii)

Therefore angle BAC = 90° − $3x$ + x = 90° − $2x$

Angle BOC = 2 × angle BAC = 180° − $4x$ (angle at centre = 2 × angle at circumference)

Angle TBO = angle TCO = 90° (angles between tangents and radii)

In quadrilateral TBOC, $y + 90 + 180 − 4x + 90 = 360°$

Therefore $y = 4x$

UNIT 6: SETS 2

EXERCISE 1

1 ▶ a 35   **b** 3   **c** 11   **d** 2   **e** 64

2 ▶ a $\mathscr{E}$

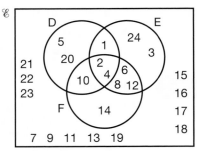

     b {1, 2, 4, 5, 6, 8, 10, 12, 14, 20}

     c 16

3 ▶ a Diagram not unique.

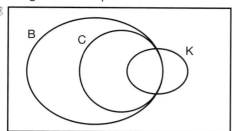

b All cards that are black or a king or both.

c All cards that are black or a king or both.

d All cards that are red or a king or both.

4 ▶ a All houses with electricity have mains water. $E \subset W$

b House p has mains water and gas but no electricity.

c and **d**

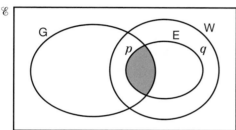

5 ▶ a

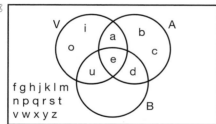

b {a, e, i, o, u, b, c, d}

c Consonants

d Yes

6 ▶ a

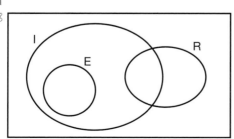

b An isosceles right-angled triangle.

c $I \cup E$ = isosceles triangles, $I \cup R$ = triangles that are isosceles or right-angled or both.

d Equilateral triangles; ∅

EXERCISE 1*   **1 ▶ a** 39   **b** 22   **c** 8   **d** 12

      e 7 different types of ice-cream

2 ▶ a

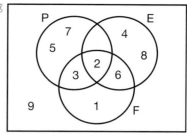

b $P' \cap E = \{4, 6, 8\}$, $E \cap F = \{2, 6\}$, $P \cap F' = \{5, 7\}$

c The even prime factor of 6

3 ▶

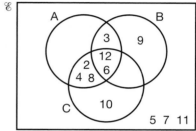

4 ▶ 17

5 ▶

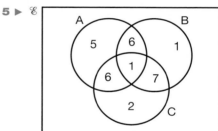

$n(A \cup B \cup C) = 28$

6 ▶ 2^n

EXERCISE 2   **1 ▶** 6   **2 ▶** 93   **3 ▶** 22   **4 ▶** 41

   5 ▶ a 10   **b** 8   **c** 5

   6 ▶ 18

EXERCISE 2*   **1 ▶ a** 4   **b** 14

   2 ▶ 11   **3 ▶** 10   **4 ▶** 100   **5 ▶** 8

   6 ▶ a 4   **b** 3

   7 ▶ $8 \leqslant x \leqslant 14$, $0 \leqslant y \leqslant 6$

   8 ▶ 40% ⩽ percentage who do both ⩽ 65%

EXERCISE 3   **1 ▶**

a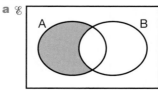

$A \cap B'$

b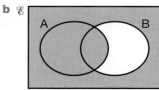

$A \cup B'$

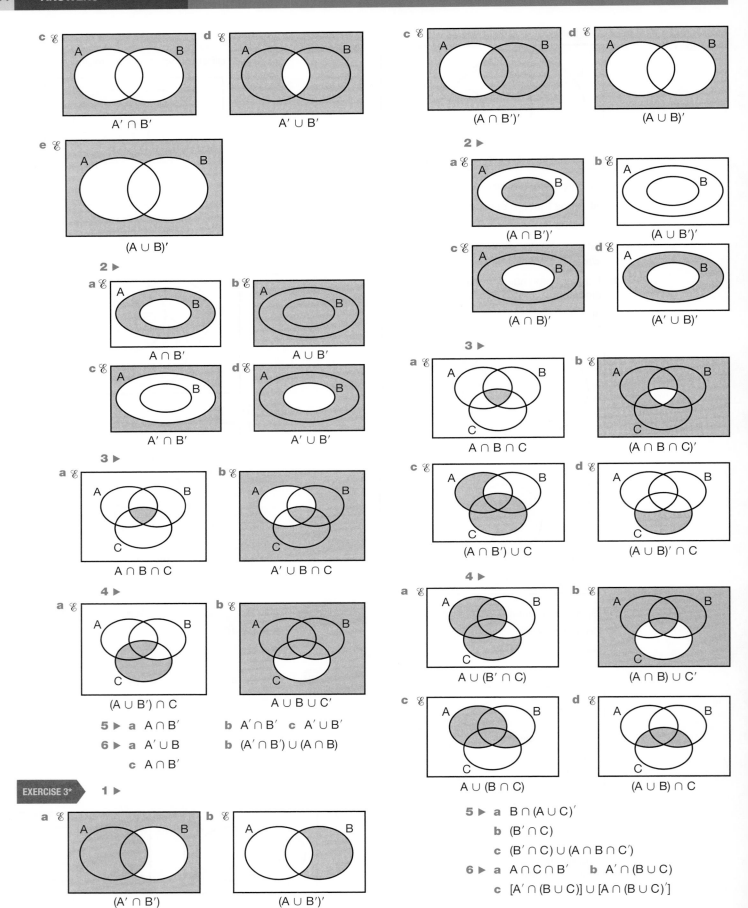

c ℰ
A′ ∩ B′

d ℰ
A′ ∪ B′

e ℰ
(A ∪ B)′

c ℰ
(A ∩ B′)′

d ℰ
(A ∪ B′)′

2 ▶

a ℰ
(A ∩ B′)′

b ℰ
(A ∪ B′)′

c ℰ
(A ∩ B)′

d ℰ
(A′ ∪ B′)′

2 ▶

a ℰ
A ∩ B′

b ℰ
A ∪ B′

c ℰ
A′ ∩ B′

d ℰ
A′ ∪ B′

3 ▶

a ℰ
A ∩ B ∩ C

b ℰ
(A ∩ B ∩ C)′

c ℰ
(A ∩ B′) ∪ C

d ℰ
(A ∪ B)′ ∩ C

3 ▶

a ℰ
A ∩ B ∩ C

b ℰ
A′ ∪ B ∩ C

4 ▶

a ℰ
A ∪ (B′ ∩ C)

b ℰ
(A ∩ B) ∪ C′

c ℰ
A ∪ (B ∩ C)

d ℰ
(A ∪ B) ∩ C

4 ▶

a ℰ
(A ∪ B′) ∩ C

b ℰ
A ∪ B ∪ C′

5 ▶ **a** A ∩ B′   **b** A′ ∩ B′   **c** A′ ∪ B′

6 ▶ **a** A′ ∪ B   **b** (A′ ∩ B′) ∪ (A ∩ B)
    c A ∩ B′

5 ▶ **a** B ∩ (A ∪ C)′
   b (B′ ∩ C)
   c (B′ ∩ C) ∪ (A ∩ B ∩ C′)

6 ▶ **a** A ∩ C ∩ B′   **b** A′ ∩ (B ∪ C)
   c [A′ ∩ (B ∪ C)] ∪ [A ∩ (B ∪ C)′]

EXERCISE 3*

1 ▶

a ℰ
(A′ ∩ B′)

b ℰ
(A ∪ B′)′

ACTIVITY 1

A ∪ B

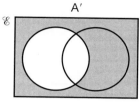

(A ∪ B)′

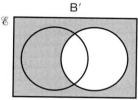

A′

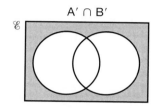

B′

B′ diagram

A′ ∩ B′

A′ ∩ B′ diagram

(A ∪ B)′ = A′ ∩ B′

c ∅   **d** {−1 + √7, −1 − √7}

5 ▶ ℰ

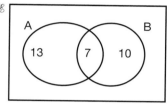

EXERCISE 5   REVISION

1 ▶ a ℰ

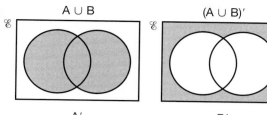

b 17   **c** 30

2 ▶ a 6   **b** 2   **c** 10

3 ▶ a 17%   **b** 52%   **c** 31%

4 ▶

a ℰ   **b** ℰ

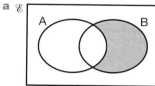

5 ▶ A′ ∪ B′

6 ▶ a {−2, −1, 0, 1, 2, 3}

     b {1, 2, 3, 4}   **c** ∅

7 ▶ a {x: x is even, x ∈ ℕ}

     b {x: x is a factor of 24, x ∈ ℕ}

     c {x: −1 ⩽ x ⩽ 4, x ∈ ℕ}

8 ▶ {x: x ⩾ −4, x ∈ ℤ}

EXERCISE 4

1 ▶ a {Tuesday, Thursday}

     b {Red, Amber, Green}

     c {1, 2, 3, 4, 5, 6}

     d {−1, 0, 1, 2, 3, 4, 5, 6}

2 ▶ a {Africa, Antarctica, Asia, Australia, Europe, North America, South America}

     b {all Mathematics teachers in the school}

     c {1, 2, 3, 4, 5}   **d** {−3, −2, −1, 0, 1, 2}

3 ▶ a {x: x < 7, x ∈ ℕ}

     b {x: x > 4, x ∈ ℕ}

     c {x: 2 ⩽ x ⩽ 11, x ∈ ℕ}

     d {x: −3 < x < 3, x ∈ ℕ}

     e {x: x is odd, x ∈ ℕ}

     f {x: x is prime, x ∈ ℕ}

4 ▶ a {x: x > 3, x ∈ ℕ}

     b {x: x ⩽ 9, x ∈ ℕ}

     c {x: 5 < x < 19, x ∈ ℕ}

     d {x: −4 ⩽ x ⩽ 31, x ∈ ℕ}

     e {x: x is a multiple of 5, x ∈ ℕ} or {x: x = 5y, y ∈ ℕ}

     f {x: x is a factor of 48, x ∈ ℕ}

EXERCISE 4*

1 ▶ a {2, 4, 6, 8, 10, 12}

     b {3, 7, 11, 15, 19, 23}   **c** {2, 4, 6}

     d {integers between 1 and 12 inclusive}

2 ▶ a {0, 1, 4}   **b** {$\frac{1}{4}$, $\frac{1}{2}$, 1, 2, 4}

     c {1}   **d** {(1, 1), (2, 2)}

3 ▶ a ∅   **b** (1, $\frac{1}{2}$, $\frac{1}{4}$, $\frac{1}{8}$, $\frac{1}{16}$, $\frac{1}{32}$}

     c {2}   **d** {−3, 2}

4 ▶ a ∅   **b** {1, 2, 4, 8, 16}

EXERCISE 5*   REVISION

1 ▶ 34   **2 ▶** 10   **3 ▶** 2

4 ▶

a ℰ   **b** ℰ

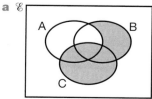

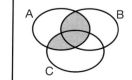

5 ▶ a (A ∪ B′) ∩ C

     b A ∪ B ∪ C′

6 ▶ a {−1, 1}

  b {0, −4}

  c ∅

7 ▶ a {$x: x > 5, x \in \mathbb{N}$}

  b {$x: 4 < x < 12, x \in \mathbb{N}$}

  c {$x: x$ is a multiple of 3, $x \in \mathbb{N}$} or
     {$x: x = 3y, y \in \mathbb{N}$}

8 ▶ {$x: -\frac{1}{4} < x \le 4, x \in \mathbb{R}$}

EXAM PRACTICE: SETS 2

1 ▶ 8

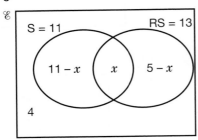

2 ▶ a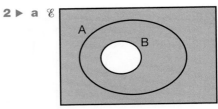

  b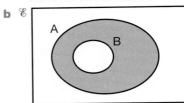

3 ▶ A′ ∩ B   **4 ▶** {−1, 0, 1, 2}

5 ▶ {$x: x = 4y, y \in \mathbb{N}$}

6 ▶ a 7   **b** 9

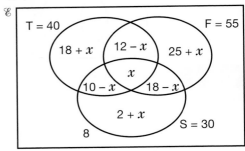

UNIT 7 ANSWERS

UNIT 7: NUMBER 7

EXERCISE 1

1 ▶ $\frac{9}{16}$ 2 ▶ $\frac{5}{32}, \frac{3}{8}$ 3 ▶ $\frac{3}{20}, \frac{5}{64}$

4 ▶ $\frac{3}{40}, \frac{7}{80}, \frac{9}{25}, \frac{9}{24}$ 5 ▶ $\frac{1}{3}$ 6 ▶ $\frac{4}{9}$

7 ▶ $\frac{5}{9}$ 8 ▶ $\frac{2}{3}$ 9 ▶ $\frac{7}{9}$ 10 ▶ 1

11 ▶ $\frac{7}{90}$ 12 ▶ $\frac{1}{90}$ 13 ▶ $\frac{1}{30}$ 14 ▶ $\frac{1}{45}$

15 ▶ $\frac{1}{18}$ 16 ▶ $\frac{1}{15}$ 17 ▶ $\frac{73}{99}$ 18 ▶ $\frac{5}{3}$

EXERCISE 1*

1 ▶ $\frac{8}{33}$ 2 ▶ $\frac{38}{99}$ 3 ▶ $\frac{10}{33}$ 4 ▶ $\frac{31}{33}$

5 ▶ $9\frac{19}{990}$ 6 ▶ $8\frac{29}{990}$ 7 ▶ $\frac{3}{110}$ 8 ▶ $\frac{2}{55}$

9 ▶ $\frac{412}{999}$ 10 ▶ $\frac{101}{999}$ 11 ▶ $\frac{128}{333}$ 12 ▶ $\frac{158}{333}$

13 ▶ $\frac{11}{90}$ 14 ▶ $\frac{13}{15}$ 15 ▶ $\frac{28}{495}$ 16 ▶ $\frac{31}{198}$

17 ▶ $0.0\dot{3}\dot{7}$ 18 ▶ $0.01\dot{6}$

EXERCISE 2

1 ▶ 7.47   2 ▶ 4.35   3 ▶ 49.6

4 ▶ 6.89×10^{-4} 5 ▶ 0.266

6 ▶ 2 200 000   7 ▶ 1.29   8 ▶ 5.83

9 ▶ 24.7   10 ▶ 0.0305   11 ▶ 31 000

12 ▶ 0.377   13 ▶ 145   14 ▶ 4.93

15 ▶ 2.38

16 ▶ a 82.4 °F   b $C = \frac{5F - 160}{9}$ c 40°C

EXERCISE 2*

1 ▶ 0.170   2 ▶ 1.55×10^{-12} 3 ▶ 115

4 ▶ 2.00   5 ▶ 2.39   6 ▶ 3.47

7 ▶ 6.04   8 ▶ 0.0322   9 ▶ 2.74

10 ▶ 0.997   11 ▶ 12.1   12 ▶ 1.46

13 ▶ $x = 0.211$ or 2.07

14 ▶ a 11 180 m/s   b 5016 m/s

15 ▶ a $r = \sqrt{\dfrac{Gm_1 m_2}{F}}$ b 1.5×10^{11} m

16 ▶ 13.3

ACTIVITY 1

1 ▶ a 120   b 720   c 5040
    d 5   e 90   f 9900

2 ▶ a 6   b 1 728 000 000   c 27
    d π e 6.72   f 2.67

3 ▶ a $x = 10$ b $x = 3$ or 4
    c $x = 4$ d $x = 8$

4 ▶ $\dfrac{n!}{(n-2)!} = \dfrac{n(n-1)(n-2)!}{(n-2)!}$
$$= n(n-1)$$
$$= n^2 - n$$

EXERCISE 3 REVISION

1 ▶ a $\frac{2}{9}$ b $\frac{7}{90}$ c $\frac{23}{99}$

2–4 ▶ Students' own answers

5 ▶ a 9.11   b 79 500   c 12.4

6 ▶ $v = 12$

7 ▶ a $T = \dfrac{D}{S}$ b 192 seconds

8 ▶ a 4654 m   b 179 107 m

EXERCISE 3* REVISION

1 ▶ a $\frac{7}{9}$ b $\frac{1}{90}$ c $\frac{67}{99}$ d $3\frac{1}{22}$

2–4 ▶ Students' own answers

5 ▶ a 0.991   b 404   c 3.30

6 ▶ 8.28

7 ▶ a £950   b $P = \dfrac{10(D - B)}{N}$ c £700

8 ▶ £1478.18

EXAM PRACTICE: NUMBER 7

1 ▶ a $\frac{8}{9}$ b $\frac{85}{99}$ c $\frac{754}{999}$

d $\frac{79}{3330}$ e $\frac{11}{15}$ f $3\frac{7}{330}$

2 ▶ a 0.464   b 0.075 2   c 13.2

3 ▶ 26.0

4 ▶ Students' own answers

UNIT 7: ALGEBRA 7

EXERCISE 1

1 ▶ −1, −2   2 ▶ 2, −3
3 ▶ −2, −5   4 ▶ 5, −3
5 ▶ 3, 3   6 ▶ 2, −6
7 ▶ 0, −1   8 ▶ 4, 0
9 ▶ 2, −2   10 ▶ 7, −12

EXERCISE 1*

1 ▶ −1, −5   2 ▶ 4, 1
3 ▶ −7, −8   4 ▶ 9, −5
5 ▶ 7, 7   6 ▶ 8, −5
7 ▶ 13, 0   8 ▶ 0, −17
9 ▶ 13, −13   10 ▶ 11, −13

EXERCISE 2

1 ▶ $-\frac{9}{4}, \frac{9}{4}$ 2 ▶ 0, −2
3 ▶ 1, 0   4 ▶ 3, 2
5 ▶ 2, $\frac{1}{2}$ 6 ▶ −1, −2
7 ▶ $\frac{5}{3}$, 0   8 ▶ 3, −2
9 ▶ $-\frac{2}{3}$, −2   10 ▶ $\frac{2}{3}$, −4

EXERCISE 2*

1 ▶ $-2\frac{2}{3}, 2\frac{2}{3}$ 2 ▶ $\frac{3}{2}, 0$

3 ▶ $2, \frac{3}{2}$ 4 ▶ $\frac{3}{2}, -\frac{1}{3}$

5 ▶ $-\frac{1}{4}, -\frac{1}{2}$ 6 ▶ $5, \frac{2}{5}$

7 ▶ $7, -\frac{4}{3}$ 8 ▶ $-5, -5$

9 ▶ $7, \frac{1}{4}$ 10 ▶ $\frac{5}{3}, \frac{5}{3}$

EXERCISE 3

1 ▶ $(x + 1)^2 + 2$ 2 ▶ $(x + 3)^2 - 13$

3 ▶ $(x - 2)^2 - 2$ 4 ▶ $(x - 5)^2 - 28$

5 ▶ $\left(x + \frac{3}{2}\right)^2 - \frac{5}{4}$ 6 ▶ $\left(x + \frac{5}{2}\right)^2 - \frac{37}{4}$

7 ▶ $\left(x - \frac{7}{2}\right)^2 - \frac{53}{4}$ 8 ▶ $\left(x - \frac{9}{2}\right)^2 - \frac{73}{4}$

EXERCISE 3*

1 ▶ $(x - 3)^2 - 8$ 2 ▶ $(x - 6)^2 - 33$

3 ▶ $\left(x + \frac{5}{2}\right)^2 - \frac{73}{4}$ 4 ▶ $\left(x - \frac{7}{2}\right)^2 - \frac{101}{4}$

5 ▶ $\left(x - \frac{9}{2}\right)^2 - \frac{85}{4}$ 6 ▶ $\left(x - \frac{11}{2}\right)^2 - \frac{33}{4}$

7 ▶ $\left(x + \frac{15}{2}\right)^2 - \frac{257}{4}$ 8 ▶ $(x + p)^2 - (p^2 - 3)$

EXERCISE 4

1 ▶ $3(x + 1)^2 - 8$ 2 ▶ $2\left(x + \frac{3}{2}\right)^2 - \frac{7}{2}$

3 ▶ $6(x - 1)^2 - 14$ 4 ▶ $2\left(x - \frac{5}{2}\right)^2 - \frac{15}{2}$

5 ▶ $-(x - 1)^2 + 5$ 6 ▶ $-\left(x + \frac{3}{2}\right)^2 + \frac{21}{4}$

EXERCISE 4*

1 ▶ $2(x + 4)^2 - 28$ 2 ▶ $2\left(x + \frac{5}{4}\right)^2 + \frac{81}{8}$

3 ▶ $5\left(x - \frac{6}{5}\right)^2 - \frac{11}{5}$ 4 ▶ $-(x - 3)^2 + 14$

5 ▶ $-4(x + 2)^2 + 19$ 6 ▶ $-6\left(x - \frac{3}{2}\right)^2 + \frac{45}{2}$

EXERCISE 5

1 ▶ 1.45, −3.45   2 ▶ $1 + \sqrt{7}, 1 - \sqrt{7}$

3 ▶ 1.46, −5.46   4 ▶ $-5 + \sqrt{10}, -5 - \sqrt{10}$

5 ▶ 0.791, −3.79   6 ▶ $\frac{7}{2} + \sqrt{\frac{29}{4}}, \frac{7}{2} - \sqrt{\frac{29}{4}}$

7 ▶ 6.90, −2.90   8 ▶ $-\frac{3}{2} + \sqrt{\frac{17}{4}}, -\frac{3}{2} - \sqrt{\frac{17}{4}}$

9 ▶ −0.184, −1.82   10 ▶ $\frac{5}{2} + \sqrt{\frac{35}{4}}, \frac{5}{2} - \sqrt{\frac{35}{4}}$

EXERCISE 5*

1 ▶ 5.83, 0.172   2 ▶ $6 + \sqrt{33}, 6 - \sqrt{33}$

3 ▶ 1.77, −6.77   4 ▶ $\frac{7}{2} + \sqrt{\frac{101}{4}}, \frac{7}{2} - \sqrt{\frac{101}{4}}$

5 ▶ 9.11, −0.110   6 ▶ $4 + \sqrt{14}, 4 - \sqrt{14}$

7 ▶ 3.39, −0.886   8 ▶ $-\frac{6}{5} + \sqrt{\frac{11}{25}}, -\frac{6}{5} - \sqrt{\frac{11}{25}}$

9 ▶ 1.09, −0.522   10 ▶ No solutions

EXERCISE 6

1 ▶ −1, −2   2 ▶ 1.45, −3.45

3 ▶ 5.45, 0.551   4 ▶ 4.45, −0.449

5 ▶ 1.46, −5.46   6 ▶ 8.16, 1.84

7 ▶ 3.33, −2   8 ▶ 0.643, −1.24

9 ▶ 1.35, 0.524   10 ▶ 0.402, −0.210

EXERCISE 6*

1 ▶ 5.65, 0.354   2 ▶ 1.77, −6.77

3 ▶ 1.46, −5.46   4 ▶ 4.44, 0.564

5 ▶ 9, −2.5   6 ▶ 1.85, −0.180

7 ▶ 2.85, −3.85   8 ▶ 5.37, 0.105

9 ▶ 1.12, −1.42   10 ▶ 1.16, −2.16

ACTIVITY 1

$ax^2 + bx + c = 0$

$x^2 + \frac{b}{a}x + \frac{c}{a} = 0$ (Divide both sides by a)

$\left(x + \frac{b}{2a}\right)^2 - \left(\frac{b}{2a}\right)^2 + \frac{c}{a} = 0$ (Completing the square)

$\left(x + \frac{b}{2a}\right)^2 = \left(\frac{b}{2a}\right)^2 - \frac{c}{a}$ (Rearranging)

$\left(x + \frac{b}{2a}\right)^2 = \frac{b^2}{4a^2} - \frac{c}{a}$ (Squaring the fraction)

$\left(x + \frac{b}{2a}\right)^2 = \frac{b^2 - 4ac}{4a^2}$ (Simplifying)

$x + \frac{b}{2a} = \pm\sqrt{\frac{b^2 - 4ac}{4a^2}}$ (Square rooting both sides)

$x + \frac{b}{2a} = \frac{\pm\sqrt{b^2 - 4ac}}{2a}$ (Square rooting fraction)

$x = \frac{-b \pm \sqrt{b^2 - 4ac}}{2a}$ (Rearranging and simplifying)

EXERCISE 7

1 ▶ b 2 s and 5s   c 7 s

2 ▶ 12 and 18, −12 and −18

3 ▶ 4.18 cm by 7.18 cm

4 ▶ 2.32

5 ▶ 14 or −15

6 ▶ 5.41 cm, 8.41 cm

7 ▶ 3 cm

8 ▶ b 7 and 8, −8 and −7

9 ▶ 25.9 cm²

10 ▶ a 1414

b $\frac{n(n + 1)}{2} = 1\,000\,000$ doesn't have an integer solution.

EXERCISE 7*

1 ▶ a 0.517 s, 3.48 s

b 4.05 s

2 ▶ 4 cm

3 ▶ b 7 and 9, −9 and −7

4 ▶ 14

5 ▶ 4.5 m by 6 m

6 ▶ 4.86 m

7 ▶ 4.83 m

8 ▶ 32 s

9 ▶ a 38

b $\frac{n(n - 3)}{2} = 406$ doesn't have an integer solution.

10 ▶ 6 days

EXERCISE 8

1 ▶ $-4 < x < 4$

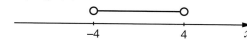

2 ▶ $x \leqslant -5$ or $x \geqslant 5$

3 ▶ $-9 \leqslant x \leqslant 9$

4 ▶ $-5 < x < 5$

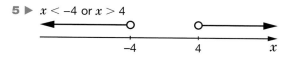

5 ▶ $x < -4$ or $x > 4$

6 ▶ $-3 \leqslant x \leqslant 1$

7 ▶ $x < -4$ or $x > -3$

8 ▶ $-1 < x < \frac{1}{2}$

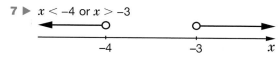

9 ▶ $x \leqslant -5$ or $x \geqslant -2$

10 ▶ $-5 \leqslant x \leqslant 3$

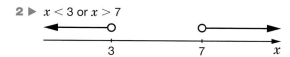

EXERCISE 8*

1 ▶ $x \leqslant -2$ or $x \geqslant 2$

2 ▶ $x < 3$ or $x > 7$

3 ▶ $x \leqslant -\frac{5}{3}$ or $x \geqslant 4$

4 ▶ $x \leqslant 0$ or $x \geqslant \frac{1}{4}$

5 ▶ $x \leqslant -4$ or $x \geqslant -1$

6 ▶ $-8 < x < 2$

7 ▶ $-13 <$ smaller number < 6

8 ▶ $x < \frac{1}{2}$ or $x > 1$

9 ▶ $6 <$ width < 8

10 ▶ $0 <$ width < 3 or width > 4

EXERCISE 9   REVISION

1 ▶ **a** $x = -5$ or $x = 5$
    b $x = -4$ or $x = 0$

2 ▶ **a** $x = 3$ or $x = -4$
    b $x = 3$ or $x = -2$
    c $x = \frac{2}{3}$ or $x = -1$

3 ▶ **a** $-2 + \sqrt{7}, -2 - \sqrt{7}$
    b $2 + \sqrt{5}, 2 - \sqrt{5}$
    c $-3 - \sqrt{13}, -3 + \sqrt{13}$

4 ▶ **a** $x = -1.24$ or $x = 3.24$
    b $x = 0.232$ or $x = 1.43$

5 ▶ 1.70

6 ▶ **b** 8 and 9, -9 and -8

7 ▶ 11.13 cm and 18.87 cm (to 2 d.p.)

8 ▶ **a** $x \leqslant -2$ or $x \geqslant 2$

    b $-5 < x < 3$

EXERCISE 9*   REVISION

1 ▶ **a** $x = -4.47$ or $x = 4.47$
    b $x = 0$ or $x = 9$

2 ▶ **a** $x = -9$ or $x = 8$
    b $x = -4$ or $x = 6$
    c $x = -2.5$ or $x = 0.75$

3 ▶ a $3 - \sqrt{11}, 3 + \sqrt{11}$

b $\dfrac{-5 + \sqrt{21}}{4}, \dfrac{-5 - \sqrt{21}}{4}$

c $\dfrac{7 + \sqrt{53}}{2}, \dfrac{7 - \sqrt{53}}{2}$

4 ▶ a $x = -0.573$ or $x = 2.91$

b $x = -4.46$ or $x = 0.459$

5 ▶ width 5, length 6

6 ▶ 3.68

7 ▶ 14.8 cm

8 ▶ a $4 < x < 8$

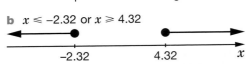

b $x \leqslant -2.32$ or $x \geqslant 4.32$

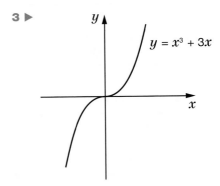

EXAM PRACTICE: ALGEBRA 7

1 ▶ a $x = 5$ or -5
b $x = \frac{2}{3}$ or -1
c $x = -\frac{1}{2}$ or 3

2 ▶ a $(x - 4)^2 - 5$
b $x = 4 + \sqrt{5}$ or $4 - \sqrt{5}$

3 ▶ a $x = 0.618$ or $x = -1.62$ (3 s.f.)
b $x = 1.71$ or $x = 0.293$ (3 s.f.)

4 ▶ a $x > 3$ or $x < -3$
b $-2 \leqslant x \leqslant 1$

5 ▶ a $x(x + 2) = 6$ **b** $x = 1.65$

6 ▶ 6.79 cm

UNIT 7: GRAPHS 6

EXERCISE 1   **1 ▶**

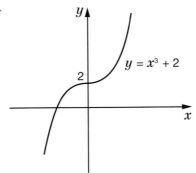

2 ▶

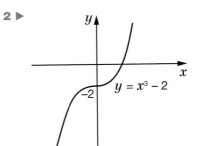

3 ▶

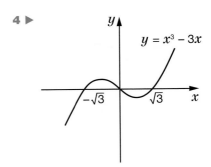

4 ▶

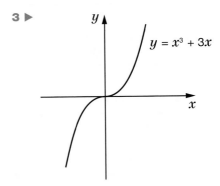

5 ▶

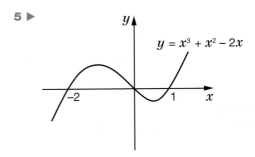

6 ▶

x	−2	−1	0	1	2	3
y	−22	−5	0	−1	−2	3

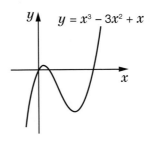

7 ▶ a $V = x^2(x - 1) = x^3 - x^2$

b

x	2	2.5	3	3.5	4	4.5	5
V	4	9.4	18	30.6	48	70.9	100

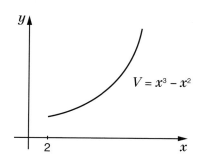

c $48\,m^2$

d 4.6 m by 4.6 m by 3.6 m

8 ▶ a

x	0	1	2	3	4	5
y	8	17	16	11	8	13

b 9.5 m

EXERCISE 1*

1 ▶

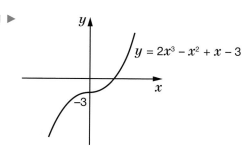

$y = 2x^3 - x^2 + x - 3$

2 ▶

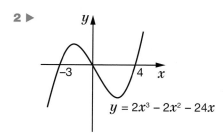

$y = 2x^3 - 2x^2 - 24x$

3 ▶

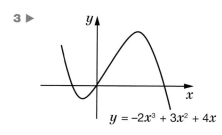

$y = -2x^3 + 3x^2 + 4x$

4 ▶

x	−4	−3	−2	−1	0	1	2	3	4
y	0	−12	−10	0	12	20	18	0	−40

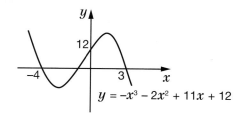

$y = -x^3 - 2x^2 + 11x + 12$

5 ▶

v	0	1	2	3	4	5
t	0	26	46	54	44	10

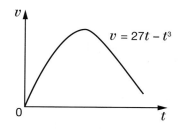

$v = 27t - t^3$

b $v_{max} = 54$ m/s and occurs at $t = 3$ s

c $v \geqslant 30$ m/s when $1.2 \geqslant t \geqslant 4.5$ so for about 3.3 s

6 ▶ a $V = \pi x^3 + \frac{1}{3}\pi x^2 \times 6 = \pi x^3 + 2\pi x^2 = \pi x^2(x + 2)$

b

x	0	1	2	3	4	5
V	0	3π	16π	45π	96π	175π

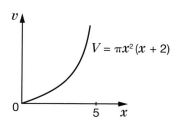

$V = \pi x^2 (x + 2)$

c When $x = 3.5$ cm, $V \simeq 212$ cm^3

d When $V = 300$ cm^3, $x \simeq 4$ cm
$\Rightarrow A \simeq 100.5$ cm^3

7 ▶ a $A = 100\pi = 2\pi r^2 + 2\pi rh$

$\Rightarrow 100\pi - 2\pi r^2 = 2\pi rh$

$\Rightarrow \dfrac{50}{r} - r = h$

$\Rightarrow V = \pi r^2 h = \pi r^2\left(\dfrac{50}{r} - r\right) = 50\pi r - \pi r^3$

b

r	0	1	2	3	4	5	6	7
V	0	153.9	289.0	386.4	427.3	392.7	263.9	22.0

c $V_{max} \approx 428$ cm^3

d When $V = 428$ cm^3, $r \approx 4$ cm
$\Rightarrow d \approx 8$ cm and $h \approx 8.5$ cm

8 ▶ b $V = (10 - 2x)(10 - 2x)x$
$= 100x - 40x^2 + 4x^3$

c

x	0	1	2	3	4	5
y	0	64	72	48	16	0

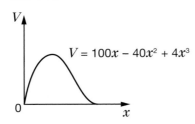

$V = 100x - 40x^2 + 4x^3$

d $V_{max} \approx 74 \, cm^3$, 1.6 cm by 6.66 cm by 6.66 cm

ACTIVITY 1

Year interval	Fox numbers	Rabbit numbers	Reason
A–B	Decreasing	Increasing	Fewer foxes to eat rabbits
B–C	Increasing	Increasing	More rabbits attract more foxes into the forest
C–D	Increasing	Decreasing	More foxes to eat rabbits so rabbit numbers decrease
D–A	Decreasing	Decreasing	Fewer rabbits to be eaten by foxes so fox numbers decrease

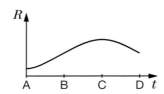

ACTIVITY 2

x	−3	−2	−1	$-\frac{1}{2}$	$-\frac{1}{4}$	$\frac{1}{4}$	$\frac{1}{2}$	1	2	3
y	$-\frac{1}{3}$	$-\frac{1}{2}$	−1	−2	−4	4	2	1	$\frac{1}{2}$	$\frac{1}{3}$

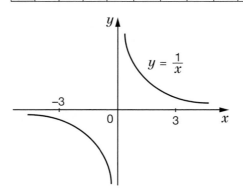

$y = \frac{1}{x}$

EXERCISE 2

1 ▶

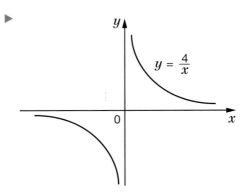

$y = \frac{4}{x}$

2 ▶

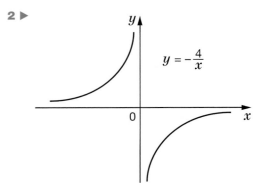

$y = -\frac{4}{x}$

3 ▶

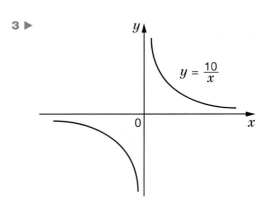

$y = \frac{10}{x}$

4 ▶

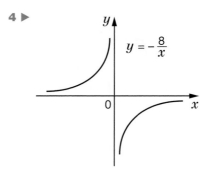

$y = -\frac{8}{x}$

5 ▶ a

t (months)	1	2	3	4	5	6
y	2000	1000	667	500	400	333

b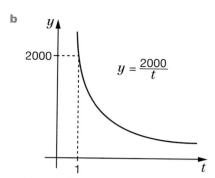

Hyperbola

c 3.3 months

d When $3.3 < t \le 4$, so about 2.7 months

6 ▶ a

t (hours)	1	5	10	15	20
v (m³)	1000	200	100	67	50

b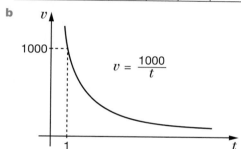

c 4 hours   d 62.5 m³

7 ▶ a $k = 400$

m (min)	5	6	7	8	9	10
t (°C)	80	67	57	50	44	40

b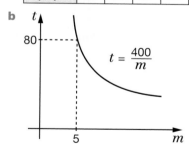

c 53 °C   d 6 min 40 s

e $5.3 \le m \le 8$

EXERCISE 2*

1 ▶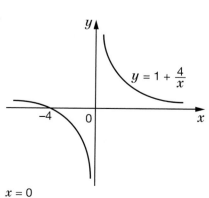

$x = 0$

2 ▶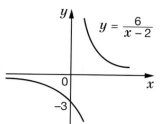

$x = 2$

3 ▶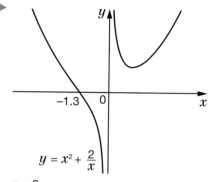

$x = 0$

4 ▶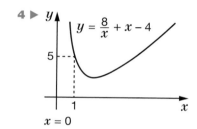

$x = 0$

5 ▶ a

x (°)	30	35	40	45	50	55	60
d (m)	3.3	7.9	10	10.6	10	8.6	6.7

b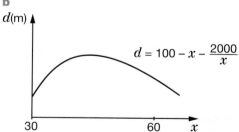

c 10.6 m at $x \approx 45°$

d $37° \le x \le 54°$

6 ▶ a

x	0	1	2	3	4	5
R	0	3	4	4.5	4.8	5

x	1	3	5
C	2	4	6

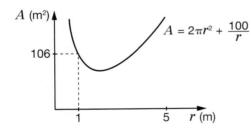

R, C(£1000's)

$C = x + 1$

$R = \dfrac{6x}{x + 1}$

b $P > 0$ when $0.27 < x < 3.7$ so between 27 and 370 boards per week

c £1100 when $x = 1.45$ so 145 boards hired out

7 ▶ a Volume = $\pi r^2 h$

$50 = \pi r^2 h$

$\dfrac{50}{\pi r^2} = h$

Area = $2\pi r^2 + 2\pi rh$

$A = 2\pi r^2 + 2\pi r\dfrac{50}{\pi r^2}$

$A = 2\pi r^2 + \dfrac{100}{r}$

b

r (m)	1	2	3	4	5
A (m²)	106	75	90	126	177

A (m²)

$A = 2\pi r^2 + \dfrac{100}{r}$

106

c $A \approx 75\,\text{m}^2$ at $r \approx 2.0\,\text{m}$

ACTIVITY 3

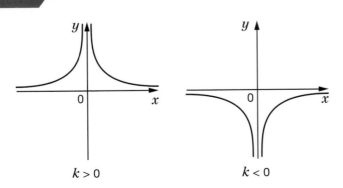

$k > 0$ $k < 0$

EXERCISE 3   **REVISION**

1 ▶

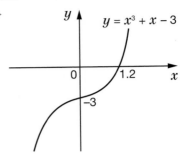

$y = x^3 + x - 3$

1.2

-3

2 ▶

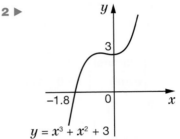

3

-1.8

$y = x^3 + x^2 + 3$

3 ▶ a

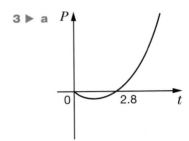

P

2.8 t

b 2.8 years   **c** 5.2 years

4 ▶ a $k = 2800$

t (weeks)	30	32	34	36	38	40
w (kg)	93	88	82	78	74	70

b

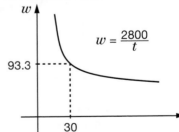

w

$w = \dfrac{2800}{t}$

93.3

30   t

c 35 weeks

d Clearly after 500 weeks, for example, Nick cannot weigh 5.6 kg. So there is a domain over which the equation fits the situation being modelled.

5 ▶ a

m	5	6	7	8	9	10
t	85	70.8	60.7	53.1	47.2	42.5

b $6 < m < 8.5$

1 ▶

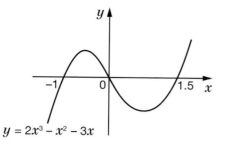

$y = 2x^3 - x^2 - 3x$

2 ▶

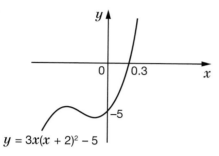

$y = 3x(x + 2)^2 - 5$

3 ▶ A: Linear B: Linear C: Quadratic
D: Reciprocal (Hyperbola) E: Cubic F: Linear

4 ▶ a $\dfrac{600}{x}$

c

x	5	10	15	20	25	30	35	40
L	130	80	70	70	74	80	87	95

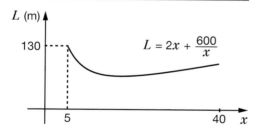

$L = 2x + \dfrac{600}{x}$

d 69.3 m at $x = 17.3$ m

e $11.6 < x < 25.9$

5 ▶ a When $x = 0$, $y = 5$

$0 - 0 + 0 + b = 5$

$b = 5$

When $x = 1$, $y = 6$

$1 - 5 + a + 5 = 6$

$a = 5$

b

x	−1	0	1	2	3	4
y	−6	5	6	3	2	9

c

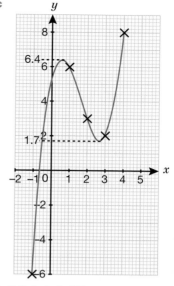

d R(0.6, 6.4), S(2.7, 1.7)

1 ▶ a

x	−5	−4	−3	−2	−1	0	1	2	3
y	−48	−15	0	3	0	−3	0	15	48

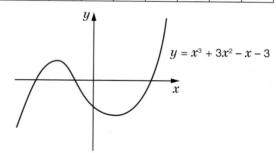

$y = x^3 + 3x^2 - x - 3$

b $x = -3, -1, 1$

2 ▶ a

t	0	1	2	3	4	5
Q	10	17	14	7	2	5

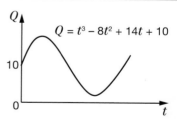

$Q = t^3 - 8t^2 + 14t + 10$

b $Q_{max} = 17.1$ m³/s at 01:06

c Between midnight and 02:35

3 ▶ a

x	60	70	80	90	100	110	120	130
$-\dfrac{x}{3}$	−20	−23.3	−26.7	−30	−333.3	−36.7	−40	−43.3
$-\dfrac{2400}{x}$	−40	−34.3	−30	−26.7	−24	−21.8	−20	−18.5
y	10	12.4	13.3	13.3	12.7	11.5	10	8.2

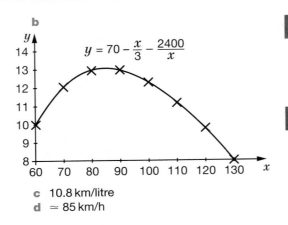

b

$$y = 70 - \frac{x}{3} - \frac{2400}{x}$$

c 10.8 km/litre
d $\simeq$ 85 km/h

UNIT 7: SHAPE AND SPACE 7

EXERCISE 1

1 ▶ 20.6 cm, 25.1 cm²
2 ▶ 33.6 cm, 58.9 cm²
3 ▶ 22.3 cm, 30.3 cm²
4 ▶ 50.8 cm, 117 cm²
5 ▶ 46.8 cm, 39.5 cm²
6 ▶ 34.3 cm, 42.4 cm²
7 ▶ Radius 0.955 cm; Area 2.86 cm²
8 ▶ Radius 2.11 cm; Circumference 13.3 cm
9 ▶ 6.03 m
10 ▶ 7960

EXERCISE 1*

1 ▶ 47.0 cm, 115 cm²
2 ▶ 37.7 cm, 92.5 cm²
3 ▶ 43.7 cm, 99.0 cm²
4 ▶ 66.8 cm, 175 cm²
5 ▶ 37.7 cm, 56.5 cm²
6 ▶ 37.7 cm, 62.8 cm²
7 ▶ $r = 3.19$ cm, $P = 11.4$ cm
8 ▶ 16.0 m
9 ▶ 2 cm
10 ▶ a 40 100 km   b 464 m/s
11 ▶ 6.28 km
12 ▶ $r = 2.41$ cm, $A = 32.3$ cm²

EXERCISE 2

1 ▶ 8.62 cm   2 ▶ 25.6 cm
3 ▶ 38.4 cm   4 ▶ 63.6 cm
5 ▶ 34.4°   6 ▶ 115°
7 ▶ 14.3 cm   8 ▶ 10.6 cm

EXERCISE 2*

1 ▶ 11 cm   2 ▶ 38.3 cm
3 ▶ 25.1°   4 ▶ 121°
5 ▶ 13.4 cm   6 ▶ 117 cm
7 ▶ 33.0 cm   8 ▶ 15.5 cm
9 ▶ 4.94 cm

EXERCISE 3

1 ▶ 12.6 cm²   2 ▶ 61.4 cm²
3 ▶ 170 cm²   4 ▶ 11.9 cm²
5 ▶ 76.4°   6 ▶ 129°
7 ▶ 5.86 cm   8 ▶ 8.50 cm

EXERCISE 3*

1 ▶ 15.8 cm²   2 ▶ 625 cm²
3 ▶ 53.3°   4 ▶ 103°
5 ▶ 4.88 cm   6 ▶ 19.7 cm
7 ▶ 11.5 cm²   8 ▶ 5.08 cm
9 ▶ 1.45 cm²   10 ▶ 6.14 cm
11 ▶ 2.58 cm²

EXERCISE 4

1 ▶ 120 cm³
2 ▶ 48 cm³, 108 cm²
3 ▶ Volume = 452 cm³ (3 s.f.);
    Area = 358 cm² (3 s.f.)
4 ▶ 800 m³
5 ▶ 0.785 m³
6 ▶ 9.9 cm (1 d.p.)

EXERCISE 4*

1 ▶ 4800 cm³
2 ▶ 1.2×10^5 cm³, 1.84×10^4 cm²
3 ▶ 229 cm³, 257 cm²
4 ▶ 1.18 cm³, 9.42 cm²
5 ▶ 405 cm³, 417 cm²
6 ▶ 0.04 mm

EXERCISE 5

1 ▶ 8779 m³
2 ▶ 9817 cm³, 2407 cm²
3 ▶ Volume = 7069 cm³; Surface area = 1414 cm²
4 ▶ 25.1 cm³
5 ▶ 396 m³, 311 m²
6 ▶ $r = 3.5$ cm; Surface area = 154 cm² (3 s.f.)
7 ▶ 61 cm
8 ▶ 0.47 m

EXERCISE 5*

1 ▶ $83\frac{1}{3}$ mm³   2 ▶ 98.2 cm³
3 ▶ 2150 cm³, 971 cm²
4 ▶ 2.92×10^5 m³   5 ▶ 1089 cm³
6 ▶ 5.12×10^8 km²   7 ▶ 0.417 cm
8 ▶ 12 cm   9 ▶ 4.5×10^{-4} mm

EXERCISE 6

1 ▶ 163 cm²
2 ▶ a Angles are the same   b 8.55 cm²
3 ▶ 213 cm²   4 ▶ 84.4 cm²
5 ▶ 6 cm   6 ▶ 3 cm
7 ▶ 3 cm   8 ▶ 24 cm

EXERCISE 6*

1 ▶ 675 cm²   2 ▶ 45 cm²
3 ▶ 7.5 cm   4 ▶ 10 cm
5 ▶ 1000 cm²   6 ▶ 44%
7 ▶ 19%   8 ▶ 75 cm²
9 ▶ 280 cm²

EXERCISE 7

1 ▶ 54 cm³   2 ▶ 23.4 cm³
3 ▶ 222 cm³   4 ▶ 25.3 cm³
5 ▶ 15.1 cm   6 ▶ 9.9 mm
7 ▶ 5.06 cm   8 ▶ 50 cm²

EXERCISE 7*

1 ▶ 86.4 cm³   2 ▶ 31.25 cm³   3 ▶ 33.4 cm
4 ▶ 18.6 cm   5 ▶ 72.8%   6 ▶ $40
7 ▶ a 270 g   b 16 cm
8 ▶ a 25 cm   b 48 g
9 ▶ a 270 g   b 180 cm²
10 ▶ a 2000 quills   b 22.5 m
     c 810 cm²   d 240 g

ACTIVITY 2

1 kg, 800 kg, 400 cm², 2 kg, 4 m

EXERCISE 8   REVISION

1 ▶ Area = 11.1 cm²; Perimeter = 15.1 cm
2 ▶ 22.3 cm², 21.6 cm
3 ▶ Volume = 288 cm³; Area = 336 cm²
4 ▶ 132 m³
5 ▶ 98 cm²
6 ▶ 27 litres

EXERCISE 8*   REVISION

1 ▶ Area = 15.3 cm²; Perimeter = 29.7 cm
2 ▶ Area = 6.98 cm²; Perimeter = 11 cm
3 ▶ Volume = 453 cm³; Area = 411 cm²
4 ▶ 335 cm³, 289 cm²
5 ▶ 54 cm
6 ▶ a $6.75
     b 12 cm

EXAM PRACTICE: SHAPE AND SPACE 7

1 ▶ 49.1 cm, 146 cm²
2 ▶ Area = 133 cm², perimeter = 51.9 cm
3 ▶ 1.4 m³ (1 d.p.)
4 ▶ a 45 cm²
     b 7.11 cm³
5 ▶ a 12.0 cm
     b 360 cm²

UNIT 7: SETS 3

EXERCISE 1

1 ▶ a $\frac{3}{35}$ b $\frac{15}{35}\left(=\frac{3}{7}\right)$ c $\frac{2}{35}$

2 ▶ a ℰ
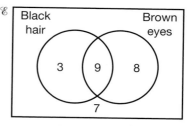
   b $\frac{9}{27}\left(=\frac{1}{3}\right)$ c $\frac{3}{27}\left(=\frac{1}{9}\right)$ d $\frac{7}{27}$

3 ▶ a ℰ
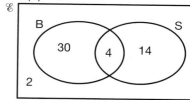
   b i $P(S) = \frac{24}{30}\left(=\frac{4}{5}\right)$ ii $P(F \cap S) = \frac{15}{30}\left(=\frac{1}{2}\right)$
     iii $P(F \cup S) = 1$ iv $P(F' \cap S) = \frac{9}{30}\left(=\frac{3}{10}\right)$

4 ▶ a $n(\mathscr{E}) = 50$
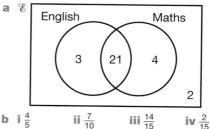
   b i $\frac{2}{25}$ ii $\frac{23}{25}$

5 ▶ a 30   b 9
   c i $\frac{1}{3}$ ii $\frac{1}{5}$ iii $\frac{2}{15}$

EXERCISE 1*

1 ▶ a ℰ
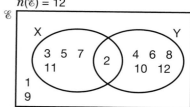
   b i $\frac{4}{5}$ ii $\frac{7}{10}$ iii $\frac{14}{15}$ iv $\frac{2}{15}$
   c Probability of failing English and passing mathematics

2 ▶ $\frac{27}{45}$

3 ▶ a $n(\mathscr{E}) = 12$
ℰ

X   3 5 7   2   4 6 8   Y
   11   10 12
1
9

   b i $\frac{1}{6}$ ii $\frac{1}{2}$ iii $\frac{2}{3}$

4 ▶ a $\frac{6}{7}$ **b** $\frac{1}{7}$

5 ▶ $\frac{1}{3}$

EXERCISE 2

1 ▶ a $n(\mathcal{E}) = 20$

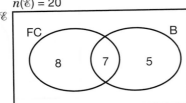

b $\frac{7}{20}$ **c** $\frac{2}{5}$ **d** $\frac{7}{15}$

2 ▶ a $\mathcal{E}$

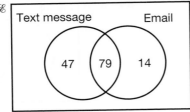

b $\frac{93}{140}$ **c** $\frac{79}{126}$

3 ▶ a $n(\mathcal{E}) = 1000$

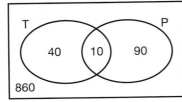

b i 0.01   **ii** 0.1

4 ▶ a $\mathcal{E}$

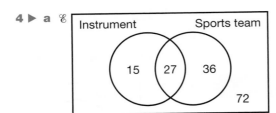

b $\frac{7}{25}$ **c** $\frac{3}{7}$

5 ▶ a 7   **b** 40

    c i $\frac{7}{40}$ **ii** $\frac{11}{40}$ **iii** $\frac{11}{13}$

EXERCISE 2*

1 ▶ a 12   **b i** $\frac{35}{70}$ **ii** $\frac{12}{70}\left(=\frac{6}{35}\right)$ **iii** $\frac{27}{38}$

2 ▶ a 120   **b i** $\frac{31}{60}$ **ii** $\frac{7}{24}$ **iii** $\frac{10}{29}$

3 ▶ a $\mathcal{E}$

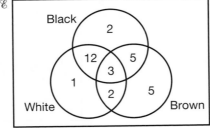

b $\frac{10}{30}\left(=\frac{1}{3}\right)$ **c** $\frac{8}{22} = \frac{4}{11}$

4 ▶ a $\mathcal{E}$

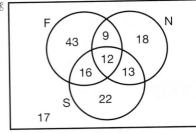

b $\frac{38}{150}\left(=\frac{19}{75}\right)$ **c** $\frac{28}{63}\left(=\frac{4}{9}\right)$

5 ▶ a $\frac{1}{15}$ **b** $\frac{1}{2}$ **c** $\frac{17}{31}$

ACTIVITY 1

$n(\mathcal{E}) = 1000$

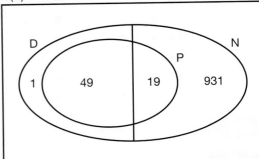

Probability $(N|P) = \frac{9}{19+49} = \frac{19}{68} \approx 28\%$.

Not fair.

EXERCISE 3   **REVISION**

1 ▶ a 6 people   **b** 5 people

    c 2 people   **d** 17 people

    e $\frac{9}{17}$ **f** $\frac{4}{17}$

2 ▶ a $\frac{10}{21}$ **b** $\frac{8}{9}$

3 ▶ a $\mathcal{E}$

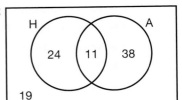

b $\frac{49}{92}$ **c** $\frac{11}{35}$

4 ▶ a $\frac{42}{80}\left(=\frac{21}{40}\right)$ **b** $\frac{6}{22}\left(=\frac{3}{11}\right)$

5 ▶ a 4   **b** 29

    c i $\frac{4}{29}$ **ii** $\frac{4}{29}$

    iii $\frac{6}{15}\left(=\frac{2}{5}\right)$

EXERCISE 3*   **REVISION**

1 ▶ a $\frac{5}{26}$ **b** $\frac{19}{26}$ **c** Only sing

2 ▶ a $\frac{1}{10}$ **b** $\frac{7}{10}$ **c** $\frac{8}{10}$ **d** 0

3 ▶ a 0.35　　　　**b** 0.875

4 ▶ a

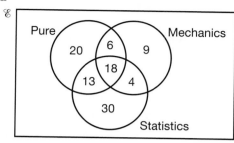

b $\frac{33}{100}$ or 33% or 0.33　　　**c** $\frac{22}{65}$

5 ▶ a

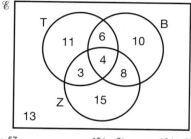

b i $\frac{57}{70}$　　　**ii** $\frac{12}{28}\left(=\frac{3}{7}\right)$　　　**iii** $\frac{16}{40}\left(=\frac{2}{5}\right)$

1 ▶ a

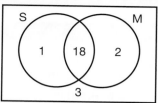

b i $\frac{19}{24}$　　**ii** $\frac{18}{24}\left(=\frac{3}{4}\right)$　　**iii** $\frac{21}{24}\left(=\frac{7}{8}\right)$

iv $\frac{2}{24}\left(=\frac{1}{12}\right)$

2 ▶ a $\frac{2}{7}$　　　　　　**b** $\frac{5}{7}$

3 ▶ a 0.7　　　　　　**b** $\frac{4}{15}$

4 ▶ a 48

b i $\frac{16}{48}\left(=\frac{1}{3}\right)$　　**ii** $\frac{1}{16}$　　**iii** $\frac{5}{16}$

5 ▶ a $\frac{1}{8}$　　　　　　**b** $\frac{2}{7}$

UNIT 8 ANSWERS

UNIT 8: NUMBER 8

EXERCISE 1

1 ▶ 0.033 km

2 ▶ 9.46×10^{17} cm

3 ▶ 6700 km

4 ▶ 8.2×10^{-3} km

5 ▶ 9×10^{-6} mm

6 ▶ 8.85×10^{9} µm

EXERCISE 1*

1 ▶ 3.99×10^{19} mm

2 ▶ 5×10^{-8} km

3 ▶ 2.72 m

4 ▶ 2.00 km

5 ▶ 21 200 km

6 ▶ 2722 feet

EXERCISE 2

1 ▶ a 2.9×10^{4} mm^2 b 0.029 m^2

2 ▶ 1×10^{-3} km^2

3 ▶ 0.0624 m^2

4 ▶ 0.055 mm^2

5 ▶ 2×10^{-6} m^2

6 ▶ 196 m^2

EXERCISE 2*

1 ▶ 1.53×10^{20} mm^2

2 ▶ 1×10^{-16} km^2

3 ▶ Minimum area is 0.418 hectares, maximum area is 1.09 hectares

4 ▶ 1.11×10^{9} cm^2/person

5 ▶ 538 ft^2 to 807 ft^2

6 ▶ 3.77×10^{7} mm^2

EXERCISE 3

1 ▶ 2×10^{8} m^3

2 ▶ 8.5×10^{5} mm^3

3 ▶ 1.2 km^3

4 ▶ 8.01×10^{-3} m^3

5 ▶ a 0.062 mm^3 b 16 100

6 ▶ 6.5×10^{8} litres

EXERCISE 3*

1 ▶ 3.75×10^{18} litres

2 ▶ 1.08×10^{12} km^3

3 ▶ a 1.6×10^{-5} cm^3 b 3.09×10^{7}

4 ▶ 2.7×10^{8} m^3

5 ▶ 1×10^{15}

6 ▶ 4.55 litres

ACTIVITY 1

a 7.1×10^{4} cm^3

b 5×10^{-3} cm^3

c Human volume is 5.25×10^{14} cm^3, ant volume is 5×10^{13} cm^3 so humans have the greater volume.

EXERCISE 4

1 ▶ a 5.4 km/hr   b 1.5 m/s

2 ▶ 900 km/h

3 ▶ Falcon is faster. Car: 350 km/h = 97.2 m/s (1 d.p.); Falcon: 388.8 km/h = 108 m/s

4 ▶ 40 cm^3

5 ▶ 208 g

6 ▶ a 6.84 kg   b 630 cm^3

7 ▶ 0.8 g/cm^3

8 ▶ 17.3 N/m^2 (1 d.p.)

9 ▶ 90 N

10 ▶

Force	Area	Pressure
60 N	2.6 m^2	23.1 N/m^2
73.0 N	4.8 m^2	15.2 N/m^2
100 N	8.33 m^2	12 N/m^2

11 ▶ 2.74 litres/day

12 ▶ a 10.8 km   b 30 km/hr

EXERCISE 4*

1 ▶ 0.0467 m^3/min

2 ▶ a 128 km   b 297 km/hr

3 ▶ Usain Bolt is faster.
Usain Bolt: 12.3 m/s = 44.3 km/h;
White shark: 11.1 m/s = 40 km/h

4 ▶ 760 m/s

5 ▶ 53.8 km/h

6 ▶ a 33.8 kg   b 0.16 m^3

7 ▶ 0.001 g/cm^3

8 ▶ 40 000 m^3

9 ▶ 1.01 g/cm^3

10 ▶ 0.153 N/cm^2

11 ▶ 12 000 N

12 ▶ a 50 N/cm^2 b 18 750 N/m^2

c Standing 1.875 N/cm^2, so when sitting.

ACTIVITY 2

a 245 N/cm^2

b 4.69 N/cm^2

c Approximately 52 : 1. The claim is supported.

EXERCISE 5

REVISION

1 ▶ 3.5×10^{-5} km

2 ▶ 1.4×10^{19} cm^2

3 ▶ 0.0165 km^2

4 ▶ 100

5 ▶ 6.75×10^{10} mm^3

6 ▶ a 2.5×10^{6} litres   b 2.5×10^{-6} km^3

7 ▶ a i 1.5 litres   ii 3.75 litres

b 40 hours

8 ▶ a 16 km/litre   b 4.1 litres (1 d.p.)

9 ▶ 40.5 km

10 ▶ 3.125 m/s

11 ▶ 50 km/h

12 ▶ 0.69 g/cm³

13 ▶ 70 200 g or 70.2 kg

14 ▶ 45 000 000 cm³ or 45 m³

15 ▶ 25 N/m²

16 ▶ 18 N

EXERCISE 5* REVISION

1 ▶ 8.95×10^{-3} km

2 ▶ 7×10^{-5} cm²

3 ▶ 7.66×10^{-3} m²

4 ▶ 10^{45}

5 ▶ a 0.02 cm³   b 4×10^{16}

6 ▶ 0.568 litres

7 ▶ a i 200 litres   ii 700 litres

    b 3.75 days

8 ▶ a 7.2 miles per litre

    b 38.2 litres (3 s.f.)

9 ▶ 30.6 m/s (3 s.f.)

10 ▶ 1.8 km/h

11 ▶ a $\dfrac{18b}{5}$ km/h or $3.6b$ km/h

    b $\dfrac{5c}{18}$ m/s or $\dfrac{c}{3.6}$ m/s

12 ▶ Dead Sea

13 ▶ 8.48 g/cm³

14 ▶ $1000y$ kg/m³

15 ▶ 0.0849 N/m²

16 ▶ 17.5 N

EXAM PRACTICE: NUMBER 8

1 ▶ 4.8×10^{-10} km²

2 ▶ 1×10^{12}

3 ▶ a 9.46×10^{12} km   b 8.47×10^{38} km³

4 ▶ 1.71 g/min

5 ▶ a 26.4 km   b 20.25 km/hr

6 ▶ a 1900 cm³   b 2.4 g/cm³

7 ▶

Force	Area	Pressure
40 N	3.2 m²	12.5 N/m²
106 N	6.4 m²	16.5 N/m²
2000 N	8 m²	250 N/m²

UNIT 8: ALGEBRA 8

EXERCISE 1

1 ▶ Many to one, function.

2 ▶ Many to many, not a function.

3 ▶ One to one, function.

4 ▶ One to many, not a function.

5 ▶ Function, any vertical line intersects at one point, one to one.

6 ▶ Function, any vertical line intersects at one point, one to one.

7 ▶ Not a function, most vertical lines intersect at two points.

8 ▶ Function, any vertical line intersects at one point, many to one

9 ▶ Function, any vertical line intersects at one point, many to one

10 ▶ Not a function, most vertical lines intersect at two points.

11 ▶ Function, any vertical line intersects at one point, many to one

12 ▶ Not a function, any vertical line intersects at two points.

EXERCISE 1*

1 ▶ $x \mapsto x^2$
many to one, function

2 ▶ $x \mapsto \sin x°$
many to one, function

3 ▶ $x \mapsto \pm\sqrt{x-1}$
one to many, not a function

4 ▶ $x \mapsto x + \sqrt{x}$
one to one, function

5 ▶ Function, any vertical line intersects at one point, many to one.

6 ▶ Not a function, most vertical lines intersect at two points.

7 ▶ Function, any vertical line intersects at one point, many to one.

8 ▶ Function, any vertical line intersects at no more than one point, one to one.

9 ▶ Not a function, most vertical lines intersect at two points.

10 ▶ Function, any vertical line intersects at no more than one point, one to one.

11 ▶ Not a function, most vertical lines intersect at more than one point.

12 ▶ Not a function, most vertical lines intersect at more than one point.

EXERCISE 2

1 ▶ a 4   b 0   c 3

2 ▶ a 15   b 5   c 20

3 ▶ a 90   b 3

4 ▶ 9

5 ▶ a 2   b 0   c 10

6 ▶ a $\dfrac{1}{5}$ b −1   c $\dfrac{1}{1+2a}$

7 ▶ a $1\tfrac{1}{2}$ b $2\tfrac{1}{2}$ c $2 - \dfrac{1}{y}$

8 ▶ 3   9 ▶ 3   10 ▶ 3

EXERCISE 2*

1 ▶ a 9   b −3   c 3
2 ▶ a 12   b −11   c −25
3 ▶ a 4   b −2
4 ▶ 7
5 ▶ a $-\frac{1}{3}$ b $\frac{1}{5}$ c $-\frac{1}{195}$
6 ▶ a $\sqrt{8}$ b 0   c $2\sqrt{a^2 + a}$
7 ▶ a −4   b $\frac{2}{3}$ c $\frac{9y + 2}{3y - 4}$
8 ▶ 1
9 ▶ −2, 3
10 ▶ $\frac{1}{2}$

EXERCISE 3

1 ▶ a $-2x + 1$ b $2x + 5$ c $2x + 3$
2 ▶ a $4x + 1$ b $8x - 3$ c $8x - 6$
3 ▶ a $3 + x$ b $3 + 3x$ c $3x - 9$
4 ▶ a $x^2 - 1$ b x^2 c $2 - x^2$
5 ▶ a $9x^2 - 3x$ b $3x^2 - 3x$ c $x^2 + x$
6 ▶ $x = \frac{2}{3}$
7 ▶ $x = -1$ or 3
8 ▶ $x = -1$

EXERCISE 3*

1 ▶ a $2 + x$ b $2 + 2x$ c $2x - 4$
2 ▶ a $x^2 + 4x + 5$ b $x^2 + 3$ c $x^2 + 1$
3 ▶ a $2x^2 + x$ b $x - 2x^2$ c $8x^2 - 2x$
4 ▶ a $3 - 9x^2$ b $9 - 3x^2$ c $9 - 3x^2$
5 ▶ a x^2 b x^2 c $\frac{1}{x^2}$
6 ▶ $x = -3$ or 6
7 ▶ $x = -2$ or $-\frac{5}{4}$
8 ▶ $x = -1$ or 3

EXERCISE 4

1 ▶ a $0 \leqslant y \leqslant 2$ b $-2 \leqslant y \leqslant 2$
2 ▶ a $0 \leqslant y \leqslant 10$ b $-10 \leqslant y \leqslant 10$
3 ▶ a $1 \leqslant y \leqslant 4$ b $-2 \leqslant y \leqslant 4$
4 ▶ a $-3 \leqslant y \leqslant 5$ b $-11 \leqslant y \leqslant 5$
5 ▶ a $-1 \leqslant y \leqslant 0$ b $-2 \leqslant y \leqslant 0$
6 ▶ a $11 \leqslant y \leqslant 20$ b $2 \leqslant y \leqslant 20$
7 ▶ a $0 \leqslant y \leqslant 4$ b $0 \leqslant y \leqslant 8$
8 ▶ a $4 \leqslant y \leqslant 6$ b $-5 \leqslant y \leqslant 15$
9 ▶ a $-0.5 \leqslant y \leqslant 1.5$ b $-4.5 \leqslant y \leqslant 5.5$
10 ▶ a $12 \leqslant y \leqslant 18$ b $-15 \leqslant y \leqslant 45$

EXERCISE 4*

1 ▶ a {8, 5, 2, −1}   b All real numbers
2 ▶ a {1, 5, 9, 13}   b All real numbers
3 ▶ a {0, 0, 8, 24}
   b $\{y : y \geqslant 0, y$ a real number$\}$
4 ▶ a {6, 2, 0, 0}
   b $\{y : y \geqslant 12, y$ a real number$\}$

5 ▶ a {27, 11, 3, 3}
   b $\{y : y \geqslant 2, y$ a real number$\}$
6 ▶ a {7, −1, −1, 7}
   b $\{y : y \geqslant -2, y$ a real number$\}$
7 ▶ a {−10, 0, 10, 68}
   b $\{y : y \geqslant 2, y$ a real number$\}$
8 ▶ a {−125, −27, −1, 1}
   b $\{y : y \geqslant -1, y$ a real number$\}$
9 ▶ a $\{1, \frac{1}{2}, \frac{1}{3}, \frac{1}{4}\}$
   b $\{y : 0 \leqslant y \leqslant 1, y$ a real number$\}$
10 ▶ a {8, 7, 8, 13}
   b $\{y : y \geqslant 8, y$ a real number$\}$

EXERCISE 5

1 ▶ $x = -1$
2 ▶ $x = 1$
3 ▶ $\{x : x < 2, x$ a real number$\}$
4 ▶ $\{x : x > 2, x$ a real number$\}$
5 ▶ $x = 0$
6 ▶ $x = 0$
7 ▶ None
8 ▶ None
9 ▶ $-2 < x < 2$
10 ▶ $x < -3$ or $x > 3$

EXERCISE 5*

1 ▶ $x = \frac{1}{2}$
2 ▶ $x = \frac{3}{4}$
3 ▶ $\{x : x > 9, x$ a real number$\}$
4 ▶ $\{x : x < -4, x$ a real number$\}$
5 ▶ $x = -1$
6 ▶ $x = 1$
7 ▶ $x = \pm 1$
8 ▶ None
9 ▶ $\{x : x \leqslant -2, x$ a real number$\}$
10 ▶ $\{x : x \geqslant 2, x$ a real number$\}$

EXERCISE 6

1 ▶ fg(3) = 6, gf(3) = 6
2 ▶ fg(1) = 9, gf(1) = 3
3 ▶ fg(4) = 5, gf(4) = $\frac{4}{5}$
4 ▶ a $x - 1$ b $x - 1$
   c $x - 8$ d $x + 6$
5 ▶ a $2x + 4$ b $2x + 2$
   c $4x$ d $x + 4$
6 ▶ a $(x + 2)^2$ b $x^2 + 2$
   c x^4 d $x + 4$
7 ▶ a x b x
   c $x - 12$ d $x + 12$
8 ▶ a 7   b 6

EXERCISE 6*

1 ▶ fg(–3) = 19, gf(–3) = 8

2 ▶ fg(2) = 82, gf(2) = 36

3 ▶ fg(–3) = –4.5, gf(–3) = $-\frac{3}{7}$

4 ▶ a $x - 2$ b $x - 4$

   c $\dfrac{x - 12}{4}$ d $4x$

5 ▶ a $2(x - 2)^2$ b $2x^2 - 2$

   c $8x^4$ d $x - 4$

6 ▶ a $\dfrac{1}{x}$ b $\dfrac{1}{x - 2} + 2$

   c $\dfrac{x - 2}{5 - 2x}$ d $x + 4$

7 ▶ a $4\sqrt{\dfrac{x}{4} + 4}$ b $\sqrt{x + 4}$

   c $16x$ d $\sqrt{\dfrac{1}{4}\sqrt{\left(\dfrac{x}{4} + 4\right)}}$

8 ▶ a $\frac{5}{4}$ b $-\frac{1}{2}$

9 ▶ a $\{x : x \neq 5,\ x \text{ a real number}\}$

   b $\{x : x \neq \pm 2,\ x \text{ a real number}\}$

10 ▶ a $\{x : x \geqslant -1,\ x \text{ a real number}\}$

   b $\{x : x \geqslant -2,\ x \text{ a real number}\}$

ACTIVITY 1

$x = y + 1,\ y = x - 1$

$x = 2y,\ y = \dfrac{x}{2}$

$x = 4 - y,\ y = 4 - x$

EXERCISE 7

1 ▶ $f^{-1}(x) = \dfrac{(x - 4)}{6}$ 2 ▶ $f^{-1}(x) = 2x - 3$

3 ▶ $f^{-1}(x) = -\dfrac{(x - 12)}{5}$ 4 ▶ $f^{-1}(x) = \dfrac{x}{3} + 6$

5 ▶ $f^{-1}(x) = \dfrac{x}{6} - 5$ 6 ▶ $g^{-1}(x) = \frac{1}{3}\left(\dfrac{1}{x} - 4\right)$

7 ▶ $p^{-1}(x) = \dfrac{3}{(4 - x)}$ 8 ▶ $f^{-1}(x) = \sqrt{(x - 7)}$

9 ▶ a 4 b $\frac{5}{2}$ c 1

10 ▶ $x = -5$

EXERCISE 7*

1 ▶ $f^{-1}(x) = \dfrac{x + 5}{3} - 4$ 2 ▶ $f^{-1}(x) = -\dfrac{(x - 12)}{4}$

3 ▶ $f^{-1}(x) = \dfrac{4}{3} - \dfrac{x}{24}$ 4 ▶ $f^{-1}(x) = 2 - \dfrac{3}{2x}$

5 ▶ $f^{-1}(x) = \dfrac{7}{(4 - x)}$ 6 ▶ $g^{-1}(x) = \sqrt{(x^2 - 7)}$

7 ▶ $p^{-1}(x) = \sqrt{\dfrac{(x - 16)}{2}}$ 8 ▶ $r^{-1}(x) = \dfrac{(4x - 3)}{(x + 2)}$

9 ▶ a 4 b 7 c 0

10 ▶ $x = 2$

11 ▶ $x = 1$ or $x = 2$

12 ▶ $x = -3$ or $x = 1$

ACTIVITY 2

a

A	B	C	D	E	F	G	H	I
D	E	F	G	H	I	J	K	L

J	K	L	M	N	O	P	Q	R
M	N	O	P	Q	R	S	T	U

S	T	U	V	W	X	Y	Z
V	W	X	Y	Z	A	B	C

b I hate maths

c and d Students' own answers

EXERCISE 8 REVISION

1 ▶ a Function as any vertical line only cuts graph once.

   b Not a function as most vertical lines cut graph twice.

2 ▶ a 13 b -2 c 7

3 ▶ a -0.5 b 1.25

4 ▶ a $5x - 1$ b $5x + 3$

5 ▶ $x = 10$

6 ▶ a 1 b $\frac{1}{2}$ c $x < -1$ d $x < 2$

7 ▶ a All y b $y \geqslant 1$ c $y \geqslant 0$ d All y

8 ▶ a i $\dfrac{1}{x^2} + 1$ ii $\dfrac{1}{x^2 + 1}$ b x

9 ▶ a $\frac{1}{2}\left(\dfrac{x}{4} - 3\right)$ b $7 - x$

   c $\dfrac{1}{x} - 3$ d $\sqrt{(x - 4)}$

10 ▶ a x b Inverse of each other

EXERCISE 8* REVISION

1 ▶ a Not a function as a vertical line can cut the graph more than once.

   b Function as any vertical line only cuts graph once.

2 ▶ a 4 b 3 c 0

3 ▶ a $-2, 3$ b $-4, 3$

4 ▶ a $4 - 2x$ b $7 - 2x$

5 ▶ $-4, 2$

6 ▶ a $\frac{4}{3}$ b -1

   c $x < -\frac{2}{5}$ d $-3 < x < 3$

7 ▶ a $y \geqslant 3$ b $y \geqslant 0$

   c $y \geqslant 0$ d All real numbers

8 ▶ a i $\dfrac{1}{(x - 8)^3}$ ii $\dfrac{1}{x^3 - 8}$

   b i 8 ii 2

   c $\dfrac{x - 8}{65 - 8x}$

9 ▶ a $\frac{1}{2}\left(1 - \dfrac{x}{4}\right)$ b $4 - \dfrac{3}{2 - x}$

   c $\dfrac{x^2 + 3}{2}$ d $2 + \sqrt{x}$

10 ▶ a x b Inverse of each other

   c $\sqrt{7}$

EXAM PRACTICE: ALGEBRA 8

1 ▶ a Function as any vertical line only cuts graph once.

b Function as any vertical line only cuts graph once.

c Not a function as any vertical line cuts graph twice.

2 ▶ a $4x^2 + 2x$ **b** $2x^2 + 2x$ **c** $x^2 - x$

3 ▶ $\{y : y \geqslant 0, y$ a real number$\}$

4 ▶ a $x < 4$ **b** None   **c** $x = 3$

5 ▶ 2

6 ▶ a $(2x + 2)^2$ **b** $4x$

7 ▶ a $2(x + 3)$ **b** $5 - \dfrac{2}{x}$

8 ▶ $a = 4$

UNIT 8: GRAPHS 7

Note: When read from a graph, answers are given to 1 dp.

EXERCISE 1

1 ▶ $x = -2.2$ or $x = 2.2$

2 ▶ $x = -1$ or $x = 2$

3 ▶ $x = -3.8$ or $x = 1.8$

4 ▶ $x = 0.6$ or $x = 3.4$

5 ▶ $x = -2.9$ or $x = 3.4$

6 ▶ $x = -0.8$ or 0.4

EXERCISE 1*

1 ▶ $x = -1.3$ or $x = 2.3$

2 ▶ $x = -2.6$ or $x = -0.4$

3 ▶ $x = 2$

4 ▶ $x = -2.7$ or $x = 2.2$

5 ▶ $x = -2.8$ or $x = 3.2$

6 ▶ $x = -1.7$ or 0.9

EXERCISE 2

1 ▶ a $x = 0$ or $x = 3$

b $x = -0.6$ or $x = 3.6$

c $x = 0.4$ or $x = 2.6$

d $x = -0.2$ or $x = 4.2$

e $x = -0.8$ or $x = 3.8$

f $x = 0.2$ or $x = 4.8$

2 ▶ a $x = 1$ or $x = 3$

b $x = -0.5$ or $x = 4.5$

c $x = 0.7$ or $x = 4.3$

d $x = -0.6$ or $x = 3.6$

3 ▶ a $2x^2 + 2x - 1 = 0$

b $x^2 + 5x - 5 = 0$

4 ▶ a $y = 2x + 2$ **b** $y = x$ **c** $y = -3x - 3$

5 ▶ (2.71, 3.5), no

6 ▶ (1.15, −2), no

EXERCISE 2*

1 ▶ a $x = 5$ or $x = 0$

b $x = 4.3$ or $x = 0.7$

c $x = 3.7$ or $x = 0.3$

d $x = 0.8$ or $x = 5.2$

2 ▶ a $x = -1.8$ or $x = 0.3$

b $x = -2.4$ or $x = 0.9$

c $x = -2.3$ or $x = -0.2$

3 ▶ a $6x^2 - 7x - 2 = 0$ **b** $5x^2 - 7x - 4 = 0$

4 ▶ a $y = x + 2$ **b** $y = -2x - 1$

5 ▶ a Yes   **b** 17.9 m, so legal

6 ▶ (72, 14)

EXERCISE 3

1 ▶ b i $x = -1.7$, $x = 0$ or $x = 1.7$

    ii $x = -1.5$, $x = -0.4$ or $x = 1.9$

    iii $x = -1.6$, $x = 0.6$ or $x = 1$

2 ▶ a

x	−3	−2.5	−2	−1.5	−1	−0.5	0
y	−11	−2.1	3	5.1	5	3.4	1

x	0.5	1	1.5	2	2.5	3
y	−1.4	−3	−3.1	−1	4.1	13

   c i $x = -2.3$, $x = 0.2$ or $x = 2.1$

     ii $x = -2$, $x = -0.4$ or $x = 2.4$

     iii $x = -2.6$, $x = -0.1$ or $x = 2.7$

3 ▶ a

x	−3	−2.5	−2	−1.5	−1	−0.5
y	−2	−2.4	−3	−4	−6	−12

x	0.5	1	1.5	2	2.5	3
y	12	6	4	3	2.4	2

   c i $x = 1.2$

     ii $x = -2$ or $x = 1.5$

4 ▶ a $y = -5$ **b** $y = 2x$ **c** $y = 6x - 8$

5 ▶ a $y = 6$ **b** $y = -2x + 7$

   c $y = x^2 - x - 1$

EXERCISE 3*

1 ▶ b i $x = -0.5$, $x = 0.7$ or $x = 2.9$

    ii $x = -1$ or $x = 2$

    iii $x = -1.1$, $x = 1.3$ or $x = 2.9$

2 ▶ b i $x = -1.9$, $x = -0.5$, $x = 0.5$ or $x = 1.9$

    ii $x = 2.1$ or $x = 0.7$

    iii $x = 0.6$ or $x = 1.9$ or $x = -0.5$ or $x = -2$

3 ▶ b i $x = 1.8$

    ii $x = -1.4$, $x = 2$ or $x = 4.4$

    iii $x = 2.7$

4 ▶ a $y = -5x + 4$

b $y = -3x - 2$

c $y = 2x^3 + 4x^2 - 3x + 2$

5 ▶ a $y = x$

b $y = 3x + 8$

ACTIVITY 1

x	0	2	4
$2x$	0	4	8
$-\frac{1}{4}x^2$	0	–1	–4
$y = 2x - \frac{1}{4}x^2$	0	3	4

x	6	8	10
$2x$	12	16	20
$-\frac{1}{4}x^2$	–9	–16	–25
$y = 2x - \frac{1}{4}x^2$	3	0	–5

x	0	4	8
$\frac{1}{4}x$	0	1	2
$y = \frac{1}{4}x - 1$	–1	0	1

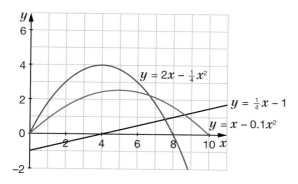

$y = 2x - \frac{1}{4}x^2$

$y = \frac{1}{4}x - 1$

$y = x - 0.1x^2$

Mary is not successful. Peter is made wet provided he is taller than around 0.5 m.

EXERCISE 4

1 ▶ $x = -3, y = -5; x = 1, y = 3$

2 ▶ $x = 2, y = 7; x = -1, y = -2$

3 ▶ $x = 2, y = 2; x = 4, y = 6$

4 ▶ $x = -1.6, y = 1.4; x = 1.9, y = 0.5$

5 ▶ $x = -1.6, y = -2.6; x = 2.6, y = 1.6$

6 ▶ $x = -2, y = 0; x = -0.7, y = 0.7;$
$x = 0.7, y = 1.4$

EXERCISE 4*

1 ▶ $x = 2, y = -3; x = -3, y = 7$

2 ▶ $x = -2, y = 8; x = 2.5, y = 3.5$

3 ▶ $x = -0.5, y = -3; x = 0.4, y = -1.2$

4 ▶ $x = -3.6, y = -0.6; x = 0.6, y = 3.6$

5 ▶ $x = -0.5, y = 4.0; x = 1.3, y = 0.4;$
$x = 3.2, y = -3.3$

6 ▶ $x = -1.2, y = -4.1; x = 1.6, y = 10.1$

EXERCISE 5   REVISION

1 ▶ (137, –50)

2 ▶ a $x = -0.4$ or $x = 2.4$

b $x = -1.2$ or $x = 3.2$

c $x = -1.3$ or $x = 2.3$

3 ▶ a $y = 3$ **b** $y = 5 - 2x$

4 ▶ a $2x^2 - 3x - 1 = 0$ **b** $y = 3x + 2$

5 ▶ a $y = 4$ **b** $y = 5x$ **c** $y = 2 - 3x$

6 ▶ $x = 0.6, y = 2.4; x = 3.4, y = -0.4$

EXERCISE 5*   REVISION

1 ▶ a $x = -0.5, x = 2$ **b** $x = -1.6, x = 2.1$

c $x = -0.7, x = 2.7$

2 ▶ a $y = 3x - 1$ **b** $y = 6 - 4x$

3 ▶ a $y = 12x$ **b** $y = 20 - 4x$

c $y = 9x - 15$

4 ▶ a $2x^3 - x^2 - 8x + 5 = 0$

b $y = 2 - 6x^2$

5 ▶ $x = -1.2, y = -1.7; x = 0.9, y = 0.7$

6 ▶ $xy = 30, x + y = 12; x = 8.5, y = 3.6$

EXAM PRACTICE: GRAPHS 7

1 ▶ a

x	–2	–1	0	1	2	3	4
y	8	3	0	–1	0	3	8

b i $x = -0.6$ or $x = 1.6$

ii $x = -0.6$ or $x = 3.4$

2 ▶ a $y = x + 6$ **b** $y = 7 - x$ **c** $y = 8 - 4x$

3 ▶ a $y = 4x$ **b** $y = 4 - 2x$

4 ▶ a

x	–2	–1	0	1	2	3	4
y	–4	1	4	5	4	1	–4

b $x = -0.3, y = 3.1; x = 3.8, y = 1.1$

UNIT 8: SHAPE AND SPACE 8

ACTIVITY 1   **b** $\begin{pmatrix} 2 \\ -2 \end{pmatrix}$; 2.83; 135°   **c** $\begin{pmatrix} 4 \\ -2 \end{pmatrix}$; 4.47; 117°

d $\begin{pmatrix} 6 \\ 1 \end{pmatrix}$; 6.08; 081°   **e** $\begin{pmatrix} -2 \\ 1 \end{pmatrix}$; 2.24; 297°

$\overrightarrow{OH} = \begin{pmatrix} 8 \\ 0 \end{pmatrix}$; $\overrightarrow{OH} = \mathbf{a} + \mathbf{b} = \mathbf{c} + \mathbf{d} + \mathbf{e}$

EXERCISE 1

1 ▶ $\mathbf{p} + \mathbf{q} = \begin{pmatrix} 6 \\ 8 \end{pmatrix}$; $\mathbf{p} - \mathbf{q} = \begin{pmatrix} -2 \\ -2 \end{pmatrix}$; $2\mathbf{p} + 3\mathbf{q} = \begin{pmatrix} 16 \\ 21 \end{pmatrix}$

2 ▶ $\mathbf{u} + \mathbf{v} + \mathbf{w} = \begin{pmatrix} -1 \\ 0 \end{pmatrix}$; $\mathbf{u} + 2\mathbf{v} - 3\mathbf{w} = \begin{pmatrix} -13 \\ 23 \end{pmatrix}$;

$3\mathbf{u} - 2\mathbf{v} = \begin{pmatrix} 9 \\ 5 \end{pmatrix}$

3 ▶ $\mathbf{p} + \mathbf{q} = \begin{pmatrix} 4 \\ 6 \end{pmatrix}$; $\mathbf{p} - \mathbf{q} = \begin{pmatrix} -2 \\ -2 \end{pmatrix}$; $2\mathbf{p} + 5\mathbf{q} = \begin{pmatrix} 17 \\ 24 \end{pmatrix}$

4 ▶ $\mathbf{s} + \mathbf{t} + \mathbf{u} = \begin{pmatrix} 7 \\ -5 \end{pmatrix}$; $2\mathbf{s} - \mathbf{t} + 2\mathbf{u} = \begin{pmatrix} 8 \\ -19 \end{pmatrix}$;

$2\mathbf{u} - 3\mathbf{s} = \begin{pmatrix} 5 \\ -1 \end{pmatrix}$

5 ▶ $\mathbf{v} + \mathbf{w} = \begin{pmatrix} 4 \\ 5 \end{pmatrix}$, $\sqrt{41}$

$2\mathbf{v} - \mathbf{w} = \begin{pmatrix} 5 \\ -2 \end{pmatrix}$, $\sqrt{29}$

$\mathbf{v} - 2\mathbf{w} = \begin{pmatrix} 1 \\ -7 \end{pmatrix}$, $\sqrt{50}$

6 ▶ $\mathbf{p} + \mathbf{q} = \begin{pmatrix} 5 \\ 4 \end{pmatrix}$, $\sqrt{41}$

$3\mathbf{p} + \mathbf{q} = \begin{pmatrix} 9 \\ 2 \end{pmatrix}$, $\sqrt{85}$

$\mathbf{p} - 3\mathbf{q} = \begin{pmatrix} -7 \\ -16 \end{pmatrix}$, $\sqrt{305}$

7 ▶ **a**

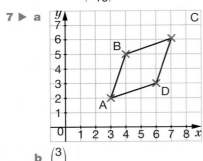

b $\begin{pmatrix} 3 \\ 1 \end{pmatrix}$

c Parallelogram   **d** They are equal.

8 ▶ **a**

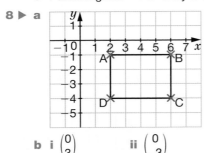

b i $\begin{pmatrix} 0 \\ 3 \end{pmatrix}$ **ii** $\begin{pmatrix} 0 \\ -3 \end{pmatrix}$

They are the inverse of one another.

c i They are equal.   **ii** They are equal.

9 ▶ Parallelogram

10 ▶ **a** (8, 7)   **b** $\begin{pmatrix} 2 \\ -2 \end{pmatrix}$ **c** 8.60

EXERCISE 1*

1 ▶ $\mathbf{p} + \mathbf{q} = \begin{pmatrix} 5 \\ 0 \end{pmatrix}$, 5, 090°

$\mathbf{p} - \mathbf{q} = \begin{pmatrix} -1 \\ 2 \end{pmatrix}$, $\sqrt{5}$, 333°

$2\mathbf{p} - 3\mathbf{q} = \begin{pmatrix} -5 \\ 5 \end{pmatrix}$, $\sqrt{50}$, 315°

2 ▶ $2(\mathbf{r} + \mathbf{s}) = \begin{pmatrix} 10 \\ -4 \end{pmatrix}$, $\sqrt{116}$, 112°

$3(\mathbf{r} - 2\mathbf{s}) = \begin{pmatrix} -21 \\ -15 \end{pmatrix}$, $\sqrt{666}$, 234°

$(4\mathbf{r} - 6\mathbf{s}) \sin 30° = \begin{pmatrix} -10 \\ -9 \end{pmatrix}$, $\sqrt{181}$, 228°

3 ▶ $m = -1$, $n = -2$

4 ▶ $m = -4\frac{1}{2}$, $n = 2\frac{2}{3}$

5 ▶ **a** Chloe $\begin{pmatrix} 5 \\ 7 \end{pmatrix}$; Leo $\begin{pmatrix} 4 \\ 5 \end{pmatrix}$; Max $\begin{pmatrix} 3 \\ 2 \end{pmatrix}$

b Chloe: $\sqrt{74}$ km ≈ 8.6 km, 2.9 km/h

Leo: $\sqrt{41}$ km ≈ 6.4 km, 2.1 km/h

Max: $\sqrt{13}$ km ≈ 3.6 km, 1.2 km/h

6 ▶ **a** Chloe: $\sqrt{20}$ km ≃ 4.5 km

Leo: 3 km

Max: $\sqrt{13}$ km ≃ 3.6 km

b Chloe: 243°; Leo 270°; Max: 326°

7 ▶ **a** $\begin{pmatrix} 10.4 \\ 6 \end{pmatrix}$ km   **b** $\begin{pmatrix} 13 \\ -7.5 \end{pmatrix}$ km

8 ▶ **a** $\begin{pmatrix} -5 \\ 8.7 \end{pmatrix}$ km   **b** $\begin{pmatrix} -8.5 \\ -3.1 \end{pmatrix}$ km

9 ▶ **a** $m = 1$, $n = 3$ **b** $p = 3$, $q = 10$

10 ▶ **a** The $t = 0$ position vector is $\begin{pmatrix} -2 \\ 1 \end{pmatrix}$, and

vector $\begin{pmatrix} 2 \\ 1 \end{pmatrix}$ is added every 1 second to

give Anne's position vector.

b $\sqrt{5} \simeq 2.2$ m/s

c $a = \frac{8}{3}$, $b = 2$ so speed = 3.3 m/s

EXERCISE 2

1 ▶ **a** $\overrightarrow{XY} = \mathbf{x}$ **b** $\overrightarrow{EO} = 4\mathbf{y}$

c $\overrightarrow{WC} = -8\mathbf{y}$ **d** $\overrightarrow{TP} = -4\mathbf{x}$

2 ▶ **a** $\overrightarrow{KC} = 2\mathbf{x} - 4\mathbf{y}$ **b** $\overrightarrow{VC} = \mathbf{x} - 8\mathbf{y}$

c $\overrightarrow{CU} = -2\mathbf{x} + 8\mathbf{y}$ **d** $\overrightarrow{AS} = 3\mathbf{x} + 6\mathbf{y}$

3 ▶ **a** $\overrightarrow{HJ}$ **b** $\overrightarrow{HN}$ **c** $\overrightarrow{HL}$ **d** $\overrightarrow{HO}$

4 ▶ **a** $\overrightarrow{HT}$ **b** $\overrightarrow{HP}$ **c** $\overrightarrow{HD}$ **d** $\overrightarrow{HY}$

5 ▶ **a** $\overrightarrow{DC} = \mathbf{x}$ **b** $\overrightarrow{DB} = \mathbf{x} + \mathbf{y}$

c $\overrightarrow{BC} = -\mathbf{y}$ **d** $\overrightarrow{AC} = \mathbf{x} - \mathbf{y}$

6 ▶ **a** $\overrightarrow{AC} = 2\mathbf{x} - \mathbf{y}$ **b** $\overrightarrow{DB} = \mathbf{x} + \mathbf{y}$

c $\overrightarrow{BC} = \mathbf{x} - \mathbf{y}$ **d** $\overrightarrow{CB} = \mathbf{y} - \mathbf{x}$

7 ▶ **a** $2\mathbf{b}$ **b** $\mathbf{a} + \mathbf{b}$ **c** $\mathbf{a} + 2\mathbf{b}$

8 ▶ **a** $\mathbf{q}$

b $\frac{1}{2}\mathbf{q}$

c $-\mathbf{p} + \frac{1}{2}\mathbf{q}$

d $-\mathbf{p} - \frac{1}{2}\mathbf{q}$

9 ▶ **a** $\overrightarrow{AB}$, $\overrightarrow{EF}$, $\overrightarrow{GH}$

b i $7\mathbf{p} - 13\mathbf{p}$

ii $4\frac{1}{2}\mathbf{a} - 4\mathbf{b}$

10 ▶ **a** $\frac{1}{2}\mathbf{b}$ **b** $-\mathbf{a} + \frac{1}{2}\mathbf{b}$ **c** $-\mathbf{a} - \frac{1}{2}\mathbf{b}$

EXERCISE 2*

1 ▶ a $\overrightarrow{DC} = \mathbf{x}$

   b $\overrightarrow{AC} = \mathbf{x} + \mathbf{y}$

   c $\overrightarrow{BD} = \mathbf{y} - \mathbf{x}$

   d $\overrightarrow{AE} = \frac{1}{2}(\mathbf{x} + \mathbf{y})$

2 ▶ a $\overrightarrow{BD} = \mathbf{y} - \mathbf{x}$

   b $\overrightarrow{BE} = \frac{1}{2}(\mathbf{y} - \mathbf{x})$

   c $\overrightarrow{AC} = \mathbf{x} + \mathbf{y}$

   d $\overrightarrow{AE} = \frac{1}{2}(\mathbf{x} + \mathbf{y})$

3 ▶ a $\overrightarrow{AB} = \mathbf{x} - \mathbf{y}$

   b $\overrightarrow{AD} = 3\mathbf{x}$

   c $\overrightarrow{CF} = 2\mathbf{y} - 3\mathbf{x}$

   d $\overrightarrow{CA} = \mathbf{y} - 3\mathbf{x}$

4 ▶ a $\overrightarrow{PQ} = \mathbf{x} - \mathbf{y}$ b $\overrightarrow{PC} = 2\mathbf{x} - \mathbf{y}$

   c $\overrightarrow{QB} = 2\mathbf{y} - \mathbf{x}$ d $\overrightarrow{BC} = 2\mathbf{x} - 2\mathbf{y}$

5 ▶ a $\mathbf{r}$ b $-\mathbf{s}$

   c $\frac{1}{2}\mathbf{r}$ d $\mathbf{s} + \frac{1}{2}\mathbf{r}$

6 ▶ a $\overrightarrow{AB} = \mathbf{y} - \mathbf{x}$ b $\overrightarrow{AM} = \frac{1}{2}(\mathbf{y} - \mathbf{x})$

   c $\overrightarrow{OM} = \frac{1}{2}(\mathbf{x} + \mathbf{y})$

7 ▶ a $\overrightarrow{AB} = \mathbf{y} - \mathbf{x}$ b $\overrightarrow{AM} = \frac{1}{3}(\mathbf{y} - \mathbf{x})$

   c $\overrightarrow{OM} = \frac{1}{3}(2\mathbf{x} + \mathbf{y})$

8 ▶ a $\overrightarrow{AB} = \mathbf{y} - \mathbf{x}$; $\overrightarrow{OD} = 3\mathbf{x}$; $\overrightarrow{DC} = 2\mathbf{y} - 3\mathbf{x}$

   b $\overrightarrow{OM} = \frac{1}{2}(3\mathbf{x} + 2\mathbf{y})$

9 ▶ $\overrightarrow{AB} = \mathbf{y} - \mathbf{x}$; $\overrightarrow{BC} = \mathbf{y} - 2\mathbf{x}$; $\overrightarrow{AD} = 2\mathbf{y} - 4\mathbf{x}$;

   $\overrightarrow{BD} = \mathbf{y} - 3\mathbf{x}$

10 ▶ a $\overrightarrow{OX} = 2\mathbf{x}$; $\overrightarrow{AB} = \mathbf{y} - \mathbf{x}$; $\overrightarrow{BP} = 2\mathbf{x} - \mathbf{y}$

   b $\overrightarrow{OM} = \mathbf{x} + \frac{1}{2}\mathbf{y}$

EXERCISE 3

1 ▶ a $2\mathbf{b}$ b $-\mathbf{a} - \mathbf{b}$ c $\mathbf{a} + 2\mathbf{b}$

2 ▶ a $\binom{1}{3}$ and $\binom{7}{5}$

   b i $\binom{6}{2}$ ii $\binom{36}{12}$

   c The lines are parallel.

3 ▶ $\binom{-4}{2}$

4 ▶ a $\overrightarrow{AB} = \mathbf{b}$. They are parallel.

   b $\overrightarrow{BC} = -2\mathbf{a} - \frac{1}{2}\mathbf{b}$. They are not parallel.

   c Trapezium

5 ▶ a i $\binom{2}{7}$ ii $\binom{4}{14}$

   b They are collinear.

6 ▶ $\overrightarrow{PQ} = \binom{4}{5}$; $\overrightarrow{QR} = \binom{8}{10}$

   $2\overrightarrow{PQ} = \overrightarrow{QR}$ so P, Q and R are collinear.

EXERCISE 3*

1 ▶ a i $6\mathbf{b} - 6\mathbf{a}$ ii $6\mathbf{a}$

   b $12\mathbf{b} - 3\mathbf{a}$

   c $\overrightarrow{EX} = 12\mathbf{b} - 3\mathbf{a} = 3(4\mathbf{b} - \mathbf{a})$;

   $\overrightarrow{EY} = 16\mathbf{b} - 4\mathbf{a} = 4(4\mathbf{b} - \mathbf{a})$

   The lines are parallel, and they both pass through point X so E, X and Y are collinear.

2 ▶ a i $2\mathbf{j}$ ii $\mathbf{j} - \mathbf{k}$ iii $-\mathbf{k} - \mathbf{j}$

   b i $\mathbf{j} - \mathbf{k}$

   ii $\overrightarrow{JX} = \overrightarrow{KJ}$, and point J is common.

3 ▶ a $\mathbf{a} - 3\mathbf{b}$

   b $\overrightarrow{NM} = \frac{1}{2}(\mathbf{a} - \mathbf{b})$; $\overrightarrow{NC} = 2(\mathbf{a} - \mathbf{b})$; $\overrightarrow{NC}$ is a

   multiple of $\overrightarrow{NM}$ and point N is common, so NMC is a straight line.

4 ▶ a i $2\mathbf{q} - 4\mathbf{p}$ ii $3(\mathbf{q} - \mathbf{p})$ iii $2(\mathbf{q} - \mathbf{p})$

   b $\overrightarrow{AB}$ and $\overrightarrow{AC}$ are multiples of $\mathbf{q} - \mathbf{p}$. Point A is common, so ABC is a straight line.

   c $9\,\text{cm}$

5 ▶ a $\mathbf{b} - \mathbf{a}$ b $2\mathbf{b} - \mathbf{a}$

6 ▶ a $6\mathbf{b} - 3\mathbf{a}$

   b $\overrightarrow{AX} = \frac{1}{3}\overrightarrow{AB} = 2\mathbf{b} - \mathbf{a}$

   $\overrightarrow{OX} = \overrightarrow{OA} + \overrightarrow{AX} = 2(\mathbf{b} + \mathbf{a})$

   $\overrightarrow{OY} = \overrightarrow{OB} + \overrightarrow{BY} = 5(\mathbf{b} + \mathbf{a})$

   $\overrightarrow{OX} = \frac{2}{5}\overrightarrow{OY}$

ACTIVITY 2

Time	$t = 0$	$t = 1$	$t = 2$	$t = 3$	$t = 4$
r	$\binom{12}{5}$	$\binom{9}{9}$	$\binom{6}{13}$	$\binom{3}{17}$	$\binom{0}{21}$

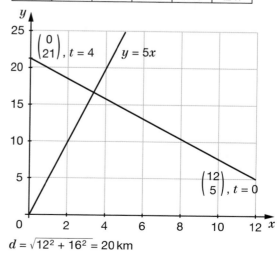

$d = \sqrt{12^2 + 16^2} = 20\,\text{km}$

$V = \dfrac{20\,\text{km}}{4\,\text{min}} = 300\,\text{km/h}, 323.1°$

About 12:03

$x = 12 - 3t$, $y = 5 + 4t$

$y = 5x$

So $5 + 4t = 5(12 - 3t)$ therefore $t = \frac{55}{19}$

Boundary is crossed at 12:02:54

EXERCISE 4 REVISION

1 ▶ $\mathbf{p} + \mathbf{q} = \begin{pmatrix} 1 \\ 5 \end{pmatrix}$, $\sqrt{26}$

   $\mathbf{p} - \mathbf{q} = \begin{pmatrix} 5 \\ 3 \end{pmatrix}$, $\sqrt{34}$

   $3\mathbf{p} - 2\mathbf{q} = \begin{pmatrix} 13 \\ 10 \end{pmatrix}$, $\sqrt{269}$

2 ▶ $2\mathbf{r} + \mathbf{s} = \begin{pmatrix} 7 \\ -6 \end{pmatrix}$, $\sqrt{85}$

   $2\mathbf{r} - \mathbf{s} = \begin{pmatrix} 1 \\ -14 \end{pmatrix}$, $\sqrt{197}$

   $\mathbf{s} - 2\mathbf{r} = \begin{pmatrix} -1 \\ 14 \end{pmatrix}$, $\sqrt{197}$

3 ▶ a $\overrightarrow{AB} = 3\mathbf{y} + \mathbf{x}$ b $\overrightarrow{AC} = 2\mathbf{y} + 2\mathbf{x}$

   c $\overrightarrow{CB} = -\mathbf{x} + \mathbf{y}$

4 ▶ a $\overrightarrow{AB} = \mathbf{w} - \mathbf{v}$ b $\overrightarrow{AM} = \frac{1}{2}(\mathbf{w} - \mathbf{v})$

   c $\overrightarrow{OM} = \frac{1}{2}(\mathbf{v} + \mathbf{w})$

5 ▶ a $\overrightarrow{AB} = \mathbf{y} - \mathbf{x}$ b $\overrightarrow{FB} = 2\mathbf{y} - \mathbf{x}$

   c $\overrightarrow{FD} = \mathbf{y} - 2\mathbf{x}$

6 ▶ a $\overrightarrow{OP} = 2\mathbf{a}$; $\overrightarrow{OQ} = 2\mathbf{b}$; $\overrightarrow{AB} = \mathbf{b} - \mathbf{a}$;

   $\overrightarrow{PQ} = 2(\mathbf{b} - \mathbf{a})$

   b PQ is parallel to AB and is twice the length of AB.

7 ▶ a $\begin{pmatrix} 4 \\ 1 \end{pmatrix}$ b $\begin{pmatrix} 6 \\ -4 \end{pmatrix}$ c $\sqrt{29}$

   d $v = 1$, $w = 2$

8 ▶ a i $\overrightarrow{FE} = \mathbf{b}$ ii $\overrightarrow{CE} = -2\mathbf{a} + \mathbf{b}$

   b $\overrightarrow{FX} = -2\mathbf{a} + 2\mathbf{b}$

   $\overrightarrow{CD} = -\mathbf{a} + \mathbf{b}$

   $\overrightarrow{FX} = -2\mathbf{a} + 2\mathbf{b} = 2(-\mathbf{a} + \mathbf{b})$

   $\overrightarrow{FX} = 2\overrightarrow{CD}$

   So FX and CD are parallel.

EXERCISE 4* REVISION

1 ▶ $m = 3$, $n = 1$

2 ▶ $m = -2$, $n = 5$

3 ▶ a $\overrightarrow{XM} = \begin{pmatrix} -1 \\ 3 \end{pmatrix}$, $\overrightarrow{XZ} = \begin{pmatrix} -10 \\ 6 \end{pmatrix}$

   b $v\begin{pmatrix} 7 \\ 3 \end{pmatrix}$ c $\begin{pmatrix} 8 \\ 0 \end{pmatrix} + w\begin{pmatrix} -10 \\ 6 \end{pmatrix}$

   d $v = \frac{2}{3}$, $w = \frac{1}{3}$

4 ▶ a $\overrightarrow{AC} = \mathbf{x} + \mathbf{y}$; $\overrightarrow{BE} = \frac{1}{3}\mathbf{y} - \mathbf{x}$

   b i $\overrightarrow{BF} = v\left(\frac{1}{3}\mathbf{y} - \mathbf{x}\right)$

   ii $\overrightarrow{AF} = \mathbf{x} + \overrightarrow{BF} = \mathbf{x} + v\left(\frac{1}{3}\mathbf{y} - \mathbf{x}\right)$

   $= (1 - v)\mathbf{x} + \frac{1}{3}v\mathbf{y}$

   iii $v = \frac{3}{4}$

5 ▶ a $\overrightarrow{MA} = \frac{3}{5}\mathbf{x}$; $\overrightarrow{AB} = \mathbf{y} - \mathbf{x}$; $\overrightarrow{AN} = \frac{3}{5}(\mathbf{y} - \mathbf{x})$;

   $\overrightarrow{MN} = \frac{3}{5}\mathbf{y}$

   b OB and MN are parallel; $\overrightarrow{MN} = \frac{3}{5}\overrightarrow{OB}$

6 ▶ a $\overrightarrow{AB} = \mathbf{y} - \mathbf{x}$; $\overrightarrow{MN} = \frac{2}{3}\mathbf{x}$

   b OA and MN are parallel and $\overrightarrow{MN} = \frac{2}{3}\overrightarrow{OA}$

7 ▶ a i $\overrightarrow{OA} = 4\mathbf{a}$

   ii $\overrightarrow{OB} = 4\mathbf{b}$

   iii $\overrightarrow{AB} = -4\mathbf{a} + 4\mathbf{b}$

   b $\overrightarrow{AX} = -6\mathbf{a} - 2\mathbf{b}$

   c $\overrightarrow{DC} = 4\mathbf{b}$

   $\overrightarrow{BY} = -4\mathbf{a} + 4\mathbf{b}$

   $\overrightarrow{CY} = \overrightarrow{CB} + \overrightarrow{BY} = 4\mathbf{a} + (-4\mathbf{a} + 4\mathbf{b}) = 4\mathbf{b}$

   DC and CY are parallel with a point in common (C) so lie on a straight line.

8 ▶ a i $\mathbf{a} = \begin{pmatrix} 3 \\ 1 \end{pmatrix}$, $\mathbf{w} = \begin{pmatrix} 0 \\ 5 \end{pmatrix}$ ii $\mathbf{a} = \begin{pmatrix} 4 \\ 3 \end{pmatrix}$, $\mathbf{w} = \begin{pmatrix} 3 \\ 6 \end{pmatrix}$

   b $\begin{pmatrix} -1 \\ 3 \end{pmatrix}$; $\sqrt{10}$

   c $\sqrt{5}$; $\sqrt{10}$

EXAM PRACTICE: SHAPE AND SPACE 8

1 ▶ a $\begin{pmatrix} 13 \\ -9 \end{pmatrix}$ b 15.8 (1 d.p.)

2 ▶ a $\begin{pmatrix} 10 \\ 25 \end{pmatrix}$ b 26.9

   c 68.2°   d $\begin{pmatrix} 25 \\ 27.5 \end{pmatrix}$

3 ▶ a $2\mathbf{n}$

   b $2\mathbf{m}$

   c $2\mathbf{m} - 2\mathbf{n}$

   d $\mathbf{m} - \mathbf{n}$

   e NM is parallel to AB and half the length

4 ▶ $\overrightarrow{XY} = \overrightarrow{XZ} + \overrightarrow{ZY} = 5\mathbf{a} - 2\mathbf{b} + 3\mathbf{a} + 4\mathbf{b} = 8\mathbf{a} + 2\mathbf{b}$

   $\overrightarrow{WZ} = 6\mathbf{a} + k\mathbf{b}$ and $\overrightarrow{XY} = 8\mathbf{a} + 2\mathbf{b}$ are parallel, so the ratio of the coefficients of $\mathbf{a}$ and $\mathbf{b}$ must be the same for both vectors.

   $\frac{k}{2} = \frac{6}{8}$, hence $k = 1.5$

UNIT 8: HANDLING DATA 5

EXERCISE 1

1 ▶ a $\frac{1}{36}$ b $\frac{25}{36}$
   c $\frac{5}{36}$ d $\frac{5}{18}$

2 ▶ a $\frac{4}{25}$ b $\frac{9}{25}$
   c $\frac{6}{25}$ d $\frac{12}{25}$

3 ▶ a $\frac{4}{25}$ b $\frac{9}{25}$
   c $\frac{6}{25}$ d $\frac{12}{25}$

4 ▶ a $\frac{9}{49}$ b $\frac{24}{49}$

5 ▶ a $\frac{4}{9}$ b $\frac{4}{9}$ c $\frac{1}{9}$

6 ▶ a $\frac{13}{28}$ b $\frac{15}{28}$

7 ▶ a $\frac{1}{169}$ b $\frac{1}{2}$ c $\frac{30}{169}$ d $\frac{1}{8}$

8 ▶ a

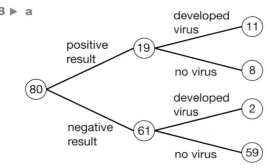

   b $\frac{13}{80}$

EXERCISE 1*

1 ▶ a

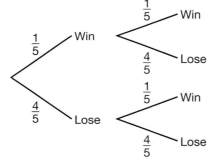

   b i $\frac{1}{25}$ ii $\frac{16}{25}$ iii $\frac{8}{25}$ iv $\frac{9}{25}$

2 ▶ a

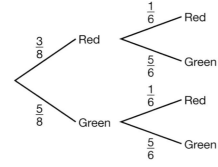

   b i $\frac{7}{12}$ ii $\frac{5}{12}$ iii $\frac{25}{48}$ iv $\frac{23}{48}$

3 ▶ a

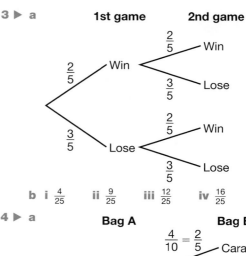

   b i $\frac{4}{25}$ ii $\frac{9}{25}$ iii $\frac{12}{25}$ iv $\frac{16}{25}$

4 ▶ a

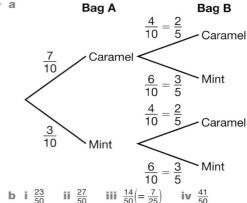

   b i $\frac{23}{50}$ ii $\frac{27}{50}$ iii $\frac{14}{50}\left(=\frac{7}{25}\right)$ iv $\frac{41}{50}$

5 ▶ a

   b i $\frac{1}{50}$ ii $\frac{2}{25}$

6 ▶ a $\frac{1}{9}$ b $\frac{4}{9}$ c $\frac{5}{9}$

7 ▶ a

	First shot	Second shot
Scores	$\frac{2}{3}$	$\frac{4}{7}$
Misses	$\frac{1}{3}$	$\frac{3}{7}$

   b i $\frac{1}{7}$ ii $\frac{10}{21}$ iii $\frac{6}{7}$

8 ▶ a $\frac{1}{4}$ b $\frac{1}{2}$ c $\frac{15}{32}$ d $\frac{3}{16}$

ACTIVITY 1

i P(1st class survive) $= \frac{203}{325} \approx 0.62$

ii P(2nd class survive) $= \frac{118}{285} \approx 0.41$

iii P(3rd class survive) $= \frac{178}{706} \approx 0.25$

P(passenger not rescued) $= \frac{817}{1316} \approx 0.62$

P(crew not rescued) $= \frac{673}{885} \approx 0.76$

Higher proportion of crew were fatalities.

EXERCISE 2

1 ▶ a Independent
  b Independent
  c Dependent
  d Independent
  e Dependent

2 ▶ a $\frac{23}{80}$ **b** $\frac{27}{80}$ **c** $\frac{8}{23}$ **d** $\frac{7}{16}$

3 ▶ a

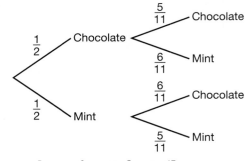

  1st sweet   **2nd sweet**

  Chocolate —$\frac{5}{11}$— Chocolate
            —$\frac{6}{11}$— Mint
  $\frac{1}{2}$

  $\frac{1}{2}$
  Mint —$\frac{6}{11}$— Chocolate
       —$\frac{5}{11}$— Mint

  b i $\frac{5}{22}$ **ii** $\frac{6}{11}$ **iii** $\frac{5}{22}$ **iv** $\frac{17}{22}$

4 ▶ b 33%

5 ▶ 0.39

6 ▶ a

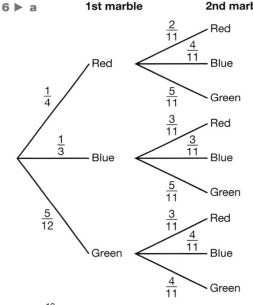

  1st marble   **2nd marble**

  Red —$\frac{2}{11}$— Red
      —$\frac{4}{11}$— Blue
      —$\frac{5}{11}$— Green
  $\frac{1}{4}$

  $\frac{1}{3}$
  Blue —$\frac{3}{11}$— Red
       —$\frac{3}{11}$— Blue
       —$\frac{5}{11}$— Green

  $\frac{5}{12}$
  Green —$\frac{3}{11}$— Red
        —$\frac{4}{11}$— Blue
        —$\frac{4}{11}$— Green

  b $\frac{19}{66}$

7 ▶ $\frac{66}{182}\left(=\frac{33}{91}\right)$

8 ▶ a $\frac{66}{336}\left(=\frac{11}{56}\right)$ **b** $\frac{96}{336}\left(=\frac{2}{7}\right)$

EXERCISE 2*

1 ▶ **b i** $\frac{4}{15}$ **ii** $\frac{8}{15}$ **iii** $\frac{3}{5}$

2 ▶ **b i** $\frac{9}{56}$ **ii** $\frac{31}{56}$ **iii** $\frac{47}{56}$

3 ▶ **b i** $\frac{2}{3}$ **ii** $\frac{1}{3}$

4 ▶ a $\frac{4}{15}$ **b** $\frac{11}{15}$ **c** 8 days

5 ▶ More likely to be in time for meeting than late; (P(in time) = 0.55, P(late) = 0.45)

6 ▶ a   **1st set**   **2nd set**   **3rd set**

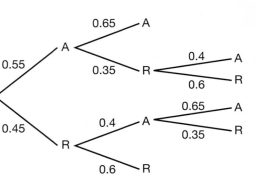

  0.55 — A —0.65— A
            —0.35— R —0.4— A
                       —0.6— R
  0.45 — R —0.4— A —0.65— A
                    —0.35— R
            —0.6— R

  b 0.194 **c** 0.4485

7 ▶ a $\frac{1}{5}$ **b** $\frac{13}{35}$ **c** $\frac{4}{5}$

8 ▶ a

	Good shot	Bad shot
Right club	$\frac{2}{3}$	$\frac{1}{3}$
Wrong club	$\frac{1}{4}$	$\frac{3}{4}$

  b i $\frac{1}{9}$ **ii** $\frac{8}{9}$

ACTIVITY 2

Look for sensible explanations especially good use of the concept of conditional probability. Careful thought about which sample space proportion is being examined should gain credit.

EXERCISE 3 REVISION

1 ▶ b i $\frac{9}{25}$ **ii** $\frac{12}{25}$

2 ▶ a $\frac{1}{16}$ **b** $\frac{3}{8}$

3 ▶ a $\frac{9}{25}$ **b** $\frac{12}{25}$

4 ▶ a $\frac{4}{5}$ **b** $\frac{6}{25}$

5 ▶ a $\frac{1}{36}$ **b** $\frac{5}{18}$ **c** $\frac{4}{9}$

6 ▶ a 0.1 **b** 0.7 **c** 0.15

EXERCISE 3* REVISION

1 ▶ a $\frac{1}{6}$ **b** $\frac{5}{18}$ **c** $\frac{13}{18}$

2 ▶ a $\frac{2}{9}$ **b** $\frac{8}{45}$ **c** $\frac{2}{45}$ **d** $\frac{43}{45}$

3 ▶ a $\frac{1}{32}$ **b** 0 **c** $\frac{3}{16}$

4 ▶ $\frac{n}{25} \times \frac{(n-1)}{24} = 0.07$
  $n^2 - n - 42 = 0 \rightarrow n = 7$
  P(diff colours) $= \frac{7}{25} \times \frac{18}{24} + \frac{18}{25} \times \frac{7}{24} = \frac{21}{50}$

5 ▶ a 1 : 3 : 5 **b** $\frac{4}{45}$
  c i $\frac{16}{2025}$ **i** $\frac{164}{2025}$ **i** $\frac{344}{2025}$

6 ▶ a 0.04
  b 0.03
  c 0.91

EXAM PRACTICE: HANDLING DATA 5

1 ▶ **a** 0.42   **b** 0.46

2 ▶ **a** 77   **b** $\frac{41}{77}$

3 ▶ **a** $\frac{2}{15}$ **b** $\frac{8}{15}$ **c** $\frac{16}{45}$

  d $\frac{64}{225}$ **e** $\frac{139}{225}$

4 ▶ **a** 0.55

  b

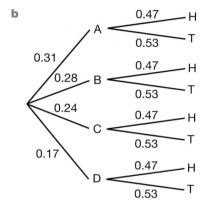

  c 0.1484

UNIT 9 ANSWERS

UNIT 9: NUMBER 9

EXERCISE 1

1 ▶ X $4.50 /litre, Y $4.20 /litre. Y better value
2 ▶ P $3.25 /kg, Q $3.30 /kg. P better value
3 ▶ A $6.50 /m, B $6.25 /m. B better value
4 ▶ I $0.36 /ml, II $0.32 /ml. II better value
5 ▶ Tin A 167 g/$, tin B 176 g/$. Tin B gives better value.

EXERCISE 1*

1 ▶ 1.5 m tree is $30 /m, 2 m is $27.50 /m, 3.5 m is $31.43 /m. The order, from worst value to best, is 3.5 m, 1.5 m, 2 m.
2 ▶ Marble $75 /m², slate $72 /m², limestone $72.22 /m². Slate is cheapest.
3 ▶ Mega-Movie $1 per day per DVD; Films R Us $1.20 per day per DVD. Films R Us gives better value.
4 ▶ Small: $1.25 /ml, Medium: $1.13 /ml, Large: $1.12 /ml. Large is best value.
5 ▶ Yellow: $195, Lime: $200, Rainbow: $207. Yellow is best value over 10 hours.

EXERCISE 2

1 ▶ $1380   2 ▶ £368
3 ▶ €28 200   4 ▶ €126
5 ▶ £862 500

EXERCISE 2*

1 ▶ £4800   2 ▶ €400
3 ▶ Overcharged by $3.75
4 ▶ 8 hrs
5 ▶ Students' own answers

EXERCISE 3

1 ▶ $8000   2 ▶ $10 000
3 ▶ $3166.67   4 ▶ $13 333.33
5 ▶ $5333.33

EXERCISE 3*

1 ▶ 38.7%   2 ▶ 23.3%
3 ▶ $50 000   4 ▶ $2 525 000
5 ▶ Students' own answers

EXERCISE 4

1 ▶ a £175   b €230   c ₱19 437.50
2 ▶ a R1258.50   b ₦14 943.75   c ₸492.75
3 ▶ a $2142.86   b $22.13   c $1630.43
4 ▶ $684 931.51
5 ▶ UK: $274 285.71; France: $271 739.13. France house cheaper by $2546.58

EXERCISE 4*

1 ▶ a $357.14   b €328.57   c ₹24 210.71
2 ▶ a R2554.03   b ₱11 834.09   c $222.22

3 ▶ £9003.22
4 ▶ Nigerian : South Africa = £800 : £1600 = ₦227 714.29 : R38 354.29
5 ▶ China : Australia : America = €600 : €1200 : €1800 = = ₸4284.78 : $(AUS) 1904.35 : $(USA) 1956.52

EXERCISE 5 REVISION

1 ▶ Green: $3.80/kg
Mint: $3.80/kg
Same value for both!
2 ▶ $315
3 ▶ $666.67
4 ▶ a $171.43   b £91.30
5 ▶ $42.05

EXERCISE 5* REVISION

1 ▶ 1st: Wessex: 4.3 litres/100 km
2nd: Fizz: 4.4 litres/100 km
3rd: Tyrol: 4.5 litres/100 km
2 ▶ 48
3 ▶ $5671.23 per day
4 ▶ UK: $64 285.71   Spain: $65 217.39
UK is cheaper by $931.68 so buy in UK.
5 ▶ Students' own answers

EXAM PRACTICE: NUMBER 9

1 ▶ Pluto: 5.5 litres/100 km
Jupiter: 4.7 litres/100 km
Jupiter more economical.
2 ▶ $380
3 ▶ $1692.31
4 ▶ a £17.50   b $35.71   c $27.17
5 ▶ January 1st: £132 000
April 1st: £100 000
£32 000 less in April!

UNIT 9: ALGEBRA 9

ACTIVITY 1

$x = 0$, $y = 5$ or $x = 4$, $y = 3$
There is only one point of intersection, $(4, -3)$; the line is a tangent to the circle.
No solutions

EXERCISE 1

1 ▶ $(-2, 4)$, $(3, 9)$
2 ▶ $x = -1$, $y = 1$ or $x = 3$, $y = 9$
3 ▶ $(-1, 1)$, $(4, 16)$
4 ▶ $x = -2$, $y = 4$ or $x = 4$, $y = 16$
5 ▶ $x = 1$, $y = 2$ or $x = 2$, $y = 3$
6 ▶ $x = 2$, $y = 1$ or $x = -3$, $y = -4$
7 ▶ $(-2, -1)$, $(1, 2)$
8 ▶ $(-1, -1)$, $(2, 2)$

EXERCISE 1*

1 ▶ $x = -3.45$, $y = -7.90$ or $x = 1.45$, $y = 1.90$

2 ▶ $x = 0.268$, $y = 1.80$ or $x = 3.73$, $y = 12.2$

3 ▶ $x = -1.54$, $y = -4.17$ or $x = 4.54$, $y = 20.2$

4 ▶ $x = 0.586$, $y = -0.757$ or $x = 3.41$, $y = -9.24$

5 ▶ $(-4, -2)$, $(2, 4)$

6 ▶ $(-3, \frac{1}{3})$, $(1, 3)$

7 ▶ $(6.85, 7.85)$, $(0.146, 1.15)$

8 ▶ $(1.73, 2.67)$, $(-1.73, 2.67)$, $(2.24, 1.6)$, $(-2.24, 1.6)$

EXERCISE 2

1 ▶ $x = -0.2$, $y = 1.4$ or $x = 1$, $y = -1$

2 ▶ $x = 0.8$, $y = 0.6$ or $x = 0$, $y = 1$

3 ▶ $x = -2$, $y = -1.5$ or $x = 3$, $y = 1$

4 ▶ $x = -\frac{2}{3}$, $y = 6$ or $x = 2$, $y = -2$

5 ▶ $x = -2.87$, $y = 4.87$ or $x = 0.87$, $y = 1.13$

6 ▶ $x = 9.74$, $y = -6.74$ or $x = 2.26$, $y = 0.74$

7 ▶ $x = 1$, $y = -1$

8 ▶ $x = \frac{1}{2}$, $y = 1$

9 ▶ $x = 2.17$, $y = 0.17$ or $x = 7.83$, $y = 5.83$

10 ▶ $x = 0.785$, $y = 3.22$ or $x = 2.55$, $y = 1.45$

11 ▶ **a** $y = 3$ **b** $x = 29.85$ and $y = 3$

     c 29.85 cm

12 ▶ 4 cm

EXERCISE 2*

1 ▶ $x = 1$, $y = 0$ or $x = 7$, $y = -12$

2 ▶ $x = -2$, $y = 2$ or $x = -1$, $y = 3$

3 ▶ $x = \frac{2}{3}$, $y = \frac{1}{3}$ or $x = \frac{1}{3}$, $y = \frac{2}{3}$

4 ▶ $(6, -6)$; tangent

5 ▶ **a** $(6, 1)$, $(-2, 7)$ **b** $AB = 10$

6 ▶ **a** $y = 1.5$

     b $A(-1.68, 1.5)$, $B(1.68, 1.5)$, $AB = 3.36$ cm

7 ▶ $(-2238, 5996)$, $(2238, 5996)$

8 ▶ **a** $y = -3$

     b $(-2.65, -3)$, $(2.65, -3)$, diameter is 5.30 cm

9 ▶ $(7.53, 0.88)$, No

10 ▶ $(23.7, 21.8)$

11 ▶ $(2.67, -1.78)$

12 ▶ $(44, -22)$, length 49.2 m

EXERCISE 3

Note: other counter-examples exist for some of these.

1 ▶ $3 + 3 = 6$

2 ▶ Rhombus

3 ▶ 2

4 ▶ $2 - (-1) > 2 + (-1)$

5 ▶ $(1 + 2)^2 \neq 1^2 + 4$

6 ▶ $41^2 + 41 + 41 = 41(41 + 1 + 1)$

7 ▶ $2^2 + 3^2 = 13$

8 ▶ $0.5^2 < 0.5$

EXERCISE 3*

Note: other counter-examples exist for some of these.

1 ▶ $4^2 - 2^2 = 12$

2 ▶ $1^3 + 3^3 = 28$

3 ▶ $0.5 \times 1 < 0.5 + 1$

4 ▶ $0.5^3 < 0.5^2$

5 ▶ $1 < 2$ but $-1 \times 1 > -1 \times 2$

6 ▶ $-2 < -1$ but $(-2)^2 > (-1)^2$

7 ▶ $(1 + 2)^2 \neq 1^2 + 2^2$

8 ▶ $101^4 + 29 \times 101^2 + 101$
$= 101(101^3 + 29 \times 101 + 1)$

EXERCISE 4

1 ▶ $n + (n + 1) + (n + 2) = 3n + 3 = 3(n + 1)$

2 ▶ **a** $n - 1, n, n + 1$

     b $\dfrac{(n - 1) + n + (n + 1)}{3} = \dfrac{3n}{3} = n$

3 ▶ $n^2 + (n + 1)^2 = 2n^2 + 2n + 1 = 2(n^2 + n) + 1$

4 ▶ $(3n + 1)^2 - (3n - 1)^2$
$= 9n^2 + 6n + 1 - (9n^2 - 6n + 1) = 12n$

5 ▶ Let the integers be $a = n - 1$, $b = n$ and $c = n + 1$ then $c^2 - a^2 = (n + 1)^2 - (n - 1)^2$
$= 4n = 4b$

EXERCISE 4*

1 ▶ $n + (n + 1) + (n + 2) + (n + 3) = 4n + 6$
$= 4(n + 1) + 2$

2 ▶ $8 \times \dfrac{n(n + 1)}{2} = 4n^2 + 4n = (2n + 1)^2 - 1$

3 ▶ Substituting c for n gives $ac^2 + bc + c$
$= c(ac + b + 1)$

4 ▶ $n(n + 1) + (n + 1)(n + 2) = 2n^2 + 4n + 2$
$= 2(n + 1)^2$

5 ▶ $(n - 1)n(n + 1) = n^3 - n$

     a At least one of $(n - 1)$, n or $(n + 1)$ is even.

     b At least one of $(n - 1)$, n or $(n + 1)$ is divisible by 3.

     c Since $n^3 - n$ is divisible by 2 and 3 then it is divisible by 6.

ACTIVITY 2

1 ▶ **a** 8   **b** 8

     c $\{(n + 4) + (n + 5)\} - \{n + (n + 1)\} = 8$

     d $\{(n + 5) + (n + 6)\} - \{n + (n + 1)\} = 10$

     e $\{(n + m) + (n + m + 1)\} - \{n + (n + 1)\} = 2m$

2 ▶ **a** $(n + 1)(n + 4) - n(n + 5) = 4$

     b $(n + 1)(n + m) - n(n + m + 1) = m$

     c Same as part **b**.

EXERCISE 5

1 ▶ $(2n + 1) + (2n + 3) = 4n + 4 = 4(n + 1)$

2 ▶ $(2n + 1) - (2m + 1) = 2n - 2m = 2(n - m)$

3 ▶ $(2n + 1)^2 = 4n^2 + 4n + 1 = 2(2n^2 + 2n) + 1$

4 ▶ $2m(2n + 1) = 2(2mn + m)$

5 ▶ $x = 2n + 5 = 2(n + 2) + 1$

6 ▶ $n^2 - 2n + 1 = (n - 1)^2$

7 ▶ $100a + 10b + 5 = 5(20a + 2b + 1)$

8 ▶ $(n - 1)^2 + n^2 = (n + 1)^2 \Rightarrow n^2 - 4n = 0 \Rightarrow$
$n(n - 4) = 0 \Rightarrow n = 4$ is only possible answer.

EXERCISE 5*

1 ▶ $2n + (2n + 2) + (2n + 4) = 6(n + 1)$

2 ▶ $(2n + 1)^3 = 8n^3 + 12n^2 + 6n + 1$
$= 2(4n^3 + 6n^2 + 3n) + 1$

3 ▶ $(2n + 3)^2 - (2n + 1)^2 = 4(2n + 2)$

4 ▶ $x = -4n + 10 = 2(-2n - 5)$

5 ▶ If n is even $= 2m$ then
$$\frac{n(n + 1)}{2} = \frac{2m(2m + 1)}{2} = m(2m + 1);$$
if n is odd $= 2m + 1$ then
$$\frac{n(n + 1)}{2} = \frac{(2m + 1)(2m + 2)}{2} =$$
$(2m + 1)(m + 1)$. (Note it is easier to say one
of n or $n + 1$ must be even so result follows.)

6 ▶ $a + b + c = 3n \Rightarrow c = 3n - a - b$ so
$100a + 10b + c = 99a + 9b + 3n =$
$3(33a + 3b + n)$

7 ▶ Let $b = 2n + 1$ then $a * b = 2a + (2n + 1) =$
$2(a + n) + 1$

8 ▶ $100a + 10b + c = 100a + 10(a + c) + c =$
$110a + 11c = 11(10a + c)$; for $abcd$ rule
is $a + c = b + d$; $1000a + 100b + 10c + d$
$= 1000a + 99b + 10c + b + d =$
$1001a + 99b + 11c = 11(91a + 9b + c)$

ACTIVITY 3

1 ▶ $(a - b) = 0$ and division by zero is not
allowed.

2 ▶ $(b - a) < 0$ and when dividing an inequality
by a negative number the inequality must
be reversed.

EXERCISE 6

1 ▶ **a** $x^2 + 4x + 4 = (x + 2)^2 \geqslant 0$

b
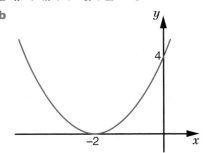

2 ▶ **a** $2x - x^2 - 1 = -(x - 1)^2 \leqslant 0$

b
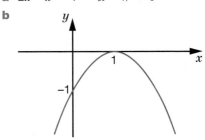

3 ▶ **a** $(x - 3)^2 - 6$ **b** -6

c
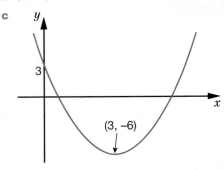

4 ▶ **a** $2(x + 1)^2 - 5$ **b** $(-1, -5)$

5 ▶ **a** $c \geqslant 25$ **b** $b \leqslant 8$

EXERCISE 6*

1 ▶ **a** $x^2 - 6x + 9 = (x - 3)^2 \geqslant 0$

b
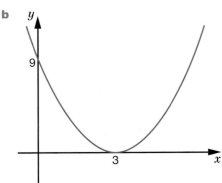

2 ▶ **a** $14x - x^2 - 52 = -(x - 7)^2 - 3 < 0$

b -3

c
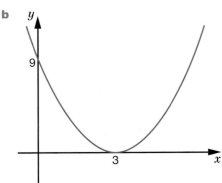

3 ▶ $2x^2 - 32x + 129 = 2(x - 8)^2 - 128 + 129 =$
$2(x - 8)^2 + 1 \geqslant 0 \Rightarrow 2x^2 + 129 > 32x$

4 ▶ $x^2 + 2bx + 4 = x + b^2 + 4 - b^2 \Rightarrow$ smallest
value is $4 - b^2$

5 ▶ **a** $(x - y)^2 \geqslant 0$
$(x - y)^2 = x^2 - 2xy + y^2$
$\Rightarrow x^2 - 2xy + y^2 \geqslant 0 \Rightarrow x^2 + y^2 \geqslant 2xy$

b $x^2 + y^2 = 2xy \Rightarrow (x - y^2) = 0 \Rightarrow x = y$

6 ▶ **a** $t^2 + 2 - t^2 = 2t^2 - 4t + 4$

b $2t^2 - 4t + 4 = 2t - 1^2 + 2$ which has
a minimum value of 2 so minimum
distance is $\sqrt{2}$

c When $t = 1$ so 13:00

EXERCISE 7 REVISION

1 ▶ a (−3, 9), (4, 16) b (−1, −7), (5, 5)

2 ▶ a $x = 1.37, y = −0.63$ or $x = −4.37, y = −6.37$

   b $x = 1.82, y = −0.82; x = −0.82, y = 1.82$ (symmetry)

3 ▶ a $x = 1.7$

   b A(1.7, 1.05), B(1.7, −1.05), 2.11 m

4 ▶ e.g. $2^2 + 2^2 = 8$ (even)

5 ▶ $(2n + 1) + (2n + 3) + (2n + 5) + (2n + 7) = 8(n + 2)$

6 ▶ $2m \times 2n = 4mn$

7 ▶ a $x^2 + 8x + 16 = (x + 4)^2 \geq 0$

   b (−4, 0)

   c
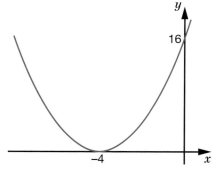

EXERCISE 7* REVISION

1 ▶ a $\left(−\frac{1}{2}, \frac{1}{2}\right)$, (3, 18) b (−2, −2), (1, 4)

2 ▶ a $x = 0.64, y = 0.27$ or $x = −3.14, y = −7.28$

   b $x = 1.85, y = 0.15; x = −0.18, y = 2.18$

3 ▶ (15, −1.658), (15, 1.658)

4 ▶ e.g. $4^3 − 2^3 = 56$ (odd)

5 ▶ Let the numbers be $a = n − 1$, $b = n$ and $c = n + 1$ then $ac = (n − 1)(n + 1) = n^2 − 1 = b^2 − 1$

6 ▶ $(2n + 1)(2n + 3) = 4n^2 + 8n + 3 = 4(n^2 + 2n + 1) − 1$

7 ▶ a $12x − x^2 − 40 = −(x − 6)^2 − 4 < 0$

   b
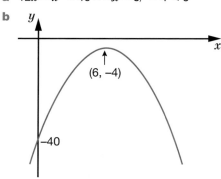

EXAM PRACTICE: ALGEBRA 9

1 ▶ (4.73, 6.46), (1.27, −0.464)

2 ▶ (0, −4), (3.2, 2.4)

3 ▶ e.g. $(1 + 4)^2 \neq 1^2 + 16$

4 ▶ Let the numbers be $n + 4$ and n, then $(n + 4)^2 − n^2 = n^2 + 8n + 16 − n^2 = 8n + 16 = 8(n + 2)$

5 ▶ $(2n + 1)(2n + 3) = 4n^2 + 8n + 3 = 2(2n^2 + 4n + 1) + 1$

6 ▶ a $x^2 − 10x + 25 = (x − 5)^2 \geq 0$

   b (5, 0)

   c
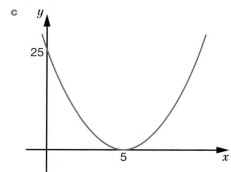

UNIT 9: GRAPHS 8

EXERCISE 1

1 ▶ a 2 b 3 c −1

   d −2 e $x = 1$

2 ▶ a 2 b 2 c 0.1

   d 5.8 e $x = 0.18$ or 1.8

3 ▶ a −0.25 b −1 c −0.44

   d −4 e $x = \pm0.71$

4 ▶ a 4 b 2 c −4 d $x = 3$

EXERCISE 1*

1 ▶ a 2 b 4 c −1

   d −3 e $x = 1.5$

2 ▶ a −2.75 b 1.25 c −0.75

   d −4 e $x = −1$ or 0.33

3 ▶ a 1 b −0.37 or 1.37

   c −1.3, 0.17 or 1.13

   d −1.13, −0.17, 1.3

4 ▶ b

x co-ordinate	−4	−3	−2	−1	0	1	2	3	4
Gradient	−8	−6	−4	−2	0	2	4	6	8

   c The value of the gradient is twice the x-co-ordinate.

5 ▶ a

x	0	1	2	3	4	5
2^x	1	2	4	8	16	32

   b 1.4, 5.5

   c $x = 4.1, y = 17$ (2 s.f.)

EXERCISE 2

1 ▶ a i 1 m/s ii 0 m/s iii 2 m/s

b 0–20 s gradually increased speed then slowed down to a stop

20–30 s stationary

30–40 s speed increasing

40–50 s travelling at a constant speed of 2 m/s

50–60 s slowing down to a stop

2 ▶ a i $\frac{1}{4}$ m/s² ii 0 m/s² iii $-\frac{1}{4}$ m/s²

b 0–20 s accelerating up to a speed of 5 m/s

20–30 s running at a constant speed of 5 m/s

30–50 s decelerating to a speed of 2.5 m/s

50–60 s running at a constant speed of 2.5 m/s

60–80 s decelerating to a stop

3 ▶ b i –9.6 °C/min ii –6.7 °C/min

iii –5.5 °C/min

4 ▶ a

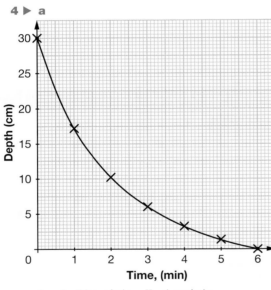

b i –12 cm/min ii –4 cm/min

iii –1.4 cm/min

5 ▶ b i 0.5 m/s² ii 1.5 m/s² iii 2.5 m/s²

6 ▶ a

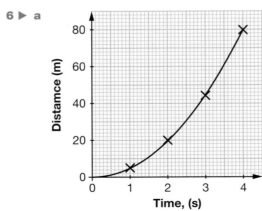

b i 10 m/s² ii 20 m/s² iii 30 m/s²

EXERCISE 2*

1 ▶ a

t (min)	0	20	40	60	80	100	120
N	10	40	160	640	2560	10 240	40 960

b i 0.7 ii 44 iii 710

2 ▶

t (months)	0	1	2	3	4
N (millions)	2	2.1	2.21	2.32	2.43

t (months)	5	6	7	8	9
N (millions)	2.55	2.68	2.81	2.95	3.1

b i 97 000 ii 119 000 iii 144 000

3 ▶ a

t (min)	0	10	20	30	40
V (cm³)	2000	1700	1445	1228	1044

t (min)	50	60	70	80	90
V (cm³)	887	754	641	545	463

b i –28 cm³/min ii –8.9 cm³/min

c t = 0, –32.5 cm³/min

4 ▶ a

t (s)	0	10	20	30	40
M (g)	120	96	76.8	61.4	49.2

t (s)	50	60	70	80	90
M (g)	39.3	31.5	25.2	20.1	16.1

b i –1.71 g/s ii –0.56 g/s

c At t = 0, –2.68 g/s

5 ▶ a i 1.67 m/h ii –1.67 m/h iii 0 m/h

b Max at t = 0, 4, 8, 12, ±2.36 m/h

6 ▶ a i –9.44 m/s ii 0 m/s iii 1.56 m/s

b Max at t = 1.75, –11.7 m/s

EXERCISE 3

1 ▶

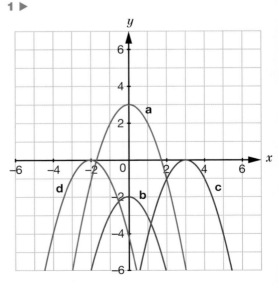

a (0, 3) b (0, –2) c (3, 0) d (–2, 0)

2 ▶ **a** $\begin{pmatrix} 0 \\ -7 \end{pmatrix}$ **b** $\begin{pmatrix} 7 \\ 0 \end{pmatrix}$

     c $\begin{pmatrix} 0 \\ 7 \end{pmatrix}$ **d** $\begin{pmatrix} -7 \\ 0 \end{pmatrix}$

3 ▶

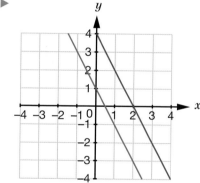

    a Red curve   **b** Blue curve
    c Purple curve   **d** Green curve

4 ▶

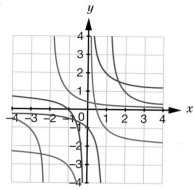

    a Red line   **b** Blue line
    c $y = 4 - 2x$

5 ▶

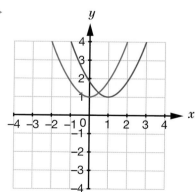

    a Red curve   **b** Blue curve
    c $y = x^2 - 2x + 2$

6 ▶ **a** $y = f(x) - 2$
     b $y = f(x - 2)$

7 ▶ $y = x^2 + 5x + 3$

8 ▶ $y = -x^2 - 5x - 3$

EXERCISE 3*

1 ▶

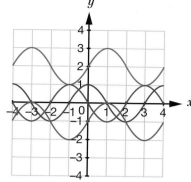

    a Red curve   **b** Blue curve
    c Purple curve   **d** Green curve

2 ▶ **a** $\begin{pmatrix} 0 \\ 5 \end{pmatrix}$ **b** $\begin{pmatrix} 13 \\ 0 \end{pmatrix}$

     c $\begin{pmatrix} -11 \\ 0 \end{pmatrix}$ **d** $\begin{pmatrix} 0 \\ -9 \end{pmatrix}$

3 ▶

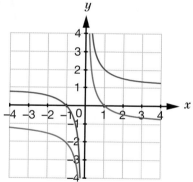

    a Red curve   **b** Blue curve
    c $y = \dfrac{1}{x} + 1$

4 ▶

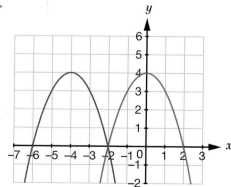

    a Red curve   **b** Blue curve
    c $y = -x^2 - 8x - 12$

5 ▶ **a** $y = f(x + 3)$ **b** $y = f(x) + 3$

6 ▶ $y = 2x^3$

7 ▶ $y = 1 + 3x - x^2$

8 ▶ a Red curve   **b** Blue curve

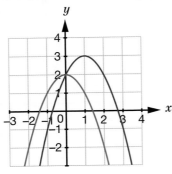

c $y = 2 + 2x - x^2$

ACTIVITY 1
$\sin(x + 90) = \cos x, \ a = 90$
$\cos(x - 90) = \sin x, \ a = -90$

EXERCISE 4

1 ▶

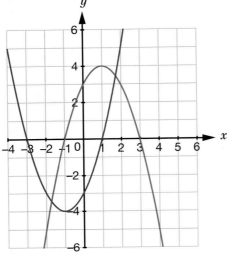

a Red curve

b Blue curve

2 ▶ a i Reflection in the x-axis
ii Reflection in the y-axis

b Red line is $y = x + 1$, blue line is **a i**,
purple line is **a ii**

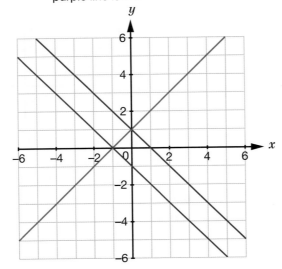

3 ▶ a $y = f(-x)$
b $-f(x)y = -x^3 - x^2 - x - 1$

4 ▶ a and **b** are the same.

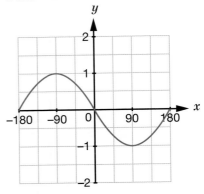

5 ▶ (0, 4), (−2, 0), (−4, −2)

EXERCISE 4*

1 ▶ a i Reflection in y-axis
ii Reflection in the x-axis.

b Red line is $y = 2 - x$, blue line is **a i**,
purple line is **a ii**

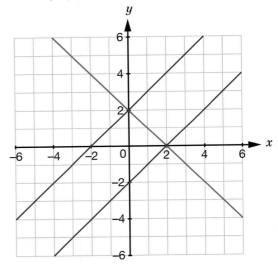

2 ▶ a Red curve   **b** Blue curve

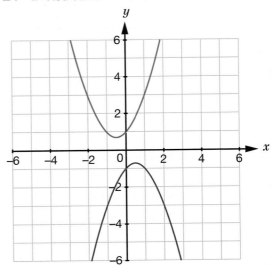

3 ▶ a $(a, 0)$ **b** $(0, -b)$ **c** $(c, -d)$

4 ▶ a $y = -x^3 - x^2 - x - 1$

b $y = -x^3 + x^2 - x + 1$

5 ▶ a

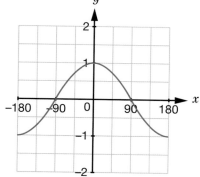

b

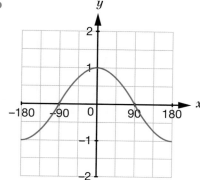

c The curves are the same, so $\cos(x) = \cos(-x)$

d

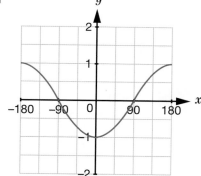

6 ▶

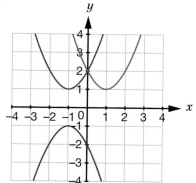

a Red curve is $y = f(x)$, blue curve is $y = f(-x)$, purple curve is $y = -f(-x)$

b Rotation of 180° about the origin

c No

EXERCISE 5 **1 ▶**

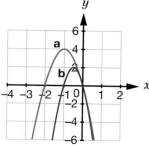

a $(-1, 4)$ **b** $(-0.5, 2)$

2 ▶ a Stretch in the x direction, scale factor $\frac{1}{3}$

b Stretch in the y direction, scale factor 3

3 ▶

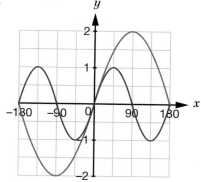

a Red curve   **b** Blue curve

4 ▶

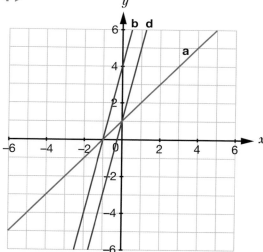

c $y = 4 + 4x$ **e** $y = 1 + 4x$

5 ▶ a Stretch in the y direction, scale factor 2

b $y = 2f(x)$

c Stretch in the x direction, scale factor $\frac{1}{2}$

d $y = f(2x)$

6 ▶ a $y = 5x^2 + 5x + 5$

b $y = x^2 - 2x + 2$

EXERCISE 5*

1 ▶

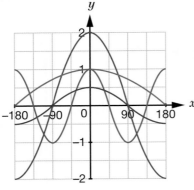

a Red curve   b Blue curve

c Purple curve   d Green curve

2 ▶ a Stretch in the y direction, scale factor $\frac{1}{4}$

b Stretch in the x direction, scale factor $\frac{1}{5}$

c Stretch in the y direction, scale factor 7

d Stretch in the x direction, scale factor 2

3 ▶

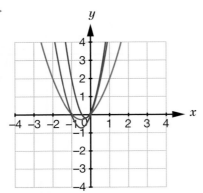

a Blue curve   b Purple curve

4 ▶

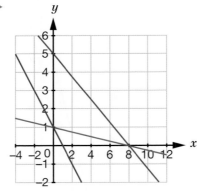

a Red line   b Blue line

c $y = \dfrac{5(8 - x)}{8}$ d Purple line

e $y = 1 - x$

5 ▶ a Stretch in the x direction, scale factor 3

b $y = f\!\left(\dfrac{x}{3}\right)$

c Stretch in the y direction, scale factor $\frac{1}{4}$

d $y = \frac{1}{4}f(x)$

6 ▶ a $y = 64x^3 + 32x^2 - 12x$

    b $y = x^2 - 4x + 8$

ACTIVITY 2

$y = (x - 1)^2, y = -(x - 1)^2, y = (x + 1)^2, y = -(x + 1)^2$

$y = x^3, y = -x^3, y = \dfrac{1}{x}, y = -\dfrac{1}{x}$

$y = \cos(2x) + 1, y = -\cos(2x) - 1$

EXERCISE 6   REVISION

1 ▶ b $-3, 0, 5$ c $y = -3x - 1$

2 ▶ b 2.6 mm/s, 6.3 mm/s

3 ▶ a Graph 1,   b Graph 2

4 ▶ a D   b C   c B   d A

5 ▶

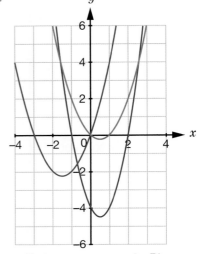

a Red curve   b Blue curve

c Purple curve

6 ▶ a $(2, -4)$ b $(-2, 4)$

7 ▶ a $y = 4 - x^2$ b $(-2, 0)$ and $(2, 0)$

    c $y = 4 - 4x^2$

EXERCISE 6*   REVISION

1 ▶ a

x	0	1	2	3	4
y	1	3	9	27	81

b 3.3, 9.9   c $y = 3.3x - 0.3$

2 ▶ a $(1, 35), (2, 64), (3, 75), (4, 80), (5, 75),$
$(6, 60), (7, 35), (8, 0)$

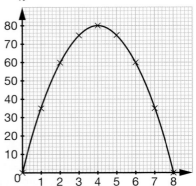

b

t (s)	0	1	2	3	4
v (m/s)	40	30	20	10	0

t (s)	5	6	7	8
v (m/s)	−10	−20	−30	−40

c Straight-line graph passing through (0, 40) and (8, −40)

d Acceleration is constant (−10 m/s), i.e. constant deceleration (10 m/s)

3 ▶ a D   **b** C   **c** B   **d** A

4 ▶ a

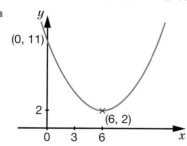

b (6, 2)   **c** (1.5, 2)

d $x = 2x = \frac{1}{2}x = 0$ hence stretching horizontally does not affect the point where the graphs intersect the y-axis.

5 ▶ a (0, 4), (3, 0), (2, −8)

b (0, 2), (−3, 0), (−2, −4)

c (−1, 2), (2, 0), (1, −4)

6 ▶

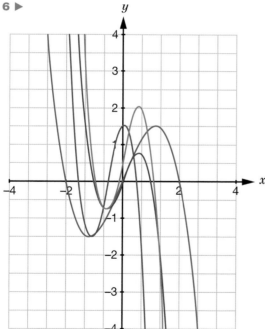

a Red curve   **b** Blue curve

c Purple curve   **d** Green curve

7 ▶ a Stretch scale factor 2 in the x direction

b $y = \sin\left(\frac{1}{2}x\right)$

c Stretch scale factor 2 in the y direction followed by a translation of $\begin{pmatrix} 0 \\ 3 \end{pmatrix}$

d $y = 2\sin\left(\frac{1}{2}x\right) + 3$

e 4.73 m (3 s.f.)

1 ▶ a i 5   **ii** −3

b $x = -0.39$ or $x = 1.72$

2 ▶ a

m	1	2	4	6	10	15	20
t	52	50	46	42	37	31	28

b

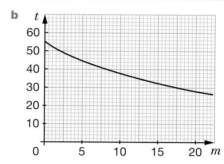

c −1.5 °C/min

3 ▶ a

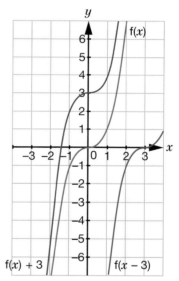

b i (0, 3)   **ii** (3, 0)

4 ▶

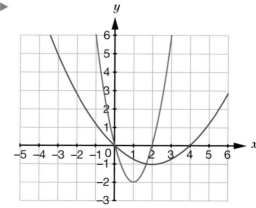

a Red curve   **b** Blue curve

5 ▶ a

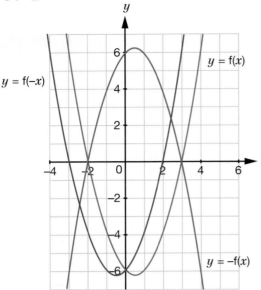

y = f(−x)

y = f(x)

y = −f(x)

b Reflection in the x-axis

c Reflection in the y-axis

UNIT 9: SHAPE AND SPACE 9

EXERCISE 1

1 ▶ a 11.7 cm   b 14.2 cm   c 34.4°

2 ▶ a 18.6 cm   b 28.1 cm   c 48.5°

3 ▶ a 14.1 cm   b 17.3 cm   c 35.3°

4 ▶ a 28.3 cm   b 34.6 cm

    c 35.3°   d 19.5°

5 ▶ a 4.47 m   b 4.58 m

    c 29.2°   d 12.6°

6 ▶ a 407 m   b 402 m

    c 8.48°   d 13.3°

7 ▶ a 43.3 cm   b 68.7 cm   c 81.2 cm

8 ▶ a 28.9 cm   b 75.7 cm   c 22.4°

EXERCISE 1*

1 ▶ a 16.2 cm   b 67.9°   c 55.3 cm²

2 ▶ a 26.5 cm   b 61.9°   c 1530 cm²

3 ▶ a 30.3°   b 31.6°   c 68.9°

4 ▶ a 36.9°   b 828 cm²

5 ▶ a 15 m   b 47.7°   c £91 300

6 ▶ a 66.4°   b 32.9°

7 ▶ 46.5 m

8 ▶ a OW = 4290 m, OS = 2760 m

    b 36.0°   c 197 km/h

ACTIVITY 1 $d^2 = a^2 + b^2 + c^2$

EXERCISE 2 **REVISION**

1 ▶ a   i 15 cm   ii 20.5 cm   iii 16.6 cm

     iv 20.5 cm

   b 43.0°   c 43.0°   d 35.8°

2 ▶ a 19.3 cm   b 21.2°

3 ▶ a AC = 42.4 cm

   b 33.9 cm   c 68.0°   d 58.0°

4 ▶ a 18.4°   b 500 m   c 11.3°

5 ▶ a 18.7 m   b 31.9°   c 20.5°

EXERCISE 2* **REVISION**

1 ▶ a 21.9 cm

   b Angle ADB = 33.2°

   c Angle DAE = 24.2°

2 ▶ a AC = 70.7 cm

   b 98.7 cm

   c 27.9°

   d 216 000 cm²

3 ▶ a 40 cm

   b 18.1 cm

   c 42.2°

   d 24.2 cm

   e 48.6°

4 ▶ 25.5°

5 ▶ a x is the length of the diagonal of the square that is the top face of the cube.

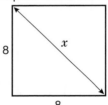

Using Pythagoras' theorem:

$x^2 = 8^2 + 8^2 = 128$

$x = \sqrt{128} = \sqrt{64 \times 2} = \sqrt{64} \times \sqrt{2} = 8\sqrt{2}$

   b 50.5°

EXAM PRACTICE: SHAPE AND SPACE 9

1 ▶ a 43.3 cm

   b 35.3°

2 ▶ a ii 21.2 cm   ii 10.6 cm   iii 14.5 cm

   b 54°

   c 27°

3 ▶ 61.1°

UNIT 9: HANDLING DATA 6

1 ▶ a, b

Age, a (years)	Frequency	Class width	Frequency density
$0 < a \leqslant 10$	3	10	0.3
$10 < a \leqslant 20$	14	10	1.4
$20 < a \leqslant 40$	17	20	0.85
$40 < a \leqslant 60$	19	20	0.95
$60 < a \leqslant 80$	7	20	0.35

2 ▶

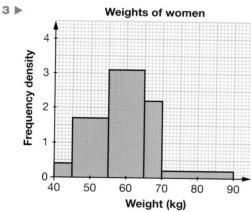

Time spent watching TV

3 ▶

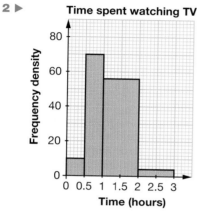

Weights of women

4 ▶ a 67.8%

b

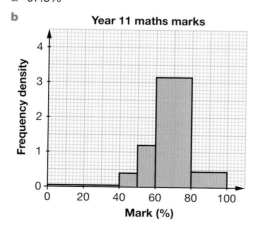

Year 11 maths marks

1 ▶ a 175 litres/day

b

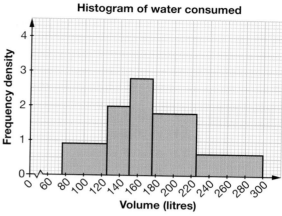

Histogram of water consumed

2 ▶ 102

3 ▶ a The number waiting for less than 20 minutes is 150. The total number of patients is 310, so Rachel is wrong.

   b 38 patients

4 ▶ a Bar for $30 < t \leqslant 60$ class drawn with a frequency density of 1.8

     Bar for $100 < t \leqslant 200$ class drawn with a frequency density of 2.5

     Frequency for the $60 < t \leqslant 100$ class = 64

   b 112 minutes

1 ▶ a 2   **b** 17   **c** 40

2 ▶ a 27   **b** 59   **c** 87

3 ▶ a

Distance, d (miles)	Frequency
$0 < d \leqslant 5$	140
$5 < d \leqslant 10$	190
$10 < d \leqslant 20$	210
$20 < d \leqslant 35$	120
$35 < d \leqslant 50$	30

b

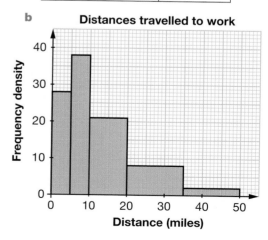

Distances travelled to work

4 ▶ **a** 25 **b** 49 **c** 3.6 kg

d

Weight, w (kg)	Frequency
$2.8 < w \leqslant 3.0$	5
$3.0 < w \leqslant 3.5$	15
$3.5 < w \leqslant 4.0$	25
$4.0 < w \leqslant 4.2$	4

e 3.54 kg **f** 27

EXERCISE 2*

1 ▶ **a** 52 **b** 172 cm **c** 38

2 ▶ **a** 90 plants **b** 22 cm **c** 16 plants

3 ▶ **a** 23.8 g

b

Mass, m (grams)	Frequency
$16 < m \leqslant 20$	8
$20 < m \leqslant 22$	14
$22 < m \leqslant 24$	20
$24 < m \leqslant 25$	19
$25 < m \leqslant 26$	13
$26 < m \leqslant 29$	6

c 23.25 g

4 ▶ **a** 250 runners **b** 40 runners

c i

Time, t (minutes)	Frequency
$35 < t \leqslant 45$	18
$45 < t \leqslant 50$	54
$50 < t \leqslant 60$	110
$60 < t \leqslant 70$	40
$70 < t \leqslant 90$	28

ii 56.7 minutes

ACTIVITY 1

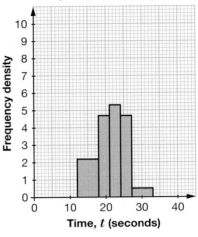

Group times before programme

Median = 23.1 s

Mean = 23.2 s

IQR = 7.3 s

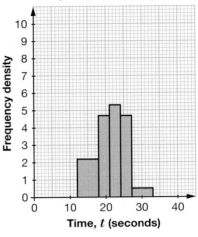

Group times after programme

Median = 21.6 s

Mean = 21.3 s

IQR = 6 s

Evidence suggests that the Brain Training programme does have a positive impact. The mean and median times are both reduced for the group and there appears to be less dispersion of ability after the programme, as the IQR is also reduced. The second histogram is shifted slightly towards the shorter times.

EXERCISE 3 **REVISION**

1 ▶

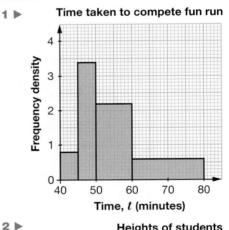

Time taken to compete fun run

2 ▶

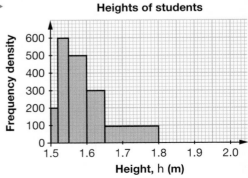

Heights of students

3 ▶ **a** 40 **b** 230 **c** 94

4 ▶ **a** 46 **b** 23.5th = 4.9375 **c** 15

1 ▶

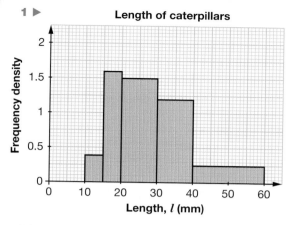

Length of caterpillars

2 ▶ a Frequency for the $30 < t \leqslant 60$ class = 33

Bar for $150 < t \leqslant 210$ class drawn with a frequency density of 1.5

b 105 vehicles

3 ▶ About 86 or 87 farms

4 ▶ a Frequency for the $160 < h \leqslant 165$ class = 18

Frequency for the $165 < h \leqslant 170$ class = 25

Bar for $140 < h \leqslant 160$ class drawn with a frequency density of 1.1

Bar for $170 < h \leqslant 185$ class drawn with a frequency density of 1.8

Bar for $185 < h \leqslant 200$ class drawn with a frequency density of 0.8

b There are 104 adults in total, so the median height is the mean of the 52nd and 53rd heights.

The number of people up to 167 cm is 22 + 18 + 10 = 50 and the number of people up to 168 cm is 22 + 18 + 15 = 55.

So the median lies between 167 cm and 168 cm, and Clare is correct.

c 168.4 cm

1 ▶ a

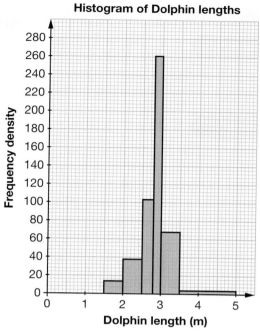

Histogram of Dolphin lengths

b i 2.85 m   **ii** 2.87 m   **iii** 24

c $2.8 < l \leqslant 3.0$ m

2 ▶ a 100   **b** 85   **c** 1.51 cm

d

Height, (m)	Frequency
$1.4 < m \leqslant 1.45$	5
$1.45 < m \leqslant 1.48$	15
$1.48 < m \leqslant 1.5$	20
$1.5 < m \leqslant 1.55$	20
$1.55 < m \leqslant 1.6$	15
$1.6 < m \leqslant 1.7$	10

e 1.524 cm   **f** 36

UNIT 10 ANSWERS

UNIT 10: NUMBER 10

ACTIVITY 1

a 4.51, 0.528, 0.602, 0.0402, 2.36×10^{-3}, 8.49×10^{-6}, 0.658, 7.18×10^{-3}, 3.21×10^{-8}

b Around 3×10^7 km (around 80 times the distance from the Earth to the Moon)

ACTIVITY 2

a
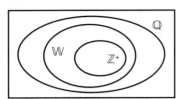

b
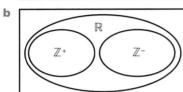

EXERCISE 1

1 ▶ $\frac{57}{10}$　　2 ▶ $\frac{47}{99}$　　3 ▶ $\frac{7}{1}$

4 ▶ Irrational　　5 ▶ $\frac{3}{1}$　　6 ▶ $\frac{1}{1}$

7 ▶ Irrational　　8 ▶ 0　　9 ▶ e.g. 2.5

10 ▶ e.g. $\sqrt{53}$　　11 ▶ $\frac{2}{\pi}$　　12 ▶ $\frac{3}{\sqrt{\pi}}$

EXERCISE 1*

1 ▶ Irrational　　2 ▶ $\frac{2}{5}$　　3 ▶ $\frac{3}{5}$

4 ▶ $\frac{3}{2}$　　5 ▶ Irrational　　6 ▶ $\frac{1}{1}$

7 ▶ Irrational　　8 ▶ e.g. 3.5　　9 ▶ e.g. $\sqrt{7}$

10 ▶ e.g. $\sqrt{2} \times \sqrt{8}$　　11 ▶ $\frac{9}{\pi}$

12 ▶ a e.g. $3:4:5$　　b e.g. $\sqrt{2}:\sqrt{2}:2$

　　　c e.g. $1:2:\sqrt{5}$　　d e.g. $1:\sqrt{3}:2$

EXERCISE 2

1 ▶ $6\sqrt{5}$　　2 ▶ $4\sqrt{3}$　　3 ▶ 32

4 ▶ 20　　5 ▶ 8　　6 ▶ 105

7 ▶ $2\sqrt{2}$　　8 ▶ 4　　9 ▶ 8

EXERCISE 2*

1 ▶ $5\sqrt{11}$　　2 ▶ $4\sqrt{7}$　　3 ▶ 99

4 ▶ 96　　5 ▶ $56\sqrt{7}$　　6 ▶ $120\sqrt{2}$

7 ▶ 9　　8 ▶ 4　　9 ▶ 6

EXERCISE 3

1 ▶ $2\sqrt{3}$　　2 ▶ $3\sqrt{2}$　　3 ▶ $8\sqrt{3}$

4 ▶ $9\sqrt{5}$　　5 ▶ $3\sqrt{3}$　　6 ▶ $2\sqrt{2}$

7 ▶ $14\sqrt{2}$　　8 ▶ $\sqrt{50}$　　9 ▶ $\sqrt{27}$

10 ▶ $\sqrt{54}$　　11 ▶ $\frac{1}{2}$　　12 ▶ $\frac{2}{5}$

13 ▶ $\frac{2}{3}$　　14 ▶ 18, $10\sqrt{3}$, $\sqrt{3}\sqrt{13}$

EXERCISE 3*

1 ▶ $2\sqrt{7}$　　2 ▶ $3\sqrt{11}$　　3 ▶ $20\sqrt{5}$

4 ▶ $9\sqrt{13}$　　5 ▶ $5\sqrt{3}$　　6 ▶ $\sqrt{3}$

7 ▶ $22\sqrt{7}$　　8 ▶ $\sqrt{75}$　　9 ▶ $\sqrt{80}$

10 ▶ $\sqrt{63}$　　11 ▶ $\frac{1}{6}$　　12 ▶ $\frac{9}{10}$

13 ▶ $\frac{7}{13}$

14 ▶ $2\sqrt{2}$ cm, $(8 + 2\sqrt{2})$ cm, $(2 + 3\sqrt{2})$ cm²

ACTIVITY 3

a i $\sqrt{2}$　　ii 45°

iii

$\sin 45° = \frac{1}{\sqrt{2}}$	$\cos 45° = \frac{1}{\sqrt{2}}$	$\tan 45° = 1$

b i $\sqrt{3}$　　ii 60°, 30°

iii

$\sin 30° = \frac{1}{2}$	$\cos 30° = \frac{\sqrt{3}}{2}$	$\tan 30° = \frac{1}{\sqrt{3}}$
$\sin 60° = \frac{\sqrt{3}}{2}$	$\cos 60° = \frac{1}{2}$	$\tan 60° = \sqrt{3}$

EXERCISE 4

1 ▶ $3 + 2\sqrt{2}$　　　　2 ▶ $4 - 2\sqrt{3}$

3 ▶ $21 + 12\sqrt{3}$　　　4 ▶ $27 - 18\sqrt{2}$

5 ▶ -4　　　　　　　　6 ▶ $5 + 2\sqrt{6}$

7 ▶ $7 - 2\sqrt{10}$　　　　8 ▶ $1 + \sqrt{2} - \sqrt{5} - \sqrt{10}$

9 ▶ $1 + \sqrt{2} - \sqrt{3} - \sqrt{6}$　　10 ▶ $4\sqrt{3}$, 2, $2\sqrt{2}$

EXERCISE 4*

1 ▶ $9 + 4\sqrt{5}$

2 ▶ $18 - 8\sqrt{2}$

3 ▶ 13

4 ▶ $36 + 16\sqrt{2}$

5 ▶ $91 - 40\sqrt{3}$

6 ▶ $10 - 2\sqrt{21}$

7 ▶ 2

8 ▶ $15 + 10\sqrt{2} - 6\sqrt{7} - 4\sqrt{14}$

9 ▶ $8 + 12\sqrt{3} - 6\sqrt{5} - 9\sqrt{15}$

10 ▶ $1 + \sqrt{2}$, $1.5 + \sqrt{2}$

ACTIVITY 4

a i -1　　ii 13　　iii 1

b Rational

c i $(2 - \sqrt{2})$　　ii $(3 + 2\sqrt{5})$

d i $(a - \sqrt{b})$　　ii $(a + c\sqrt{b})$

EXERCISE 5

1 ▶ $\frac{\sqrt{5}}{5}$　　　　　2 ▶ $\sqrt{3}$

3 ▶ $2\sqrt{2}$　　　　　4 ▶ $\frac{\sqrt{2}}{2}$

5 ▶ $\sqrt{5}$　　　　　6 ▶ $\sqrt{3}$

7 ▶ $\frac{3\sqrt{2}}{4}$　　　　8 ▶ $\frac{2 + \sqrt{2}}{2}$

9 ▶ $2 + \sqrt{5}$　　　　10 ▶ $\frac{-1 + \sqrt{5}}{4}$

11 ▶ $2 + \sqrt{3}$　　　　12 ▶ $\frac{5 - 4\sqrt{2}}{7}$

EXERCISE 5*

1 ▶ $\dfrac{\sqrt{13}}{13}$

2 ▶ $\sqrt{a}$

3 ▶ $2\sqrt{3} - 1$

4 ▶ $\dfrac{\sqrt{6}}{3}$

5 ▶ $3 + 2\sqrt{7}$

6 ▶ $\sqrt{2}$

7 ▶ $\dfrac{2 - \sqrt{7}}{-3}$

8 ▶ $\dfrac{12 + 2\sqrt{3}}{3}$

9 ▶ $7 + 4\sqrt{3}$

10 ▶ $4 + 6\sqrt{2}$

11 ▶ $5(\sqrt{5} - \sqrt{3})$

12 ▶ $a = -3,\ b = 4$

ACTIVITY 6

a $\dfrac{1}{\sqrt{1} + \sqrt{2}} = \sqrt{2} - \sqrt{1},\ \dfrac{1}{\sqrt{2} + \sqrt{3}} = \sqrt{3} - \sqrt{2},$

$\dfrac{1}{\sqrt{3} + \sqrt{4}} = \sqrt{4} - \sqrt{3},$

so sum is $-\sqrt{1} + \sqrt{4} = 2 - 1 = 1$

b   Sum is $-\sqrt{1} + \sqrt{9} = 3 - 1 = 2$

c   Sum is $-\sqrt{1} + \sqrt{n} \Rightarrow n = 100$

EXERCISE 6   REVISION

1 ▶ $0.\dot{3}$ and $\sqrt{25}$

2 ▶ 2, for example (answers may vary)

3 ▶ e.g. $\sqrt{11}$

4 ▶ $\sqrt{45}$

5 ▶ $5\sqrt{3}$

6 ▶ $\sqrt{3}$

7 ▶ 18

8 ▶ 1.5

9 ▶ $3\sqrt{7}$

10 ▶ $12\sqrt{2}$

11 ▶ $3\sqrt{2}$

12 ▶ $2\sqrt{2}$

13 ▶ $59 + 30\sqrt{2}$

14 ▶ -1

15 ▶ $2\sqrt{3}$

16 ▶ $\dfrac{3}{2}$

17 ▶ $1 + \sqrt{3}$

18 ▶ $\sqrt{7} - 1$

19 ▶ $4\sqrt{5} + 8$

20 ▶ $16\sqrt{2},\ 30,\ 2\sqrt{17}$

EXERCISE 6*   REVISION

1 ▶ $(\sqrt{3})^2$ and $0.\dot{2}\dot{3}$

2 ▶ 3, for example (answers may vary)

3 ▶ e.g. $\sqrt{40}$

4 ▶ $\sqrt{176}$

5 ▶ $2\sqrt{5}$

6 ▶ 75

7 ▶ $11\sqrt{2}$

8 ▶ $19\sqrt{6}$

9 ▶ $\dfrac{\sqrt{3}}{6}$

10 ▶ 8

11 ▶ $-5 + \sqrt{14}$

12 ▶ $37 - 20\sqrt{3}$

13 ▶ $4 + 4\sqrt{2}$

14 ▶ $\dfrac{\sqrt{5}}{10}$

15 ▶ $2\sqrt{6}$

16 ▶ $2 + \sqrt{6}$

17 ▶ 2

18 ▶ $\dfrac{5 - \sqrt{7}}{2}$

19 ▶ $\dfrac{7(\sqrt{11} - \sqrt{7})}{2}$

20 ▶ $\dfrac{2 + 3\sqrt{2}}{4}$

21 ▶ $5\sqrt{3},\ 12\sqrt{3},\ 18$

22 ▶ $\sqrt{3},\ \dfrac{1}{2},\ \dfrac{\sqrt{3}}{2}$

EXAM PRACTICE: NUMBER 10

1 ▶ a $2\sqrt{5}$ b $\dfrac{7}{8}$

2 ▶ a $8\sqrt{3}$ b $7\sqrt{2}$

3 ▶ a $16 - 8\sqrt{3}$ b -2

4 ▶ a $1 + \dfrac{\sqrt{5}}{5}$ b $2 + \sqrt{11}$

5 ▶ a $\dfrac{7\sqrt{2}}{2}$ b $\dfrac{-\sqrt{2}}{6}$

6 ▶ a $\sin 45° = \dfrac{\sqrt{2}}{2},\ \cos 45° = \dfrac{\sqrt{2}}{2}$ b $18\sqrt{2}$ km

7 ▶ $x = 2$

UNIT 10: ALGEBRA 10

EXERCISE 1

1 ▶ $\dfrac{3}{2}$

2 ▶ $\dfrac{5}{2x}$

3 ▶ $\dfrac{y}{x}$

4 ▶ $x + 2$

5 ▶ $\dfrac{x - 3}{5}$

6 ▶ $\dfrac{1}{x + 2}$

7 ▶ $x + 2$

8 ▶ $\dfrac{2x}{x + 1}$

9 ▶ $\dfrac{x + y}{x - y}$

10 ▶ $\dfrac{4x}{x - 6}$

11 ▶ $\dfrac{x + 1}{x - 3}$

12 ▶ $\dfrac{x + 3}{x + 2}$

EXERCISE 1*

1 ▶ $\dfrac{3}{5}$

2 ▶ $\dfrac{2 - x}{x + 2}$

3 ▶ $\dfrac{1}{y}$

4 ▶ $\dfrac{x - 4}{x + 3}$

5 ▶ $\dfrac{x - 3}{x + 5}$

6 ▶ $\dfrac{x + 3}{x + 4}$

7 ▶ $\dfrac{x}{3x + 2}$

8 ▶ $\dfrac{r - 3}{r + 1}$

9 ▶ $\dfrac{t - 2}{t + 2}$

10 ▶ $\dfrac{a - b}{a + b}$

11 ▶ $\dfrac{2(x + 8)}{x}$

12 ▶ $\dfrac{3(x + 2)}{x}$

EXERCISE 2

1 ▶ $\dfrac{5x + 3}{6}$

2 ▶ $\dfrac{x - 2}{4}$

3 ▶ $\dfrac{8x + 12}{15}$

4 ▶ $\dfrac{5x + 3}{4}$

5 ▶ $\dfrac{1 - 4x}{5}$

6 ▶ $\dfrac{7x + 2}{12}$

7 ▶ $\dfrac{3x - 17}{10}$

8 ▶ $\dfrac{5x + 8}{6}$

9 ▶ $\dfrac{3x - 10}{18}$

10 ▶ $\dfrac{3x + 2}{6}$

11 ▶ $\dfrac{39 - 2x}{12}$

12 ▶ $\dfrac{23x - 11}{10}$

EXERCISE 2*

1 ▶ $\dfrac{9x + 13}{10}$

2 ▶ $\dfrac{1 - 2x}{12}$

3 ▶ $\dfrac{-3x + 16}{14}$

4 ▶ $\dfrac{6x - 2}{35}$

5 ▶ $\dfrac{-x + 5}{2}$

6 ▶ $\dfrac{x - 230}{15}$

7 ▶ $\dfrac{16x - 21}{6}$

8 ▶ $\dfrac{x - 6}{72}$

9 ▶ $\dfrac{47x - 22}{60}$

10 ▶ $\dfrac{5x - 8}{6}$

11 ▶ $\dfrac{23x + 7}{18}$

12 ▶ $\dfrac{59 - 78x}{10}$

EXERCISE 3

1 ▶ $\dfrac{5}{6x}$

2 ▶ $\dfrac{1}{4x}$

3 ▶ $\dfrac{x-4}{2(x-2)}$

4 ▶ $\dfrac{4y-2x}{x^2y}$

5 ▶ $\dfrac{2x}{(x-1)(x+1)}$

6 ▶ $\dfrac{x+10}{(x-4)(x+3)}$

7 ▶ $\dfrac{x+8}{(x-1)(x+2)}$

8 ▶ $\dfrac{x^2+x-3}{x(x-3)}$

9 ▶ $\dfrac{2}{(x-1)}$

10 ▶ $\dfrac{3}{(x+3)}$

EXERCISE 3*

1 ▶ $\dfrac{5}{12x}$

2 ▶ $\dfrac{3(x+6)}{(x+1)(x+4)}$

3 ▶ $\dfrac{x}{1+x}$

4 ▶ $\dfrac{3x^2-2y^2}{4xy}$

5 ▶ $\dfrac{1}{x+1}$

6 ▶ $\dfrac{2x}{(x+2)(x-2)}$

7 ▶ $\dfrac{1}{x+1}$

8 ▶ $\dfrac{7x-11}{2(x-1)(x+3)}$

9 ▶ $\dfrac{8}{(x-3)(x+1)}$

10 ▶ $\dfrac{8-x}{(x-4)(x+1)(x-2)}$

EXERCISE 4

1 ▶ $\dfrac{(x+1)^2}{2}$ **2** ▶ $(x+2)(x-1)$ **3** ▶ $\dfrac{3}{4}(x+3)$

4 ▶ $2x$ **5** ▶ 6 **6** ▶ $\dfrac{y}{x}$

7 ▶ $\dfrac{2a}{b}$ **8** ▶ $\dfrac{p-1}{p-2}$ **9** ▶ $\dfrac{r+2}{r-1}$

10 ▶ $\dfrac{x-2}{x-4}$ **11** ▶ $\dfrac{x-3}{x-5}$ **12** ▶ $\dfrac{x+3}{x+4}$

EXERCISE 4*

1 ▶ $\dfrac{2(x-3)}{x+2}$ **2** ▶ $\dfrac{2(x-4)}{x-3}$ **3** ▶ $\dfrac{2(x-3)}{x+2}$

4 ▶ $\dfrac{1}{x+1}$ **5** ▶ $\dfrac{(x+1)^2}{(x+2)(x+3)}$

6 ▶ $x+2$ **7** ▶ $\dfrac{x+2}{x-2}$ **8** ▶ $\dfrac{x-4}{x+2}$

9 ▶ $\dfrac{p+4}{p-5}$ **10** ▶ $\dfrac{q+2}{q+6}$ **11** ▶ $\dfrac{y}{x+3y}$

12 ▶ $\dfrac{x-y}{x}$

EXERCISE 5

1 ▶ 21   **2** ▶ 2   **3** ▶ $\dfrac{1}{2}$

4 ▶ 3   **5** ▶ −8   **6** ▶ $-\dfrac{2}{3}$

7 ▶ $\dfrac{1}{2}$ **8** ▶ $21\dfrac{1}{2}$ **9** ▶ $\dfrac{3}{2}$

10 ▶ −6   **11** ▶ 2   **12** ▶ 0

EXERCISE 5*

1 ▶ $\dfrac{2}{5}$ **2** ▶ 15   **3** ▶ $\dfrac{1}{3}$

4 ▶ 0   **5** ▶ 15   **6** ▶ 7

7 ▶ 1   **8** ▶ $\dfrac{7}{3}$ **9** ▶ 4

10 ▶ 3   **11** ▶ $-\dfrac{8}{3}$ **12** ▶ $-\dfrac{5}{13}$

13 ▶ 6 km   **14** ▶ 30 km

EXERCISE 6

1 ▶ −7, 2   **2** ▶ 6   **3** ▶ 2

4 ▶ $-\dfrac{5}{3}$, 4   **5** ▶ $-\dfrac{2}{3}$ **6** ▶ $\dfrac{3}{4}$

7 ▶ 2   **8** ▶ −4, 5   **9** ▶ −3, 7

10 ▶ −1, 4   **11** ▶ −6, 3   **12** ▶ $-\dfrac{1}{3}$, 2

EXERCISE 6*

1 ▶ −3.5, 1   **2** ▶ −2, 5   **3** ▶ 4

4 ▶ 3   **5** ▶ −6, 6   **6** ▶ $-\dfrac{7}{3}$, 2

7 ▶ −0.768, 0.434   **8** ▶ −8.28, 0.785

9 ▶ −2, 6   **10** ▶ 3.2, 5   **11** ▶ 60

12 ▶ 10.47   **13** ▶ −2.5   **14** ▶ $\dfrac{2}{5}$, 2

15 ▶ $-\dfrac{2}{3}$, 5

EXERCISE 7   REVISION

1 ▶ 3   **2** ▶ $x+2$ **3** ▶ $\dfrac{x+3}{x-3}$

4 ▶ $\dfrac{x-1}{2x+3}$ **5** ▶ $\dfrac{7x-8}{12}$ **6** ▶ $\dfrac{3x-10}{18}$

7 ▶ $\dfrac{5x+1}{(x-1)(x+1)}$ **8** ▶ $\dfrac{4x+10}{(x+2)(x+4)}$

9 ▶ −2   **10** ▶ $\dfrac{1}{2}$ **11** ▶ −8, 2

12 ▶ 1

EXERCISE 7*   REVISION

1 ▶ $\dfrac{2}{3}$ **2** ▶ $\dfrac{x-11}{x+5}$ **3** ▶ $\dfrac{x+4}{x-7}$

4 ▶ $\dfrac{x-1}{3x+1}$ **5** ▶ $\dfrac{13x-5}{18}$ **6** ▶ $\dfrac{5x+7}{12}$

7 ▶ $\dfrac{2x+3}{(x+1)(x+2)}$ **8** ▶ $\dfrac{-1}{(x-4)(x+1)}$

9 ▶ 5   **10** ▶ $-\dfrac{2}{3}$ or 1

11 ▶ −0.464 or 6.46   **12** ▶ −9.16 or 3.16

EXAM PRACTICE: ALGEBRA 10

1 ▶ a x b $\dfrac{x+2}{2}$ c $\dfrac{x+3}{x+6}$

2 ▶ a $\dfrac{22-5x}{12}$ b $\dfrac{-1}{(x-4)(x+1)}$

3 ▶ a $\dfrac{2(x+1)}{x}$ b $(x-2)(2x-1)$

4 ▶ a 2   b $\dfrac{1}{3}$ or 2

c −2.24 or 6.24

UNIT 10: GRAPHS 9

EXERCISE 1

1 ▶ $\dfrac{dy}{dx}=0$ **2** ▶ $\dfrac{dy}{dx}=2$

3 ▶ $\dfrac{dy}{dx}=3x^2$ **4** ▶ $\dfrac{dy}{dx}=4x^3$

5 ▶ $\dfrac{dy}{dx}=5x^4$ **6** ▶ $\dfrac{dy}{dx}=10x^9$

7 ▶ $\dfrac{dy}{dx} = 6x^2$ **8** ▶ $\dfrac{dy}{dx} = 8x^3$

9 ▶ $\dfrac{dy}{dx} = 10x^4$ **10** ▶ $\dfrac{dy}{dx} = 20x^9$

11 ▶ $\dfrac{dy}{dx} = 0$ **12** ▶ $\dfrac{dy}{dx} = 0$

13 ▶ $\dfrac{dy}{dx} = 2$ **14** ▶ $\dfrac{dy}{dx} = 12$

15 ▶ $\dfrac{dy}{dx} = 24$ **16** ▶ $\dfrac{dy}{dx} = 20$

EXERCISE 1*

1 ▶ $\dfrac{dy}{dx} = -2x$ **2** ▶ $\dfrac{dy}{dx} = -3x^2$

3 ▶ $\dfrac{dy}{dx} = -x^{-2}$ **4** ▶ $\dfrac{dy}{dx} = -2x^{-3}$

5 ▶ $\dfrac{dy}{dx} = -3x^{-4}$ **6** ▶ $\dfrac{dy}{dx} = -4x^{-5}$

7 ▶ $\dfrac{dy}{dx} = -\dfrac{1}{x^2}$ **8** ▶ $\dfrac{dy}{dx} = -\dfrac{2}{x^3}$

9 ▶ $\dfrac{dy}{dx} = \dfrac{1}{2}x^{-\frac{1}{2}}$ **10** ▶ $\dfrac{dy}{dx} = \dfrac{1}{2\sqrt{x}}$

11 ▶ $\dfrac{dy}{dx} = \dfrac{1}{3}x^{-\frac{2}{3}}$ **12** ▶ $\dfrac{dy}{dx} = \dfrac{1}{3\sqrt[3]{x^2}}$

13 ▶ $\dfrac{dy}{dx} = -1$ **14** ▶ $\dfrac{dy}{dx} = -\dfrac{1}{4}$

15 ▶ $\dfrac{dy}{dx} = \dfrac{1}{4}$ **16** ▶ $\dfrac{dy}{dx} = \dfrac{1}{12}$

EXERCISE 2

1 ▶ $\dfrac{dy}{dx} = 2x + 1$

2 ▶ $\dfrac{dy}{dx} = 2x + 2$

3 ▶ $\dfrac{dy}{dx} = 3x^2 + 2x$

4 ▶ $\dfrac{dy}{dx} = 4x^3 + 3x^2 + 2x + 1$

5 ▶ $\dfrac{dy}{dx} = 6x^2 - 6x$

6 ▶ $\dfrac{dy}{dx} = 50x^4 + 5$

7 ▶ $\dfrac{dy}{dx} = -2x^{-3} - x^{-2}$

8 ▶ $\dfrac{dy}{dx} = -4x^{-3} - 3x^{-2}$

9 ▶ $\dfrac{dy}{dx} = -6x^{-3} + 2x^{-2}$

10 ▶ $\dfrac{dy}{dx} = -6x^{-4} + 12x^{-5}$

11 ▶ $\dfrac{dy}{dx} = 100x^9 + 25x^4$

12 ▶ $\dfrac{dy}{dx} = -100x^{-11} + 25x^{-6}$

13 ▶ $\dfrac{dy}{dx} = 3$ **14** ▶ $\dfrac{dy}{dx} = 6$

15 ▶ $\dfrac{dy}{dx} = 16$ **16** ▶ $\dfrac{dy}{dx} = 125$

17 ▶ $\dfrac{dy}{dx} = 9$ **18** ▶ $\dfrac{dy}{dx} = -9$

EXERCISE 2*

1 ▶ $\dfrac{dy}{dx} = 2x + 3$ **2** ▶ $\dfrac{dy}{dx} = 3x^2 + 6x$

3 ▶ $\dfrac{dy}{dx} = 2x + 8$ **4** ▶ $\dfrac{dy}{dx} = 4x + 11$

5 ▶ $\dfrac{dy}{dx} = 2x + 8$ **6** ▶ $\dfrac{dy}{dx} = 8x - 12$

7 ▶ $\dfrac{dy}{dx} = 18x - 6$ **8** ▶ $\dfrac{dy}{dx} = 27x^2 - 12x + 1$

9 ▶ $\dfrac{dy}{dx} = 18x - 2x^{-3}$ **10** ▶ $\dfrac{dy}{dx} = 18x - 2x^{-3}$

11 ▶ $\dfrac{dy}{dx} = -\dfrac{1}{x^2} - \dfrac{2}{x^3}$ **12** ▶ $\dfrac{dy}{dx} = -\dfrac{1}{x^2} + \dfrac{2}{x^3}$

13 ▶ $\dfrac{dy}{dx} = -\dfrac{2}{x^2} - \dfrac{6}{x^3}$ **14** ▶ $\dfrac{dy}{dx} = 4x + 4$

15 ▶ $\dfrac{dy}{dx} = 3 + \dfrac{3}{x^2}$ **16** ▶ $\dfrac{dy}{dx} = 1 + \dfrac{1}{2\sqrt{x}}$

17 ▶ $\dfrac{dy}{dx} = 1 - \dfrac{1}{2\sqrt{x^3}}$ **18** ▶ $\dfrac{dy}{dx} = 1 + \dfrac{1}{2\sqrt{x^3}}$

19 ▶ $\dfrac{dy}{dx} = 5$ **20** ▶ $\dfrac{dy}{dx} = 15$

21 ▶ $\dfrac{dy}{dx} = 4$ **22** ▶ $\dfrac{dy}{dx} = -8$

23 ▶ $\dfrac{dy}{dx} = 5$ **24** ▶ $\dfrac{dy}{dx} = 6$

EXERCISE 3

1 ▶ $y = 7x - 3$

2 ▶ $y = 2x + 4$

3 ▶ **a** $\dfrac{dy}{dt} = t - 3$

    b i $\dfrac{dy}{dt} = -2$ m/s

    ii $\dfrac{dy}{dt} = 0$ m/s

    iii $\dfrac{dy}{dt} = 3$ m/s

4 ▶ **a** $\dfrac{dP}{dt} = t + 1$

    b i $\dfrac{dP}{dt} = 2$ millions/day

    ii $\dfrac{dP}{dt} = 4$ millions/day

    iii $\dfrac{dP}{dt} = 5$ millions/day

5 ▶ a $\dfrac{dT}{dt} = 6t + 5$

    b i $\dfrac{dT}{dt} = 11\,°C/min$

      ii $\dfrac{dT}{dt} = 35\,°C/min$

      iii $\dfrac{dT}{dt} = 65\,°C/min$

6 ▶ a $\dfrac{dh}{dt} = 11 - 4t$

    b i $\dfrac{dh}{dt} = 7\,m/hr$

      ii $\dfrac{dh}{dt} = -1\,m/hr$

      iii $\dfrac{dh}{dt} = -7\,m/hr$

EXERCISE 3*

1 ▶ $y = 7x - 15$

2 ▶ $y = -20x - 14$

3 ▶ a $\dfrac{dN}{dt} = 40t + 80$

    b i $\dfrac{dN}{dt} = 80$ people/hr

      ii $\dfrac{dN}{dt} = 160$ people/hr

      iii $\dfrac{dN}{dt} = 230$ people/hr

4 ▶ a $\dfrac{dQ}{dt} = 3t^2 - 16t + 24$

    b i $\dfrac{dQ}{dt} = 11\,m^3/s$

      ii $\dfrac{dQ}{dt} = 2.75\,m^3/s$

      iii $\dfrac{dQ}{dt} = 3.6875\,m^3/s$

5 ▶ a $\dfrac{dT}{dm} = -\dfrac{400}{m^2}$

    b i $\dfrac{dT}{dm} = -16\,°C/min$

      ii $\dfrac{dT}{dm} = -4\,°C/min$

6 ▶ a $\dfrac{dP}{dt} = 4t - \dfrac{180}{t^2}$

    b i $\dfrac{dP}{dt} = -176$ spiders/month

      ii $\dfrac{dP}{dt} = 46.75$ spiders/month

EXERCISE 4

1 ▶ $\dfrac{dy}{dx} = 2x - 2$, $(1, 2)$ min.

2 ▶ $\dfrac{dy}{dx} = 2x + 4$, $(-2, -5)$ min.

3 ▶ $\dfrac{dy}{dx} = 6 - 2x$, $(3, 14)$ max.

4 ▶ $\dfrac{dy}{dx} = -8 - 2x$, $(-4, 28)$ max.

5 ▶ $\dfrac{dy}{dx} = 4x - 4$, $(1, 5)$ min.

6 ▶ $\dfrac{dy}{dx} = -12 - 4x$, $(-3, 26)$ max.

7 ▶ $\dfrac{dy}{dx} = 2x - 2$, $(1, -4)$ min.

8 ▶ $\dfrac{dy}{dx} = -8x$, $(0, 1)$ max.

9 ▶ a $\dfrac{dN}{dt} = 100t - 300$

    b $t = 3$, 8:03 pm

10 ▶ a $\dfrac{dN}{dt} = 200 - 20t$

    b 2000 leaves, when $t = 10$, on September 10th

EXERCISE 4*

1 ▶ $\dfrac{dy}{dx} = 3x^2 - 12x$, $(0, 0)$ max, $(4, -32)$ min.

2 ▶ $\dfrac{dy}{dx} = 3x^2 + 6x$, $(0, 0)$ min, $(-2, 4)$ max.

3 ▶ $\dfrac{dy}{dx} = 3x^2 - 18x$, $(6, -105)$ min, $(0, 3)$ max.

4 ▶ $\dfrac{dy}{dx} = -6x - 3x^2$, $(-2, 0)$ min, $(0, 4)$ max.

5 ▶ $\dfrac{dy}{dx} = 3x^2 + 6x - 9$, $(1, 0)$ min, $(-3, 32)$ max.

6 ▶ $\dfrac{dy}{dx} = -18 - 24x - 6x^2$, $(-3, -11)$ min, $(-1, 19)$ max.

7 ▶ $\dfrac{dy}{dx} = 6x^2 + 18x - 24$, $(1, -13)$ min, $(-4, 112)$ max.

8 ▶ $\dfrac{dy}{dx} = 12x^2 - 8x + 1$, $\left(\dfrac{1}{2}, 0\right)$ min, $\left(\dfrac{1}{6}, \dfrac{2}{27}\right)$ max.

9 ▶ a $V = x(10 - 2x)^2 = 100x - 40x^2 + 4x^3$

    b $\dfrac{dV}{dx} = 100 - 80x + 12x^2$

    c $\dfrac{dV}{dx} = 0 = (3x - 5)(x - 5)$, V_{max} at $x = \dfrac{5}{3}$,

      $V_{max} = \dfrac{2000}{27}$; $\dfrac{20}{3} \times \dfrac{20}{3} \times \dfrac{5}{3}$

      $x = 5$ is not in the domain for the model.

10 ▶ a $\dfrac{dT}{dt} = 5 - \dfrac{20}{t^2}$

    b T_{min} at $t = 2$ from graph, $T = 15\,°C$ March 1st

EXERCISE 5

1 ▶ a $v = 20t - 30\,m/s$, $10\,m/s$

    b $a = 20\,m/s^2$

2 ▶ a $v = 7 - 2t\,m/s$, $1\,m/s$

    b $a = -2\,m/s^2$

3 ▶ a $v = 3t^2 + 4t - 3\,m/s$, $17\,m/s$

    b $a = 6t + 4\,m/s^2$, $16\,m/s^2$

4 ▶ a $v = 12 + 6t - 3t^2$ m/s, 12 m/s

     b $a = 6 - 6t$ m/s^2, −6 m/s^2

5 ▶ $v = 16 - 8t$, $v = 0$ at $t = 2$, $s_{max} = 16$ m

6 ▶ a $v = 24$ m/s

     b $a = 12 - 2t$ m/s^2, $a = 8$ m/s^2

     c $t = 6$, $v_{max} = 40$ m/s

EXERCISE 5*

1 ▶ a $v = 10t + \dfrac{4}{t^2}$ m/s, 21 m/s

     b $a = 10 - \dfrac{8}{t^3}$ m/s^2, 9 m/s^2

2   **a** $v = \dfrac{5}{\sqrt{t}}$ m/s, $\dfrac{5}{2}$ m/s

     b $a = -\dfrac{5}{2\sqrt{t^3}}$ m/s^2, $a = -\dfrac{5}{16}$ m/s^2

3   **a** $s = 50$ m   **b** $t = \sqrt{10}$ s

     c $-10\sqrt{10}$ m/s

4   **a** $v = 40 - 10t$ m/s, $t = 0$, 40 m/s

     b $v = 0$, $t = 4$, $s = 80$ m

5   **a** $a = -1 + \dfrac{25}{t^2}$ km/s^2

     b $t = 5$, $v_{max} = 10$ km/s

6   **a** $t = 11$ s   **b** $v = -60$ m/s

     c Mean speed = $\dfrac{305}{11}$ m/s

ACTIVITY 1

a   Volume = $\pi r^2 h$

     $50 = \pi r^2 h$

     $\dfrac{50}{\pi r^2} = h$

     Area = $2\pi r^2 + 2\pi r h$

     $A = 2\pi r^2 + \dfrac{2r\pi \times 50}{\pi r^2}$

     $A = \pi r^2 + \dfrac{100}{r}$

c $r = 2.00$ m (3 s.f.), $h = 3.99$ m (3 s.f.),
$A_{min} = 75.1$ m^2 (3 s.f.)

EXERCISE 6   **REVISION**

1 ▶ $\dfrac{dy}{dx} = 4$

2 ▶ a $\dfrac{dp}{dt} = \$40\,000$/yr (profit increasing)

     b $\dfrac{dp}{dt} = -\$40\,000$/yr (profit decreasing)

3 ▶ (0, −2) minimum point

4 ▶ a $v = 25$ m/s   **b** $a = 6$ m/s^2

EXERCISE 6*   **REVISION**

1 ▶ $y = 11x - 25$

2 ▶ $t = 50$, $n_{max} \simeq 104\,167$

---

3 ▶ (4, −13) minimum point, (1, 14) maximum
point

4 ▶ a $a = -1 + \dfrac{144}{t^2}$ m/s^2

     b $a = 0$ at $t = 12$, $v_{max} = 6$ m/s

EXAM PRACTICE: GRAPHS 9

1 ▶ a 19   **b** 14   **c** 4

2 ▶ (2, 3) minimum point, (1, 4) maximum point

3 ▶ a $\dfrac{dp}{dx} = -\dfrac{1}{3} + \dfrac{2400}{x^2}$

     b $x = 60\sqrt{2} = 84.9$ km/hr,
$y_{max} = 13.4$ km/litre

4 ▶ a $v = 6$ m/s   **b** $a = -\dfrac{5}{2}$ m/s^2

UNIT 10: SHAPE AND SPACE 10

ACTIVITY 1

$\theta(°)$	0	30	60	90	120	150	180
$\sin \theta$	0	0.5	0.87	1	0.87	0.5	0

$\theta(°)$	210	240	270	300	330	360
$\sin \theta$	−0.5	−0.87	−1	−0.87	−0.5	0

The graph of $y = \sin \theta$ is given in the text following Activity 1.

ACTIVITY 2

$\theta(°)$	0	30	60	90	120	150	180
$\cos \theta$	1	0.87	0.5	0	−0.5	−0.87	−1

$\theta(°)$	210	240	270	300	330	360
$\cos \theta$	−0.87	−0.5	0	0.5	0.87	1

The graph of $y = \cos \theta$ is given in the text following Activity 2.

ACTIVITY 3

$\theta(°)$	0	30	60	90	120	150	180
$\tan \theta$	0	0.58	1.7	−	−1.7	−0.58	0

$\theta(°)$	210	240	270	300	330	360
$\tan \theta$	0.58	1.7	−	−1.7	−0.58	0

The graph of $y = \tan \theta$ is given in the text following Activity 3.

EXERCISE 1

1 ▶ a $\theta = 0°, 180°, 360°$ **b** $\theta = 90°, 270°$

     c $\theta = 0°, 180°, 360°$ **d** $\theta = 90°$

     e $\theta = 0°, 360°$ **f** $\theta = 45°, 225°$

2 ▶ a $\theta = 30°, 150°$ **b** $\theta = 60°, 300°$

     c $\theta = 30°, 210°$ **d** $\theta = 45°, 135°$

     e $\theta = 45°, 315°$ **f** $\theta = 60°, 240°$

EXERCISE 1*

1 ▶ a $\theta = -360°, -180°, 0°, 180°, 360°, 540°,$
720°

     b $\theta = -270°, -90°, 90°, 270°, 450°, 630°$

     c $\theta = -360°, -180°, 0°, 180°, 360°, 540°,$
720°

d $\theta = -270°, 90°, 450°$

e $\theta = -360°, 0°, 360°, 720°$

f $\theta = -315°, -135°, 45°, 225°, 405°, 585°$

2 ▶ a $\theta = -150°, -30°, 210°, 330°, 570°, 690°$

b $\theta = -240°, -120°, 120°, 240°, 480°, 600°$

c $\theta = -210°, -30°, 150°, 330°, 510°, 690°$

d $\theta = -135°, -45°, 225°, 315°, 585°, 675°$

e $\theta = -225°, -135°, 135°, 225°, 495°, 585°$

f $\theta = -240°, -60°, 120°, 300°, 480°, 660°$

EXERCISE 2

1 ▶ $x = 5.94$ 2 ▶ $y = 11.1$

3 ▶ MN = 39.0 cm   4 ▶ RT = 8.75 cm

5 ▶ AC = 37.8 cm   6 ▶ YZ = 33.0 cm

7 ▶ $x = 37.3°$ 8 ▶ $y = 37.8°$

9 ▶ ∠ABC = 38.8°   10 ▶ ∠XYZ = 26.0°

11 ▶ ∠ACB = 62.2°   12 ▶ ∠DCE = 115°

EXERCISE 2*

1 ▶ $x = 29.7$ 2 ▶ $y = 8.35$

3 ▶ ∠LMN = 67.4°   4 ▶ ∠RST = 71.9°

5 ▶ EF = 10.4 cm, ∠DEF = 47.5°, ∠FDE = 79.0°

6 ▶ MN = 10.8 cm, ∠MLN = 68.3°, ∠LNM = 49.7°

7 ▶ 13.5 km   8 ▶ 1089 m

9 ▶ BC = 261 m

10 ▶ YT = 53.3 m, 17.35 m

11 ▶ a 11.3 cm   b 38.7°

12 ▶ 59.0° or 121.0°

EXERCISE 3

1 ▶ $x = 7.26$ 2 ▶ $b = 8.30$

3 ▶ AB = 39.1 cm   4 ▶ AB = 32.9 cm

5 ▶ RT = 24.2 cm   6 ▶ MN = 6.63 cm

7 ▶ $X = 73.4°$ 8 ▶ $Y = 70.5°$

9 ▶ ∠ABC = 92.9°   10 ▶ ∠XYZ = 110°

EXERCISE 3*

1 ▶ $x = 9.34$ 2 ▶ $y = 13.3$

3 ▶ ∠XYZ = 95.5°   4 ▶ ∠ABC = 59.0°

5 ▶ ∠BAC = 81.8°   6 ▶ ∠RST = 27.8°

7 ▶ a 30.4 km   b 092°

8 ▶ a 10.3 km and 205°   b 8.60 km

9 ▶ 11.6 km

10 ▶ a VWU = 36.3°   b 264°

EXERCISE 4

1 ▶ a 9.64   b 38.9°

2 ▶ a 6.62   b 49.0°

3 ▶ a 54.8°   b 92.1°   c 33.1°

4 ▶ a 24.1°   b 125.1°   c 30.8°

5 ▶ a 7.88   b 6.13

6 ▶ a 4.13   b 4.88

7 ▶ a 79.1°   b 7.77

8 ▶ a 44.7°   b 4.11

9 ▶ a 16.8 km   b 168°

10 ▶ a 8.89 km   b 063.0°

EXERCISE 4*

1 ▶ 055.1°

2 ▶ a 52.7 km   b 076.9°

3 ▶ 247 km, 280°

4 ▶ 14.7 km/h, 088.9°

5 ▶ a 50.4°   b 7.01 m   c 48.4°

6 ▶ $x = 5.29$ cm, $y = 8.72$ cm

7 ▶ ∠BXA = 75.9°

8 ▶ CS = 2.64 km, 040.2°

9 ▶ a 38.1°   b 29.4 cm

10 ▶ a 16.8 cm   b 9.23 cm   c 98.3°

EXERCISE 5

1 ▶ 7.39 cm²   2 ▶ 29.7 cm²

3 ▶ 36.2 cm²   4 ▶ 8.46 cm²

5 ▶ 121 cm²   6 ▶ 173 cm²

EXERCISE 5*

1 ▶ 48.1 cm   2 ▶ 16.5 cm

3 ▶ 51.4°   4 ▶ 53.5 cm²

5 ▶ 65.8 cm   6 ▶ 15 600 m² (3 s.f.)

ACTIVITY 5

a $\frac{1}{2}ab \sin C$

b $\frac{1}{2}ac \sin B, \frac{1}{2}bc \sin A$

c Each expression must have the same value so,
$\frac{1}{2}ab \sin C = \frac{1}{2}ac \sin B$
$ab \sin C = ac \sin B$
$b \sin C = c \sin B$
$\dfrac{b}{\sin B} = \dfrac{c}{\sin C}$

ACTIVITY 6

a $a^2 = h^2 + b^2 - 2bx + x^2$

b $c^2 = h^2 + x^2$

c Substituting for $h^2 + x^2$ in part **a** gives
$a^2 = b^2 + c^2 - 2bx$

d $x = c \cos A$, so $a^2 = b^2 + c^2 - 2bc \cos A$

EXERCISE 6 REVISION

1 ▶ A(90, 1), B(180, 0), C(270, −1), D(540, 0)

2 ▶ A(90, 0), B(180, −1), C(360, 1)

3 ▶ a Every 180°   b i 1.7   ii −1.7

c Rotational symmetry of order 2 about (180, 0)

d i 240°   ii 280°   iii 300°

4 ▶ a 22.9°   b 22.9

5 ▶ 148 cm²

6 ▶ a 50°, 60°, 70°   b AB = 4.91 km

7 ▶ a 8.04 cm   b 82.0°   c 104 cm²

8 ▶ a 061.2°   b 224 km²

EXERCISE 6* REVISION

1 ▶ A(90, 1), B(180, 0), C(−90, −1), D(−180, 0)
2 ▶ A(0, 1), B(180, −1), C(−90, 0), D(−180, −1)
3 ▶ a

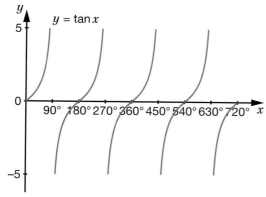

    b 60°, 240°, 420°, 600°
4 ▶ BC = 506 m
5 ▶ 6.32 cm and 9.74 cm
6 ▶ 4.68 m²
7 ▶ a 15.2 cm   b 68.0°
8 ▶ 79.5° or 100.5°

EXAM PRACTICE: SHAPE AND SPACE 10

1 ▶ a

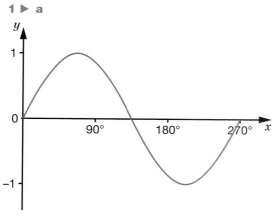

    b $\theta = 60°, 120°$
2 ▶ 10.4 cm, 35.5°
3 ▶ a 47.1°   b 131 cm²
4 ▶ a 31.0 km   b 078°

UNIT 10: HANDLING DATA 7

EXERCISE 1

1 ▶ a $\frac{1}{9}$ b $\frac{4}{9}$
2 ▶ a $\frac{1}{5}$ b $\frac{1}{5}$
    c Let X be the number of kings dealt in the first three cards:
$$P(X \geqslant 1) = 1 − P(X = 0)$$
$$= 1 − \frac{16}{20} \times \frac{15}{19} \times \frac{14}{18} = \frac{29}{57}$$

3 ▶ a $\frac{43}{63}$ b $\frac{20}{63}$ c $\frac{2}{7}$
    d Let X be the number of beads added to the box:
$$P(W_2) = \frac{2}{7} \times \frac{(2 + X)}{(7 + X)} + \frac{5}{7} \times \frac{2}{(7 + X)}$$
$$= \frac{2}{[7(7 + X)]} \times [(2 + X) + 5] = \frac{2}{7}$$
    Therefore true!

4 ▶ a 0.0459   b 0.3941
5 ▶ a 0.655   b 0.345
6 ▶ a 0.1   b 0.7   c 0.36
    d 0.147   e 0.441
7 ▶ a $\frac{1}{36}$ b $\frac{5}{18}$
8 ▶ a i $\frac{1}{15}$ ii $\frac{1}{15}$
    b $\frac{104}{105}$
9 ▶ a $\frac{2}{9}$ b $\frac{5}{9}$ c $\frac{1}{9}$
10 ▶ a $\frac{2}{15}$ b $\frac{5}{21}$ c $\frac{2}{21}$

EXERCISE 1*

1 ▶ a 0.0034   b 0.0006   c 0.0532
2 ▶ a 0.6   b 0.025   c 0.725
3 ▶ a $\frac{1}{8}$ b $\frac{8}{15}$ c $\frac{13}{60}$
4 ▶ a $\frac{5}{18}$
    b Let event X be 'clock is slow at noon on Wednesday':
$$P(X) = 1 − P(X) = 1 − \frac{7}{54} = \frac{47}{54}$$
5 ▶ a $\frac{1}{11}$ b $\frac{1}{3}$ c $\frac{3}{11}$ d $\frac{9}{55}$
6 ▶ a $\frac{9}{16}$ b $\frac{27}{64}$ c $\frac{29}{128}$
7 ▶ a $\frac{19}{66}$ b $\frac{13}{33}$ c $\frac{15}{22}$
8 ▶ a $\frac{1}{16}$ b $\frac{1}{4}$ c $\frac{15}{16}$
9 ▶ a $P(H_1) = \frac{1}{4}$
    b $P(H_2) = P(HH) + P(H'H) = \frac{1}{4} \times \frac{12}{51} + \frac{3}{4} \times \frac{13}{51}$
$$= \frac{1}{4}$$
    c $P(H_3) = P(HHH) + P(H'H'H) +$
$$P(H'HH) + P(HH'H)$$
$$= \frac{1}{4} \times \frac{12}{51} \times \frac{11}{50} + \frac{3}{4} \times \frac{38}{51} \times \frac{13}{50} +$$
$$\frac{3}{4} \times \frac{13}{51} \times \frac{12}{50} + \frac{1}{4} \times \frac{39}{51} \times \frac{12}{50}$$
$$= \frac{1}{4}$$
10 ▶ a $P(RR) = \frac{3}{5} \times \frac{5}{9} = \frac{1}{3}$
    b

Outcome		Probability
Bag X	Bag Y	
4R + 4W	5R + 5W	$\frac{1}{3}$
5R + 3W	4R + 6W	$\frac{8}{15}$
6R + 2W	3R + 7W	$\frac{2}{15}$

    c i $P(WY \rightarrow X) = \frac{1}{3} \times \frac{1}{2} = \frac{1}{6}$
       ii $P(RY \rightarrow X) = \frac{2}{15} \times \frac{3}{10} = \frac{1}{25}$

iii P(RY → X or WY → X)

$$= \frac{8}{15} \times \frac{4}{10} + \frac{2}{15} \times \frac{7}{10} = \frac{23}{75}$$

iv P(WY → X or WY → X)

$$= \frac{8}{15} \times \frac{6}{10} + \frac{1}{3} \times \frac{1}{2} = \frac{73}{150}$$

ACTIVITY 1

Possible questions:

Have you ever broken a school rule?

Do you enjoy Mathematics?

Do you think the number 13 is unlucky?

Do you know the birthday of your parents?

Do you think school meals are delicious?

ACTIVITY 2

Let Y be the event that Yosef wins the game.

P(O) + P(Y) = 1

P(Y) = 1 − P(O) = $1 - \frac{6}{11} = \frac{5}{11}$

(Clearly it is an advantage to go first!)

EXERCISE 2 REVISION

1 ▶ a i $\frac{4}{9}$ ii $\frac{4}{9}$ b $\frac{7}{27}$

2 ▶ a 0.36 b 0.42

c 0.256 (3 s.f.)

3 ▶ a $\frac{6}{25}$ b $\frac{19}{25}$

c i $\frac{12}{43}$ ii $\frac{31}{43}$ d $\frac{191}{597}$

4 ▶ a 0.1 b 0.1 c 0.69

5 ▶ a 0.9 b 0.3 c 0.35

EXERCISE 2* REVISION

1 ▶ a $\frac{48}{125}$ b $\frac{12}{125}$ c $\frac{61}{125}$

2 ▶ a 0.614 b 0.0574 c 0.0608

3 ▶ a $\frac{19}{45}$ b i $\frac{2}{15}$ ii $\frac{8}{105}$

4 ▶ $p = \frac{1}{2}$

5 ▶ b i $\frac{9}{16}$ ii $\frac{7}{16}$ c $\frac{1}{4}$

EXAM PRACTICE: HANDLING DATA 7

1 ▶ a $\frac{38}{132}\left(= \frac{19}{66}\right)$ b $\frac{60}{132}\left(= \frac{5}{11}\right)$ c $\frac{70}{132}\left(= \frac{35}{66}\right)$

2 ▶ a

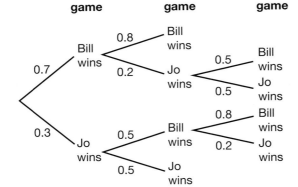

b 0.75

3 ▶ a

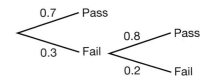

b 0.24

c P(3rd attempt) = 0.048, P(4th attempt) = 0.0096, so P(3rd or 4th) = 0.0576

OTHER ANSWERS

FACT FINDER: GOTTHARD BASE TUNNEL

EXERCISE 1

1 ▶ 1999

2 ▶ 2.01×10^6 per day

3 ▶ 2.68×10^3 m

4 ▶ 6.20×10^9 kg

5 ▶ $k = \frac{750}{79}$

EXERCISE 1*

1 ▶ a 8.12×10^6 m³ b 32 480 classrooms

2 ▶ 5.05×10^2 m³/s

3 ▶ 08:20:19

4 ▶ 2400 kg/m³

5 ▶ $p = 24.85$, $q = 20$, $p^q = 8.06 \times 10^{27}$

FACT FINDER: MOUNT VESUVIUS

EXERCISE 1

1 ▶ 1736 years 2 ▶ 14.4%

3 ▶ $\simeq 53\,300$ years 4 ▶ 57.3%

5 ▶ 932°F 6 ▶ 0.524 m/s

EXERCISE 1*

1 ▶ a 3.20×10^7 m³ b 7.17×10^7 m³

2 ▶ a 35 years approx. b 1979!

3 ▶ 4.63×10^4 m³/s

4 ▶ a 2.08×10^5 tonnes/s

b Approx. 208 000 cars per second!

5 ▶ 3.69×10^9 m³ 6 ▶ 9.93 km²

FACT FINDER: THE SOLAR SYSTEM

EXERCISE 1

1 ▶ Diameter = $\frac{12\,800 \times 8}{1\,390\,000}$ = 0.07 cm

Distance = $\frac{1.5 \times 10^8 \times 8}{1\,390\,000}$ = 863 cm

2 ▶

Body	Diameter (mm)	Distance from the orange OR 'Sun' (m)
Sun	80	
Earth	0.737	8.63
Mars	0.391	13.1
Jupiter	8.23	44.8

3 ▶ 14.4 m

4 ▶ 8.56 years

5 ▶ 8.3 min

EXERCISE 1*

1 ▶ 9.46×10^{12} km

2 ▶ 2287 km (approx. London–Athens)

3 ▶ 2 270 000 years

4 ▶ 2.08×10^{19} km

5 ▶ 2.84×10^3 revolutions around Earth

FACT FINDER: THE WORLD'S POPULATION

EXERCISE 1

1 ▶ **a** 1927   **b** 1960

2 ▶ **a** 8.25×10^7 people

    b 2.62 people/sec

    c 1.93×10^9

3 ▶ 3.6%

4 ▶ 1.85×10^7 km^2

5 ▶ 398 people/km^2

EXERCISE 1*

1 ▶ **a** 1.0113

    b **i** 7.38×10^9

      ii 7.47×10^9

      iii 7.64×10^9

2 ▶ Multiplying factor is 1.0113 each year, so n years after 2016: $P = 7.3 \times 10^9 \times 1.0113^n$

3 ▶ **a**

Year	2016	2020	2040	2060	2080	2100
n	0	4	24	44	64	84
$P\,(\times 10^9)$	7.3	7.6	9.6	12	15	19

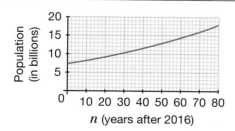

    b **i** Population at start of 2016 is 7.3×10^9

        Population is 14.6×10^9 at about start of 2078.

      ii UN estimate for 2011 is 11.2 billion.

        Model estimate is 19 billion, so a huge difference of 7.8 billion.

        UN hopes that education about contraception will spread faster around the world.

4 ▶ Population density of habitable land in the year 2100 is 1012 people/km^2.

    Approximately 2.5 times the population density in 2016.

5 ▶ Mass of Earth 6.6×10^{24} kg

    Mass of humans in year 4733 = $7.3 \times 10^9 \times 50 \times (1.0113)^{(4733 - 2016)} = 6.6 \times 10^{24}$ kg

    The mass of the Earth will equal the mass of humans in year 4733 assuming a rate of growth of 1.113% p.a. is constant.

FACT FINDER: THE TOUR DE FRANCE 2015

EXERCISE 1

1 ▶ 198

2 ▶ 43.2 km/hr

3 ▶ 90 hrs 2 mins 48 secs

4 ▶ **a** 53.4%   **b** 0.164%

5 ▶ 107 000 burgers (3 s.f.)

EXERCISE 1*

1 ▶ 0.0743

2 ▶ **a** 6.51 m/s   **b** 0.483 m/s

3 ▶ 1690%

4 ▶ 22.4 pedal strokes/dollar

5 ▶ 347 revs/min

CHALLENGES

1 ▶ 204

2 ▶ $\frac{1}{4}$

3 ▶ 39°

4 ▶ $2\pi x$ m

5 ▶ 1089

6 ▶ 333 333 332 666 666 667

7 ▶ 225

8 ▶ 59 (43 and 16)

9 ▶ $\frac{11}{24}$

10 ▶ 100π

11 ▶ 4.8 m

12 ▶ 6 cm^2

13 ▶ Dividing by zero $(a - b)$ in the third line

14 ▶ 37

15 ▶ 7.5 cm

16 ▶

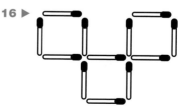

17 ▶ 9 cm^2

18 ▶ **a** 70.7 m^2 **b** 34.4 m^2

19 ▶ 10 extra bars

20 ▶ $\dfrac{h}{n} = \dfrac{p}{100}$ and $\dfrac{(h + 1)}{(n + 1)} = \dfrac{(p + 1)}{100} \Rightarrow n + p = 99$

21 ▶ 24 cm

22 ▶ **a** $2.7 \times 10^{-29}\, m^3$

b 7.4×10^{24}

c 5.6×10^7

23 ▶ **a**

b No. When the 7×4 rectangle is coloured as a chess board, 14 squares will be one colour, and 14 another colour. When the tetrominoes are coloured, the 3rd shape from the left above can only be coloured with 3 squares one colour and 1 square another colour. All the rest when coloured have 2 squares one colour and 2 another colour. The 7 tetrominoes will have 15 squares one colour and 13 another colour, so they cannot fit together as required.

24 ▶ 778.75 days

25 ▶ 60

26 ▶ $6(\sqrt{3} - 1)\, cm$

27 ▶ Students' proof

28 ▶ $2\, cm^2$

29 ▶ **a** $305\, m^2$ **b** $218\, m^2$

30 ▶ $y = 1 - x,\ y^4 = x^4 - 5$

$5 = x^4 - y^4 = (x^2 - y^2)(x^2 + y^2)$ [1]

$x + y = 1$ [2]

$5 = (x + y)(x - y)[(x + y)^2 - 2xy]$
$\quad = (x - y)[1 - 2xy]$

$5 = (2x - 1)[1 - 2x(1 - x)]$
$\quad = (2x - 1)[2x^2 - 2x + 1]$

$0 = 2x^3 - 3x^2 + 2x - 3$

$x = \frac{3}{2},\ y = \frac{1}{2},$

and from Pythagoras' Theorem

$d^2 = \left(\frac{3}{2}\right)^2 + \left(\frac{1}{2}\right)^2 = \frac{5}{2}$, so $d = \sqrt{\frac{5}{2}}$

EXAMINATION PRACTICE PAPER ANSWERS

Paper 1

1 a Both 10^6 factors cancel out.

$\dfrac{1.8}{5.1} \times 100 = 35.29... = 35.3\%$ (3 s.f.)

b Divide both sides by 2 when ratio is $2:5$
or divide both sides by 1.4 when ratio is $1.4:3.5$
$1.4:3.5 = 14:35 = 2:5 = 1:2.5$

2 a Multiply each term in the bracket by 5.
$5(2y - 3) = 10y - 15$

b 'FOIL' expansion and simplification.
$(2x - 1)(x + 5) = 2x^2 + 10x - x - 5$
$\qquad\qquad\qquad = 2x^2 + 9x - 5$

c Common factor $4y$ produces the given product of $4y(1 + 6z)$.
$4y + 24yz = 4y(1 + 6z)$

d Common factor $2x$ produces the given product of $2x(x - 11)$.
$2x^2 - 22x = 2x(x - 11)$

e Divide both sides by 3.
$3x - 2 = 7$

Add 2 to both sides.
$3x = 9$

Divide both sides by 3.
$x = 3$

3 a The mid-point is the mean of the x-values and the y-values.

$M = \left(\dfrac{2 + 8}{2}, \dfrac{1 + 5}{2}\right) = (5, 3)$

b Pythagoras' theorem.
$AB = \sqrt{(8 - 2)^2 + (5 - 1)^2} = \sqrt{52} = 7.211...$
$\qquad = 7.21$ cm (3 s.f.)

c Equation of line perpendicular to AB through M:

gradient $= \dfrac{\text{rise}}{\text{run}} = \dfrac{y_2 - y_1}{x_2 - x_1}$

Gradient of AB $= m_1 = \dfrac{5 - 1}{8 - 2} = \dfrac{4}{6} = \dfrac{2}{3}$

The condition for perpendicular gradients m_1 and m_2 is given by $m_1 \times m_2 = -1$
Gradient of line perpendicular to AB $= -\dfrac{3}{2}$

Use the equation of a straight line.
$y = mx + c$

M is on the line so must satisfy the equation.

$3 = -\dfrac{3}{2} \times 5 + c$

$c = 10\dfrac{1}{2}$

Multiply through by 2 to express equation as desired.

$y = -\dfrac{3}{2}x + 10\dfrac{1}{2}$

$2y = -3x + 21$

$3x + 2y - 21 = 0$

4 a $7x - 11 > 3$

Add 11 to both sides and then divide by 7.
$7x > 14$
$\ x > 2$

b

5 a Median is at $\dfrac{1}{2}(n + 1)$th position.
Median score is at the 8th position.
Median score $= 2$

b mean $= \dfrac{\Sigma fx}{\Sigma f}$

Mean $= \dfrac{5 \times 1 + 4 \times 2 + \cdots + 1 \times 5}{15}$

$\qquad = \dfrac{37}{15} = 2.47$ (3 s.f.)

6 a p(A or B) = p(A) + p(B) if A,B independent.
Note that the answer should be given in full as no rounding has been requested.
p(bb or gg) = p(bb) + p(gg) = $0.45 \times 0.45 + 0.12 \times 0.12$
$\qquad\qquad\qquad\qquad = 0.2169$

b Expected number of events = no. of trials × probability of the event
Number of green-eyed people $= 500 \times 0.12$
$\qquad\qquad\qquad\qquad\qquad = 60$

7 a area of trapezium $= \dfrac{1}{2}(a + b)h$

Area $= \dfrac{1}{2}(50 + 90)60 = 4.2 \times 10^3\,\text{mm}^2$

b volume of prism = cross-sectional area × length
Volume of prism $= 150 \times 4200 = 6.3 \times 10^5\,\text{mm}^3$

8 a i Elements in both A and B.
$A \cap B = \{1, 3\}$

ii Elements in A or B.
$A \cup B = \{1, 2, 3, 4, 5\}$

b A' is the complement of set A.
5 is a member of the set that is not A.

9 a

Age (years)	Frequency f	Mid-points x	fx	Cumulative frequency
$10 < t \leqslant 20$	8	15	$8 \times 15 = 120$	8
$20 < t \leqslant 30$	28	25	$28 \times 25 = 700$	36
$30 < t \leqslant 40$	30	35	$30 \times 35 = 1050$	66
$40 < t \leqslant 50$	10	45	$10 \times 45 = 450$	76
$50 < t \leqslant 60$	4	55	$4 \times 55 = 220$	80

$$\Sigma f = 80 \qquad \Sigma fx = 2540$$

Mean estimate $= \dfrac{\Sigma fx}{\Sigma f}$, x are mid-points

Mean estimate $= \dfrac{\Sigma fx}{\Sigma f} = \dfrac{2540}{80} = 31.75$ years

b Check that final value is the total sum of 80. Above.

c

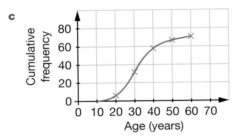

d n 'large' so median at $\frac{1}{2}n = 40$
Median ≈ 32

Lower quartile at $\frac{1}{4}n = 20$
Lower quartile (LQ) ≈ 24.3 years

Upper quartile at $\frac{3}{4}n = 60$
Upper quartile (UQ) ≈ 38 years

IQR = UQ − LQ
Interquartile range (IQR) ≈ 13.7 years

10 a 32 capsules equally spaced around the wheel.
$$\angle AOB = \frac{360}{32} = 11.25°$$
$$\angle BOC = 180° − 11.25° = 168.75°$$

Triangle OBC is isosceles.
$$\angle OCB = \frac{180° − 168.75°}{2} = 5.625°$$

AC is a vertical line.
Angle required $= 90° − 5.625° = 84.375°$

b $C = \pi d$
Distance $= \pi \times 135 = 424.115...\,\text{m} = 424\,\text{m}$ (3 s.f.)

c speed $= \dfrac{\text{distance}}{\text{time}}$

time $= \dfrac{\text{distance}}{\text{speed}} = \dfrac{424.115}{0.26} = 1631.2115...\,\text{s}$
$$= 27 \text{ mins } 11 \text{ secs}$$

11 a $3x − 4y = 15$
$4y = 3x − 15$
$y = \dfrac{3}{4}x − \dfrac{15}{4}$

$y = mx + c$; m is the gradient
Gradient $= \dfrac{3}{4}$

b Solve by eliminating y.

$3x − 4y = 15$	[1] × 3
$5x + 6y = 6$	[2] × 2
$9x − 12y = 45$	[3]
$10x + 12y = 12$	[4]

[3] + [4] $19x = 57$
$$x = \frac{57}{19} = 3 \text{ sub in [2]}$$
[2] $15 + 6y = 6$, so $y = -1\frac{1}{2} \Rightarrow (3, -1\frac{1}{2})$

12 $f(x) = 3x − 2$, $g(x) = \dfrac{1}{x}$

a i Find g(3) to use as the input into f(x).
$$fg(3) = f\left(\frac{1}{3}\right) = 1 − 2 = −1$$

ii Find f(3) to use as the input into g(x).
$$gf(3) = g(7) = \frac{1}{7}$$

b i $g(x)$ is the input into f(x).
$$fg(x) = f\left(\frac{1}{x}\right) = 3\left(\frac{1}{x}\right) − 2 = \frac{3 − 2x}{x}$$

ii if $x = 0$ there is no output for fg(x).

c $f(x) = fg(x)$
$$3x − 2 = \frac{3 − 2x}{x}$$

Multiply both sides by x.
$$(3x − 2)x = 3 − 2x$$
$$3x^2 − 2x = 3 − 2x$$
$$x^2 = 1 \rightarrow x = ±1$$

13 a i

Number of calories (n)	Frequency	Frequency density = frequency ÷ class width
$0 < n \leqslant 1000$	90	0.09
$1000 < n \leqslant 2000$	130	0.13
$2000 < n \leqslant 2500$	140	0.28
$2500 < n \leqslant 4000$	120	0.08

ii Histogram can now be completed.

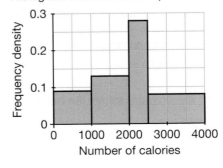

b UQ is at $\frac{3}{4}$ of area of histogram.

UQ is $\frac{3}{4} \times 480 = $ 360th person = 2500 calories.

14 a i Inversely implies that r^2 is on the denominator.

$R\alpha \frac{1}{r^2}$, so $R = \frac{k}{r^2}$, where k is a constant.

Substitute $R = 0.9$ and $r = 2$ into the formula.

$0.9 = \frac{k}{2^2}$, so $k = 3.6$

ii

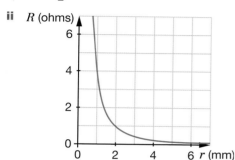

b Substitute values into formula then solve for r.

$R = \frac{3.6}{r^2}$, $0.1 = \frac{3.6}{r^2}$, $r^2 = 36$, $r = 6\,mm$

15 shaded area = large rectangle − small rectangle = 35

$(x + 4)(x + 1) - 15 = 35$

$(x^2 + 5x + 4) - 15 = 35$

$x^2 + 5x - 46 = 0$

Use the quadratic formula $x = \frac{-b \pm \sqrt{b^2 - 4ac}}{2a}$

$a = 1$, $b = 5$, $c = -46$

$x = \frac{-5 \pm \sqrt{5^2 - 4(1)(-46)}}{2(1)}$, $x = 4.73\,cm$ (3 s.f.)

16 a $x = -1.2$, 3.2 (1 d.p.)

b $y = x^2 - 2x - 4 = x + 2$ so draw $y = x + 2$

Solutions are $x = 4.4$, −1.4 (1 d.p.)

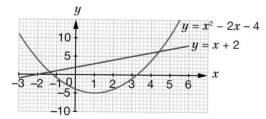

c i $\frac{dy}{dx} = 2x - 2$

ii Gradient = 0 at minimum point.

$\frac{dy}{dx} = 0 = 2x - 2$, $x = 1$, $y = -5$

Min point (1, −5)

17

A

110°

42°

B   7.8 cm   C

a sine rule: $\dfrac{a}{\sin A} = \dfrac{b}{\sin B} = \dfrac{c}{\sin C}$

$\angle ACB = 28°$, sine rule $\dfrac{AB}{\sin 28°} = \dfrac{7.8}{\sin 110°}$

$AB = \dfrac{7.8}{\sin 110°} \times \sin 28° = 3.89688...cm$

$AB = 3.90\,cm$ (3 s.f.)

b Note 3.90 is not used!

Area $= \frac{1}{2} \times 7.8 \times 3.89688 \times \sin 42° = 10.1693...cm^2$

area $= \frac{1}{2} ab \sin C$

Area $= 10.2\,cm^2$ (3 s.f.)

18 a p(X and Y) = p(X) × p(Y)

p(AN) $= \frac{3}{6} \times \frac{2}{5} = \frac{6}{30} = \frac{1}{5}$

b p(X) + p(X′) = 1

p(not same) = 1 − p(same)

p(same) = p(AA) + p(NN)

p(X or Y) = p(X) + p(Y)

$= \frac{3}{6} \times \frac{2}{5} + \frac{2}{6} \times \frac{1}{5} = \frac{8}{30} = \frac{4}{15}$

p(not same) $= 1 - \frac{4}{15} = \frac{11}{15}$

19 Volume of the thin cylinder of water increase after the sphere is submerged = volume of the sphere.

$V = \pi r^2 h$ (volume of cylinder)

$V = \frac{4}{3} \pi r^3$ (volume of sphere)

$\pi \times 30^2 \times 5 = \frac{4}{3} \times \pi \times r^3$ where r is the radius of the sphere.

Rearrange to make r the subject.

$r^3 = \dfrac{3 \times 30^2 \times 5}{4}$

$r = \sqrt[3]{\dfrac{3 \times 30^2 \times 5}{4}} = 15\,cm$

$A = 4\pi r^2$ (surface area of sphere)

$A = 4\pi \times 15^2 = 2827.43...cm^2 = 2830\,cm^2$ (3 s.f.)

20 $\dfrac{x}{x + 2} + \dfrac{x + 17}{x^2 + x - 6} = 1$

Factorise the denominator.

$\dfrac{3}{(x + 2)} + \dfrac{x + 17}{(x + 2)(x - 3)} = 1$

Multiply first fraction by $\dfrac{(x - 3)}{(x - 3)}$ which is 1.

$\dfrac{(x - 3)}{(x - 3)} \times \dfrac{3}{(x + 2)} + \dfrac{x + 17}{(x + 2)(x - 3)} = 1$

Express LHS as a single fraction.

$\dfrac{3(x - 3) + x + 17}{(x - 3)(x + 2)} = 1$

$\dfrac{3x - 9 + x + 17}{(x - 3)(x + 2)} = 1$

Factorise numerator and cancel common factor $(x + 2)$.

$\dfrac{4x + 8}{(x - 3)(x + 2)} = \dfrac{4(x + 2)}{(x - 3)(x + 2)} = \dfrac{4}{(x - 3)} = 1$

Add 3 to both sides.

$4 = (x - 3)$, $x = 7$

21 Consider the sum of the first odd numbers:

$1 + 3 + 5 + 7 + \cdots$

This is an A.P. with $a = 1$ and $d = 2$

$t_n = a + (n - 1)d = 1 + (n - 1) \times 2$

$t_n = 2n - 1$

Sum of the A.P. $1 + 3 + 5 + 7 + \cdots + (2n - 1)$

Sum of an A.P. $S_n = \dfrac{n}{2}[2a + (n - 1)d]$

$S_n = \dfrac{n}{2}[2 \times 1 + (n - 1) \times 2] = \dfrac{n}{2}[2n] = n^2$

22

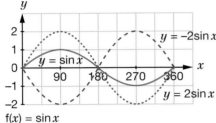

$f(x) = \sin x$

Stretch of $f(x)$ SF = 2 parallel to y-axis.

$2f(x) = 2\sin x$

$-2f(x) = -2\sin x$

Reflection of $2f(x)$ in x-axis.

Min point (90, −2) Max point (270, 2)

Paper 2

1

180 units at £0.0825 per unit	£14.85
440 units at £0.0705 per unit	£31.02
Total amount	£45.87
Tax at 15% of the total amount	£6.88
Amount to pay	£52.75

$620 - 180 = 440$ units

15% of £45.87 = $0.15 \times £45.87 = £6.88$

2 (triangle ABD is isosceles)

$\angle ABD = \dfrac{180 - 38}{2} = 71°$

(corresponding angle to $\angle ABD$)

$p = 71°$

(angle sum of a straight line = 180°)

$\angle BDE = 180 - 71 = 109°$

3 Sweets are the same in number.

Arul   Nikos

$x + 6 = 4x - 6$

$12 = 3x$

$x = 4$

4 mean = $\dfrac{\text{sum of scores}}{\text{number of scores}}$

$158 = \dfrac{S_4}{4}$, $S_4 = 156 \times 4 = 632 \, \text{cm}$

$156 = \dfrac{632 + y}{5}$, where y is Sienna's height in cm.

$156 \times 5 = 632 + y$, $y = 148 \, \text{cm}$

5 a tin:lead = 1:2

240 g ÷ sum of parts = 1 part

240 = 3 parts, 1 part = 80

tin:lead = 80 g:160 g

b tin:lead = 75:m

$m = 75 \times 2 = 150 \, \text{g}$

mass of solder = 75 + 150 = 225 g

6 $48 = 2^4 \times 3$

First find prime factors of both numbers.

$180 = 2^2 \times 3^2 \times 5$

a Product of common factors of both numbers.

HCF = $2^2 \times 3 = 12$

b LCM = $2^4 \times 3^2 \times 5 = 720$

7 a Round one number up and one down as this is a product.

$g = 10$, $h = 0.8$

b Square both sides.

$v = \sqrt{2gh}$

Divide both sides by $2h$.

$v^2 = 2gh$

$g = \dfrac{v^2}{2h}$

8 a Kitchen chairs

b i $P \cup Q = \{1, 2, 3, 4, 5, 6, 7, 8, 9\}$

ii Yes, an integer cannot be both odd and even. ϕ means that the set is empty.

9 a Let angle PQR = θ

$\sin\theta = \dfrac{\text{opposite side}}{\text{hypotenuse}}$

$\sin(\theta) = \dfrac{4.7}{7.6}$, $\theta = 38.2009\ldots$

Use the inverse sin 'button' used to find angle.

$\theta = 38.2°$ (1 d.p.)

b i $\cos\theta = \dfrac{\text{adjacent side}}{\text{hypotenuse}}$

$QR_{max} = 7.65 \times \cos(38.2009\ldots)$

Upper bound of 7.6 = 7.65

= 6.01 cm (3 s.f.)

ii $QR_{min} = 7.55 \times \cos(38.2009\ldots)$

Lower bound of 7.6 = 7.55

= 5.93 cm (3 s.f.)

10 a $60 \times 1.20 = 72 \, \text{s}$

b cost per sec = $\dfrac{12}{72}$, cost per min = $\dfrac{12}{72} \times 60 = 10$ c/min

c old rate: 12 c/min; new rate: 10 c/min

percentage change = $\dfrac{\text{change}}{\text{original}} \times 100$

% decrease = $\dfrac{2}{12} \times 100 = 16\dfrac{2}{3}$%

11 $2x^2 = 8x - 7$

Rearrange in form $ax^2 + bx = c$.
$2x^2 - 8x = -7$

Divide both sides by 2.
$x^2 - 4x = -\dfrac{7}{2}$

Complete the square.
$(x - 2)^2 = -\dfrac{7}{2} + 4 = \dfrac{1}{2}$

Take square root of both sides.
$x - 2 = \pm\sqrt{\dfrac{1}{2}}$

$x = 2 \pm \dfrac{1}{\sqrt{2}} = \dfrac{2\sqrt{2} \pm 1}{\sqrt{2}}$

12 a Ratios of lengths are $8:12 = 2:3$
$5:x = 2:3$, $\dfrac{5}{x} = \dfrac{2}{3}$, $5 \times 3 = 2 \times x$, $x = \dfrac{15}{2}$, $x = 7.5$

b $y:15 = 2:3$, $\dfrac{y}{15} = \dfrac{2}{3}$, $3 \times y = 15 \times 2$, $y = \dfrac{30}{3}$, $y = 10$

c Length are in ratio $x:y$; areas are in ratio $x^2:y^2$.
Ratios of areas are $4:9$
Let area of Q be A
$80:A = 4:9$, $\dfrac{80}{A} = \dfrac{4}{9}$, $80 \times 9 = 4 \times A$,
$A = 180\,\text{cm}^2$

13 Intersecting chords theorem.
$AX \times XB = CX \times DX$
$2.8 \times 1.6 = 1.2 \times DX$
$\qquad DX = 3.7333...\text{cm}$

Pythagoras' theorem
$AD^2 = AX^2 + DX^2$
$\qquad = 2.8^2 + (3.7333...)^2$
$AD = 4.67\,\text{cm}$ (3 s.f.)
Let angle DAX $= \theta$

$\tan\theta = \dfrac{\text{opposite side}}{\text{adjacent side}}$

$\tan\theta = \dfrac{DX}{AX} = \dfrac{3.7333...}{2.8}$

$\theta = 53.1°$ (3 s.f.)

14 a i $\mathbf{OX} = k\mathbf{OB} = k\begin{pmatrix}5\\2\end{pmatrix}$

ii $\mathbf{AX} = \mathbf{AO} + \mathbf{OX} = k\begin{pmatrix}5\\2\end{pmatrix} - \begin{pmatrix}1\\2\end{pmatrix}$

iii $\mathbf{XC} = \mathbf{XO} + \mathbf{OC} = \begin{pmatrix}4\\0\end{pmatrix} - k\begin{pmatrix}5\\2\end{pmatrix}$

b If $\mathbf{AX} = \mathbf{XC}$, $5k - 1 = 4 - 5k$, $k = \dfrac{1}{2}$

c X is halfway along OB as $k = \dfrac{1}{2}$.

$\mathbf{XC} = \dfrac{1}{2}\mathbf{AX}$ implies $\mathbf{AX} = \mathbf{XC}$ and vectors are parallel, so AXC is a straight line with X as its mid-point.

15 a Multiplying a quantity by 1.03 increases it by 3%...
Investment value = $\$1200 \times (1.03)^5 = \$1391.13$

b Let original cost be $\$y$,
$y \times (0.70) = \$98$

Multiplying a quantity by 0.70 decreases it by 30%.
$y = \dfrac{98}{0.70} = \$140$

16 a Area of rectangle = base × width
$y = x(28 - 2x) = 28x - 2x^2$

b Graph of y against x will be an inverted parabola which has a maximum point where $\dfrac{dy}{dx} = 0$

If $y = ax^n$, $\dfrac{dy}{dx} = nax^{n-1}$

$\dfrac{dy}{dx} = 28 - 4x = 0$ at max point

Solve for x.

$x = 7$

Substitute x into width $= 28 - 2x$
$y = 28 - 2 \times 7 = 14$
$y_{\text{max}} = 7 \times 14 = 98\,\text{m}^2$

17 a area of circle $= \pi r^2$
Area $= \dfrac{110}{360} \times \pi \times 12^2 = 138\,\text{cm}^2$ (3 s.f.)

b Total perimeter $= 2r + \dfrac{120}{360} \times 2\pi r = 2r + \dfrac{2}{3}\pi r = \dfrac{2}{3}r(3 + \pi)$

Perimeter $= 2 \times 12 + \left(\dfrac{110}{360}\right) \times 2 \times \pi \times 12$
$\qquad = 47.0$ (3 s.f.)

circumference of circle $= 2\pi r$

$200 = \dfrac{2}{3}r(3 + \pi)$

Rearrange to make r the subject.
$600 = 2r(3 + \pi)$
$300 = r(3 + \pi)$
$r = \dfrac{300}{3 + \pi}$ cm

18 a $p(A) + p(A') = 1$

First attempt　　　Second attempt

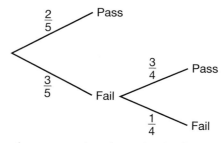

b p(pass course) = p(pass 1st time) + p(pass 2nd time)

p(pass 2nd time) = fail 1st test and pass 2nd test
$= \dfrac{2}{5} + \dfrac{3}{5} \times \dfrac{3}{4} = \dfrac{17}{20}$

19 Solve by substitution.
$y = x - 1$　　　[1]
$2x^2 + y^2 = 2$　　[2]

subs [1] into [2]
[2] $2x^2 + (x - 1)^2 = 2$
$2x^2 + x^2 - 2x + 1 = 2$

Quadratic factorisation.

$3x^2 - 2x - 1 = 0$

$(3x + 1)(x - 1) = 0$

$x = -\dfrac{1}{3}$ or 1 sub into [1]

$y = -\dfrac{4}{3}$ or 0

$(1, 0), \left(-\dfrac{1}{3}, -\dfrac{4}{3}\right)$

20 a $x = 3t - t^2 + \dfrac{8}{t} = 3t - t^2 + 8t^{-1}$

If $x = f(t)$, $v = \dfrac{dx}{dt}$, $a = \dfrac{dv}{dt}$

$v = 3 - 2t - 8t^{-2} = 3 - 2t - \dfrac{8}{t^2}$ m/s

b $a^{-m} = \dfrac{1}{a^m}$

$a = \dfrac{dv}{dt} = -2 + 16t^{-3} = -2 + \dfrac{16}{t^3}$ m/s²

c $a = 0 = -2 + \dfrac{16}{t^3}$, $t^3 = 8$, $t = 2$ s

21 $(2x - 1)(x + 1)(x + 2) - (x + 1)(x + 2)(x + 3) = 0$

$(x + 1)(x + 2)[(2x - 1) - (x + 3)] = 0$

$(x + 1)(x + 2)(x - 4) = 0$

$x = -1, -2$ or 4

Be careful not to divide by $(x + 1)(x + 2)$ first as solutions can be lost!

22   area of triangle $= \dfrac{1}{2} \times$ base $\times$ perpendicular height

Let height be y. Consider equilateral triangle of side $4n$.

$A = \dfrac{1}{2} \times 4n \times y = 2ny$

Pythagoras' theorem

$(4n)^2 = y^2 + (2n)^2$

$16n^2 = y^2 + 4n^2$

$12n^2 = y^2$

$\sqrt{12} = \sqrt{4 \times 3} = 2\sqrt{3}$

$2\sqrt{3}\,n = y$

$A = 2n \times 2\sqrt{3}\,n = 4\sqrt{3}\,n^2$

Paper 3

1 $\dfrac{3.23 \times 10^4}{1.8 \times 10^6} = 24.0749\ldots = 2.41 \times 10^1$ (3 s.f.)

2   width = height

$6x = 4(x + 7)$

Solve for x.

$6x = 4x + 28$

$2x = 28$, $x = 14$ cm

3

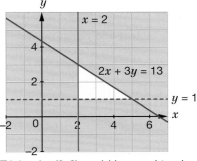

Trial point $(0, 0)$ could be used to check region $2x + 3y \leqslant 13$

4   mean $= \dfrac{\Sigma fx}{\Sigma f}$, where x are mid-points.

$5.5 = \dfrac{4 \times 1 + 15 \times 4 + 18 \times 7 + p \times 10}{37 + p}$

Solve for p.

$5.5(37 + p) = 190 + 10p$

$203.5 + 5.5p = 190 + 10p$

$13.5 = 4.5p$, $p = 3$

5   (ABCD is a cyclic quadrilateral, so opposite angles sum to 180°)

$x + (2x + 12) = 180°$

$3x = 168°$

$x = 56°$

(angle at centre of circle = 2 × angle at circumference)

$y = 2 \times 56°$

$y = 112°$

6 a i   'FOIL'

$(x - 3)(x + 7) = x^2 + 7x - 3x - 21 = x^2 + 4x - 21$

ii   'FOIL'

$(5x - 3)^2 = (5x - 3)(5x - 3) = 25x^2 - 30x + 9$

iii   Expand last two brackets by 'FOIL' then expand this by $(3x - 1)$

$(3x - 1)(2x - 1)(x - 1) = (3x - 1)(2x^2 - 3x + 1)$

$= 6x^3 - 9x^2 + 3x - 2x^2 + 3x - 1$

$= 6x^3 - 11x^2 + 6x - 1$

iv $(\sqrt{7} - 1)(2\sqrt{7} + 1) = 2 \times 7 + \sqrt{7} - 2\sqrt{7} - 1 = 13 - \sqrt{7}$

b i $\dfrac{\sqrt{7}}{\sqrt{7}} = 1$

$\dfrac{\sqrt{7} + 1}{\sqrt{7}} \times \dfrac{\sqrt{7}}{\sqrt{7}} = \dfrac{7 + \sqrt{7}}{7}$

ii $\dfrac{\sqrt{7} - 1}{\sqrt{7} - 1} = 1$, $a^2 - b^2 = (a + b)(a - b)$

$\dfrac{\sqrt{7}}{\sqrt{7} + 1} \times \dfrac{\sqrt{7} - 1}{\sqrt{7} - 1} = \dfrac{7 - \sqrt{7}}{\sqrt{7}^2 - 1^2} = \dfrac{7 - \sqrt{7}}{6}$

7   Let A_1 = area of $\triangle ABC$.

Let A_2 = total surface area of pyramid.

Base area + four triangular faces.

$A_2 = 10 \times 10 + 4A_1$

area $= \dfrac{1}{2} \times$ base $\times$ height

$A_1 = \dfrac{1}{2} \times 10 \times$ AM

Pythagoras' theorem

$AC^2 = AM^2 + MC^2$

$AM^2 = 13^2 - 5^2 = 144$, AM = 12

$A_1 = 5 \times 12 = 60$, $A_2 = 100 + 4 \times 60$

$A_2 = 340$ cm²

8 n-sided regular polygon

exterior angle, $e = 180° - 120° = 60°$

$e = \dfrac{360}{n}$ for a regular n-sided polygon

$60 = \dfrac{360}{n}$, $n = 6$

$4n$-sided polygon

$e = \dfrac{360}{4 \times 6} = 15°$

$e + i = 180°$ for a regular n-sided polygon

interior angle, $i = 180° - 15° = 165°$

9

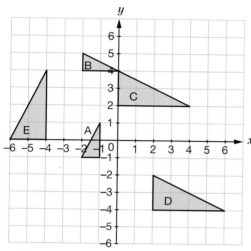

e Draw construction lines from vertices from E to A shows centre of enlargement is where they meet.

E mapped onto A by an enlargement SF = $\dfrac{1}{2}$ about centre (2, –2)

10 a

Age (years)	Cumulative frequency
$0 < t \leqslant 5$	41
$0 < t \leqslant 10$	67
$0 < t \leqslant 15$	87
$0 < t \leqslant 20$	97
$0 < t \leqslant 25$	100

b Plot 'endpoints' (5, 41)…(25, 100)

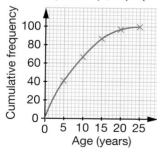

c Lower quartile = 3 years; upper quartile = 12 years.

LQ found at 25th value, UQ found at 75th value

IQR = UQ – LQ

Interquartile range = 12 – 3 = 9 years

11 a $y = mx + c$; m is gradient, c is y-intercept

gradient = $\dfrac{\text{rise}}{\text{run}}$

$m = 2$

$y = 2x - 1$

b Gradient of line perpendicular to line L is m_1.

Product of the gradients of perpendicular lines = –1

$$2 \times m_1 = -1, \ m_1 = -\dfrac{1}{2}$$

$$y = -\dfrac{1}{2}x + c$$

(2, 3) $3 = -\dfrac{1}{2} \times 2 + c, \ c = 4$

$$y = -\dfrac{1}{2}x + 4$$

12 a $p^{\frac{1}{2}} = (3^8)^{\frac{1}{2}} = 3^4, \ k = 4$

b $(a^m)^n = a^{m \times n}, \ (ab)^m = a^m \times b^m$

$q^{-\frac{1}{3}} = (2^9 \times 5^{-6})^{-\frac{1}{3}} = (2^9)^{-\frac{1}{3}} \times (5^{-6})^{-\frac{1}{3}} = 2^{-3} \times 5^2$,

$m = -3, n = 2$

13 a $y = 5000x - 625x^2$

$\dfrac{dy}{dx} = 0$ at turning points

$\dfrac{dy}{dx} = 5000 - 1250x = 0$

$x = 4, y = 10\,000$

b (4, 10 000) is a maximum point as the curve is an inverted parabola ($y = -ax^2 \ldots$)

c £4 as this produces the greatest profit of £10 000.

14 a $3600 = \pi r^2 + \dfrac{1}{2} \times 4\pi r^2 = 3\pi r^2$

area of circle = πr^2; surface area of sphere = $4\pi r^2$

$r^2 = \dfrac{3600}{3\pi}, \ r = \sqrt{\dfrac{1200}{\pi}} = 19.5441\ldots, \ d = 39.1$ cm (3 s.f.)

b volume of sphere = $\dfrac{4}{3}\pi r^3$

$V = \dfrac{1}{2} \times \dfrac{4}{3}\pi \times 19.5441 = 15\,635.2\ldots,$

$v = 15\,600$ cm^3 (3 s.f.)

15 a Similar figures of corresponding lengths l and L.

$l \times k = L$ where k is the length scale factor.

$l \times 3 = L$

$v \times k^3 = V$

6000×3^3 = volume of large drum, V

$V = 1.62 \times 10^5$ cm^3

b Number of large drums = $\dfrac{3240 \times 10^6}{1.62 \times 10^5} = 20\,000$

$a \times k^2 = A$

2000×3^2 = Area of a large drum = $18\,000$ cm^2

$10\,000$ cm^2 = 1 m^2

Area of all large drums = $20\,000 \times \dfrac{18\,000}{10^4}$ m^2 = $36\,000$ m^2

Total cost = $1.20 \times 36\,000 = \$43\,200$

16 a If $f(x) = \dfrac{x}{x - 1}$, $f(3) = \dfrac{3}{3 - 1} = \dfrac{3}{2}$

b If $x = 1$, the denominator $= 0$
x cannot be 1

c i $ff(x) = f\left(\dfrac{x}{x - 1}\right) = \dfrac{\dfrac{x}{x - 1}}{\dfrac{x}{x - 1} - 1} = \dfrac{\dfrac{x}{x - 1}}{\dfrac{1}{x - 1}} = \dfrac{x}{x - 1} \times \dfrac{x - 1}{1} = x$

ii The input is the output as $ff(x) = ff^{-1}(x) = x$
$f(x)$ is its own inverse

17 Triangle BAG has $\angle A = 30°$, $\angle B = 20°$, $\angle G = 130°$,
AB $= 1500\,$m

'SASA' so use the sine rule.

$\dfrac{BG}{\sin 30°} = \dfrac{1500}{\sin 130°}$, $BG = \dfrac{1500}{\sin 130°} \times \sin 30° = 979.055...$

BG $= 979\,$m (3 s.f.)

18 a $7 - x$

b All cells sum to 50.
Total must sum to 50:
$50 = x + 10 + 9 + 13 + (7 - x) + (8 - x) + 6$
$50 = 53 - x$, $x = 3$

19 a $3n$

b p(A and B) = p(A) × p(B) if A, B are independent.
$\dfrac{n}{3n} \times \dfrac{n - 1}{3n - 1} = \dfrac{1}{10}$
Solving for n by multiplying across by $30n(n - 1)$
gives $n^2 - 7n = 0$
$n(n - 7) = 0$, so $n = 7$, number of people in club $= 3n = $
21 people.

20 a $x^2 - 8x + 21 = (x - 4)^2 - 16 + 21 = (x - 4)^2 + 5$
$a = 4$, $b = 5$

b If $f(x) = x^2 - 8x + 21 = (x - 4)^2 + 5$

$(x - 4)^2$ will always be positive so, when $x = 4$,
this value is 0.
$f(x)_{min} = 5$ when $x = 4$
Also, the curve is a U-shaped parabola with minimum
point $(4, 5)$

21 Let the first integer be n, the next integer is $n + 1$.
$(n + 1)^2 - n^2 = (n^2 + 2n + 1) - n^2 = 2n + 1$
$n + (n + 1) = 2n + 1$

22 a

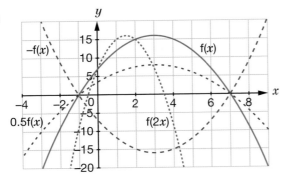

b,c i $-f(x) = -(7 + 6x - x^2) = x^2 - 6x - 7$: Reflection of f(x)
in x-axis

ii $0.5f(x) = 0.5(7 + 6x - x^2) = 3.5 + 3x - 0.5x^2$: Stretch
parallel to y-axis, SF = 0.5

iii $f(2x) = 7 + 6(2x) - (2x)^2 = 7 + 12x - 4x^2$: Stretch
parallel to x-axis, SF = 0.5

Paper 4

1 Bearings are measured clockwise from North.
Bearing of Q from P = 250°.

2 a i $(3x - 4)(3x + 4) = 9x^2 + 12x - 12x - 16 = 9x^2 - 16$

ii 'FOIL' used to expand the brackets.
$(3x - 4)^2 = (3x - 4)(3x - 4) = 9x^2 - 12x - 12x + 16$
$= 9x^2 - 24x + 16$

b i $2xy - 12x^2y^3 = 2xy(1 - 6xy^2)$

ii $3x^2y - 9x^3y^2 + 15x^4y^3 = 3x^2y(1 - 3xy + 5x^2y^2)$

c i $a^m \times a^n = a^{m+n}$
$xy \times x^3y^5 = x^4y^6$

ii $a^m \div a^n = a^{m-n}$
$\dfrac{x^2y^7z^5}{xy^2z^3} = xy^5z^2$

3 a $37\,000\,000\,000 = 3.7 \times 10^{10}$

b $7.5 \times 10^{-5} = 0.000075$

c Standard form is $a \times 10^n$ where $1 \leqslant a < 10$, n is an
integer.
$\dfrac{2.5 \times 10^{-3}}{1.25 \times 10^7} = 2 \times 10^{-10}$

4 $\angle CBA = 73°$ (alternate segment theorem)
$\angle OBC = 34°$ ($\triangle OBC$ is isosceles)
$\angle OBA = 73° - 34° = 39°$

5 Increase by 2%: multiplying factor = 1.02
Big Bank: €1200 × (1.02)³ = €1273.45

Increase by 3%: multiplying factor= 1.03
Small Bank: €1273.45 × (1.03)² = €1351.00

percentage profit $= \dfrac{change}{original} \times 100$

Percentage profit $= \dfrac{1351 - 1200}{1200} \times 100 = 12.6\%$ (3 s.f.)

6 $3x + 4y = 5$ [1] × 5 ⇒ [3]
$2x - 5y = 11$ [2] × 4 ⇒ [4]
$15x + 20y = 25$ [3]
$8x - 20y = 44$ [4]
$23x = 69$ [3]+[4]
$x = 3$ ⇒ [1]
[1] $9 + 4y = 5$, $y = -1$ ⇒ $(3, -1)$

7 $\mathbf{OR} = \mathbf{OP} + \mathbf{PR} = \mathbf{OP} + \frac{5}{2}\mathbf{PQ} = \mathbf{OP} + \frac{5}{2}(\mathbf{OQ} - \mathbf{OP})$

$\qquad = \frac{5}{2}\mathbf{OQ} - \frac{3}{2}\mathbf{OP} = \frac{1}{2}(5\mathbf{OQ} - 3\mathbf{OP})$

Set up the vector path equation first, and then substitute in the vector elements.

$\qquad = \frac{1}{2}[5(4\mathbf{a} + \mathbf{b}) - 3(2\mathbf{a} + 5\mathbf{b})]$

$\qquad = \frac{1}{2}[14\mathbf{a} - 10\mathbf{b}] = 7\mathbf{a} - 5\mathbf{b}$

8 **a** $P(T \cap C \cap G') = \frac{6}{80} = \frac{3}{40}$

  b $P((T \cap G \cap C)/T) = \frac{11}{42}$

  c $P(T/(G \cup C)) = \frac{6 + 11 + 13}{58} = \frac{30}{58} = \frac{15}{29}$

  d $P(A/B) = \frac{P(A \cap B)}{P(B)}$

$\qquad P(G/C') = \frac{13 + 15}{50} = \frac{28}{50} = \frac{14}{25}$

9 **a,b** $t_n = a + (n - 1)d$ for an A.P.

$\qquad t_{10} = 39 = a + 9d \qquad [1]$

$\qquad t_5 = 19 = a + 4d \qquad [2]$

$\qquad 20 = 5d \qquad\qquad [1] - [2]$

$\qquad d = 4 \Rightarrow [1],\ a = 3$

  c $Sn = \frac{n}{2}(a + l)$ for an A.P.

$\qquad S_{10} = \frac{10}{2}(3 + 39) = 210$

10 **a** $y_{max} = \frac{\text{max value}}{\text{min value}}$

$\qquad y_{max} = \frac{8.3}{6.6 - 2.6} = 2.1$ (2 s.f.)

  b $y_{min} = \frac{\text{min value}}{\text{max value}}$

$\qquad y_{min} = \frac{7.3}{7.6 - 1.6} = 1.2$ (2 s.f.)

11 **a** $g(x) = \frac{2}{3x - 1}$, $g(2) = \frac{2}{6 - 1} = \frac{2}{5}$

  b $f^2(x) = ff(x)$; note that $f^2(x) \neq [f(x)]^2$

$\qquad g^2(x) = gg(x) = g(\frac{2}{5}) = \frac{2}{\frac{6}{5} - 1} = \frac{2}{\frac{1}{5}} = 10$

  c $\qquad x = \frac{2}{3x - 1}$

$\qquad x(3x - 1) = 2$

$\qquad 3x^2 - x - 2 = 0$

$\qquad (3x + 2)(x - 1) = 0$

$\qquad x = -\frac{2}{3},\ 1$

  d Let $y = g(x)$

Switch y and x for the inverse of $g(x)$.

$\qquad y = \frac{2}{3x - 1}$

Make y the subject.

$\qquad x = \frac{2}{(3y - 1)}$, $x(3y - 1) = 2$, $3xy - x = 2$

Use proper notation for inverse function.

$\qquad 3xy = x + 2$, $y = \frac{x + 2}{3x}$, $g^{-1}(x) = \frac{x + 2}{3x}$

12 RHS regular polygon, exterior angle $e = \frac{360}{10} = 36°$

$\qquad e = \frac{360}{n}$, for an n-sided regular polygon.

LHS regular polygon, $e = 60° - 36° = 24°$

$\qquad 24° = \frac{360°}{n}$, $n = \frac{360}{24} = 15$ sides

13 Sum of parts $= 4 + 5 + 6$

150 cm $\Rightarrow$ 15 parts

1 part $=$ 10 cm, so triangle sides are 40 cm, 50 cm and 60 cm.

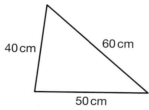

Let angle between 40 and 50 sides be A:

cosine rule: $a^2 = b^2 + c^2 - 2bc\cos A$

$\cos A = \frac{40^2 + 50^2 - 60^2}{2 \times 40 \times 50}$, $A = 82.819\ldots°$

Area of triangle $= \frac{1}{2} \times 40 \times 50 \times \sin(82.819°)$

area of triangle $= \frac{1}{2}ab\sin C$

$= 992.156\ldots\text{cm}^2 = 992\,\text{cm}^2$ (3 s.f.)

14 **a** arc length $= \frac{\theta}{360} \times 2\pi r$

Perimeter, $p = \frac{60 \times 3}{360} \times 2 \times \pi \times 4 = 12.6$ cm (1 d.p.)

  b area of triangle $= \frac{1}{2}ab\sin C$; area of sector $= \frac{\theta}{360} \times \pi r^2$

Area $= \frac{1}{2} \times 8 \times 8 \times \sin 60° - \frac{180}{360} \times \pi \times 4^2 = 2.6\,\text{cm}^2$ (1 d.p.)

15 **a**

x	0.5	1	1.5	2	3	4	5
y	−5.75	−2	0.25	2.5	8	15.25	24.4

  b

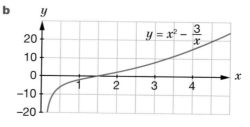

  c Read off graph where curve cuts the x-axis.

$\qquad x \approx 1.4$

d $x^2 - 2x - \dfrac{3}{x} = 0$, $x^2 - \dfrac{3}{x} = 2x$, so draw $y = 2x$.

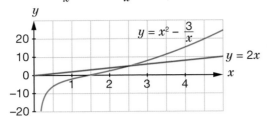

Read off x-axis where curve cuts $y = 2x$
$x \approx 2.5$

16 a frequency density = frequency ÷ class width

Weight (w kg)	Frequency	Frequency density
$0 < w \leq 2$	128	$128 \div 2 = 64$
$2 < w \leq 3.5$	150	$150 \div 1.5 = 100$
$3.5 < w \leq 4.5$	136	$136 \div 1 = 136$
$4.5 < w \leq 6$	72	$72 \div 1.5 = 48$

b

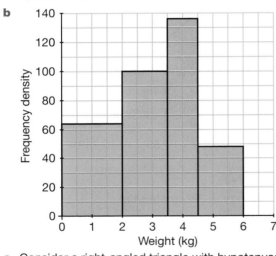

17 a Consider a right-angled triangle with hypotenuse
$AB = 4 + x$

Pythagoras' theorem
$$(x + 4)^2 = (5 - x)^2 + (6 - x)^2$$
$$x^2 + 8x + 16 = 25 - 10x + x^2 + 36 - 12x + x^2$$
$$0 = x^2 - 30x + 45, \text{ as required.}$$

b Use the quadratic formula
$a = 1$, $b = -30$, $c = 45$

$$x = \frac{-b \pm \sqrt{b^2 - 4ac}}{2a}$$

$$x = \frac{-(-30) \pm \sqrt{(-30)^2 - 4(1)45}}{2(1)} = 1.58\ldots \text{ or } 28.4\ldots,$$

so $x = 1.58\,\text{cm}$ (3 s.f.)

18 a $v = \dfrac{\mathrm{d}x}{\mathrm{d}t} = 2t^2 - 9t + 4 = 0$ when stationary.

$(2t - 1)(t - 4) = 0$, $t = \dfrac{1}{2}$ or 4

b $a = \dfrac{\mathrm{d}v}{\mathrm{d}t} = 4t - 9$, at $t = 3$, $a = 12 - 9 = 3$

19 a $p(R) = \dfrac{3}{8} = \dfrac{x}{48}$, $x = 18$ red beads

b $p(R) = \dfrac{18 + r}{48 + r} = \dfrac{1}{2}$, $36 + 2r = 48 + r$, $r = 12$ red beads

20 a Let length of diagonal of square base be y
Pythagoras' theorem
$y^2 = x^2 + x^2 = 2x^2$, $y = \sqrt{2}\,x$, $\dfrac{1}{2}y = \dfrac{x}{\sqrt{2}}$

Consider right-angled triangle of height h and base $\dfrac{1}{\sqrt{2}}x$ and hypotenuse x.

$x^2 = h^2 + \dfrac{1}{2}x^2$, $h^2 = \dfrac{1}{2}x^2$, $h = \dfrac{x}{\sqrt{2}}$

b Let angle between an edge and the base be θ.

$\tan \theta = \dfrac{\text{opposite side}}{\text{adjacent side}}$

$\tan \theta = \dfrac{\frac{x}{\sqrt{2}}}{\frac{x}{\sqrt{2}}} = 1$, $\theta = 45°$

21 Let A be the area of the field to be maximized.
Width of field = $150 - 2x$
$A = x(150 - 2x) = 150x - 2x^2$
$\dfrac{\mathrm{d}A}{\mathrm{d}x} = 150 - 4x = 0$ at turning point,
solving $x = 37.5$, $y = 75$, $A_{max} = 2812.5$

Maximum value as shape of A vs x graph is an inverted parabola yielding a maximum point.

22 volume of a sphere = $\dfrac{4}{3}\pi r^3$; volume of cone = $\dfrac{1}{3}\pi r^2 h$

curved surface area of cone = $\pi r l$; slant height is l
where $l^2 = r^2 + h^2$
volume of sphere = volume of cone
$\dfrac{4}{3}\pi r^3 = \dfrac{1}{3}\pi r^2 h$, so $4r = h$
Total surface area of cone, $A = \pi r^2 + \pi r l$
$l^2 = r^2 + h^2 = r^2 + 16r^2 = 17r^2$,
so $A = \pi r^2 + \pi r \sqrt{17}\,r = \pi r^2(1 + \sqrt{17})$

INDEX

...ank the following individuals and organisations for

...nt; t-top)

...6, 87, 138tr, 169, 393, scottff72 125;

...nikishiyev 127, Rodolfo Arpia 180, Aurora Photos 44, Sergio ...21cr, bilwissedition Ltd. & Co. KG 374tr, Martin Carlsson ...Cultura Creative (RF) 253, David Tipling Photo Library 131tl, ...KE 275br, Elena Elisseeva 177, EyeEm 343, Bella Falk 293, Petro ...Historical Picture Archive 112, 228tr, Grapheast 62, Roberto Herrett ...34tl, Image Source Plus 174, Interface Images - Human Interest 438, ...Jason Smalley Photography 299, JohanH 267br, Joy 383, LatitudeStock ...ic and Arts Photo Library 219, 374tl, Chris Lofty 233tr, Melvyn Longhurst ...ray - CC 233br, James Nesterwitz 187, Steve Nichols 440br, PACIFIC ..., Pictorial Press Ltd 30cl, 421tr, Prixpics 195, REUTERS 276tr, 436t, 444t, ...hotolibrary 31, Stocktrek Images, Inc. 440t, Tetra Images 286, Erik Tham 228tl, ...ley 131tr, Peter Titmuss 265, Tuul and Bruno Morandi 178tr, Universal Art Archive ..., Rogan Ward 184bl, WaterFrame 182cr, Westend61 GmbH 63, World History Archive ..., 323, Xinhua 211b;

...idgeman Art Library Ltd: Arctic Sunset with Rainbow, 1877 (oil on canvas), Bradford, William (1823-92) / Private Collection / Photo © Christie's Images 102, Composition No.10, 1939-42 (oil on canvas), Mondrian, Piet (1872-1944) / Private Collection / Bridgeman Images / 2017 / Mondrian/Holtzman Trust 348;

Fotolia.com: Africa Studio 98cr, Alexander 320, alfexe 88, Artistic Endeavor 432, ChrisVanLennepPhoto 172cr, ck001 11tr, dbrnjhrj 274br, deaddogdodge 10tr, Dimitrius 268tl, golandr 259, isuaneye 6cr, Juulijs 140, krasnevsky 138cr, Erik Lam 254, logos2012 383br, Monkey Business 426b, nyul 175, Steve Oehlenschlager 188, photogoodwin 274tl, rozakov 170tr, Roman Samokhin 164cr, Roman Sigaev 271, Piotr Skubisz 266, Sam Spiro 58, sutiporn 190tr, Taigi 164tl, 164tr, Tupungato 273, tyneimage 431br, vectorfusionart 442b, viiwee 64, Kim Warden 107, zhu difeng 387;

Getty Images: De Agostini Picture Library 162, Image Source 6br, jameslee1 443, Mark Kolbe 134, JONATHAN NACKSTRAND / AFP 220cr, SAM PANTHAKY / AFP 8br, Popperfoto 255, Jan Sleurink / NiS / Minden Pictures 260, Jaimie D. Travis 105br, Stephanie Zieber 391;

Ian Potts: 111;

Mary Evans Picture Library: INTERFOTO / Sammlung Rauch 247;

Pearson Education Ltd: Lord and Leverett 442tr, Debbie Rowe 397, Tudor Photography 136, Coleman Yuen 382;

Rob Ijbema: Rob Ijbema 445b;

Science Photo Library Ltd: NATIONAL LIBRARY OF MEDICINE 333tr, ROYAL ASTRONOMICAL SOCIETY 421tl;

Shutterstock.com: 59783 244, 96, AdamEdwards 161br, aerogondo2 59b, Andy-pix 166, Arieliona 19, badahos 196, Bayanova Svetlana 172br, bluecrayola 448, cherries 275tl, adithep chokrattanakan 173, Croisy 181, Steve Cukrov 133, Raphael Daniaud 60, Dchauy 384cl, Janaka Dharmasena 21tl, Dudarev Mikhail 389, entropic 301, Juergen Faelchle 182tl, Mette Fairgrieve 122, gkuna 10br, Anton Gvozdikov 268br, Mau Horng 33, imdb 98br, JPagetRFPhotos 297, Graeme Knox 267tr, Jakub Krechowicz 21tr, John Leung 61, Lucky Business 220tr, Marzolino 276, Petinov Sergey Mihilovich 178cl, Monkey Business Images 97, 170br, neelsky 120, Vitalii Nesterchuk 25tr, Nicku 43, nico99 385, Tyler Olson 55, Pichugin Dmitry 9br, Leonid Shcheglov 17, sivilla 333br, Solent News / REX 18, Stubblefield Photography 341, Studio 37 346, T.Fabian 211, Thor Jorgen Udvang 241, Taras Vyshnya 396, Yeko Photo Studio 28;

Werner Forman Archive Ltd: 59cr

We are grateful to the following for permission to reproduce copyright material: Limerick on page 281 by Rebecca Siddall and used by permission of Rebecca Siddall.

Cover images: *Front*: **Shutterstock.com:** Grey Carnation
Inside front cover: **Shutterstock.com:** Dmitry Lobanov

All other images © Pearson Education

Glossary terms have been taken from *The Longman Dictionary of Contemporary English Online*.